LONDON MATHEMATICAL SOCIETY LECTURE NOTE SERIES

Managing Editor: Professor J.W.S. Cassels, Department of Pure Mathematics and Mathematical Statistics, University of Cambridge, 16 Mill Lane, Cambridge CB2 1SB, England

The titles below are available from booksellers, or, in case of difficulty, from Cambridge University Press.

London Mathematical Society Lecture Note Series. 247

Analytic Number Theory

Edited by

Y. Motohashi
College of Science and Technology, Nihon University

CAMBRIDGE
UNIVERSITY PRESS

CAMBRIDGE UNIVERSITY PRESS
Cambridge, New York, Melbourne, Madrid, Cape Town, Singapore, São Paulo

Cambridge University Press
The Edinburgh Building, Cambridge CB2 8RU, UK

Published in the United States of America by Cambridge University Press, New York

www.cambridge.org
Information on this title: www.cambridge.org/9780521625128

© Cambridge University Press 1997

First published 1997

A catalogue record for this publication is available from the British Library

ISBN 978-0-521-62512-8 paperback

Transferred to digital printing 2008

Contents

Foreword

This volume is an outcome of

The 39th Taniguchi International Symposium on Mathematics
Analytic Number Theory
May 13–17, 1996, Kyoto

organized by myself, and of its forum, May 20–24, organized by N. Hirata-Kohno, L. Murata and myself as a conference at the Research Institute for Mathematical Sciences, Kyoto University.

I am deeply indebted to the Taniguchi Foundation for the generous support that made the symposium and the conference possible. The organizers of the conference acknowledge sincerely that the speakers were supported in part by the Inoue Science Foundation, the Kurata Foundation, Saneyoshi Scholarship Foundation, the Sumitomo Foundation; College of Science and Technology of Nihon University, the Research Institute for Mathematical Sciences of Kyoto University; and the Grant-Aid for General Scientific Research from the Ministry of Education, Science and Culture (through the courtesy of Prof. M. Koike, Kyushu University).

My special thanks are due to Profs. Hirata-Kohno and Murata for their unfailing collaboration during the three years of difficult preparation for the meetings.

Tokyo
May, 1997   Y.M.

Speakers

RIMS, Kyoto University, May 20–24, 1996

Plenary Sessions

E. Bombieri	A. Ivić	P. Sarnak
P.D.T.A. Elliott	H. Iwaniec	H.-P. Schlickewei
J.-H. Evertse	M. Jutila	R. Tijdeman
J. Friedlander	Y. Kitaoka	M. Waldschmidt
A. Granville	Y. Motohashi	T.D. Wooley
K. Hashimoto	M.R. Murty	

Parallel Sessions

S. Akiyama	N. Hirata-Kohno	A. Perelli
M. Amou	M.N. Huxley	J. Pintz
R.C. Baker	S. Kanemitu	B. Ramakrishnan
R. Balasubramanian	M. Katsurada	W. Schwarz
J. Brüdern	K. Kawada	T.N. Shorey
H. Diamond	A. Laurinčikas	C.L. Stewart
S. Egami	E. Manstavičius	K.-M. Tsang
S. Gonek	K. Matsumoto	C. Viola
G. Greaves	H. Mikawa	D. Wolke
M. Hata	L. Murata	E. Yoshida
A.J. Hildebrand	V.K. Murty	J. Yu
M. Hindry	M. Nagata	T. Zhan

Subvarieties of Linear Tori and the Unit Equation
A Survey

ENRICO BOMBIERI

0. Introduction Let K be a number field. A classical and important problem is that of determining the units u of K such that $1 - u$ is also a unit. More generally, let Γ be a finitely generated subgroup of $(K^*)^2$. The unit equation in Γ is the equation

$$x + y = 1$$

to be solved with $(x, y) \in \Gamma$. A basic result, going back to Siegel, Mahler and Lang, asserts that this equation has only finitely many solutions. Moreover, a fundamental result of Baker provides us with effective bounds for their height.

One may also consider a linear torus G over a number field K, a finitely generated subgroup Γ of $G(K)$ and study the set $C \cap \Gamma$, where C is an irreducible algebraic curve in G. Lang [8, Ch.8, Th.3.2] proved that if $C \cap \Gamma$ is an infinite set then C is a translation of a subtorus of G, and conjectured that the same conclusion holds if we replace Γ by its division group, that is the group Γ' consisting of all points $y \in G$ such that $y^n \in \Gamma$ for some n (we use multiplicative notation in G). This conjecture of Lang was later proved by Liardet [8, Ch.8, Th.7.4]. Similar statements can be made for G a commutative algebraic group with no $\mathbb{G}_a$ components (that is, a semiabelian variety) and replacing C by a subvariety X of G, but they are far more difficult to prove; indeed, even the simplest case of a curve in an abelian variety turns out to be equivalent to Mordell's Conjecture.

In this lecture, we shall report on some recent results on the distribution of small algebraic points on subvarieties of $\mathbb{G}_m^n$ and their application to the study of the unit equation in groups of finite rank. The exposition in section §2 is a simplification of a method of Y. Bilu and the proof of Theorem 3 follows closely the proof of F. Beukers and H.P. Schlickewei.

We shall not comment here about results for the case of abelian varieties similar to those in sections §1 and §2; instead, we refer the interested reader to the papers [14], [15], [16], and [13] listed in the bibliography.

1. Zhang's Theorem In 1992 S. Zhang [14] obtained a surprising result on the height of algebraic points on curves in $\mathbb{G}_m^n$. He showed that for any curve $C \subset \mathbb{G}_m^n$ there is a positive lower bound for the height of non-torsion algebraic points in $C(\overline{\mathbb{Q}})$. This was new even in the simplest case of the unit equation

$x+y = 1$. Later, Zhang [15] extended this to general subvarieties of $\mathbb{G}_m^n$. As we shall see, there are interesting applications to the problem of obtaining good upper bounds for the number of solutions of the generalized unit equation.

We identify $\mathbb{G}_m$ with the affine line punctured at the origin, together with the usual multiplication. The *standard height* on $\mathbb{G}_m^n$ is

$$\widehat{h}(P_1,\ldots,P_n) = \sum_{i=1}^{n} h(P_i)$$

with h the absolute Weil height on $\mathbb{P}^1$.

The height $\widehat{h}$ has the following properties:

(i) Homogeneity and symmetry: $\widehat{h}(P^m) = |m| \cdot \widehat{h}(P)$ for $m \in \mathbb{Z}$.

(ii) Non-degeneracy: $\widehat{h}(P) = 0$ if and only if P is a torsion point of G.

(iii) Triangle inequality: $\widehat{h}(PQ^{-1}) \leq \widehat{h}(P) + \widehat{h}(Q)$.

(iv) Finiteness: There are only finitely many points in $\mathbb{G}_m^n(\overline{\mathbb{Q}})$ of bounded degree and bounded height.

These properties are clear from the corresponding properties of the Weil height, with (iv) following from Northcott's Theorem [8, Ch.3, Th.2.6]. Thus we see that $\widehat{h}(PQ^{-1})$ defines a translation invariant semidistance $d(P,Q)$ on $\mathbb{G}_m^n(\overline{\mathbb{Q}})$ and actually a translation invariant distance on $\mathbb{G}_m^n(\overline{\mathbb{Q}})/\text{tors}$.

By a *subtorus H* of $\mathbb{G}_m^n$ we mean a geometrically irreducible closed algebraic subgroup; H is non-trivial if $\dim(H) \geq 1$. A *torsion coset* εH is a translation of H by a torsion point ε of $\mathbb{G}_m^n$. We have:

Zhang's Theorem *Let X/K be a closed subvariety of $\mathbb{G}_m^n$ defined over a number field K and let X^* be the complement in X of the union of all torsion cosets $\varepsilon H \subseteq X$. Then:*

(a) The number of maximal torsion cosets $\varepsilon H \subseteq X$ is finite.

(b) The height of points $P \in X^*(\overline{K})$ has a positive lower bound.

The results (a), (b) above are effective.

The following uniform version of Zhang's Theorem is due to E. Bombieri and U. Zannier [3].

Theorem 1 *Let $\widehat{h}$ be the standard height on $\mathbb{G}_m^n(\overline{\mathbb{Q}})$ and let $d(P,Q) = \widehat{h}(PQ^{-1})$ be the associated semidistance.*

Let X/K be a closed subvariety of $\mathbb{G}_m^n$ defined over a number field K by polynomial equations of degree at most d and let X° be the complement in X of the union of all cosets $gH \subseteq X$ with non-trivial H. Then:

(a) X° is Zariski open in X. Moreover, the number and degrees of the irreducible components of $X - X^\circ$ are bounded in terms of d and n.

(b) There are a positive constant $\gamma(d, n)$ and a positive integer $N(d, n)$, depending only on d and n, with the following property. Let $Q \in \mathbb{G}_m^n(\overline{K})$. Then

$$\{P : P \in X^\circ(\overline{K}), \quad d(P, Q) \le \gamma(d, n)\}$$

is a finite set of cardinality at most $N(d, n)$. Moreover, for every point P in this set we have $[K(P, Q) : K(Q)] \le N(d, n)$.

Remark The K-irreducible components of $X - X^\circ$ have a special structure. Let Z be a K-irreducible component of $X - X^\circ$.

Then we can find an isomorphism $\varphi : \mathbb{G}_m^n \xrightarrow{\sim} \mathbb{G}_m^n$, given by a monomial change of coordinates of degree bounded in terms of d and n, such that $\varphi(Z) = \mathbb{G}_m^k \times Y$, where $k \ge 1$ and Y is defined by polynomial equations of degree effectively bounded in terms of d and n and height not exceeding the height of the polynomial equations used to define X. By Northcott's Theorem, Y can be effectively determined.

The constants $\gamma(d, n)$ and $N(d, n)$ are effective and the finite set of points in (b) can be effectively determined for every $Q \in \mathbb{G}_m^n(\overline{K})$.

The inductive proof in [3] yields extraordinarily small values for $\gamma(d, n)$. W.M. Schmidt [9] introduced new ideas and obtained explicit good values for $\gamma(d, n)$ and $N(d, n)$.

Remark There is no uniform version of Zhang's Theorem. For example, if a, b are not roots of unity the equation $1 + ax + by = 0$ in $\mathbb{G}_m^2$ has a non-torsion solution $\xi = (a^{-1}\rho, b^{-1}\rho^2)$ with ρ a primitive cubic root of unity. We have $\widehat{h}(\xi) = h(a) + h(b) > 0$, and we can make it arbitrarily small by choosing a and b.

Lemma 1 *Let $f(x_1, \ldots, x_n)$ be a polynomial with integer coefficients, with degree at most d and height $H(f)$ and let $p > e\binom{d+n}{n}H(f)$ be a prime number. Let $\boldsymbol{\xi} = (\xi_1, \ldots, \xi_n)$ be an algebraic point with $f(\xi_1, \ldots, \xi_n) = 0$ and $f(\xi_1^p, \ldots, \xi_n^p) \ne 0$. Then $\widehat{h}(\boldsymbol{\xi}) \ge 1/(pd)$.*

Proof We repeat the proof in [3]. We may assume that the coefficients of f have no common divisor greater than 1. Let K be a number field containing all coordinates ξ_i.

By Fermat's Little Theorem we have

$$f^p(x_1, \ldots, x_n) = f(x_1^p, \ldots, x_n^p) + p\, g(x_1, \ldots, x_n),$$

where $g(x_1, \ldots, x_n) \in \mathbb{Z}[x_1, \ldots, x_n]$ has degree at most pd. Since by hypothesis $f(\xi_1, \ldots, \xi_n) = 0$, we get

$$f(\xi_1^p, \ldots, \xi_n^p) = -p\, g(\xi_1, \ldots, \xi_n). \tag{1}$$

For any $\zeta \in K^*$ the product formula yields

$$\sum_{v \in M_K} \log |\zeta|_v = 0.$$

We apply this with $\zeta = f(\xi_1^p, \ldots, \xi_n^p)$ and estimate terms as follows.

If $v|p$ we have by (1) and the fact that g has integer coefficients:

$$\log |\zeta|_v = \log |p\, g(\xi_1, \ldots, \xi_n)|_v \le \log |p|_v + pd \sum_{i=1}^{n} \log^+ |\xi_i|_v + \log |f|_v,$$

because $\log |f|_v = 0$ (the coefficients of f have no non-trivial common divisor and v is finite).

For the other v's we have, with $\varepsilon_v = 0$ if v is finite, and $\varepsilon_v = [K_v : \mathbb{Q}_v]/[K : \mathbb{Q}]$ if v is infinite:

$$\log |\zeta|_v = \log |f(\xi_1^p, \ldots, \xi_n^p)|_v \le pd \sum_{i=1}^{n} \log^+ |\xi_i|_v + \log |f|_v + \varepsilon_v \log \binom{d+n}{n},$$

because the number of monomials in f does not exceed $\binom{d+n}{n}$.

Summing over all $v \in M_K$ and using $\sum_{v|p} \log |p|_v = -\log p$ we infer

$$0 = \sum_{v \in M_K} \log |\zeta|_v \le -\log p + pd\, \widehat{h}(\xi) + h(f) + \log \binom{d+n}{n},$$

which ends the proof.

Corollary (W.M. Schmidt)   *Let $f(\mathbf{x})$ be as in Lemma 1 and let m be a positive integer all of whose prime factors are greater than $e\binom{d+n}{n} H(f)$. Suppose $f(\boldsymbol{\xi}) = 0$. Then either $f(\boldsymbol{\xi}^m) = 0$ or $\widehat{h}(\xi) > 1/(md)$.*

Proof   The easy proof is by induction on the number of prime factor of m, writing $m = pm'$. If $f(\boldsymbol{\xi}^{m'}) \ne 0$ we apply the corollary inductively with m' in place of m. If instead $f(\boldsymbol{\xi}^{m'}) = 0$ we apply Lemma 1 to the point $\boldsymbol{\xi}^{m'}$ in place of $\boldsymbol{\xi}$ and note that $\widehat{h}(\boldsymbol{\xi}^{m'}) = m'\widehat{h}(\xi)$.

Proof of Zhang's Theorem

Step 1: By taking the union of X with all its conjugates over $\mathbb{Q}$, one needs only consider the case in which X is defined over $\mathbb{Q}$.

Step 2: By applying an isomorphism $\mathbb{G}_m^n \xrightarrow{\sim} \mathbb{G}_m^n$ followed by a projection $\mathbb{G}_m^n \longrightarrow \mathbb{G}_m^k$ we may assume that X is a hypersurface, defined by a polynomial $f(\mathbf{x})$ of degree d, with integer coefficients.

Step 3 (W.M. Schmidt): Let q be the product of all primes $p \leq e\binom{d+n}{n}H(f)$ and apply the Corollary to Lemma 1 with $m = qj+1$, for $j = 0, 1, \ldots, \binom{d+n}{n}-1$.
If $f(\boldsymbol{\xi}^{qj+1}) \neq 0$ for some j, we obtain a lower bound for $\widehat{h}(\boldsymbol{\xi})$.
If instead $f(\boldsymbol{\xi}^{qj+1}) = 0$ for all such j, we get

$$\sum_{\mathbf{m}} a_{\mathbf{m}}(\boldsymbol{\xi}^{\mathbf{m}})^{qj+1} = 0,$$

with $f(\mathbf{x}) = \sum a_{\mathbf{m}} \mathbf{x}^{\mathbf{m}}$.

We view this as a homogeneous linear system with coefficients $(\boldsymbol{\xi}^{\mathbf{m}})^{qj}$ and unknowns $a_{\mathbf{m}}\boldsymbol{\xi}^{\mathbf{m}}$, so that its determinant must be 0. This is a Vandermonde determinant, and looking at its factorization we see that

$$\boldsymbol{\xi}^{q\mathbf{m}} = \boldsymbol{\xi}^{q\mathbf{m}'}$$

for some $\mathbf{m} \neq \mathbf{m}'$. Now $\mathbf{x}^{q\mathbf{m}} = \mathbf{x}^{q\mathbf{m}'}$ is a finite union of torsion cosets of codimension 1 and none of them is contained in X^* by hypothesis. Thus the intersection Z of X with the coset containing $\boldsymbol{\xi}$ has codimension 2, and we may apply induction on n, replacing X by Z and going to step 2. This ends the proof.

Sketch of proof of Theorem 1

Step 1: Statement (a) follows fairly easily from the fact that $\mathbb{G}_m^n$ has only finitely many subtori of given degree.

Step 2: The proof of (b) is by induction on n.

We may assume that Q is the origin of G, so that $d(P,Q) = \widehat{h}(P)$.

Let X/K be an irreducible subvariety of $\mathbb{G}_m^n$, of degree d; then X can be defined by polynomials $f_i(\mathbf{x}) \in K[\mathbf{x}]$ of degree at most d. Let $\mathcal{M}$ denote the set of all monomials $\mathbf{x}^{\mathbf{m}}$ of degree at most d, of cardinality $r = \binom{d+n}{n}$.

Consider the $r \times r$ matrix

$$\mathcal{X}_{\mathcal{M}} = \begin{pmatrix} \mathbf{x}_1^{\mathbf{m}} \\ \mathbf{x}_2^{\mathbf{m}} \\ \cdots \\ \mathbf{x}_r^{\mathbf{m}} \end{pmatrix}_{\mathbf{m} \in \mathcal{M}}$$

where $\mathbf{x}_1, \ldots, \mathbf{x}_r$ are points of X. Its determinant is 0, because any equation $f_i(\mathbf{x}_j) = 0$ yields a linear relation among the columns of the matrix. This means that $\det(\mathcal{X}_{\mathcal{M}})$ vanishes on X^r.

Now we apply Zhang's Theorem to the variety $\mathcal{X}_{\mathcal{M}} \subset (\mathbb{G}_m^n)^r$ defined by $\det(\mathcal{X}_{\mathcal{M}}) = 0$. By (a), the maximal torsion cosets εH in $\mathcal{X}_{\mathcal{M}}$ are finite in number. For such a coset, H will be properly contained in $(\mathbb{G}_m^n)^r$, so there exists some factor

$$G_\nu = \underbrace{e \times \cdots \times (\mathbb{G}_m)^n \times \cdots \times e}_{(\mathbb{G}_m)^n \text{ is the } \nu\text{-th factor}} \tag{2}$$

not contained in H. Define H' as the subtorus of $\mathbb{G}_m^n$ determined by the obvious projection of $G_\nu \cap H$ on $\mathbb{G}_m^n$. Since $n' = \dim(H') < n$ and $H' \xrightarrow{\sim} \mathbb{G}_m^{n'}$, the induction assumption applies to $(gX) \cap H' \subset H'$ with any $g \in \mathbb{G}_m^n$.

Thus applying inductively Theorem 1 (which is trivial if $n = 1$) we obtain that there are γ' and an integer N' (depending only on d and n) such that if $\boldsymbol{\xi}$ is an algebraic point on $(gX) \cap H'$ then:

  (i) either $\widehat{h}(\boldsymbol{\xi}) \geq \gamma'$,

or

  (ii) $\boldsymbol{\xi}$ belongs to some coset of positive dimension contained in $(gX) \cap H'$,

or

  (iii) $\boldsymbol{\xi}$ belongs to a finite set of at most N' elements.

Clearly we may suppose that the same γ' and N' work for the finitely many H' involved here. Let t be their number, let $N'' = tN' + 1$ and take any r-tuple formed from any given N'' distinct points $\boldsymbol{\xi}_i \in X^\circ(\overline{\mathbb{Q}})$. If some such r-tuple does not lie in any of the finitely many relevant torsion cosets of H contained in $X_{\mathcal{M}}$, then Zhang's Theorem applies to $X_{\mathcal{M}}$ and we are done.

Otherwise, each r-tuple corresponds to some torsion coset of H and thus to some H', as described before. Now at least $(N'')^r / t$ distinct r-tuples will lie in a same torsion coset of H'. Let G_ν be the factor as in (2) not contained in H, and let us associate to each such r-tuple the $(r-1)$-tuple obtained by projection on the $r-1$ trivial factors of G_ν. The number of $(r-1)$-tuples is $(N'')^{r-1}$, so at least $l \geq N''/t > N'$ of the r-tuples will have the same components save for the ν-th component. Therefore, dividing any such r-tuple by a fixed one we obtain, after renumbering, that

$$\boldsymbol{\xi}_i \boldsymbol{\xi}_1^{-1} \in H' \quad \text{for} \quad i = 1, \dots, l.$$

It follows that $\boldsymbol{\xi}_i \boldsymbol{\xi}_1^{-1} \in (\boldsymbol{\xi}_1^{-1} X) \cap H'$ for all $i = 1, \dots, l$ and (setting $g = \boldsymbol{\xi}_1^{-1}$) we may apply one of (i), (ii), (iii) to the points $\boldsymbol{\xi}_i \boldsymbol{\xi}_1^{-1}$.

Since $l > N'$, alternative (iii) cannot occur. Also, alternative (ii) gives that $\boldsymbol{\xi}_i \boldsymbol{\xi}_1^{-1}$ belongs to some coset of positive dimension contained in $\boldsymbol{\xi}_1^{-1} X$, which is excluded because $\boldsymbol{\xi}_i \in X^\circ$. Thus only alternative (i) remains and

$$\widehat{h}(\boldsymbol{\xi}_i) + \widehat{h}(\boldsymbol{\xi}_1) \geq \widehat{h}(\boldsymbol{\xi}_i \boldsymbol{\xi}_1^{-1}) \geq \gamma'.$$

Hence out of any $tN' + 1$ distinct points in $X^\circ(\overline{\mathbb{Q}})$ one of them has height bounded below by $\gamma'/2$.

The last statement of (b) is clear, because conjugation over K does not change height. This ends the proof.

2. The equidistribution theorem   Another approach is due to L. Szpiro, E. Ullmo and S. Zhang [13] and Y. Bilu [1]. The idea, due to S. Zhang [16], is that points of small height under the action of Galois conjugation tend to be equidistributed with respect to a suitable measure.

For $a \in \mathbb{C}^*$ let δ_a be the usual Dirac measure at a and for $\xi \in K$ let

$$\delta_\xi = \frac{1}{[K:\mathbb{Q}]} \sum_{\sigma:K\to\mathbb{C}} \delta_{\sigma\xi}$$

be the probability measure supported at all complex conjugates of ξ, with equal mass at each point.

Theorem 2 (Y. Bilu)   *Let $\{\xi_i\}$ be an infinite sequence of distinct non-zero algebraic numbers such that $h(\xi_i) \to 0$ as $i \to \infty$. Then the sequence δ_{ξ_i} converges in the weak* topology to the uniform probability measure $\mu_{\mathbb{T}}$ on the unit circle $\mathbb{T}$ in $\mathbb{C}^*$.*

Proof   Let μ be a weak* limit of the measures δ_{ξ_i}. Let a_{0i} and d_i be the leading coefficient and degree of a minimal equation for ξ_i. Since the ξ_i are distinct, Northcott's Theorem shows that $d_i \to \infty$.

By hypothesis,

$$h(\xi_i) = \frac{1}{d_i} \log|a_{0i}| + \frac{1}{d_i} \sum_\sigma \log^+ |\sigma\xi_i| \longrightarrow 0 \qquad (3)$$

as $i \to \infty$; this implies $\log|a_{0i}| = o(d_i)$ and that μ is still a probability measure.

By weak* convergence we have

$$\frac{1}{d_i} \sum_\sigma f(\sigma\xi_i) \log^+ |\sigma\xi_i| \longrightarrow \int_\mathbb{C} f(z) \log^+ |z| d\mu(z)$$

for any continuous function $f(z)$ with compact support in $\mathbb{C}$. Thus (3) shows that

$$\int_\mathbb{C} f(z) \log^+ |z| d\mu(z) = 0$$

and μ must be supported in the unit disk $|z| \leq 1$. Using the fact that $h(1/\xi_i) = h(\xi_i) \longrightarrow 0$, we deduce in similar fashion that μ is supported in $|z| \geq 1$. Hence any limit measure μ has support in the unit circle $\mathbb{T}$.

Let D_i be the discriminant of a minimal equation for ξ_i. By writing D_i as the product of $a_{0i}^{2d_i-2}$ and the square of a Vandermonde determinant and estimating the determinant using Hadamard's inequality, we have

$$\frac{1}{d_i} \log|D_i| \leq \log d_i + (2d_i - 2)h(\xi_i),$$

whence

$$0 \leq \log|D_i| = (2d_i - 2)\log|a_{0i}| + \sum_{\sigma \neq \sigma'} \log|\sigma\xi_i - \sigma'\xi_i| = o(d_i^2). \qquad (4)$$

By (4) we easily deduce that μ is a continuous measure and, taking a weak* limit, we get

$$\int_{\mathbb{T}^2} \log\frac{1}{|z - \zeta|} d\mu(z)\, d\mu(\zeta) = 0.$$

Either directly by symmetrization, or by appealing to well-known results of potential theory (see, e.g., [7], Exercise 16.2.2, Th.16.4.3 and Th.16.4.5), we see that this holds only if $\mu(z)$ is the uniform measure on $\mathbb{T}$, completing the proof.

Another way of concluding the proof, which in its discrete version is Bilu's argument, consists in noting that the integral in question is the inner product $\langle \mu * \mu', -\log|1 - e^{i\theta}|\rangle$ with μ' the composition of μ with complex conjugation $z \mapsto \bar{z}$. The n-th Fourier coefficient of $-\log|1 - e^{i\theta}|$ is 0 if $n = 0$ and $1/(2|n|)$ if $n \neq 0$ (expand in Taylor series), so the integral is

$$\sum_{n \neq 0} \frac{|\hat{\mu}(n)|^2}{2|n|} \geq 0.$$

Equality holds only if $\hat{\mu}(n) = 0$ for $n \neq 0$, or in other words only if μ is the uniform measure on $\mathbb{T}$.

Second proof of Zhang's Theorem

We proceed by induction on n. The result is trivial if $n = 1$, because height 0 characterises roots of unity (Kronecker's Theorem).

We may assume that X is defined over $\mathbb{Q}$. Suppose we have an infinite sequence of distinct points $\xi_i \in X^*$ with $\hat{h}(\xi_i) \to 0$. Since X is defined over $\mathbb{Q}$, we may assume that the set of points in this sequence is stable by Galois conjugation in $\overline{\mathbb{Q}}$. For any non-trivial character $\chi(\mathbf{x}) = x_1^{m_1} \cdots x_n^{m_n}$ of $(\mathbb{C}^*)^n$ consider the sequence $\{\chi(\xi_i)\}$. Clearly, $h(\chi(\xi_i)) \leq (\max|m_j|)\, \hat{h}(\xi_i) \to 0$.

Case 1: All sequences $\{\chi(\xi_i)\}$ ultimately consist of distinct elements. In this case, Theorem 2 shows that $\{\chi(\xi_i)\}$ determines the uniform measure on $\chi(\mathbb{T}^n)$. Since this holds for any non-trivial χ, the closure of $\{\xi_i\}$ in $(\mathbb{C}^*)^n$ contains $\mathbb{T}^n$. Hence $\mathbb{T}^n \subset X(\mathbb{C})$, a contradiction because X is a proper algebraic subvariety of $\mathbb{G}_m^n$.

Case 2: There is a non-trivial character χ such that the sequence $\{\chi(\xi_i)\}$ has an element ε_0 occurring infinitely many times. Since $h(\chi(\xi_i)) \to 0$, we have $h(\varepsilon_0) = 0$ and therefore ε_0 is a root of unity by Kronecker's Theorem. If we replace X by $\varepsilon^{-1}X$ where ε is a torsion point such that $\chi(\varepsilon) = \varepsilon_0$ and note that $(\varepsilon^{-1}X)^* = \varepsilon^{-1}(X^*)$, we may assume that $\varepsilon = 1$ and, possibly by taking an infinite subsequence of $\{\xi_i\}$, we may also assume that $\{\xi_i\}$ is contained in the connected component of the identity of the kernel of χ, say H. Now H is a proper subtorus of $\mathbb{G}_m^n$ and we may replace X, $\mathbb{G}_m^n$ by $X \cap H$, H and then use induction.

Remark As such, this proof does not lead to an effective form of Zhang's Theorem.

Remark The following immediate consequence of Theorem 2 is worth noting. If $\xi \neq 0$ is algebraic not a root of unity and has at least $\eta \deg(\xi)$ real conjugates, then $h(\xi) \geq c(\eta) > 0$ with $c(\eta)$ independent of ξ. In particular, totally real algebraic numbers other than ± 1 have height bounded below by an absolute positive constant.

Examples of totally real ξ with small height can be obtained by noticing that if ξ is totally real then $\eta - \eta^{-1} = \xi$ yields a totally real η of degree not exceeding $2 \deg(\xi)$. C.J. Smyth [11], using a result of Schinzel [10], obtained the sharp lower bound $\frac{1}{2}\log((1 + \sqrt{5})/2) \cong 0.2406059$ for $h(\xi)$, attained for $\xi = (1 + \sqrt{5})/2$. He also showed that values of $h(\xi)$ for totally real ξ are dense in a half-line (λ, ∞), with $\lambda \cong 0.2732831$. Here $\lambda = \lim h(\xi_n)$ where $\xi_0 = 1$ and $\xi_{n+1} - \xi_{n+1}^{-1} = \xi_n$.

The minimum above is isolated, and subsequently Smyth [12] determined the first four smallest values of $h(\xi)$ for totally real ξ.

3. The number of solutions of the unit equation We have the following nice result of F. Beukers and H.P. Schlickewei [1].

Theorem 3 *There are absolute computable constants C_1, C_2 with the following property.*

Let Γ be a subgroup of $(\overline{\mathbb{Q}}^*)^2$ with $\mathrm{rank}_\mathbb{Q}(\Gamma) = r < \infty$, where $\mathrm{rank}_\mathbb{Q}(\Gamma)$ is the maximum number of multiplicatively independent elements in Γ. Then the equation

$$x + y = 1, \qquad (x, y) \in \Gamma$$

has at most $C_1 \times C_2^r$ solutions.

This result improves bounds C^{r^2} and $(Cr)^r$ previously obtained by Schlickewei and Schmidt. Beukers and Schlickewei have shown that one may take $C_1 = C_2 = 256$ in Theorem 3, and K.K. Choi (unpublished) has refined their result to $C_1 = 241$, $C_2 = 70$.

It is an interesting problem to determine the maximum number of solutions of the equation $x + y = 1$ with (x, y) in a group Γ of rank r. In this connection, we may remark the following. If we take cosets of $(\Gamma/\mathrm{tors})^4$ in Γ/tors, we are led to finding rational points in K on curves $ax^4 + by^4 = 1$, which have genus 3. It is widely conjectured [4] that the number of rational points in K on a curve of genus $g \geq 2$ is bounded solely in terms of K and g therefore, since we have 4^r cosets, this argument suggests that $C_2 \leq 4$.

Example 1 The following simple argument, due to D. Zagier, yields an example of a subgroup of $\mathbb{Q}^*$ with a large number of solutions. It simplifies a more precise calculation due to P. Erdös, C.L. Stewart and R. Tijdeman [6].

Let $N \geq 2$ be a positive integer. Let M be the number of integers up to x whose prime factors do not exceed $x^{1/N}$. It is clear that

$$M \geq \frac{\pi(x^{1/N})^N}{N!}.$$

Since $\pi(y) > y/\log y$ for $y \geq 17$ and $N! \leq \frac{1}{2}N^N$ we see that $M > 2x/(\log x)^N$ if $x \geq 17^N$. Consider the M^2 sums $n' + n$ where n', n have only prime factors not exceeding $x^{1/N}$. Since $n' + n \leq 2x$ one sum must occur at least $M^2/(2x) > x/(\log x)^{2N}$ times. In other words, there is an integer b such that the equation $n' + n = b$ has at least $x/(\log x)^{2N}$ solutions.

It follows that if Γ is the subgroup of $(\mathbb{Q}^*)^2$ generated on each factor by all primes up to $x^{1/N}$ and by b then the unit equation in Γ has at least $x/(\log x)^{2N}$ solutions, provided $x \geq 17^N$. This group Γ has rank r equal to either $2\pi(x^{1/N})$ or $2\pi(x^{1/N}) + 1$, hence $r \sim 2Nx^{1/N}/\log x$ as $x^{1/N}$ tend to ∞. If we make the asymptotically optimal choice

$$N = \left[\frac{\log x}{2\log\log x} - \frac{\log x}{2(\log\log x)^2} \right]$$

then we verify that the number of solutions is at least

$$\frac{x}{(\log x)^{2N}} = e^{(\frac{\sqrt{2}}{e} + o(1)) \frac{\sqrt{r}}{\sqrt{\log r}}}.$$

Example 2 Consider the equation $ax^m + by^m = 1$ for varying m, corresponding to a group $\Gamma = (x,y)^{\mathbb{Z}}$ of rank 1. We want to find a, b, x, y such that it has the maximum number of solutions for $m \in \mathbb{Z}$. We may assume that $m = 0$ is a solution. Suppose that $m = 1$ is also a solution, so the equation becomes $(y - 1)x^m + (1 - x)y^m - (y - x) = 0$. If we fix two other solutions, say m_1 and m_2, we can eliminate y and obtain an equation for x. Taking $m_1 = 2$ or 3 always leads to roots of unity for x and y. However, taking $m_1 = 4$ and $m_2 = 6$ gives the equation $x^6 + x^5 + 2x^4 + 3x^3 + 2x^2 + x + 1 = 0$. For any root ξ of this equation we see that taking $\eta = -1/(1 + \xi + \xi^3)$, which is another root of the same equation, we have

$$\frac{\eta - 1}{\eta - \xi} \xi^m + \frac{1 - \xi}{\eta - \xi} \eta^m = 1$$

for the six values $m = 0, 1, 4, 6, 13, 52$.

Example 3 The following example gives an equation $u + v = 1$ with at least 2532 solutions u, $v \in \Gamma$, and rank$(\Gamma) = 5$. Let $K = \mathbb{Q}(\alpha)$ with α the real root $\alpha > 1$ of the Lehmer equation

$$x^{10} + x^9 - x^7 - x^6 - x^5 - x^4 - x^3 + x + 1 = 0.$$

This equation has another real root $1/\alpha$ and 8 complex roots all of absolute value 1; we shall refer to the map $\alpha \mapsto 1/\alpha$ as real conjugation in $\mathbb{Q}(\alpha)$. The Mahler height of α is $M(\alpha) = \alpha = 1.176280818259917^+$ and is widely conjectured to be the infimum of the Mahler height of an algebraic number not a root of unity – the so-called Lehmer Conjecture. The group Γ of units of K has rank 5: $\Gamma = \{\pm 1\} \times \; < \alpha, 1 - \alpha, 1 + \alpha, 1 + \alpha + \alpha^2, 1 + \alpha - \alpha^3 >$. Now an extensive computer search for solutions of the corresponding unit equation produced a remarkable total of 2532 solutions.

The following is a plot of the 2532 points $(\log |u|, \log |u'|)$ where u is a real unit and u' is the real conjugate of u.

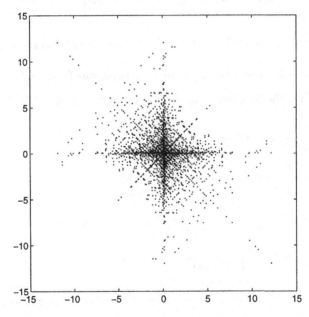

Example 4 The following remark is due to H.W. Lenstra. For a prime p, consider the cyclotomic field $\mathbb{Q}(\sqrt[p]{1})$ and the corresponding unit equation. If $u + v = 1$ and u, v are not real then $\bar{u} + \bar{v} = 1$ is another solution of the unit equation. By Kronecker's Theorem, $\varepsilon = \bar{u}/u$ and $\varepsilon' = \bar{v}/v$ are roots of unity in $\mathbb{Q}(\sqrt[p]{1})$. Solving the system $u + v = 1$, $\varepsilon u + \varepsilon' v = 1$ we get $u = (\varepsilon' - 1)/(\varepsilon' - \varepsilon)$,

$v = (1 - \varepsilon)/(\varepsilon' - \varepsilon)$. Conversely, given distinct roots of unity ε, ε' in $\mathbb{Q}(\sqrt[p]{1})$, not equal to 1, we obtain a solution u, v of the unit equation. Thus the number of complex solutions of the unit equation in $\mathbb{Q}(\sqrt[p]{1})$ is $(p - 1)(p - 2)$.

Example 5 The number of solutions in the maximal real subfield K_p of $\mathbb{Q}(\sqrt[p]{1})$ is much larger. A computer search using cyclotomic units produced 3 solutions in K_5, 42 solutions in K_7, 570 solutions in K_{11}, 1830 solutions in K_{13}, 11700 solutions in K_{17} and 28398 solutions in K_{19}.

The proof of Theorem 3 is obtained by means of a Padé approximation method which originates in the work of Thue, Siegel and Baker. We need a basic result.

Theorem 4 *Let n be a positive integer. Then we can find a non-zero polynomial $M_n(x)$ of degree n with positive integral coefficients without common divisor such that*

$$x^{2n+1} M_n(1 - x) + (-1)^n (1 - x)^{2n+1} M_n(x) - x^n M_n(1 - 1/x) = 0.$$

The polynomial $M_n(x)$ is unique and we have $M_n(1) \leq (\frac{27}{4})^n$.
Moreover, any linear combination

$$\alpha x^{2n+1} M_n(1 - x) + \beta(1 - x)^{2n+1} M_n(x) + \gamma x^n M_n(1 - 1/x)$$

is either identically 0 or has only simple roots outside 0, 1, and ∞.

Proof We have the following transformation of a hypergeometric integral:

$$\int_0^1 x^n (x - 1)^n (x - z)^n \, dx$$

$$= \int_0^{1/z} (zy)^n (zy - 1)^n (zy - z)^n \, d(zy)$$

$$= z^{2n+1} \left(\int_0^1 + \int_1^{1/z} \right) y^n (1 - y)^n (1 - zy)^n \, dy$$

$$= z^{2n+1} \int_0^1 y^n (1 - y)^n (1 - zy)^n \, dy$$

$$+ (-1)^n (1 - z)^{2n+1} \int_0^1 u^n (1 - u)^n (1 - u + zu)^n \, du$$

as one sees using the change of variable $zy = 1 - (1 - z)u$ in the integral from 1 to $1/z$. Thus we obtain three polynomial $A(z)$, $B(z)$, $C(z)$ of degree n such that

$$A(z) + (1 - z)^{2n+1} B(z) + z^{2n+1} C(z) = 0.$$

In terms of hypergeometric functions, we have

$$\frac{(2n+1)!}{n!n!} A(z) = -z^n F(n+1, -n; 2n+2; 1/z),$$

$$\frac{(2n+1)!}{n!n!} B(z) = (-1)^n F(n+1, -n; 2n+2; 1-z),$$

$$\frac{(2n+1)!}{n!n!} C(z) = F(n+1, -n; 2n+2; z),$$

and the above linear relation is one of the 20 three-term linear relations for hypergeometric series found by Kummer [5, I, (29), p. 106].

Rather than pursuing the hypergeometric function connection here, we proceed geometrically to obtain uniqueness. Consider the rational function

$$\varphi(z) = -\frac{z^{2n+1} C(z)}{A(z)}$$

and the associated covering $\varphi : \mathbb{P}^1 \longrightarrow \mathbb{P}^1$. The rational function φ has degree $3n + 1$, fixes $\{0, 1, \infty\}$ and is ramified of order at least $2n$ at $\{0, 1, \infty\}$. By Hurwitz's genus formula we have

$$-2 = -2(3n+1) + 2n + 2n + 2n + \delta$$

where δ is the sum of the orders of ramification of φ at all points outside $\{0, 1, \infty\}$, plus the order of ramification in excess of $2n$ at $\{0, 1, \infty\}$. Hence $\delta = 0$ and φ is unramified outside $\{0, 1, \infty\}$. This proves uniqueness, because if p_1/q_1 and p_2/q_2 were two distinct such functions then $(p_1 + \mu p_2)/(q_1 + \mu q_2)$ would be a one-parameter family with the same properties and we could impose additional ramification by choosing μ, a contradiction.

By uniqueness, we deduce that $(A(z), B(z), C(z))$, $(A(1-z), C(1-z), B(1-z))$ and $(z^n C(1/z), -z^n B(1/z), z^n A(1/z))$ are proportional by a factor ± 1. It follows that $A(z) = (-1)^{n+1} z^n B(1-1/z)$ and $C(z) = (-1)^n B(1-z)$ and $M_n(z)$ is proportional to

$$B(z) = \int_0^1 u^n (1-u)^n (1-u+zu)^n \, du$$

$$= \sum_{m=0}^n \binom{n}{m} \left(\int_0^1 u^{n+m} (1-u)^{2n-m} du \right) z^m$$

$$= \frac{1}{3n+1} \sum_{m=0}^n \left\{ \binom{n}{m} \Big/ \binom{3n}{n+m} \right\} z^m.$$

In order to bound $M_n(1)$, it suffices to remark that

$$\binom{2n}{n}\binom{3n}{2n}\binom{n}{m} \Big/ \binom{3n}{n+m} = \frac{(n+m)!(2n-m)!}{m!(n-m)!(n!)^2} = \binom{n+m}{n}\binom{2n-m}{n}$$

is an integer. Thus

$$M_n(z) = D^{-1} \sum_{m=0}^{n} \binom{n+m}{n}\binom{2n-m}{n} z^m,$$

where D is the greatest common divisor of the integers $\binom{n+m}{n}\binom{2n-m}{n}$. Finally

$$DM_n(1) = \sum_{m=0}^{n} \binom{n+m}{n}\binom{2n-m}{n} = \binom{3n+1}{n} \leq \left(\frac{27}{4}\right)^n.$$

The last statement of the theorem is equivalent with $\varphi(z)$ being unramified outside $\{0, 1, \infty\}$.

Remark One can show that

$$\frac{1}{n}\log(M_n(1)) \sim \frac{\pi}{2\sqrt{3}} + \frac{1}{2}\log(\frac{27}{16}).$$

Now we return to the proof of Theorem 3, following Beukers and Schlickewei. Let $h((x_0 : x_1 : x_2))$ be the height on the projective plane $\mathbb{P}^2$.

Lemma 2 *Suppose that $ax + by = 1$ and $a'x + b'y = 1$ with $\Delta = ab' - a'b \neq 0$. Then we have*

$$h(1 : x : y)) \leq \log 2 + h((1 : a : b)) + h((1 : a' : b')).$$

Proof By Cramer's rule, $x = (b' - b)/(ab' - a'b)$ and $y = (a - a')/(ab' - a'b)$. Hence

$$\begin{aligned}
h((1 : x : y)) &= h((ab' - a'b : b' - b : a - a')) \\
&= \sum_v \max(\log|b' - b|_v, \log|a - a'|_v, \log|ab' - a'b|_v) \\
&\leq \log 2 + \sum_v \max(\log|a|_v, \log|b|_v, 0) + \sum_v \max(\log|a'|_v, \log|b'|_v, 0) \\
&= \log 2 + h((1 : a : b)) + h((1 : a' : b')),
\end{aligned}$$

proving what we want.

Let us write $\mathbf{x} = (x, y)$ and $h(\mathbf{x}) = h((1 : x : y))$. The following corollary of Lemma 2 will be useful.

Corollary *Suppose that $x_1 + y_1 = 1$ and $x_2 + y_2 = 1$ with $x_1, y_1, x_2, y_2 \neq 0$. Then*

$$h(\mathbf{x}_1) \leq \log 2 + h(\mathbf{x}_2 \mathbf{x}_1^{-1}).$$

Proof Apply Lemma 2 with $(x,y) = (x_1, y_1)$, $a = b = 1$, $a' = x_2/x_1$, $b' = y_2/y_1$.

The next lemma is the key to the proof.

Lemma 3 *Suppose that $x_1 + y_1 = 1$ and $x_2 + y_2 = 1$ with $x_1, y_1, x_2, y_2 \neq 0$ and let $n \geq 2$ be an integer. Then*

$$h(\mathbf{x_1}) \leq \log(42) + \frac{1}{n-1}\, h(\mathbf{x_2 x_1^{-2n}}).$$

Proof We set $z = x_1$ in the identity provided by Theorem 4, obtaining

$$M_n(y_1)\, x_1^{2n+1} + (-1)^n M_n(x_1)\, y_1^{2n+1} = x_1^n M_n(-y_1/x_1).$$

We also have $x_2 + y_2 = 1$, hence setting

$$a = x_2/x_1^{2n}, \quad b = y_2/y_1^{2n},$$

$$a' = x_1 M_n(y_1)/(x_1^n M_n(-y_1/x_1)), \quad b' = (-1)^n y_1 M_n(x_1)/(x_1^n M_n(-y_1/x_1)),$$

we have $a x_1^{2n} + b y_1^{2n} = 1$ and $a' x_1^{2n} + b' y_1^{2n} = 1$.

Suppose first that $ab' - a'b \neq 0$. Lemma 2 gives

$$2n\, h(\mathbf{x_1}) = h(\mathbf{x_1^{2n}}) \leq \log 2 + h((1:a:b)) + h((1:a':b')).$$

Now $h((1:a:b)) = h(\mathbf{x_2 x_1^{-2n}})$ and

$$\begin{aligned}
h((1:a':b')) &= h((x_1^n M_n(-y_1/x_1) : x_1 M_n(y_1) : (-1)^n y_1 M_n(x_1)))\\
&= \sum_v \max(\log|x_1^n M_n(-y_1/x_1)|_v, \log|x_1 M_n(y_1)|_v, \log|y_1 M_n(x_1)|_v)\\
&\leq \log|M_n|_{L^1} + (n+1)\, h(\mathbf{x_1}).
\end{aligned}$$

Therefore, using the estimate of the norm of M_n given in Theorem 4, we get

$$2n\, h(\mathbf{x_1}) \leq \log 2 + \log\binom{3n+1}{n} + h(\mathbf{x_2 x_1^{-2n}}) + (n+1)\, h(\mathbf{x_1});$$

the result follows from

$$\log 2 + \log\binom{3n+1}{n} \leq (n-1)\log(42).$$

Now suppose that $ab' - a'b = 0$. This is the same as saying that

$$f(z) = M_n(1-z)\,z^{2n+1} - (x_2/y_2)(-1)^n M_n(z)\,(1-z)^{2n+1}$$

vanishes at $z = x_1$. By the last statement of Theorem 4, we see that $z = x_1$ is a simple zero of $f(z)$ and therefore its derivative does not vanish there. Now we differentiate the identity provided by Theorem 4 and set $z = x_1$. With $U_n(z) = (2n+1)M_n(z) - (1-z)M_n'(z)$, $V_n(z) = nM_n(z) + (1-z)M_n'(z)$, we obtain

$$U_n(y_1)x_1^{2n} + (-1)^{n+1}U_n(x_1)y_1^{2n} = x_1^{n-1}V_n(-y_1/x_1).$$

This time we take

$$a'' = U_n(y_1)/(x_1^{n-1}V_n(-y_1/x_1)), \quad b'' = (-1)^{n+1}U_n(x_1)/(x_1^{n-1}V_n(-y_1/x_1))$$

and apply Lemma 2 to $ax_1^{2n} + by_1^{2n} = 1$ and $a''x_1^{2n} + b''y_1^{2n} = 1$. Note that $ab'' - a''b \neq 0$, since otherwise $f'(z)$ would vanish at $z = x_1$. Lemma 3 now follows in very much the same way as before.

Let Γ be a finitely generated subgroup of $(\overline{\mathbb{Q}}^*)^2$ of rank r. By picking generators of Γ we identify Γ/tors with $\mathbb{Z}^r$. Let $\mathcal{Z}$ be the set of solutions of $x + y = 1$ in Γ and let $\mathcal{Z}_0$ be its image in $\mathbb{Z}^r$ under the projection $\Gamma \longrightarrow \mathbb{Z}^r$. We claim that

$$|\mathcal{Z}| \le 2\,|\mathcal{Z}_0|.$$

Indeed, elements of $\mathcal{Z}$ with same image in $\mathcal{Z}_0$ can be written as $(a\varepsilon, b\zeta)$ with a, b fixed and ε and ζ roots of unity such that $a\varepsilon + b\zeta = 1$. This equation gives us a triangle in the complex plane with vertices at 0, $a\varepsilon$ and 1 and sides of length 1, $|a|$, $|b|$. There are at most two such triangles (intersect a circle of radius $|a|$ and center 0 and a circle of radius $|b|$ and center 1), showing that the projection of $\mathcal{Z}$ onto $\mathcal{Z}_0$ is at most two-to-one.

We define a distance function $\|\ \|$ on $\mathbb{R}^r$ as follows. Let $\mathbf{x} = (x, y) \in \Gamma$ be any representative of $\mathbf{u} \in \mathbb{Z}^r \cong \Gamma/\text{tors}$, and set

$$\|\mathbf{u}\| = \widehat{h}(\mathbf{x}) = h(x) + h(y);$$

this is well defined because changing x or y by a root of unity does not change their height. Next, we extend this to $\mathbb{Q}^r$ by setting $\|\lambda\mathbf{u}\| = |\lambda| \cdot \|\mathbf{u}\|$, which is consistent with the definition of $\|\ \|$, because $h(x^\lambda) = |\lambda| \cdot h(x)$ for $\lambda \in \mathbb{Q}$. Finally, we extend this to $\mathbb{R}^r$ by continuity.

The triangle inequality $\|\mathbf{u} + \mathbf{v}\| \le \|\mathbf{u}\| + \|\mathbf{v}\|$ is clear. It follows that the region

$$B_t = \{\mathbf{u} \in \mathbb{R}^r : \|\mathbf{u}\| \le t\}$$

is a closed, convex, symmetric set. It is also bounded, otherwise Minkowski's theorem would give infinitely many lattice points in C_t, and hence infinitely many elements $\mathbf{x} = (x, y) \in \Gamma$ with $\widehat{h}(\mathbf{x}) = h(x) + h(y) \le t$. Since Γ is finitely generated, this would contradict Northcott's Theorem. Hence $\| \ \|$ is a distance function and B_t is the associated ball of radius t.

It is clear that

$$\tfrac{1}{2}\widehat{h}(\mathbf{x}) \le \max(h(x), h(y)) \le h((1 : x : y)) \le \widehat{h}(\mathbf{x}).$$

In view of this inequality, Lemma 3 shows that if $\mathbf{u}$ and $\mathbf{v}$ are in $\mathcal{Z}_0$ then for any integer $n \ge 2$ we have

$$\|\mathbf{u}\| \le 2 \log(42) + \frac{2}{n-1} \|\mathbf{v} - 2n\mathbf{u}\| . \tag{5}$$

In the same way, the Corollary to Lemma 2 shows that

$$\|\mathbf{u}\| \le \log 4 + 2 \|\mathbf{v} - \mathbf{u}\| . \tag{6}$$

For a vector $\mathbf{u} \in \mathbb{R}^r$ let $\nu(\mathbf{u}) = \mathbf{u}/\|\mathbf{u}\|$ be the associated unit vector with respect to the distance function $\| \ \|$. Suppose that the vectors $\nu(\mathbf{u})$ and $\nu(\mathbf{v})$ are nearly the same, so that $\mathbf{u}$ and $\mathbf{v}$ point about in the same direction. If $\|\mathbf{v}\|$ is much larger than $\|\mathbf{u}\|$, then we can find an integer n such that $\|\mathbf{v} - 2n\mathbf{u}\|$ is small compared with $n\|\mathbf{u}\|$, and now (5) can be used to get an upper bound for $\|\mathbf{u}\|$.

The details are as follows. Let $\varepsilon > 0$ be a small positive constant and let $\mathbf{u}, \mathbf{v} \in \mathcal{Z}_0$ be two points with

$$\|\nu(\mathbf{v}) - \nu(\mathbf{u})\| \le \varepsilon, \qquad \|\mathbf{v}\| \ge 4 \|\mathbf{u}\| .$$

Let $n = [\|\mathbf{v}\|/(2\|\mathbf{u}\|)]$, so that $n \ge 2$ and $0 \le \|\mathbf{v}\| - 2n\|\mathbf{u}\| < 2\|\mathbf{u}\|$. Then (5) gives

$$\|\mathbf{u}\| \le 2 \log(42) + \frac{2}{n-1} \|\mathbf{v} - 2n\mathbf{u}\|$$

$$= 2 \log(42) + \frac{2}{n-1} \big\| \|\mathbf{v}\| \cdot \nu(\mathbf{v}) - 2n \|\mathbf{u}\| \cdot \nu(\mathbf{u}) \big\|$$

$$\le 2 \log(42) + \frac{2}{n-1} \big(\|\mathbf{v}\| - 2n\|\mathbf{u}\| \big) + \frac{4n}{n-1} \|\mathbf{u}\| \cdot \|\nu(\mathbf{v}) - \nu(\mathbf{u})\|$$

$$\le 2 \log(42) + \frac{4 + 4n\varepsilon}{n-1} \|\mathbf{u}\| .$$

We take $\varepsilon = \frac{1}{10}$ and note that $(4 + 4n\varepsilon)/(n-1) \le \frac{1}{2}$ if $n \ge 45$. In this case the above chain of inequalities yields $\|\mathbf{u}\| \le 2\log(42) + \frac{1}{2}\|\mathbf{u}\|$ and $\|\mathbf{u}\| \le$

$4 \log(42) < 15$. If instead $n < 45$, we note that $\|\mathbf{v}\| - 2n \|\mathbf{u}\| < 2 \|\mathbf{u}\|$, hence $\|\mathbf{v}\| \leq 90 \|\mathbf{u}\|$. We have shown:

Lemma 4 *Let* $\mathbf{u}, \mathbf{v} \in Z_0$ *and suppose that* $\|\nu(\mathbf{v}) - \nu(\mathbf{u})\| \leq \frac{1}{10}$ *and* $15 \leq \|\mathbf{u}\| < \|\mathbf{v}\|$. *Then*

$$\|\mathbf{u}\| \leq \|\mathbf{v}\| \leq 90 \|\mathbf{u}\|.$$

Let us call a solution $\mathbf{x}$ of $x + y = 1$ in Γ *large* if $\widehat{h}(\mathbf{x}) = h(x) + h(y) \geq 15$ and *small* otherwise.

The counting of large solutions is done in two steps, first by providing an upper bound for the number of points $\mathbf{u} \in Z_0$ such that $H \leq \|\mathbf{u}\| \leq AH$ lying in a fixed cone

$$C(\varepsilon; a) = \{\mathbf{w} \in \mathbb{R}^r : \|\nu(\mathbf{w}) - a\| \leq \varepsilon\}$$

and then by covering all of $\mathbb{R}^r$ by means of finitely many cones $C(\varepsilon; a_i)$.

For the first step we use (6). Suppose we have two points $\mathbf{u}, \mathbf{v} \in Z_0 \cap C(\varepsilon; a)$ with

$$15 \leq \|\mathbf{u}\| \leq \|\mathbf{v}\| \leq (1 + \delta)\|\mathbf{u}\|.$$

Then (6) gives

$$\begin{aligned}
\|\mathbf{u}\| &\leq \log 4 + 2\|\mathbf{v} - \mathbf{u}\| \\
&= \log 4 + 2 \left\| \|\mathbf{v}\| \cdot \nu(\mathbf{v}) - \|\mathbf{u}\| \cdot \nu(\mathbf{u}) \right\| \\
&\leq \log 4 + 2 \left(\|\mathbf{v}\| - \|\mathbf{u}\| \right) + 2 \|\mathbf{u}\| \cdot \|\nu(\mathbf{v}) - \nu(\mathbf{u})\| \\
&\leq \log 4 + (2\delta + 4\varepsilon) \|\mathbf{u}\|.
\end{aligned}$$

If we take for example $\delta = \frac{1}{4}$ and $\varepsilon = \frac{1}{10}$ we obtain $\|\mathbf{u}\| \leq 10 \log 4 < 14$, contradicting the assumption $\|\mathbf{u}\| \geq 15$. Thus we have a *gap principle*:

$$\|\mathbf{v}\| > \tfrac{5}{4} \|\mathbf{u}\|.$$

Suppose we have m large solutions in a cone $C(\frac{1}{20}; a)$, say $\mathbf{u}_i \in Z_0 \cap C(\frac{1}{20}; a)$ with $15 \leq \|\mathbf{u}_1\| \leq \|\mathbf{u}_2\| \leq \dots$. Then $\|\nu(\mathbf{u}_m) - \nu(\mathbf{u}_1)\| \leq \frac{1}{10}$, therefore by Lemma 4 we have $\|\mathbf{u}_m\| \leq 90 \|\mathbf{u}_1\|$. On the other hand, the preceding gap principle shows that $\|\mathbf{u}_m\| \geq (\frac{5}{4})^{m-1} \|\mathbf{u}_1\|$. Hence $m - 1 \leq \log(90)/\log(5/4) < 21$ and, by a preceding remark, we cannot have more than 42 large solutions with image in any given cone $C(\frac{1}{20}; a)$.

We can cover the unit sphere $\|\mathbf{w}\| = 1$ with not more than $(1 + 2/\varepsilon)^r$ translates of the ball B_ε, with centers on the unit sphere. Indeed, consider a maximal set of non-overlapping balls of radius $\varepsilon/2$ with centers on the unit sphere. Since they are contained in a ball of radius $1 + \varepsilon/2$ and they are disjoint, their number does not exceed $(1 + 2/\varepsilon)^r$. On the other hand, maximality shows that doubling the radius we obtain a covering of the unit sphere. Taking $\varepsilon = \frac{1}{20}$, we infer that we can cover all of $\mathbb{R}^r$ with not more than 41^r cones $C(\frac{1}{20}; a)$.

We have already shown that any such cone determines at most 42 large solutions and we conclude that the total number of large solutions does not exceed 42×41^r.

It remains to give a bound for the number of small solutions, and this is a consequence of the uniform version of Zhang's Theorem, namely Theorem 1. We apply this theorem with $d = 1$ and $n = 2$, and deduce that there are two constants $\gamma = \gamma(1, 2) > 0$ and $N = N(1, 2) < \infty$ such that, for any $a, b \in \overline{\mathbb{Q}}^*$, we have at most N solutions $\mathbf{x} = (x, y) \in (\overline{\mathbb{Q}}^*)^2$ of

$$ax + by = 1, \qquad \widehat{h}(\mathbf{x}) \leq \gamma.$$

Let Γ' be the division group of Γ. By the same argument given before, we can cover the ball B_t with not more than $(1 + 2t/\gamma)^r$ translates of the ball B_γ, and we may assume that these translates are centered at rational points. This means that we can find points $(a_i, b_i) \in \Gamma'$, numbering not more than $(1 + 2t/\gamma)^r$, such that every $\mathbf{x} = (x, y) \in \Gamma$ with $x + y = 1$ and $\widehat{h}(\mathbf{x}) \leq t$ can be written, for some i, as $x = a_i \xi$, $y = b_i \eta$ with $a_i \xi + b_i \eta = 1$ and $h(\xi) + h(\eta) \leq \gamma$. Since there are at most N such (ξ, η) we deduce that the number of (x, y) in question does not exceed $N \times (1 + 2t/\gamma)^r$.

For our purpose of counting small solutions we can take $t = 15$, hence the number of small solutions does not exceed $N \times (1 + 30/\gamma)^r$.

Thus the total number of solutions does not exceed $42 \times 41^r + N \times (1 + 30/\gamma)^r$. This completes the proof of Theorem 3 if Γ is finitely generated.

In the general case, we may suppose that Γ is the division group of a finitely generated group Γ_0 of rank r. Then it suffices to notice that Γ is the direct limit of subgroups $\{g : g^{m!} \in \Gamma_0\}$, all of them finitely generated and of rank r.

Acknowledgements The author is grateful to Y. Bilu and H.P. Schlickewei for sharing preprints of their papers. Also the author wishes to thank D. Bertrand and U. Zannier for several useful comments and criticisms.

References

[1] Y. Bilu. Limit distribution of small points on algebraic tori. Preprint, 1996.

[2] F. Beukers and H.P. Schlickewei. The equation $x + y = 1$ in finitely generated groups. *Acta Arith.*, **78** (1996), 189–199.

[3] E. Bombieri and U. Zannier. Algebraic points on subvarieties of $\mathbb{G}_m^n$. *International Math. Res. Notices*, **7** (1995), 333–347.

[4] L. Caporaso, J. Harris and B. Mazur. Uniformity of rational points. *J. Amer. Math. Soc.*, **10** (1997), 1–35.

[5] A. Erdélyi, W. Magnus, F. Oberhettinger and F.G. Tricomi. *Higher Transcendental Functions*, vol. I. McGraw-Hill, New York, 1953.

[6] P. Erdös, C. L. Stewart and R. Tijdeman. Some diophantine equations with many solutions. *Compositio Math.*, **66** (1988), 37–56.

[7] E. Hille. *Analytic Function Theory*, vol. II. Ginn & Company, Boston, 1962.

[8] S. Lang. *Fundamental of Diophantine Geometry*. Springer-Verlag, Berlin, 1983.

[9] W.M. Schmidt. Heights of points on subvarieties of $\mathbb{G}_m^n$. Preprint, 1994.

[10] A. Schinzel. Addendum to the paper "On the product of the conjugates outside the unit circle of an algebraic number" (Acta Arith. 24 (1973), 385–399). *Acta Arith.*, **26** (1975), 329–331.

[11] C.J. Smyth. On the measure of totally real algebraic integers. *J. Austral. Math. Soc. Ser. A*, **30** (1980), 137–149.

[12] —. On the measure of totally real algebraic integers. II. *Math. of Comp.*, **37** (1981), 205–208.

[13] L. Szpiro, E. Ullmo and S. Zhang. Equidistribution des petits points. *Invent. math.*, **127** (1997), 337–347.

[14] S. Zhang. Positive line bundles on arithmetic surfaces. *Ann. Math.*, (2) **136** (1992), 569–587.

[15] —. Positive line bundles on arithmetic varieties. *J. Amer. Math. Soc.*, **8** (1995), 187–221.

[16] —. Small points and adelic metrics. *J. Alg. Geom.*, **4** (1995), 281–300.

Enrico Bombieri
School of Mathematics, Institute for Advanced Study
Princeton, New Jersey 08540, U.S.A.

2

Remarks on the Analytic Complexity of Zeta Functions

Enrico Bombieri

0. Introduction   This lecture will survey some recent results obtained in collaboration with John Friedlander [2] and discuss some problems arising from our research.

The Dirichlet series for the Riemann zeta-function

$$\zeta(s) = \sum_{n=1}^{\infty} \frac{1}{n^s}, \qquad \text{valid for} \quad \sigma > 1,$$

where $s = \sigma + it$, can be used to compute numerically $\zeta(s)$ for $\sigma > 1$. By absolute convergence one sees that, even for x not very large, the Dirichlet polynomial $\sum_{n \leq x} n^{-s}$ gives a rather good approximation to $\zeta(s)$ with a remainder $o(1)$ as $x \to \infty$. This convergence can be accelerated somewhat by expanding the remainder by means of the Euler–MacLaurin summation formula and by using Dirichlet polynomial approximations with smoothed coefficients.

On the other hand, it is the behaviour of $\zeta(s)$ in the critical strip which is of most interest to us. Here the above polynomial still provides [4, §4.11] (at least away from the pole) a useful approximation to $\zeta(s)$. The smoothed polynomials

$$\sum_{n \leq x} \frac{1}{n^s} \left(1 - \frac{n}{x}\right)^k$$

do an even better job, but all of these only for $x > (1 + o(1))|t|/(2\pi)$. The length of these polynomials is of order t and if say $t > 10^{50}$ the computation of $\zeta(\frac{1}{2} + it)$ becomes unfeasible by all methods known to us. Thus the question arises of determining the *analytic complexity* of $\zeta(s)$ in the critical strip.

In broad terms, the question can be stated as follows. Suppose we are given a small $\varepsilon > 0$ and a computer (perhaps with some constraints on storage). Then we may ask how long it will take to compute $\zeta(\sigma + it)$ within an error ε, as a function of $s = \sigma + it$. If this time is $\ll_\varepsilon (\log |t|)^A$ for some $A = A(\varepsilon)$ one talks about *polynomial complexity*, while if this time is $\gg_\varepsilon |t|^\delta$ for some $\delta = \delta(\varepsilon) > 0$, one talks about *exponential complexity*. In an intermediate situation where this time is $\ll_\varepsilon |t|^{o(1)}$ as t tends to ∞, one talks about *subexponential complexity*.

If $\sigma > 1$ we have

$$\zeta(\sigma + it) = \sum_{n=1}^{x} \frac{1}{n^{\sigma + it}} + O\left(\frac{x^{1-\sigma}}{\sigma - 1}\right)$$

uniformly in t and σ. Thus in order to compute $\zeta(\sigma+it)$ by this formula within an error of ε we need, for any fixed $\sigma > 1$, about $O(\varepsilon^{-1/(\sigma-1)})$ terms. Since $n^{-\sigma-it}$ can be computed in time $\ll_\varepsilon \log^3 |t|$, we see that $\zeta(s)$ has polynomial complexity in any strip $A \le \sigma \le B$, for any fixed $B > A > 1$.

The problem of computing $\zeta(s)$ inside the critical strip is more difficult. On the assumption of the Lindelöf hypothesis one can show that $\zeta(s)$ has subexponential complexity in any strip $\frac{1}{2} + \delta \le \sigma \le B$, for any fixed $\delta > 0$ and B. By the functional equation, we again get subexponential complexity in any strip $A \le \sigma \le \frac{1}{2} - \delta$, provided $\delta > 0$. Thus it appears that the real difficulties in the computation of $\zeta(s)$ occur precisely when σ is very close to $\frac{1}{2}$.

We suggest that *the function $\zeta(\frac{1}{2} + it)$ has exponential complexity and moreover that it requires at least $\gg_\varepsilon |t|^{\frac{1}{2} - o(1)}$ bit operations for its calculation within an error ε.* A similar complexity is also expected for Dirichlet series $L(\frac{1}{2}+it, \chi)$.

It seems quite difficult to attack in full generality the problem of determining the analytic complexity of the zeta function, and a plausible first start consists in noting that all methods so far proposed for computing the zeta function rely on either Dirichlet polynomial approximations or Mellin transform approximations (which we may consider as a continuous version of Dirichlet polynomials). This is our justification for restricting our initial investigation to approximations of this type.

The paper [2] studied such approximations to L-functions of a fairly general type, and showed in many cases that it is not possible to achieve a very good level of approximation using polynomials essentially shorter than the known approximations. The so-called approximate functional equation formulas were not treated in [2], and in this paper we shall sketch a promising way of studying them. The results obtained support the conjecture that the Riemann zeta function $\zeta(\frac{1}{2} + it)$ has exponential complexity $|t|^{\frac{1}{2}+o(1)}$.

1. Results

We shall consider L-functions $L(s)$ having the following properties (compare, for example, [3]):

(H1) $L(s)$ is given by an absolutely convergent Dirichlet series

$$L(s) = \sum_{n=1}^{\infty} a_n\, n^{-s}$$

in the half-plane $\sigma > 1$, with coefficients a_n satisfying $a_1 = 1$ and $a_n \ll n^{o(1)}$.

(H2) $L(s)$ is meromorphic of finite order in the whole complex plane, has only finitely many poles and satisfies a functional equation[1]

$$\Phi(s)L(s) = w\overline{\Phi}(1 - s)\,\overline{L}(1 - s)$$

---

[1]For a function $f(s)$ we define $\overline{f}(s) = \overline{f(\bar{s})}$.

where $\Phi(s) = Q^s \prod_{j=1}^{J} \Gamma(\lambda_j s + \mu_j)$ *and*

$$|w| = 1, \quad Q > 0, \quad \lambda_j > 0, \quad \mathrm{Re}\,\mu_j \geq 0.$$

From the fact that $L(s)$ is of finite order with finitely many poles and satisfies a functional equation of the above type and from the Phragmén -Lindelöf principle, it follows that $L(s)$ has, away from the poles, polynomial growth in any fixed vertical strip. Moreover $L(s)$, for $\sigma < \frac{1}{2}$, has order not less than $|t|^{2\Lambda(\frac{1}{2}-\sigma)}$ and for $\sigma < 0$ has order precisely $|t|^{2\Lambda(\frac{1}{2}-\sigma)}$, where $\Lambda = \sum_{j=1}^{J} \lambda_j$.

Let

$$\theta(t) = \arg \Phi(\tfrac{1}{2} + it), \qquad S(t) = \arg L(\tfrac{1}{2} + it),$$

with the usual conventions about the argument. By a well-known calculation ([4, §9.4]) if $T + \mathrm{Im}\,\mu_j > 0$ for every j the number $N([T, 2T]; L)$ of non-trivial zeros (that is, those not located at the poles of the Γ factors) of $L(s)$ satisfying $T \leq t \leq 2T$ is asymptotically given by

$$
\begin{aligned}
N([T, 2T]; L) &= \frac{1}{\pi} \theta(2T) - \frac{1}{\pi} \theta(T) + S(2T) - S(T) \\
&= \sum_{j=1}^{J} \int_{\lambda_j T + \mathrm{Im}\,\mu_j}^{\lambda_j 2T + \mathrm{Im}\,\mu_j} \log t \, dt + \frac{T}{\pi} (\log Q) + S(2T) - S(T) + O\!\left(\frac{1}{T}\right) \\
&= \frac{\Lambda}{\pi} (2T \log 2T - T \log T) + c_L T + O(\log T),
\end{aligned}
$$

where c_L is a constant depending on L. A similar formula holds for negative T. In our case where $a_1 \neq 0$ we have

$$S(T) = O(\log T)$$

and the constant c_L is explicitly given by

$$c_L = \frac{1}{\pi} \left\{ \log Q + \sum_{j=1}^{J} \lambda_j \log \lambda_j \right\}; \tag{1.1}$$

more generally, assuming $a_{n_0} \neq 0$ and $a_n = 0$ for $n < n_0$ we have

$$S(T) = -\frac{1}{\pi} (\log n_0) T + O(\log T)$$

and (1.1) still holds provided we replace Q by Q/n_0.

One should remark that the choice of the parameter Q and the Gamma factors in the above decomposition of $\Phi(s)$ are not uniquely determined due to the multiplication formula for the Gamma function. However the key quantities Λ and c_L used here are uniquely determined by $L(s)$.

The next important assumption we make about our L-function is that it satisfies a weak zero-density estimate. Let $N(\sigma, T; L)$ denote the number of non-trivial zeros $\rho = \beta + i\gamma$ of L with $0 < \gamma \leq T$ and $\beta \geq \sigma$. Then we assume:

(H3) For any fixed $\delta > 0$, we have

$$N\left(\tfrac{1}{2} + \delta, T; L\right) = o(T \log T).$$

One of the results in [2] obtains a lower bound on the length of the Dirichlet polynomial

$$D_x(s) = \sum_{n \leq x} a_n(x) n^{-s}, \quad |a_1(x)| > \tfrac{1}{2}, \quad a_n(x) \ll n^{o(1)} \tag{1.2}$$

(actually, $\tfrac{1}{2}$ may be replaced by any fixed positive constant) if it is to be a useful approximation to $L(s)$. Specifically, it is proved there

Theorem 1 *Let $L(s)$ satisfy assumptions (H1)–(H3), and let $\varepsilon, \varepsilon' > 0$. Suppose that we have*

$$L(s) = D_x(s) + O(T^{-\varepsilon}) \tag{1.3}$$

on the segment $\{\sigma = \tfrac{1}{2} - \varepsilon', \quad T \leq t \leq 2T\}$. Then $x > T^{2\Lambda - o(1)}$.

Examples of functions to which Theorem 1 applies are given by Dirichlet L-series and Hecke L-series with Grössencharacters, by L-series associated to holomorphic forms on GL(2), and by products of such functions. In particular, Theorem 1 applies to Dedekind zeta functions of abelian extensions of $\mathbb{Q}$, because they can be expressed as finite products of such L-series. It is an interesting question to verify the condition $N(\tfrac{1}{2} + \delta, T) = o(T \log T)$ in other cases, such as Dedekind zeta functions of arbitrary number fields.

We may also remark that if the L-series factorises as a linear combination of products of simpler L-series then it may pay off to compute separately the factors to a high degree of approximation, multiply them together and only then form the desired linear combination. Thus Theorem 1 is most significant only in the case in which the L-series also admits an Euler product and is irreducible (or expected to be irreducible) in Selberg's sense [3].

2. Examples of Dirichlet polynomial approximations It is well known that smoothed truncations of a Dirichlet series can provide good approximations. Let $u(x)$ be a C^∞ function with compact support in $(0, 1]$, such that

$$\int_0^\infty u(t)\, dt = 1$$

and let

$$v(x) = \int_x^\infty u(t)\, dt.$$

Thus $v(0) = 1$ and v has compact support in $[0,1]$. The Mellin transform $\widetilde{u}(s)$ of $u(x)$ is entire of exponential type and rapidly decreasing at ∞ (i.e., faster than any negative power of s) in any fixed vertical strip, and $\widetilde{u}(1) = 1$. The Mellin transform of $v(x)$ is $s^{-1}\widetilde{u}(s+1)$.

We have the integral formula for inverse Mellin transforms

$$v(x) = \frac{1}{2\pi i} \int_{(c)} \widetilde{u}(w+1)\, x^{-w}\, \frac{dw}{w}$$

valid for any $c > 0$ and $x > 0$; here (c) stands for the vertical line $\operatorname{Re} w = c$. Now consider the Mellin transform

$$\frac{1}{2\pi i} \int_{(c)} L(s+w)\, \widetilde{u}(w+1)\, x^w\, \frac{dw}{w}.$$

For $c > 1$, we may integrate term by term, getting

$$\frac{1}{2\pi i} \int_{(c)} L(s+w)\, \widetilde{u}(w+1)\, x^w\, \frac{dw}{w} = \sum_{n \le x} \frac{a_n}{n^s}\, v\left(\frac{n}{x}\right).$$

We may also compute the integral by the calculus of residues, shifting the line of integration to the left. Now an easy estimate gives (see [2])

Theorem 2 *If $x > T^{2\Lambda+\varepsilon}$ then for every fixed N and any fixed strip $A \le \sigma \le B$ we have*

$$L(s) = \sum_{n \le x} \frac{a_n}{n^s}\, v\left(\frac{n}{x}\right) + O(T^{-N})$$

as T tends to ∞.

This result shows that Theorem 1 is sharp. It also shows that if $L(s)$ is expressible in terms of new L-series with smaller Λ's bounded by Λ', then the analytic complexity of $L(s)$ is $\ll_\varepsilon T^{2\Lambda'+o(1)}$.

The situation for approximations to the right of the line $\sigma = \frac{1}{2}$ is quite different. We have

Theorem 3 *On the Lindelöf hypothesis for $L(s)$, in any fixed strip $\frac{1}{2} + \delta \le \sigma \le B$ we have*

$$L(s) = \sum_{n \le x} \frac{a_n}{n^s}\, v\left(\frac{n}{x}\right) + O(x^{-\delta} T^{o(1)})$$

as T tends to ∞.

A result of this type for the Riemann zeta function can be found in Titch-marsh [4, Th.13.3]. The proof follows the same pattern as for Theorem 2, except that this time we shift the integral only to the line $\frac{1}{2} - \sigma$ rather than $-c$ with c large.

This result justifies our remark in the introduction that on the Lindelöf hypothesis $\zeta(s)$ has subexponential complexity away from the critical line $\sigma = \frac{1}{2}$.

3. Approximate functional equation As is well known, it is possible to approximate the Riemann zeta function, using two Dirichlet polynomials rather than one, in a way which allows shorter polynomials, namely:

$$\zeta(s) = \sum_{n \leq x} \frac{1}{n^s} + \chi(s) \sum_{n \leq y} \frac{1}{n^{1-s}} + O(x^{-\sigma}) + O(|t|^{\frac{1}{2}-\sigma} y^{\sigma-1})$$

for $xy = |t|/(2\pi)$, $x, y \geq \frac{1}{2}$, and s in any fixed vertical strip, away from the pole at $s = 1$. Here $\chi(s) = \pi^{\frac{1}{2}-s}\Gamma(\frac{1}{2}(1-s))\Gamma(\frac{1}{2}s)^{-1}$ appears in the functional equation as $\zeta(s) = \chi(s)\zeta(1-s)$.

This shows that the Riemann zeta function has complexity at most $\ll_\varepsilon T^{\frac{1}{2}+o(1)}$ in the critical strip, whence

Theorem 4 *The function $\zeta(s)$ has complexity at most $T^{\frac{1}{2}+o(1)}$ in any fixed vertical strip $A \leq \sigma \leq B$, away from the pole at $s = 1$.*

The same result holds for Dirichlet L-series. An approximate functional equation of length $O(T)$ is also known for L-series associated to cusp forms for $GL(2)$, which have $\Lambda = 1$. Thus L-series associated to $GL(1)$ and $GL(2)$ have complexity at most $T^{\Lambda+o(1)}$ in every fixed vertical strip. It is an interesting open problem to extend this result to L-series on $GL(n)$ with $n \geq 3$, even on the assumption of the appropriate Riemann hypothesis; the difficulty lies with the approximate functional equation.

It is likely that there should be an analogue to Theorem 1 for approximate functional equations of this type stating that such an approximate functional equation in the range $T \leq t \leq 2T$ requires $xy > T^{2\Lambda-o(1)}$. We shall give in the next section a sketch of a proof of this statement in the case in which $\Lambda = \frac{1}{2}$.

4. Mean values A simple minded approach to Theorem 1 in the case of $\zeta(s)$ and more generally L-series associated to $GL(1)$ (what matters here is $\Lambda \leq \frac{1}{2}$) consists in comparing mean-value estimates of the L-series and of the Dirichlet polynomial approximation. This is done as follows.

By a well-known mean-value estimate (e.g., take $Q = 1$ in [1, Théorème 10]) we have, for fixed $\delta > 0$,

$$\int_T^{2T} \left| \sum_{n \leq x} \frac{a_n(x)}{n^{\frac{1}{2}-\delta+it}} \right|^2 dt \ll T \sum_{n \leq x} \frac{|a_n(x)|^2}{n^{1-2\delta}} + \sum_{n \leq x} n^{2\delta}|a_n(x)|^2$$

$$\ll T x^{2\delta+o(1)} + x^{1+2\delta+o(1)}. \qquad (4.1)$$

On the other hand, applying first our assumed approximation, and then the functional equation, we get, for each $t \in [T, 2T]$, the lower bound

$$\left| \sum_{n \leq x} \frac{a_n(x)}{n^{\frac{1}{2} - \delta + it}} \right| + O(T^{-\varepsilon}) \gg \left| L(\tfrac{1}{2} - \delta + it) \right| \gg |t|^\delta \left| \overline{L}(\tfrac{1}{2} + \delta - it) \right|. \qquad (4.2)$$

Squaring and integrating (4.2) over t, we note that the left-hand side of (4.1) is

$$\gg T^{2\delta} \int_T^{2T} |L(\tfrac{1}{2} + \delta + it)|^2 dt \gg T^{1+2\delta}. \qquad (4.3)$$

Now (4.1) and (4.3) give $x \gg T^{1-o(1)}$, which is the conclusion of Theorem 1.

Next we quickly sketch an argument showing that if we have an approximate functional equation of type

$$\zeta(\tfrac{1}{2} + it) = \sum_{m=1}^x \frac{a_m}{m^{\frac{1}{2}+it}} + \chi(\tfrac{1}{2} + it) \sum_{n=1}^y \frac{b_n}{n^{\frac{1}{2}-it}} + o(1) \qquad (4.4)$$

with x, y positive integers and $a_n, b_n = O(n^{o(1)})$ and valid for $t \in [T, 2T]$ as $T \to \infty$, then we must have $xy \gg T^{1-o(1)}$. The argument extends easily to the general case in which $\Lambda = \tfrac{1}{2}$, using the approximations given by Theorem 2.

Let $\varphi(z)$ be a smooth function with compact support in $[1, 2]$ and consider the integral

$$I(T) = \int_{-\infty}^\infty \left| \zeta(\tfrac{1}{2} + it) - \sum_{m=1}^x \frac{a_m}{m^{\frac{1}{2}+it}} - \chi(\tfrac{1}{2} + it) \sum_{n=1}^y \frac{b_n}{n^{\frac{1}{2}-it}} \right|^2 \varphi\left(\frac{t}{T}\right) dt$$
$$= o(T). \qquad (4.5)$$

In the range $[T, 2T]$, we replace $\zeta(\tfrac{1}{2} + it)$ by a partial sum $\sum_{n=1}^T n^{-\frac{1}{2}-it}$, which we can do introducing an error $\ll T^{-\frac{1}{2}}$, see for example [4, Th.4.11]. Now $\zeta(\tfrac{1}{2} + it) = \chi(\tfrac{1}{2} + it)\zeta(\tfrac{1}{2} - it)$, therefore we may write the integral (4.5) as

$$\int_{-\infty}^\infty \left(\sum_{n=x+1}^T \frac{1}{n^{\frac{1}{2}+it}} + \sum_{m=1}^x \frac{(1 - a_m)}{m^{\frac{1}{2}+it}} - \chi(\tfrac{1}{2} + it) \sum_{n=1}^y \frac{b_n}{n^{\frac{1}{2}-it}} \right)$$
$$\times \left(\chi(\tfrac{1}{2} - it) \sum_{n=y+1}^T \frac{1}{n^{\frac{1}{2}+it}} - \sum_{m=1}^x \frac{\overline{a}_m}{m^{\frac{1}{2}-it}} + \chi(\tfrac{1}{2} - it) \sum_{n=1}^y \frac{(1 - \overline{b}_n)}{n^{\frac{1}{2}+it}} \right) \varphi\left(\frac{t}{T}\right) dt.$$

Now we expand the integrand and note that $\chi(\tfrac{1}{2} + it)\chi(\tfrac{1}{2} - it) = 1$, thus expressing $I(T)$ as a sum of nine integrals

$$I(T) = \sum_{i,j=1}^3 I_{ij} \qquad (4.6)$$

where the integrand in I_{ij} is the product of the i-th summand in the first three sums into parentheses and the j-th summand in the second three sums into parentheses. Then the crux of the argument can be explained as follows.

We may assume $x, y \ll T^{1-\varepsilon}$ for some $\varepsilon > 0$. We have

$$I_{12} = -\int_{-\infty}^{\infty} \left(\sum_{n=x+1}^{T} \frac{1}{n^{\frac{1}{2}+it}} \right) \times \left(\sum_{m=1}^{x} \frac{\bar{a}_m}{m^{\frac{1}{2}-it}} \right) \varphi\left(\frac{t}{T}\right) dt,$$

therefore this integral is majorized by

$$\sum_{m=1}^{x} \sum_{n=x+1}^{T} (mn)^{-\frac{1}{2}+o(1)} \left| \int_{-\infty}^{\infty} (n/m)^{-it} \varphi\left(\frac{t}{T}\right) dt \right|. \tag{4.7}$$

Since φ is a smooth function, the integrals in (4.7) are $\ll (x/T)^N$ for any fixed positive N, thus showing that the contribution of I_{12} to $I(T)$ is negligible. The same argument applies to I_{31}.

Next we verify that

$$\int_{-\infty}^{\infty} \chi(\tfrac{1}{2}+it) A^{it} \varphi\left(\frac{t}{T}\right) dt \ll T^{-N} \tag{4.8}$$

for $2\pi A$ outside the interval $[(1-\varepsilon)T, (2+\varepsilon)T]$. This shows that the contribution of I_{32} is also negligible on the assumption that $xy \ll T^{1-\varepsilon}$.

The contribution of the integrals I_{2j} and I_{i3} is $o(T \log T)$, because a_m and b_n must be close to 1 on average if (4.4) holds. Thus the main contribution to $I(T)$ in the decomposition (4.6) comes from the integral I_{11}. By (4.8) this contribution is

$$\sum_{\substack{x<m, y<n \\ (1-\varepsilon)T<2\pi mn<(2+\varepsilon)T}} (mn)^{-\frac{1}{2}} \int_{-\infty}^{\infty} \chi(\tfrac{1}{2}-it)(mn)^{-it} \varphi\left(\frac{t}{T}\right) dt,$$

and we can evaluate the integrals accurately using stationary phase. It turns out that for $2\pi mn \in [T, 2T]$ the integral equals

$$2\pi (mn)^{\frac{1}{2}} \varphi(2\pi mn/T)$$

with a rather small error. This yields

$$I(T) = I_{11} + o(T \log T) \sim 2\pi \sum_{x<m, y<n} \varphi(2\pi mn/T)$$

$$\sim \left(\int_{-\infty}^{\infty} \varphi(z)dz \right) T \log\left(\frac{T}{xy}\right).$$

Since $xy \ll T^{1-\varepsilon}$ this contradicts $I(T) = o(T)$, concluding the sketch of the proof.

A refinement of this argument should give the more precise result that an approximation (4.4) with $t \sim T$ requires $xy \geq (1 + o(1))T/(2\pi)$; this would be best possible in view of the approximate functional equation.

5. Counting zeros

The above arguments based on square mean-values fail if $\Lambda > \frac{1}{2}$. However, an alternative method based on counting zeros works in general, provided we assume (H3). This may be considered as comparing the integral of $\log |L(s)|$ and $\log |D_x(s)|$.

The strategy of our proof consists in counting the number of zeros of $L(s)$ and $D_x(s)$ in a suitable rectangle. We expect this number to be about the same if D_x is a good approximation to L. We have

Proposition *Let $D_x(s)$ given by (1.2) satisfy $|a_1(x)| > \frac{1}{2}$ and $a_n(x) \ll n$. Then, uniformly for $\alpha < \sigma < \infty$ we have*

$$\arg D_x(s) \ll (|\alpha| + 1) \log x.$$

Let also $N(\alpha, T, T + H; D_x)$ denote the number of zeros of $D_x(s)$ satisfying $\sigma \geq \alpha$, $T \leq t \leq T + H$, where $H \leq T$. Then, uniformly for $-H < \alpha < -1$, we have

$$N(\alpha, T, T + H; D_x) \leq \frac{H}{2\pi} \log x + O\big(|\alpha|^{\frac{1}{2}} H^{\frac{1}{2}} \log x\big), \qquad (5.1)$$

where the implied constant is absolute.

The proof is a standard application of Littlewood's lemma.

We note in this respect that the finite Euler product

$$f(s) = \prod_{p \leq \log x} \big(1 - p^{-s}\big)$$

has length

$$\exp\Big(\sum_{p \leq \log x} \log p \Big) = x^{1+o(1)}$$

by the prime number theorem and has, for $T < t \leq 2T$, zeros on the imaginary axis at $t = 2n\pi / \log p$. Their number is

$$\sum_{p \leq \log x} \Big(\Big\lceil \frac{2T \log p}{2\pi} \Big\rceil - \Big\lceil \frac{T \log p}{2\pi} \Big\rceil \Big) \sim \frac{T}{2\pi} \log x,$$

again by the prime number theorem. Thus the bound given in (5.1) is asymptotically sharp.

Now we briefly sketch the proof of Theorem 1, referring to [2] for details. The idea is to use hypothesis (H3) to show that $L(s)$ behaves most of the time almost as if one had a Riemann hypothesis at our disposal, save for an exceptional set of small measure. This can now be used in two ways, one to use the approximation (1.3) to show that there is a 'good' set $E \subset [T, 2T]$ of measure $m(E) = (1 - O(\delta))T$, consisting of intervals I_ν of length 1, such that

$$\int_E \log |L(\tfrac{1}{2} - \delta + it)| \, dt$$
$$\leq \int_E \log |D_x(\tfrac{1}{2} - \delta + it)| \, dt + O\left(\left(\delta \log \tfrac{1}{\delta}\right)^2 T \log T\right). \qquad (5.2)$$

Since $L(s)$ behaves in a horizontal strip at a good interval almost as if one had a Riemann hypothesis, application of Littlewood's lemma shows that

$$\int_E \log |L(\tfrac{1}{2} - \delta + it)| \, dt \geq (\delta - O(\delta^2))\Lambda T \log T. \qquad (5.3)$$

Finally the Proposition can be used to obtain

$$\int_E \log |D_x(\tfrac{1}{2} - \delta + it)| \, dt \leq \tfrac{1}{2}(\delta + O(\delta^2))T \log x, \qquad (5.4)$$

and Theorem 2 follows from (5.2), (5.3), and (5.4) as $\delta \to 0$.

References

[1] E. Bombieri. *Le Grand Crible dans la Théorie Analytique des Nombres* (2nd ed). Astérisque **18**, Société Math. de France, Paris, 1987.
[2] E. Bombieri and J.B. Friedlander. Dirichlet polynomial approximations to zeta functions. *Ann. Sc. Norm. Sup. Pisa (IV)*, **22** (1995), 517–544.
[3] A. Selberg. Old and new conjectures and results about a class of Dirichlet series. *Proc. Amalfi Conf. Analytic Number Theory* (E. Bombieri, A. Perelli, S. Salerno and U. Zannier, Eds.), Univ. di Salerno, Salerno, 1992, pp. 367–385; *Collected Papers*, vol. II, Springer-Verlag, Berlin 1991, pp. 47–63.
[4] E.C. Titchmarsh. *The Theory of the Riemann Zeta-Function* (2nd ed., 2nd prnt., revised by D.R. Heath-Brown). Oxford Univ. Press, Oxford, 1986.

Enrico Bombieri
School of Mathematics, Institute for Advanced Study
Princeton, New Jersey 08540, U.S.A.

3

Normal Distribution of Zeta Functions and Applications

ENRICO BOMBIERI and ALBERTO PERELLI

0. Introduction   A fundamental result of Selberg obtains the existence of a distribution function for $\log \zeta(\frac{1}{2} + it)$ and more generally of $\log L(\frac{1}{2} + it)$ for a wide class of L-functions. Equally importantly, Selberg showed how the logarithms $\log L(\frac{1}{2} + it)$ of 'independent' (in a sense to be clarified later on) L-functions are also statistically independent. These results have applications to the study of the distribution of zeros of certain classes of Dirichlet series, which will be examined in this paper; detailed proofs can be found in [1] and [2].

1. Main result   We work in the moderately general setting of the paper [1] of Bombieri and Hejhal, and consider N functions $L_1(s), \ldots, L_N(s)$ satisfying the following basic hypotheses [1, §3]†:

Hypothesis $\mathcal{B}$

(I) Each L_j will be assumed to have an Euler product of the form

$$L_j(s) = \prod_p \prod_{i=1}^{d}(1 - \alpha_{ip}p^{-s})^{-1}$$

with $|\alpha_{ip}| \leq p^{\theta}$ for some fixed $0 \leq \theta < \frac{1}{2}$ and $i = 1, \ldots, d$.

(II) We shall also suppose that

$$\sum_{p \leq X} \sum_{i=1}^{d} |\alpha_{ip}|^2 = O(X^{1+\varepsilon})$$

holds for every $\varepsilon > 0$.

(III) Each L_j is assumed to have an analytic continuation to all of $\mathbb{C}$ as a meromorphic function of finite order with a finite number of poles, all on the line $\sigma = 1$, and to satisfy a functional equation of the form

$$\Phi_j(s) = \varepsilon_j \overline{\Phi}_j(1 - s),$$

---

† With respect to [1], we do not ask here that the L_j satisfy the same functional equation in *(III)*.

where $\Phi_j(s) = Q_j^s \prod_{i=1}^m \Gamma(\lambda_{ij}s + \mu_{ij})L_j(s)$, $Q_j > 0$, $\lambda_{ij} > 0$, $\mathrm{Re}\,\mu_{ij} \geq 0$ *and* $|\varepsilon_j| = 1$.

(IV) The coefficients of $L_1, \ldots, L_N$ will be assumed to satisfy

$$\sum_{p \leq X} \frac{a_j(p)\overline{a_k(p)}}{p} = \delta_{jk}\,n_j\,\log\log X + c_{jk} + O\left(\frac{1}{\log X}\right)$$

for $X \geq 2$, certain positive constants $n_1, \ldots, n_N$ and Kronecker's δ_{jk}.

We explicitly remark that all data involved in hypothesis $\mathcal{B}$ concerning a function $L_j(s)$ may depend on j.

We refer to §3 of [1] for a thorough discussion of hypothesis $\mathcal{B}$, of its standard consequences and of several examples of functions satisfying it. Here we point out only that $\mathcal{B}(II)$ implies that both the Dirichlet series and the Euler product of $L_j(s)$ converge absolutely for $\sigma > 1$, $\mathcal{B}(I)$ ensures that $L_j(s) \neq 0$ for $\sigma > 1$ and $\mathcal{B}(III)$ gives rise to the familiar notions of critical strip, critical line and trivial and non-trivial zeros. Moreover, writing

$$\Lambda_j = \sum_{i=1}^m \lambda_{ij}, \tag{1}$$

$$N_j(t) = \#\{\rho : L_j(\rho) = 0,\ 0 \leq \mathrm{Re}\,\rho \leq 1 \text{ and } 0 \leq \mathrm{Im}\,\rho \leq t\}$$

and

$$S_j(t) = \frac{1}{\pi}\mathrm{arg}L_j(\tfrac{1}{2} + it),$$

for sufficiently large t we have

$$N_j(t) = \frac{\Lambda_j}{\pi} t\log t + c_j t + c_j' + S_j(t) + O\left(\frac{1}{t}\right) \tag{2}$$

with certain constants c_j and c_j'.

Condition $\mathcal{B}(IV)$, introduced by Selberg [12], plays a special role, since it provides a form of 'near-orthogonality' of the functions $L_j(s)$; the 'independence' alluded to at the beginning of the section is this property of near-orthogonality. One can show that $\mathcal{B}(IV)$ implies the linear independence over $\mathbb{C}$ of $L_1(s), \ldots, L_N(s)$; see Bombieri–Hejhal [1] and Kaczorowski–Perelli [8] for further results in this direction.

We expect that the functions $L_j(s)$ satisfy the Generalized Riemann Hypothesis, GRH. As a substitute of it in our arguments, we will instead assume the following density estimate which trivially follows from GRH. Let

$$N_j(\sigma, T) = \#\{\rho : L_j(\rho) = 0,\ \mathrm{Re}\,\rho \geq \sigma \text{ and } |\mathrm{Im}\,\rho| \leq T\}.$$

Hypothesis $\mathcal{D}$

  There exists $0 < a < 1$ *such that*

$$N_j(\sigma, T) \ll T^{1-a(\sigma-1/2)} \log T$$

for $j = 1, \ldots, N$, *uniformly for* $\sigma \geq \frac{1}{2}$.

The main point in introducing hypothesis $\mathcal{D}$ is that, unlike GRH, it can be verified in many interesting cases. In fact, it has been proved by Selberg [11] for the Riemann zeta function, by Fujii [5] for Dirichlet L-functions, and by Luo [10] in the more difficult case of L-functions attached to holomorphic modular forms for GL(2).

The following result is stated in [12] and proved‡ in detail in [1].

Theorem B   *Suppose the L_j satisfy hypotheses $\mathcal{B}$ and $\mathcal{D}$. Then the functions*

$$\frac{\log|L_1(\frac{1}{2}+it)|}{\sqrt{\pi n_1 \log\log t}}, \frac{\arg L_1(\frac{1}{2}+it)}{\sqrt{\pi n_1 \log\log t}}, \ldots, \frac{\log|L_N(\frac{1}{2}+it)|}{\sqrt{\pi n_N \log\log t}}, \frac{\arg L_N(\frac{1}{2}+it)}{\sqrt{\pi n_N \log\log t}}$$

become distributed, in the limit of large t, like independent random variables, each having gaussian density $\exp(-\pi u^2)du$.

The proof of Theorem B in [1] is a variant of Selberg's [11] moments method, which leads in a more direct way to the distribution function for the $\log L_j(\frac{1}{2} + it)$. The study of moments gives stronger results, in particular about the rate of convergence. The best exposition of Selberg's method is still [11]; moments of $\log\zeta(\sigma + it)$ with $\sigma > \frac{1}{2}$ are studied in the book [9] by A. Laurinčikas, but we will not need those results here.

For the applications we have in mind the following 'short interval' version of Theorem B is more appropriate.

Let $M \geq e$, write $h = M/\log T$ and

$$V_j(t) = \frac{\log L_j(\frac{1}{2} + i(t+h)) - \log L_j(\frac{1}{2} + it)}{\sqrt{2\pi n_j \log M}},$$

and let μ_T denote the associated probability measure on $\mathbb{C}^N$, defined by

$$\mu_T(\Omega) = \frac{1}{T}\left|\{t \in [T, 2T] : (V_1(t), \ldots, V_N(t)) \in \Omega\}\right| \tag{3}$$

for every open set $\Omega \subset \mathbb{C}^N$. Moreover, let $e^{-\pi\|z\|^2}$ denote the gaussian measure on $\mathbb{C}^N$ and let $d\omega$ be the euclidean density on $\mathbb{C}^N$.

---

‡ Hypothesis $\mathcal{B}$ in [1] requires the L_j to satisfy the same functional equation. However, inspection of [1] shows that the functional equation is never used in the proof of Theorem B below and therefore the result holds with hypothesis $\mathcal{B}$ as stated here.

Theorem 1 *Let $L_1(s), \ldots, L_N(s)$ satisfy hypotheses $\mathcal{B}$ and $\mathcal{D}$ and suppose $M = M(T) \leq \log T / \log \log T$ tends to ∞ as $T \to \infty$. Then, as $T \to \infty$, μ_T converges weakly to the gaussian measure with associated density $e^{-\pi \|\mathbf{z}\|^2} d\omega$.*

The next section contains a sketch of the proof of Theorem 1, and the following two sections describe two applications, one [1] about zeros of linear combinations of the L_j, and the other [2] about distinct zeros of the L_j.

2. Sketch of proof of Theorem 1 In this section we follow the arguments in §5 of [1]. For $\sigma > 1$ and $j = 1, \ldots, N$ we write

$$\log L_j(s) = \sum_{n=1}^{\infty} c_j(n) \Lambda_1(n) n^{-s}, \qquad \Lambda_1(n) = \begin{cases} 0 & n = 1, \\ \dfrac{\Lambda(n)}{\log n} & n \geq 2, \end{cases}$$

and denote by $u(x)$ a real positive C^∞ function with compact support in $[1, e]$ and by $\tilde{u}(s)$ its Mellin transform. We also write

$$v(x) = \int_x^{\infty} u(t) dt$$

and assume that u is normalized so that $v(0) = 1$.

The first step, an exercise in explicit formulas, proves

Lemma 1 *Assume (I), (II), and (III) of hypothesis $\mathcal{B}$. Then*

$$\log L_j(\tfrac{1}{2} + it) = \sum_{n=1}^{\infty} \frac{c_j(n) \Lambda_1(n)}{n^{\frac{1}{2}+it}} v(e^{\log n / \log X})$$

$$+ \sum_{\rho} \int_{\frac{1}{2}}^{\infty} \frac{1}{\rho - s} \tilde{u}(1 + (\rho - s) \log X) d\sigma + O(1), \qquad (4)$$

where $|t|$ is sufficiently large and not the ordinate of a zero of $L_j(s)$, where $2 \leq X \leq t^2$ and where ρ runs over zeros of $L_j(s)$ with $0 \leq \operatorname{Re} \rho \leq 1$.

This gives us a decomposition of (4) as

$$\log L_j(\tfrac{1}{2} + it) = D_j(\tfrac{1}{2} + it, X) + R_j(\tfrac{1}{2} + it, X)$$

where $D_j(\tfrac{1}{2} + it, X)$ is the Dirichlet series in the right-hand side of (4).

The idea behind this decomposition is the following. Suppose X is a function of T which goes to ∞ slower than any positive power of T, hence $X = T^{o(1)}$. The $D_j(s, X)$ are rather short Dirichlet series and their mixed moments of any positive integral order can be calculated easily, because the contribution of the diagonal dominates over the rest, due to (IV) of hypothesis $\mathcal{B}$. As $T \to \infty$ these moments, after normalization of D_j, agree with the corresponding moments for the gaussian measure. In contrast with Selberg's approach, there is no need to control the higher moments of the remainders R_j, and control of the L^1-norm is all we need to ensure weak convergence to gaussian measure of the probability measure associated to L_j. This proves Theorem B. Hypothesis $\mathcal{D}$ is used precisely in this last step, to achieve a saving of $1/\log X$ with respect to the trivial estimate.

For the proof of Theorem 1 we need asymptotics for the mixed moments of short interval differences of the $D_j(\frac{1}{2} + it, X)$.

For sufficiently large M, write $h = M/\log T$ and

$$\Sigma_j(t) = D_j(\tfrac{1}{2} + i(t + h), X) - D_j(\tfrac{1}{2} + it, X).$$

Moreover, let $k_j \geq 0$ and $l_j \geq 0$, $j = 1, \ldots, N$ be integers and let us abbreviate $\mathbf{k} = (k_1, \ldots, k_N)$, $K_j = k_1 + \cdots + k_j$, $K = K_N$ and similarly for $\mathbf{l}$, L_j and L. We also write $\mathbf{k}! = \prod_{j=1}^{N} k_j!$.

Lemma 2 *Assume hypothesis $\mathcal{B}$ and let*

$$X \leq T^{1/(K+L+1)}, \quad M \leq \log T/\log\log X.$$

Write

$$\Sigma_j(t) = \sum_{n=1}^{\infty} \frac{b_j(n)}{n^{\frac{1}{2}+it}}, \quad b_j(n) = c_j(n)\Lambda_1(n)v(e^{\log n/\log X})(e^{-ih\log n} - 1).$$

Then

$$\int_T^{2T} \prod_{j=1}^{N} \left(\Sigma_j(t)\right)^{k_j} \left(\overline{\Sigma_j(t)}\right)^{l_j} dt = \delta_{\mathbf{k},\mathbf{l}}\, \mathbf{k}!\, T \prod_{j=1}^{N} \left(2n_j \log^+ \left(\tfrac{1}{2}M\frac{\log T}{\log X}\right)\right)^{k_j}$$

$$+ O\left(T\left(\log^+ \left(\tfrac{1}{2}M\frac{\log T}{\log X}\right)\right)^{\frac{1}{2}(K+L-1)}\right).$$

The proof is an adaptation of the proof of Lemma 6 of [1], which gives the asymptotics of the mixed moments of the $D_j(\frac{1}{2} + it, X)$ rather than of the differences $D_j(\frac{1}{2} + i(t + h), X) - D_j(\frac{1}{2} + it, X)$.

Lemma 2 yields the asymptotic estimate

$$\int_T^{2T} \prod_{j=1}^{N} \left(\Sigma_j(t)\right)^{k_j} \left(\overline{\Sigma_j(t)}\right)^{l_j} dt \sim \delta_{\mathbf{k},\mathbf{l}}\, \mathbf{k}!\, T \prod_{j=1}^{N} (2n_j \log M)^{k_j}$$

if $M = M(T) \to \infty$ with $M \ll \log T / \log \log T$, which we shall suppose.

The rest of the proof follows closely that of Theorem B of [1]. Let $M = M(T) \ll \log T / \log \log T$ tend to ∞ as $T \to \infty$ and choose

$$\log X = \frac{\log T}{(\log M)^{1/4}},$$

so that

$$\log^+\left(\tfrac{1}{2}M\frac{\log T}{\log X}\right) \sim \log M, \quad \frac{\log X}{\log T} = (\log M)^{-1/4},$$

and $X = T^{o(1)}$. Moreover, let

$$U_j(t) = (2\pi n_j \log M)^{-1/2}\Sigma_j(t)$$

and define $\tilde{\mu}_T$ to be the associated probability measure on $\mathbb{C}^N$, as in (3).

Then, assuming that $M \leq \log T / \log \log T$ and arguing exactly as in the proof of Theorem B of [1], from Lemma 2 we see that $\tilde{\mu}_T$ converges weakly, as $T \to \infty$, to the gaussian measure $e^{-\pi\|\mathbf{z}\|^2}$.

Finally, the required L^1-norm estimate for R_j is given in [1], Corollary to Lemma 3; it is a rather easy consequence of (4) and of Hypothesis $\mathcal{D}$. With the above choice of X we get the bound

$$\frac{1}{T}\int_T^{2T} |V_j(t) - U_j(t)|\, dt \ll (\log M)^{-1/4},$$

which tends to 0 as T tends to ∞. Hence μ_T converges to the same gaussian measure, completing the sketch of our proof.

3. Applications: Zeros of linear combinations of L-functions

In this section, we assume that the functions L_j satisfy a mild condition of well-spacing for zeros, namely:

Hypothesis $\mathcal{H}_0$

For each Euler product $L_j(s)$, we have

$$\lim_{\varepsilon \to 0+}\left\{\varlimsup_{T \to \infty}\frac{\#(T \leq \gamma \leq 2T \;:\; \gamma' - \gamma \leq \varepsilon/\log T)}{T \log T}\right\} = 0.$$

The following result is proved in [1]. Consider a linear combination

$$f(s) = \sum_{j=1}^{N} b_j L_j(s)$$

where the L_j satisfy a same functional equation and the phases of the coefficients b_j have been adjusted so that

$$b_j Q^s \prod_{i=1}^{m} \Gamma(\lambda_i s + \mu_i) L_j(s) \quad \text{is real on the critical line} \quad \sigma = \tfrac{1}{2}$$

for $j = 1, \ldots, N$.

Theorem A *Suppose the L_j satisfy hypotheses $\mathcal{B}$ with a same functional equation, GRH and $\mathcal{H}_0$. Let also the b_j be as before. Then, in the limit of large T, almost all zeros of $f(s) = \sum b_j L_j(s)$ are simple and lie on the critical line $\sigma = \tfrac{1}{2}$.*

One may ask what can be done here with weaker hypotheses or no hypotheses at all. First of all, one can dispense with GRH and replace it with hypothesis $\mathcal{D}$ and that almost all zeros of L_j are on the critical line, with the same conclusion.

If $L_j(s) = L(s, \chi_j)$ are Dirichlet L-series for a set of primitive even (or odd) characters to a same modulus q, then one can prove, assuming only that almost all zeros of L_j are on the critical line, that there are $\gg T \log T$ zeros of $f(s)$ on the critical line if $N = 2$ or 3, and in fact $\gg T \log T$ simple zeros if $N = 2$, thus dispensing entirely with hypothesis $\mathcal{H}_0$. The same conclusion holds if we assume that a sufficiently large percentage of zeros of L_j are on the critical line, rather than almost all zeros. For the time being, existing lower bounds for the number of zeros of Dirichlet L-series on the critical line are too low for proving unconditionally that if $N = 2$ then $f(s)$ has $\gg T \log T$ zeros on the critical line.

Unconditionally, A.A. Karatsuba [7] has shown that if the L_j are Dirichlet L-series for primitive characters, then $f(s)$ has at least $\gg T(\log T)^{\beta - \varepsilon}$ zeros on the critical line, for a certain $\beta > 0$ (this is also mentioned* in [1]). The proof, which is quite intricate, uses 'simultaneous mollifiers' for the Dirichlet series L_j.

We refer to [1] for a detailed proof of Theorem A. However, the basic idea in the proof of Theorem A is quite simple and can be described as follows. Let $F_j(t) = b_j e^{i\varphi} L_j(\tfrac{1}{2} + it)$ where $e^{i\varphi}$ is the phase factor arising from the functional equation of the L_j, so every $F_j(t)$ is real. According to Theorem B, the quantities $\log |F_j(t)| / \sqrt{\pi n_j \log \log t}$ behave like independent random variables. Thus one expects $F(t) = \sum_j F_j(t)$ to behave most of the time like $(1 + o(1)) F_j(t)$ where j is determined by $|F_j(t)| > |F_{j'}(t)|$ for $j' \neq j$. Let us

---

* The paper [7] misattributes to Selberg a result, stated there as Theorem 5. What Selberg showed, and what the first author communicated to Karatsuba in Amalfi, is the lower bound $\gg T \sqrt{\log T}$ and not the lower bound $\gg T \sqrt{\log \log T}$ stated there. This is particularly misleading in view of the statement of Theorem 6 in [7]. Compare [1], p. 822, footnote.

say that in this case F_j dominates at t. Since all F_j are real, there will be a zero of $F(t)$ between t_1 and t_2 whenever F_j dominates at t_1 and t_2 and $F_j(t_1)$ and $F_j(t_2)$ have opposite signs. On the other hand, one expects $F_j(t)$ to have $(\Lambda/\pi)M + o(M)$ zeros in any interval $[T, T + M/\log T]$. On the assumption of GRH, this is indeed the case for most such intervals. If these zeros are reasonably well spaced, and it is here that one uses hypothesis $\mathcal{H}_0$, one can then show that $F(t)$ has about the same number of sign changes as there are zeros of $F_j(t)$ in that interval.

The main difficulty in carrying out a rigorous proof along these lines arises from the fact that all the required properties can be established only in a weak measure theoretic setting. Theorem B and Theorem 1 provide the necessary tools in the proofs.

4. Applications: Distinct zeros of L-functions Another application of Theorem 1 shows that independent L-functions (in the sense of $\mathcal{B}(IV)$) satisfying functional equations with the same 'number' of Γ factors have a positive proportion of distinct zeros, counting multiplicities. In order to state our result, we define the counting function $D(T, L_1, L_2)$ of distinct non-trivial zeros, counted with multiplicity, of two functions $L_1(s)$ and $L_2(s)$ to be

$$D(T, L_1, L_2) = \sum_{\substack{0 \le \operatorname{Re} \rho \le 1 \\ 0 \le \operatorname{Im} \rho \le T}} \max(m_1(\rho) - m_2(\rho), 0),$$

where ρ runs over zeros of $L_1(s)L_2(s)$ without multiplicity and $m_j(\rho)$ is the multiplicity of ρ as a zero of L_j.

Let also $\Lambda_j = \sum_h \lambda_{hj}$ be defined by (1), so that $\prod_h \Gamma(\lambda_{hj}s + \mu_{hj})$ is the product of the gamma factors appearing in the functional equation for L_j.

Theorem 2 Let $L_1(s)$ and $L_2(s)$ satisfy hypotheses $\mathcal{B}$ and $\mathcal{D}$ and suppose that $\Lambda_1 = \Lambda_2$. Then

$$D(T, L_1, L_2) \gg T \log T.$$

As mentioned before, hypothesis $\mathcal{D}$ has been proved for large classes of automorphic GL(1) and GL(2) L-functions.

The first result of this type has been obtained by Fujii [6] in the case of two primitive Dirichlet L-functions, by means of Selberg's moments method. The problem of counting *strongly-distinct* zeros, i.e., zeros placed at different points, is more difficult. The best results are due to Conrey–Ghosh–Gonek [3], [4]. They deal with this problem, in the case of two primitive Dirichlet L-functions, by considering the more difficult question of getting simple zeros of $L(s, \chi_1)L(s, \chi_2)$, and obtain that there are $\gg T^{6/11}$ such zeros up to T. Moreover, if the Riemann Hypothesis is assumed for one of the two functions, then a positive proportion of such zeros is obtained. Note however that the analytic techniques in [3] and [4] do not extend beyond the case of GL(1) L-functions, mainly because the product of two GL(n) L-functions is a GL($2n$) L-function, and all analytic techniques based on square mean-values break down for GL(m) L-functions as soon as $m \geq 3$.

Our method of proof of Theorem 2 does not seem to be capable of refinement to yield the expected result that almost all zeros of L_1 and L_2 are distinct, let alone strongly-distinct. Moreover, since our proof is by contradiction, we do not obtain an estimate for the constant implicit in the Vinogradov symbol $\gg$ appearing in Theorem 2. This however can be done, at the cost of introducing substantial additional complications in the proof.

We give a sketch of the proof of Theorem 2.

The idea is to consider the number of zeros (counted with multiplicity) of L_1 and L_2 in rather short intervals of type $[t, t+h]$ with $t \in [T, 2T]$ and $h = M/\log T$, with $M = M(T)$ tending to ∞ arbitrarily slowly as $T \to \infty$.

Since we assume $\Lambda_1 = \Lambda_2$, using (2) we verify that the excess $\Delta(t, h)$ of zeros of L_1 over L_2 in such an interval is given by

$$\Delta(t, h) = \big(S_1(t+h) - S_1(t)\big) - \big(S_2(t+h) - S_2(t)\big) + O(h). \tag{5}$$

We also have

$$\operatorname{Im} V_j(t) = \frac{\pi}{\sqrt{2\pi n_j \log M}} \big(S_j(t+h) - S_j(t)\big) \tag{6}$$

by definition of V_j and S_j, therefore if

$$\operatorname{Im} V_2(t) < 0 \quad \text{and} \quad \operatorname{Im} V_1(t) > 1 \tag{7}$$

we obtain from (5) and (6) the estimate

$$\Delta(t, h) = \frac{1}{\pi}(2\pi n_1 \log M)^{\frac{1}{2}} \operatorname{Im} V_1(t) - \frac{1}{\pi}(2\pi n_2 \log M)^{\frac{1}{2}} \operatorname{Im} V_2(t) + O(h)$$

$$\geq \frac{1}{\pi}(2\pi n_1 \log M)^{\frac{1}{2}} + O(h).$$

Denote by E the set of $t \in [T, 2T]$ for which (7) holds. In order to get a lower bound for $|E|$, we consider the set

$$\Omega = \{(z_1, z_2) \in \mathbb{C}^2 : \operatorname{Im} z_1 > 1 \quad \text{and} \quad \operatorname{Im} z_2 < 0\},$$

so that

$$|E| = T \,\mu_T(\Omega).$$

Now from Theorem 1 we get

$$\lim_{T\to\infty} \mu_T(\Omega) = \int_\Omega e^{-\pi\|z\|^2} d\omega \gg 1,$$

thus showing that $|E| \gg T$. Hence we deduce the existence of $\gg T/h$ values $t_r \in [T, 2T]$, with $|t_r - t_s| \geq h$ if $r \neq s$, such that

$$\Delta(t_r, h) \geq \frac{1}{\pi}(2\pi n_1 \log M)^{\frac{1}{2}} + O(h).$$

It follows that

$$D(2T, L_1, L_2) - D(T, L_1, L_2) \geq \sum_r \Delta(t_r, h) \gg \frac{\sqrt{\log M}}{M} T \log T.$$

Since $M = M(T)$ tends to infinity arbitrarily slowly, we must have

$$D(2T, L_1, L_2) - D(T, L_1, L_2) \gg T \log T,$$

completing the proof of Theorem 2.

References

[1] E. Bombieri and D.A. Hejhal. On the distribution of zeros of linear combinations of Euler products. *Duke Math. J.*, **80** (1995), 821–862.

[2] E. Bombieri and A. Perelli. Distinct zeros of *L*-functions, to appear.

[3] J.B. Conrey, A. Ghosh and S.M. Gonek. Simple zeros of the zeta function of a quadratic number field. I. *Invent. math.*, **86** (1986), 563–576.

[4] —. Simple zeros of the zeta function of a quadratic number field. II. *Analytic Number Theory and Diophantine Problems* (A.C. Adolphson, J.B. Conrey, A. Ghosh and R.I. Yager, Eds.), Birkhäuser, Boston, 1987, pp. 87–114.

[5] A. Fujii. On the zeros of Dirichlet's *L*-functions. I. *Trans. A.M.S.*, **196** (1974), 225–235.

[6] —. On the zeros of Dirichlet's *L*-functions. V. *Acta Arith.*, **28** (1976), 395–403.

[7] A.A. Karatsuba. A new approach to the problem of zeros of some Dirichlet series. *Trudi Mat. Inst. Steklova*, **207** (1994); A.M.S. English translation in *Proc. Steklov Inst. Math.*, **207** (1995), 163–177.

[8] J. Kaczorowski and A. Perelli. Functional independence of the singularities of a class of Dirichlet series. *American J. Math.*, to appear.

[9] A. Laurinčikas. *Limit Theorems for the Riemann Zeta-Function*. Mathematics and Its Applications, vol. **352**, Kluwer Academic Publishers, Dordrecht, 1996.

[10] W. Luo. Zeros of Hecke *L*-functions associated with cusp forms. *Acta Arith.*, **71** (1995), 139–158.

[11] A. Selberg. Contributions to the theory of the Riemann zeta-function. *Archiv Math. Naturvid.*, **48** (1946), 89–155; *Collected Papers*, vol. I, Springer-Verlag, Berlin, 1989, pp. 214–280.

[12] —. Old and new conjectures and results about a class of Dirichlet series. *Proc. Amalfi Conf. Analytic Number Theory* (E. Bombieri, A. Perelli, S. Salerno and U. Zannier, Eds.), Univ. di Salerno, Salerno, 1992, pp. 367–385; *Collected Papers*, vol. II, Springer, Berlin, 1991, pp. 47–63.

Enrico Bombieri
School of Mathematics, Institute for Advanced Study
Princeton, New Jersey 08540, U.S.A.

Alberto Perelli
Dipartmento di Matematica, Universita di Genova
Via Dodecaneso 35, I-16146 Genova, Italy

4

Goldbach Numbers and Uniform Distribution mod 1

JÖRG BRÜDERN and ALBERTO PERELLI

1. Introduction   As an illustrative example of their celebrated circle method, Hardy and Littlewood were able to show that subject to the truth of the Generalized Riemann Hypothesis, almost all even natural numbers are the sum of two primes, the yet unproven hypothesis being removed later as a consequence of Vinogradov's work. Natural numbers which are representable as the sum of two primes are called Goldbach numbers, and it is still not known whether all, or at least all but finitely many, even positive integers ≥ 4 are of this form. The best estimate for the number of possible exceptions is due to Montgomery and Vaughan [4]. They showed that all but $O(X^{1-\delta})$ even natural numbers not exceeding X are Goldbach numbers, for some small $\delta > 0$.

More information about possible exceptions can be obtained by considering thin subsequences of the even numbers, with the aim of showing that almost all numbers in the subsequence are Goldbach numbers. In this direction, short intervals have been treated by various authors. It is now known that almost all even numbers in the interval $[X, X + X^{11/160+\epsilon}]$ are Goldbach numbers (see Baker, Harman and Pintz [1]). Perelli [5] has shown that almost all even positive values of an integer polynomial satisfying some natural arithmetical conditions are Goldbach numbers.

In this paper we give further examples of sequences with this property. They arise, roughly speaking, as integer approximations to values of real-valued functions at integers points whose fractional parts are uniformly distributed modulo one. We need some notation to make this precise.

Let $\mathcal{V} \subset \mathbb{R}$ be a set with $|v_1 - v_2| \geq 2$ for any distinct values $v_1, v_2 \in \mathcal{V}$. We write $V(X)$ for the number of $v \in \mathcal{V}$ with $v \leq X$, and associate with $\mathcal{V}$ the exponential sum

$$T(\alpha) = \sum_{v \in \mathcal{V}, \, v \leq X} e(\alpha v). \qquad (1)$$

The set $\mathcal{V}$ is called *admissible sequence* if for any $A > 0$ there is a $B = B(A) > 0$ with $B \to \infty$ as $A \to \infty$, and such that uniformly for $X^{-1}(\log X)^A \leq |\alpha| \leq (\log X)^A$ one has

$$T(\alpha) \ll V(X)(\log X)^{-B}. \qquad (2)$$

We remark that integer sequences are not admissible since (2) clearly fails for $\alpha = 1$. For an admissible sequence $\mathcal{V}$, let $E(\mathcal{V}, X)$ denote the number of all

$v \in \mathcal{V}$ with $v \leq X$ for which the inequality $|v - p_1 - p_2| < 1$ has no solution in odd primes p_1, p_2.

Theorem *Let $\mathcal{V}$ be an admissible sequence. Then we have, for any $C > 0$,*

$$E(\mathcal{V}, X) \ll V(X)(\log X)^{-C}.$$

For any $v \in \mathcal{V}$ the interval $[v-1, v+1)$ contains exactly one even integer, which we denote by $\tilde{v}$. Since $\mathcal{V}$ is admissible, the $\tilde{v}$ are all distinct. Our Theorem implies, in particular, that almost all $\tilde{v}$ are Goldbach numbers, since $\mathcal{V}$ is not an integer sequence.

Among the numerous examples of admissible sequences, the simplest is $\{n^c : n \in \mathbb{N}\}$ where $c > 1$ is a fixed non-integral real number. In this case the exponential sum estimate (2) is a simple consequence of van der Corput's method (see Graham and Kolesnik [3], Theorem 2.8, for example). A more interesting feature of our result is that it opens a road to test Goldbach's conjecture on average over sequences which increase more rapidly than any polynomial. In fact, Vinogradov's method can be used to show that $n^{\log \log n}$ is admissible, and an even thinner admissible sequence is $\exp((\log n)^\gamma)$, where $1 < \gamma < \frac{3}{2}$ is fixed. This also follows from Vinogradov's method, but lack of space does not permit to discuss the details here. In this particular example, the exponential sum estimate (2) can be considerably improved, and a variant of the argument below yields a better bound for the exceptional set. We refer to our forthcoming paper [2].

We have already remarked that integer sequences are never admissible. However, as will be clear from the argument below, it is easy to write down an analogue of our Theorem for integer sequences. But, as we shall see in §4, the numbers $\tilde{v}$ are uniformly distributed in residue classes, and this is in marked contrast with many interesting integer sequences which usually obey a more complicated distribution law. In such a case, the exponential sum $T(\alpha)$ will have more than one peak in $[0, 1]$. It appears difficult to imagine a variant of our Theorem for integer sequences which covers all situations to which the underlying principle might be applied. Perelli [5] can serve as a model of what is possible in this direction.

2. A simple lemma We require an easy auxiliary formula. For a real number v we write $v = [v] + \{v\}$ with $[v] \in \mathbb{Z}$ and $\{v\} \in [0, 1)$, as usual.

Lemma 1 *Let $f : \mathbb{R} \to \mathbb{C}$ be a function of period 1 which is integrable over $[0, 1]$, and let*

$$\hat{f}(n) = \int_0^1 f(\beta)e(-\beta n)\, d\beta$$

denote its n-th Fourier coefficient. Then we have, for any $v \in \mathbb{R}$,

$$\int_{-\infty}^{\infty} f(\alpha)e(-\alpha v)\left(\frac{\sin \pi\alpha}{\pi\alpha}\right)^2 d\alpha = (1 - \{v\})\hat{f}([v]) + \{v\}\hat{f}([v] + 1).$$

Proof For a function $g \in L^1(\mathbb{R})$ we write

$$\hat{g}(\alpha) = \int_{-\infty}^{\infty} g(\beta)e(-\alpha\beta)\,d\beta$$

for its Fourier transform. Now let

$$K(\alpha) = \left(\frac{\sin \pi\alpha}{\pi\alpha}\right)^2.$$

Then

$$\hat{K}(\alpha) = \max(1 - |\alpha|, 0). \tag{3}$$

We now observe that the left hand side of the proposed formula equals

$$\sum_{n=-\infty}^{\infty} \int_{n}^{n+1} f(\alpha)e(-\alpha v)K(\alpha)\,d\alpha$$

$$= \int_{0}^{1} f(\beta) \sum_{n=-\infty}^{\infty} e(-v(\beta+n))K(\beta+n)\,d\beta. \tag{4}$$

Since the Fourier transform of $G(t) = e(-v(\beta+t))K(\beta+t)$ is $\hat{G}(s) = e(\beta s)\hat{K}(v + s)$, from the Poisson summation formula and (3) we get

$$\sum_{n=-\infty}^{\infty} e(-v(\beta+n))K(\beta+n) = \sum_{n=-\infty}^{\infty} e(\beta n)\hat{K}(v+n)$$

$$= e(-\beta[v])\hat{K}(\{v\}) + e(-\beta[v] - \beta)\hat{K}(\{v\} - 1).$$

Inserting this into (4), the lemma follows.

3. The Fourier transform method We now embark on the main argument. The letter p, with or without subscripts, always denotes an *odd* prime. With this convention understood, we then consider the sum

$$r(v) = \sum_{|p_1+p_2-v|<1} (1 - |p_1 + p_2 - v|)(\log p_1)(\log p_2)$$

where $v > 1$ is real. Then, on writing

$$S(\alpha) = \sum_{p\leq X} (\log p)e(\alpha p),$$

we note that for $1 \le v \le X$ one has

$$r(v) = \int_{-\infty}^{\infty} S(\alpha)^2 e(-\alpha v) K(\alpha) \, d\alpha. \tag{5}$$

Now let $A > 100$ and put $P = (\log X)^{5A}$. Let $\mathfrak{M}$ denote the the union of all intervals $|\alpha - a/q| \le P/X$ with $(a,q) = 1$ and $1 \le q \le P$. Write $\mathfrak{M}_1 = \mathfrak{M} \cap [0,1]$. By Lemma 1 with $f(\alpha) = S(\alpha)^2$ for $\alpha \in \mathfrak{M}$ and $f(\alpha) = 0$ otherwise, we deduce that

$$\int_{\mathfrak{M}} S(\alpha)^2 e(-\alpha v) K(\alpha) \, d\alpha = (1 - \{v\}) R([v]) + \{v\} R([v] + 1), \tag{6}$$

where

$$R(m) = \int_{\mathfrak{M}_1} S(\alpha)^2 e(-\alpha m) \, d\alpha.$$

The integral $R(m)$ is the familiar major arc contribution in the classical treatment of the binary Goldbach problem. For integers m with $1 \le m \le X + 1$ one has

$$R(m) = m \sum_{q \le P} \frac{\mu(q)^2}{\phi(q)^2} c_q(m) + O(X (\log X)^{-4A}), \tag{7}$$

where

$$c_q(m) = \sum_{\substack{a=1 \\ (a,q)=1}}^{q} e\left(\frac{am}{q}\right) = \phi(q) \frac{\mu(q/(q,m))}{\phi(q/(q,n))}$$

is Ramanujan's sum. For a proof of (7) see Vaughan [6, (3.22)], for example.

The natural next step would be to complete the sum in (7) to the singular series. Unfortunately this does not work in our case for individual m. A standard argument based on the bound

$$|c_q(m)| \le (q, m) \tag{8}$$

shows that

$$\sum_{q > Q} \left| \frac{\mu(q)^2}{\phi(q)^2} c_q(m) \right| \ll Q^{-1} m^\epsilon. \tag{9}$$

Therefore, we can indeed complete the sum in (7), and the resulting series is absolutely convergent and can be written as an Euler product. This yields

$$\sum_{q \le P} \frac{\mu(q)^2}{\phi(q)^2} c_q(m) = \mathfrak{S}(m) - \sum_{q > P} \frac{\mu(q)^2}{\phi(q)^2} c_q(m) \tag{10}$$

with

$$\mathfrak{S}(m) = \prod_{p \nmid m} \left(1 - \frac{1}{(p-1)^2}\right) \prod_{p \mid m} \left(1 + \frac{1}{p-1}\right),$$

where the products extend over all primes including 2 (contrary to our overall convention, in this formula only). A straightforward use of (9) with $Q = P$ to estimate the error in (10) would produce an unacceptable loss. We circumvent this difficulty and apply (9) with $Q = X^{1/3}$. Then, for $m \le X + 1$, we can rearrange (10) as

$$\sum_{q \le P} \frac{\mu(q)^2}{\phi(q)^2} c_q(m) = \mathfrak{S}(m) - \Delta(m) + O(X^{-1/4}) \tag{11}$$

where

$$\Delta(m) = \sum_{P < q \le X^{1/3}} \frac{\mu(q)^2}{\phi(q)^2} c_q(m). \tag{12}$$

By (6), (7), and (11), we now conclude as follows:

Lemma 2 *For real v with $1 \le v \le X$ one has*

$$\int_{\mathfrak{M}} S(\alpha)^2 e(-\alpha v) K(\alpha) \, d\alpha = (1 - \{v\})[v](\mathfrak{S}([v]) + \Delta([v]))$$
$$+ \{v\}([v] + 1)(\mathfrak{S}([v] + 1) + \Delta([v] + 1)) + O(X(\log X)^{-4A}).$$

We end our preliminary analysis of the major arc contribution by extracting a lower bound from the asymptotic relation in Lemma 2. This will only be possible with some control on $\{v\}$ and Δ. With this in mind we first observe that $\mathfrak{S}(m) = 0$ if m is odd, and if m is even, then $\mathfrak{S}(m) \gg 1$. Let B be the positive number defined via (1) and (2). We may assume, without loss of generality, that $B < A$. If we also suppose that $\frac{1}{2}X < v \le X$ and

$$\{v\} \ge (\log X)^{-B/8}, \quad 1 - \{v\} \ge (\log X)^{-B/8}, \tag{13}$$

$$|\Delta([v])| \le (\log X)^{-B/6}, \quad |\Delta([v] + 1)| \le (\log X)^{-B/6}, \tag{14}$$

then Lemma 2 gives

$$\int_{\mathfrak{M}} S(\alpha)^2 e(-\alpha v) K(\alpha) \, d\alpha \gg X(\log X)^{-B/8}.$$

In the next section, we show that (13) and (14) hold for almost all v in an admissible sequence.

4. Counting exceptions   We remarked already that admissible sequences have uniformly distributed fractional parts. The following simple upper bound is enough for our purposes.

Lemma 3   *Let $\mathcal{V}$ be an admissible sequence. Then we have, for any $B > 0$,*

$$\#\{v \in \mathcal{V}: v \le X, a \le \{\alpha v\} \le b\} \ll V(X)(b - a + (\log X)^{-B}) \qquad (15)$$

uniformly for $0 \le a < b \le 1$ and $X^{-1/2} \le |\alpha| \le 1$.

Proof   This is well known and may be refined to an asymptotic formula. We give a quick proof for completeness. Let N be a natural number. The Féjer kernel

$$\Psi_N(t) = \frac{1}{N}\left(\frac{\sin \pi Nt}{\sin \pi t}\right)^2 = \sum_{|k| \le N} \left(1 - \frac{|k|}{N}\right) e(kt)$$

is a non-negative function of period 1, and one has $\Psi_N(t) \ge \frac{1}{4}N$ for $|t| \le 1/(2N)$.

Now fix $B > 0$ and write $\delta = b - a$. We first prove the lemma for $(\log X)^{-B} \le \delta \le \frac{1}{4}$. Take $N = [\delta^{-1}]$. Then, the left hand side of (15) does not exceed

$$\frac{4}{N} \sum_{v \in \mathcal{V}, v \le X} \Psi_N\left(\alpha v - \frac{\delta}{2}\right) = \frac{4}{N^2} \sum_{|k| \le N} (N - |k|) e\left(\frac{k\delta}{2}\right) T(k\alpha).$$

The term with $k = 0$ contributes $4N^{-1}T(0) \ll \delta V(X)$. For $0 < |k| \le N$ we have $|k\alpha| \ll (\log X)^B$. Choose A in (2) so large that $B \le B(A)$. Since we may suppose that $B(A) < A$, we deduce that $T(k\alpha) \ll V(X)(\log X)^{-B}$, and Lemma 3 follows in this case.

If $0 < \delta \le (\log X)^{-B}$ we increase the left hand side of (15) and replace b with $a + (\log X)^{-B}$. We then appeal to the previous case, and again confirm (15). For $\delta > \frac{1}{4}$ the bound (15) is trivial, and the proof of Lemma 3 is complete.

As an immediate corollary of Lemma 3 (with $\alpha = 1$) we deduce that (13) holds for all but $O(V(X) (\log X)^{-B/8})$ values of $v \le X$ in an admissible sequence $\mathcal{V}$.

We now count exceptions to (14). We shall establish the bound

$$\sum_{v \in \mathcal{V}, v \le X} |\Delta([v])| \ll V(X)(\log X)^{-B/2}. \qquad (16)$$

From this we infer that $|\Delta([v])| \ge (\log X)^{-B/6}$ can hold for at most $O(V(X) \times (\log X)^{-B/3})$ values of $v \le X, v \in \mathcal{V}$. Now note that if $\mathcal{V}$ is admissible, then the sequence $v + 1$, $v \in \mathcal{V}$, is also admissible; this follows from (1) and (2).

Hence, (16) also holds with $[v] + 1$ in place of $[v]$. Therefore, the number of $v \leq X, v \in \mathcal{V}$, for which (14) does not hold is bounded by $O(V(X)(\log X)^{-B/3})$. We summarize these results in the following

Lemma 4 *Let $\mathcal{V}$ be an admissible sequence. Then, for all but*

$$O(V(X)(\log X)^{-B/8})$$

values of $v \in \mathcal{V}$ in the range $\frac{1}{2}X < v \leq X$, we have

$$\int_{\mathfrak{M}} S(\alpha)^2 e(-\alpha v) K(\alpha) \, d\alpha \gg X(\log X)^{-B/8}.$$

Here B is defined via (1) and (2).

It remains to verify (16). For real v and $d \in \mathbb{N}$ one has $[v] \equiv 0 \,(\text{mod } d)$ if and only if $\{v/d\} < 1/d$. In Lemma 3 we choose $a = 0$, $b = \alpha = 1/d$. Then, uniformly for $1 \leq d \leq X^{1/2}$, we find that

$$\sum_{[v]\equiv 0 \,(\text{mod } d)} 1 \ll \frac{V(X)}{d} + V(X)(\log X)^{-B} \tag{17}$$

where, here and in the following argument, the summation is restricted to $v \in \mathcal{V}, v \leq X$.

By (8), (12), and (17) we get

$$\sum_v |\Delta([v])| \leq \sum_{P<q\leq X^{1/3}} \sum_v \frac{\mu(q)^2}{\phi(q)^2}(q, [v])$$

$$\ll \sum_{P<q\leq X^{1/3}} \frac{\mu(q)^2}{\phi(q)^2} \sum_{d|q}(V(X) + dV(X)(\log X)^{-B})$$

$$\ll V(X) \sum_{q>P} \frac{\mu(q)^2 d(q)}{\phi(q)^2} + \frac{V(X)}{(\log X)^B} \sum_{q\leq X^{1/3}} \frac{qd(q)}{\phi(q)^2},$$

and (16) follows immediately.

5. The minor arcs We complete the proof of the theorem by estimating the contribution of the minor arcs $\mathfrak{m} = \mathbb{R} \setminus \mathfrak{M}$ to the integral (5).

We begin by cutting off a tail from $\mathfrak{m}$. For any $k \in \mathbb{N}$,

$$\left| \int_{k<|\alpha|\leq k+1} S(\alpha)^2 e(-\alpha v) K(\alpha) \, d\alpha \right|$$

$$\leq \int_{k<|\alpha|\leq k+1} \frac{|S(\alpha)|^2}{|\alpha|^2} \, d\alpha$$

$$\leq \frac{2}{k^2} \int_0^1 |S(\alpha)|^2 \, d\alpha \ll k^{-2} X \log X$$

and hence

$$\int_{|\alpha|\geq(\log X)^A} S(\alpha)^2 e(-\alpha v)K(\alpha)\,d\alpha \ll X(\log X)^{-A}, \tag{18}$$

uniformly in v.

It now suffices to consider the truncated minor arcs

$$\mathfrak{n} = \mathfrak{m}\cap[-(\log X)^A,(\log X)^A].$$

We apply the familiar method of estimating the variance

$$\Xi = \sum_{v\in\mathcal{V},v\leq X}\left|\int_{\mathfrak{n}} S(\alpha)^2 e(-\alpha v)K(\alpha)\,d\alpha\right|^2.$$

Squaring out produces

$$\Xi = \int_{\mathfrak{n}}\int_{\mathfrak{n}} S(\alpha)^2 S(-\beta)^2 K(\alpha)K(\beta)T(\alpha-\beta)\,d\alpha\,d\beta. \tag{19}$$

Invoking the simple bound

$$\int_{\mathfrak{n}}|S(\alpha)|^2 K(\alpha)\,d\alpha \leq \int_{-\infty}^{\infty}|S(\alpha)|^2 K(\alpha)\,d\alpha$$

$$= \sum_{p\leq X}(\log p)^2 \ll X\log X, \tag{20}$$

we deduce from (2) that the contribution to (19) arising from pairs $(\alpha,\beta)\in\mathfrak{n}\times\mathfrak{n}$ with $|\alpha-\beta| > X^{-1}(\log X)^A$ does not exceed

$$\ll V(X)(\log X)^{-B}\left(\int_{\mathfrak{n}}|S(\alpha)|^2 K(\alpha)\,d\alpha\right)^2$$

$$\ll X^2 V(X)(\log X)^{2-B}.$$

In the remaining set, we have $\beta = \alpha + \zeta$ with $|\zeta| \leq X^{-1}(\log X)^A$. After a change of variable we see that this set contributes to (19) at most

$$\ll V(X)X\log X \sup_{\alpha}\int_{\substack{|\zeta|\leq X^{-1}(\log X)^A\\ \alpha+\zeta\in\mathfrak{n}}}|S(\alpha+\zeta)|^2\,d\zeta\,;$$

here we have used (20) and the trivial bound for $T(\alpha)$. By Vinogradov's estimate (see Theorem 3.1 of Vaughan [6]) we have $S(\beta) \ll X(\log X)^{-2A}$ for $\beta\in\mathfrak{m}$. Hence, the quantity under consideration is

$$\ll V(X)X^2(\log X)^{1-3A},$$

and since $A > B$ by our overall assumption, this yields

$$\Xi \ll V(X)X^2(\log X)^{2-B}.$$

A standard argument now shows that the inequality

$$\left| \int_{\mathfrak{n}} S(\alpha)^2 e(-v\alpha) K(\alpha)\, d\alpha \right| \leq X(\log X)^{-B/4}$$

holds for all but $O(V(X)(\log X)^{-B/3})$ values of $v \leq X, v \in \mathcal{V}$. Combining this with (18), (5) and Lemma 4, we have now shown that for all but $O(V(X)(\log X)^{-B/8})$ values of $v \in \mathcal{V}$ with $\frac{1}{2}X \leq v \leq X$ we have $r(v) \gg X(\log X)^{-B/8}$. The theorem follows by applying this result with $2^{-j}X$, where $0 \leq j \leq 2\log X$.

References

[1] R.C. Baker, G. Harman and J. Pintz. The exceptional set for Goldbach's problem in short intervals. *Sieve Methods, Exponential Sums, and their Applications in Number Theory* (G.R.H. Greaves, G. Harman and M.N. Huxley, Eds.), Cambridge Univ. Press, London, 1996, pp. 1–54.

[2] J. Brüdern and A. Perelli. Goldbach numbers in sparse sequences. To appear.

[3] S.W. Graham and G. Kolesnik. *Van der Corput's Method of Exponential Sums*. Cambridge Univ. Press, London, 1991.

[4] H.L. Montgomery and R.C. Vaughan. The exceptional set in Goldbach's problem. *Acta Arith.*, **27** (1975), 353–370.

[5] A. Perelli. Goldbach numbers represented by polynomials. *Rev. Mat. Iberoamericana*, **12** (1996), 477–490.

[6] R.C. Vaughan. *The Hardy–Littlewood Method*. Cambridge Univ. Press, London, 1981.

Jörg Brüdern
Mathematisches Institut A, Universität Stuttgart
Pfaffenwaldring 57, D-70550 Stuttgart, Germany

Alberto Perelli
Dipartimento di Matematica, Universita di Genova
Via Dodecaneso 35, I-16146, Genova, Italy

5

The Number of Algebraic Numbers of Given Degree Approximating a Given Algebraic Number

JAN-HENDRIK EVERTSE

1. Introduction In 1955, Roth [15] proved his celebrated theorem, that for every real algebraic number α and every real $\kappa > 2$ the inequality

$$\left| \alpha - \frac{x}{y} \right| < \{\max(|x|, |y|)\}^{-\kappa} \text{ in } x, y \in \mathbb{Z} \text{ with } \gcd(x, y) = 1 \qquad (1.1)$$

has only finitely many solutions. Roth's proof is by contradiction. Assuming that (1.1) has infinitely many solutions, Roth constructed an auxiliary polynomial in a large number of variables, k say, of which all low order partial derivatives vanish in a point $(x_1/y_1, \ldots, x_k/y_k)$ for certain solutions $(x_1, y_1), \ldots,$ (x_k, y_k) of (1.1), and then showed, using a non-vanishing result now known as Roth's lemma, that this is not possible.

Assume that $2 < \kappa < 3$. By making explicit Roth's arguments, Davenport and Roth [3] determined an explicit upper bound for the number of solutions of (1.1) and this was improved later by Mignotte [12]. Bombieri and van der Poorten [1] obtained a much better upper bound by using instead of Roth's lemma a non-vanishing result for polynomials of Esnault and Viehweg [4]. Recently, Corvaja [2] gave an alternative proof of the result of Bombieri and van der Poorten, in which he replaced the construction of an auxiliary polynomial by the use of interpolation determinants as introduced by Laurent in transcendence theory.

We recall the result of Bombieri and van der Poorten. The Mahler measure $M(\alpha)$ of an algebraic number α (always assumed to belong to $\mathbb{C}$) is defined by

$$M(\alpha) := |a_0| \prod_{i=1}^{r} \max(1, |\alpha^{(i)}|) \, ,$$

where $r = \deg \alpha$, $\alpha^{(1)}, \ldots, \alpha^{(r)}$ are the conjugates of α over $\mathbb{Q}$ and a_0 is a rational integer such that the coefficients of the polynomial $f(X) = a_0 \prod_{i=1}^{r}(X - \alpha^{(i)})$ are rational integers with gcd 1. In particular, $M(x/y) = \max(|x|, |y|)$ for $x, y \in \mathbb{Z}$ with $\gcd(x, y) = 1$. Now let $\kappa = 2 + \delta$ with $0 < \delta < 1$, and α an algebraic number of degree r. Bombieri and van der Poorten proved that (1.1) has at most

$$c_1 \cdot \delta^{-5} (\log r)^2 \log \left(\frac{\log r}{\delta} \right)$$

solutions with $M(x/y) \geq c_2 M(\alpha)$ and at most

$$c_3 \delta^{-1} \log \left(1 + \log M(\alpha)\right)$$

solutions with $M(x/y) < c_2 M(\alpha)$, where c_1, c_2, c_3 are explicitly computable absolute constants. We mention that recently Schmidt [21] gave an explicit upper bound for the number of solutions of (1.1) in the complementary case $\kappa \geq 3$.

We deal with the analogue of (1.1) in which the unknowns are algebraic numbers of given degree, i.e. we consider the inequality

$$|\alpha - \xi| < M(\xi)^{-\kappa} \text{ in algebraic numbers } \xi \text{ of degree } t, \qquad (1.2)$$

where α is an algebraic number, κ a positive real, and $t \geq 1$. In 1921, Siegel [22], [23] showed that (1.2) has only finitely many solutions if κ exceeds some bound depending on t and the degree of α. In 1966, Ramachandra [14] proved the same with a smaller lower bound for κ, but still depending on the degree of α. In 1971, Wirsing [24] succeeded in proving Roth's conjecture that (1.2) has only finitely many solutions if

$$\kappa > 2t . \qquad (1.3)$$

Independently, Schmidt [17] (Theorem 3) proved that the number of solutions of (1.2) is finite if

$$\kappa > t + 1 . \qquad (1.4)$$

In fact, the latter can be derived from Schmidt's Subspace theorem, cf. [19], p. 278. The lower bound $t + 1$ can be shown to be best possible.

It is our purpose to derive an explicit upper bound for the number of solutions of (1.2). For this, one needs, apart from the Diophantine approximation arguments of Wirsing or Schmidt in an explicit form, a "gap principle," which states that solutions of (1.2) are far away from each other. In §2 we derive a simple gap principle for $\kappa > 2t$ which is similar to one which appeared already in Ramachandra's paper [14]. The proof of this gap principle uses a Liouville-type inequality for differences of algebraic numbers. For obtaining a gap principle for $t + 1 < \kappa \leq 2t$ one would need an effective improvement of this Liouville-type inequality which, if existing, seems to be very difficult to prove.

We derive an upper bound for the number of solutions of (1.2) with $\kappa > 2t$ by combining the gap principle in §2 with Wirsing's arguments. Another possible approach is to use ideas which are used in the proof of the quantitative Subspace theorem, e.g., in [20] or [6], but this would lead to a larger bound. One of Wirsing's main tools was Leveque's generalization of Roth's lemma to number fields ([10], Chap. 4). Instead, we use the sharpening of this from [5]. Our result is as follows:

Theorem 1 *Let α be an algebraic number of degree r, t an integer ≥ 1, and $\kappa = 2t + \delta$ with $0 < \delta < 1$.*
(i). (1.2) has at most

$$2 \times 10^7 \cdot t^7 \delta^{-4} \cdot \log 4r \cdot \log \log 4r$$

solutions ξ with $M(\xi) \geq \max \left(4^{t(t+1)/\delta}, M(\alpha) \right)$.
(ii). (1.2) has at most

$$2^{(t+3)^2} \delta^{-1} \log(2 + \delta^{-1}) + t^2 \delta^{-1} \cdot \log \log 4M(\alpha)$$

solutions ξ with $M(\xi) < \max \left(4^{t(t+1)/\delta}, M(\alpha) \right)$.

We derive a result more general than Theorem 1. For every algebraic number ξ of degree t we fix an ordering of its conjugates $\xi^{(1)}, \ldots, \xi^{(t)}$. Let $\alpha_1, \ldots, \alpha_t$ be algebraic numbers. Further, let $\varphi_1, \ldots, \varphi_t$ be non-negative reals. We introduce the notation

$$|x, y| := \max(|x|, |y|) \text{ for } x, y \in \mathbb{C}.$$

Consider the system of inequalities

$$\frac{|\alpha_i - \xi^{(i)}|}{2|1, \alpha_i| \cdot |1, \xi^{(i)}|} \leq M(\xi)^{-\varphi_i} \quad (i = 1, \ldots, t)$$

$$\text{in algebraic numbers } \xi \text{ of degree } t. \quad (1.5)$$

The denominators have been inserted for technical convenience. Wirsing [24] proved that (1.5) has only finitely many solutions if

$$\max_I \ (\#I)^2 \Big(\sum_{i \in I} \varphi_i^{-1} \Big)^{-1} > 2t , \quad (1.6)$$

where the maximum is taken over all non-empty subsets I of $\{i \in \{1, \ldots, t\} : \varphi_i \neq 0\}$ and where $\#$ is used to denote the cardinality of a set. In [24], §3, Wirsing showed that

$$\Big(\sum_{j=1}^{t} \frac{1}{2j-1} \Big)^{-1} \leq \frac{\max_I \ (\#I)^2 \big(\sum_{i \in I} \varphi_i^{-1} \big)^{-1}}{\varphi_1 + \cdots + \varphi_t} \leq 1 \quad (1.7)$$

for all non-negative reals $\varphi_1, \ldots, \varphi_t$ with $\varphi_1 + \cdots + \varphi_t > 0$, and that the upper and lower bound are best possible. In fact, the upper bound is assumed if and only if all non-zero numbers among $\varphi_1, \ldots, \varphi_t$ are equal. So condition (1.6) is in general stronger than

$$\varphi_1 + \cdots + \varphi_t > 2t . \tag{1.8}$$

We prove the following quantitative version of Wirsing's result:

Theorem 2   *Let* $\alpha_1, \ldots, \alpha_t$ *be algebraic numbers with*

$$\max_{i=1,\ldots,t} M(\alpha_i) = M, \quad [\mathbb{Q}(\alpha_1, \ldots, \alpha_t) : \mathbb{Q}] = r \tag{1.9}$$

and $\varphi_1, \ldots, \varphi_t$ *non-negative reals for which*

$$\max_I (\#I)^2 \big(\sum_{i \in I} \varphi_i^{-1}\big)^{-1} \geq 2t + \delta \ \text{with } 0 < \delta < 1. \tag{1.10}$$

Put $\kappa := \varphi_1 + \cdots + \varphi_t$.
(i). (1.5) *has at most*

$$2 \times 10^7 \cdot t^7 \delta^{-4} \cdot \log 4r \cdot \log \log 4r$$

solutions with $M(\xi) \geq \max \big(4^{t(t+1)/(\kappa-2t)}, M\big)$.
(ii). (1.5) *has at most*

$$2^{t^2+t+\kappa+4}\Big(1 + \frac{\log(2 + \frac{1}{\kappa-2t})}{\log(1 + \frac{\kappa-2t}{t})}\Big) + t \cdot \frac{\log \log 4M}{\log(1 + \frac{\kappa-2t}{t})}$$

solutions ξ *with* $M(\xi) < \max \big(4^{t(t+1)/(\kappa-2t)}, M\big)$.

It is due to a limitation of Wirsing's method that we have to impose condition (1.6) on $\varphi_1, \ldots, \varphi_t$. In §2, we shall derive a gap principle for system (1.5) which is non-trivial if the weaker condition (1.8) holds. It is conceivable that by combining this gap principle with techniques used in the proof of the quantitative Subspace theorem, one can derive an explicit (but larger) upper bound for the number of solutions of (1.5) with (1.8) replacing (1.6).

Theorem 1 follows at once from Theorem 2 with $\alpha_1 = \alpha$, $\varphi_1 = \kappa$ and $\alpha_i = 0$, $\varphi_i = 0$ for $i = 2, \ldots, t$, on observing that in that case we have $\max_I (\#I)^2 \big(\sum_{i \in I} \varphi_i^{-1}\big)^{-1} = \kappa$, $\kappa - 2t = \delta < 1$, and $\log(1 + (\kappa - 2t)/t) = \log(1 + \delta/t) \geq \delta/2t$.

Another application of Theorem 2 is to an inequality involving resultants. The resultant $R(f, g)$ of two polynomials $f(X) = a_0 X^r + a_1 X^{r-1} + \cdots + a_r$ and $g(X) = b_0 X^t + b_1 X^{t-1} + \cdots + b_t$ with $a_0 \neq 0$, $b_0 \neq 0$ is defined by the determinant of order $r + t$,

$$R(f,g) = \begin{vmatrix} a_0 & a_1 & \cdots & \cdots & a_r & & & \\ & a_0 & a_1 & \cdots & \cdots & a_r & & \\ & & \ddots & & & & \ddots & \\ & & & a_0 & a_1 & \cdots & \cdots & a_r \\ b_0 & b_1 & \cdots & b_t & & & & \\ & b_0 & b_1 & \cdots & b_t & & & \\ & & \ddots & & & \ddots & & \\ & & & & b_0 & b_1 & \cdots & b_t \end{vmatrix}, \qquad (1.11)$$

of which the first t rows consist of coefficients of f and the last r rows of coefficients of g. If $f(X) = a_0 \prod_{i=1}^{r}(X - \alpha_i)$ and $g(X) = b_0 \prod_{j=1}^{t}(X - \xi_j)$, then

$$R(f,g) = a_0^t b_0^r \prod_{i=1}^{r} \prod_{j=1}^{t} (\alpha_i - \xi_j) . \qquad (1.12)$$

Hence $R(f,g) = 0$ if and only if f and g have a common zero (cf. [9], Chap. V, §10).

We define the Mahler measure of a polynomial $f(X) = a_0 \prod_{i=1}^{r}(X - \alpha_i) \in \mathbb{C}[X]$ by

$$M(f) := |a_0| \prod_{i=1}^{r} \max(1, |\alpha_i|) .$$

We fix a polynomial $f(X) \in \mathbb{Z}[X]$ of degree r and a positive real κ and consider the inequality in unknown polynomials g,

$$0 < |R(f,g)| < M(g)^{r-\kappa}$$
in polynomials $g(X) \in \mathbb{Z}[X]$ of degree t. \qquad (1.13)

In [24], Wirsing proved that (1.13) has only finitely many solutions if f has no multiple zeros and if

$$\kappa > 2t\left(1 + \frac{1}{3} + \cdots + \frac{1}{2t-1}\right). \qquad (1.14)$$

Schmidt [18] showed that (1.14) can be relaxed to

$$\kappa > 2t \qquad (1.15)$$

if f has no multiple zeros and no irreducible factors in $\mathbb{Z}[X]$ of degree $\leq t$. Finally, from a result of Ru and Wong ([16], Thm. 4.1), which is a consequence

of the Subspace theorem, it follows that (1.14) can be relaxed to (1.15) for
every polynomial f without multiple zeros.

In contrast to Wirsing et al. who obtained their results on (1.13) for
arbitrary polynomials g, we are only able to determine an upper bound for the
number of *irreducible* polynomials g satisfying (1.13). We obtain this bound
by reducing (1.13) to a finite number of systems of inequalities (1.5) to which
we apply Theorem 2. We have to impose condition (1.14) on κ because of
condition (1.10) in Theorem 2.

A polynomial in $\mathbb{Z}[X]$ is said to be primitive if its coefficients have gcd 1.

Theorem 3 *Let f be a primitive polynomial in $\mathbb{Z}[X]$ of degree r with no
multiple zeros. Suppose that*

$$\kappa = (2t + \delta)\left(1 + \frac{1}{3} + \cdots + \frac{1}{2t - 1}\right) \text{ with } 0 < \delta < 1. \tag{1.16}$$

Then there are at most

$$10^{15}(\delta^{-1})^{t+3} \cdot (100r)^t \log 4r \cdot \log\log 4r$$

primitive, irreducible polynomials $g(X) \in \mathbb{Z}[X]$ of degree t with

$$0 < |R(f, g)| < M(f)^t \cdot M(g)^{r-\kappa}, \tag{1.17}$$

$$M(g) \geq \left(2^{8r^2 t} M(f)^{4(r-1)t}\right)^{\delta^{-1}(1 + \frac{1}{3} + \cdots + \frac{1}{2t-1})^{-1}}. \tag{1.18}$$

In (1.17), we have inserted the factor $M(f)^t$ to make the inequality homogeneous in f; without this factor, our bound would not have been better.

2. A gap principle In this section, we derive a gap principle for the system
(1.5) of inequalities in the case where $\alpha_1, \ldots, \alpha_t$ are algebraic numbers, and
$\varphi_1, \ldots, \varphi_t$ are reals with

$$\varphi_i \geq 0 \text{ for } i = 1, \ldots, t; \quad \kappa := \varphi_1 + \cdots + \varphi_t > 2t. \tag{2.1}$$

After that, we prove part (ii) of Theorem 2. Our gap principle is as follows:

Lemma 1 *(i). Let $\xi_1, \ldots, \xi_{t+1}$ be distinct solutions of (1.5) with $M(\xi_{t+1}) \geq
M(\xi_t) \geq \cdots \geq M(\xi_1)$. Then*

$$U^{-1} M(\xi_{t+1}) \geq \left(U^{-1} M(\xi_1)\right)^{1+(\kappa-2t)/t} \text{ where } U := 2^{t(t+1)/(\kappa-2t)}. \tag{2.2}$$

*(ii). Put $C := [t \cdot 2^{t^2 + \kappa + 1}]$. Let $\xi_1, \ldots, \xi_{C+1}$ be distinct solutions of (1.5) with
$M(\xi_{C+1}) \geq M(\xi_C) \geq \cdots \geq M(\xi_1)$. Then*

$$2M(\xi_{C+1}) \geq \left(2M(\xi_1)\right)^{1+(\kappa-2t)/t}. \tag{2.3}$$

Proof The assertion (i): Since solutions of (1.5) are assumed to have degree t, at least two numbers among $\xi_1, \ldots, \xi_{t+1}$ are not conjugate to each other, $\xi := \xi_i$, $\eta := \xi_j$ with $i < j$, say. Denote the minimal polynomials (in $\mathbb{Z}[X]$ with coefficients having gcd 1) of ξ, η by f, g, respectively. Then f and g have no common zeros, i.e., their resultant $R(f, g)$ is a non-zero integer. Let $f(X) = a_0 \prod_{k=1}^{t}(X - \xi^{(k)})$, $g(X) = b_0 \prod_{l=1}^{t}(X - \eta^{(l)})$. Then by (1.12) (on noting that a_0, b_0 are cancelled) we have

$$\frac{|R(f,g)|}{M(\xi)^t M(\eta)^t} = \prod_{k=1}^{t} \prod_{l=1}^{t} \frac{|\xi^{(k)} - \eta^{(l)}|}{|1, \xi^{(k)}| \cdot |1, \eta^{(l)}|} \,, \qquad (2.4)$$

and since $R(f, g)$ is a non-zero integer, this implies the Liouville-type inequality,

$$\prod_{k=1}^{t} \prod_{l=1}^{t} \frac{|\xi^{(k)} - \eta^{(l)}|}{|1, \xi^{(k)}| \cdot |1, \eta^{(l)}|} \geq \frac{1}{M(\xi)^t M(\eta)^t} \,. \qquad (2.5)$$

We estimate the left-hand side from above. For $k \neq l$ we use the trivial estimate

$$\frac{|\xi^{(k)} - \eta^{(l)}|}{|1, \xi^{(k)}| \cdot |1, \eta^{(l)}|} \leq 2 \,. \qquad (2.6)$$

Let $k = l \in \{1, \ldots, t\}$. We apply the following variation on the triangle inequality:

$$\frac{|x - y|}{|1, x| \cdot |1, y|} \leq \frac{|x - z|}{|1, x| \cdot |1, z|} + \frac{|z - y|}{|1, z| \cdot |1, y|} \quad \text{for } x, y, z \in \mathbb{C}. \qquad (2.7)$$

Thus, using that ξ, η satisfy (1.5),

$$\frac{|\xi^{(k)} - \eta^{(k)}|}{|1, \xi^{(k)}| \cdot |1, \eta^{(k)}|} \leq \frac{|\xi^{(k)} - \alpha_k|}{|1, \xi^{(k)}| \cdot |1, \alpha_k|} + \frac{|\eta^{(k)} - \alpha_k|}{|1, \eta^{(k)}| \cdot |1, \alpha_k|}$$
$$\leq 2M(\xi)^{-\varphi_k} + 2M(\eta)^{-\varphi_k} \leq 4M(\xi)^{-\varphi_k} \,.$$

Together with (2.5), (2.6) this implies

$$\frac{1}{M(\xi)^t M(\eta)^t} \leq 2^{t^2+t} M(\xi)^{-(\varphi_1 + \cdots + \varphi_t)} = U^{\kappa - 2t} M(\xi)^{-\kappa} \,, \qquad (2.8)$$

whence

$$U^{-1} M(\eta) \geq \left(U^{-1} M(\xi) \right)^{1 + (\kappa - 2t)/t} \,.$$

Together with $M(\xi_1) \leq M(\xi) \leq M(\eta) \leq M(\xi_{t+1})$ this implies (2.2).

The assertion (ii): Let p be a prime number which will be chosen later. We partition the solutions of (1.5) into equivalence classes as follows. Let ξ and η be solutions of (1.5) with minimal polynomials f, g, respectively. By definition, both f and g have $t+1$ integer coefficients without a common factor. We call ξ, η equivalent if there is an integer λ, not divisible by p, such that $(f - \lambda g)/p$ has its coefficients in $\mathbb{Z}$, in other words, if the reductions modulo p of the vectors of coefficients of f, g, respectively, represent the same point in the t-dimensional projective space $\mathbb{P}^t(\mathbb{F}_p)$. Clearly, the number of equivalence classes is at most the number of points in $\mathbb{P}^t(\mathbb{F}_p)$, which is

$$\frac{p^{t+1} - 1}{p - 1} \leq 2p^t . \tag{2.9}$$

Now for equivalent ξ, η with minimal polynomials f, g and with λ as above we have, by (1.11), that

$$R(f,g) = R(f - \lambda g, g) = p^t R((f - \lambda g)/p, g) \equiv 0 \pmod{p^t}. \tag{2.10}$$

Choose p such that $2^{t-1+\kappa/t} \leq p < 2^{t+\kappa/t}$. Then by (2.9), the number of equivalence classes is at most $2p^t < 2^{t^2+\kappa+1}$. So among the solutions $\xi_1, \ldots, \xi_{C+1}$ there must be at least $t + 1$ belonging to the same equivalence class. Among these $t + 1$ solutions we can choose two, $\xi := \xi_i$ and $\eta := \xi_j$, say, with $i < j$, which are not conjugate to each other. Now if f, g are the minimal polynomials of ξ, η, then in view of (2.4), (2.10), we can replace (2.5) by

$$\prod_{k=1}^{t}\prod_{l=1}^{t} \frac{|\xi^{(k)} - \eta^{(l)}|}{|1,\xi^{(k)}| \cdot |1,\eta^{(l)}|} \geq \frac{p^t}{M(\xi)^t M(\eta)^t} \geq \frac{2^{t^2-t+\kappa}}{M(\xi)^t M(\eta)^t} .$$

By repeating the argument of (i) we obtain instead of (2.8),

$$\frac{2^{t^2-t+\kappa}}{M(\xi)^t M(\eta)^t} \leq 2^{t^2+t} M(\xi)^{-(\varphi_1+\cdots+\varphi_t)} = 2^{t^2+t} M(\xi)^{-\kappa}$$

and so

$$2M(\eta) \geq \bigl(2M(\xi)\bigr)^{1+(\kappa-2t)/t} .$$

Together with $M(\xi_1) \leq M(\xi) \leq M(\eta) \leq M(\xi_{C+1})$, this implies (2.3).

We need the following simple consequence of Lemma 1:

Lemma 2 (i). *Let A, B be reals with $B \geq A \geq U^2 = 4^{t(t+1)/(\kappa-2t)}$. Then the number of solutions ξ of (1.5) with $A \leq M(\xi) < B$ is at most*

$$t \cdot \left(1 + \frac{\log(2 \log B/ \log A)}{\log(1 + (\kappa - 2t)/t)}\right) .$$

(ii). Let A, B be reals with $B \geq A \geq 1$. Then the number of solutions ξ of (1.5) with $A \leq M(\xi) < B$ is at most

$$C \cdot \left(1 + \frac{\log(\log 2B/\log 2A)}{\log(1 + (\kappa - 2t)/t)}\right).$$

Proof The assertion (i): Put $\theta := 1 + (\kappa - 2t)/t$. Let k be the smallest integer with

$$\left(U^{-1}A\right)^{\theta^k} \geq U^{-1}B.$$

Part (i) of Lemma 1 implies that for each $i \in \{0, \ldots, k-1\}$, (1.5) has at most t solutions ξ with $(U^{-1}A)^{\theta^i} \leq U^{-1}M(\xi) < (U^{-1}A)^{\theta^{i+1}}$. Hence (1.5) has at most $t \cdot k$ solutions with $A \leq M(\xi) < B$. Now part (i) follows since in view of our assumption $A \geq U^2$ we have

$$k \leq 1 + \frac{\log(\log U^{-1}B/\log U^{-1}A)}{\log \theta} \leq 1 + \frac{\log(2\log B/\log A)}{\log \theta}.$$

The assertion (ii): Use part (ii) of Lemma 1 and repeat the argument given above with 2 replacing U^{-1} and $C \cdot k$ replacing $t \cdot k$.

Proof of part (ii) of Theorem 2 Put $\theta := 1 + (\kappa - 2t)/t$. We first estimate the number of solutions ξ of (1.5) with

$$4^{t(t+1)/(\kappa-2t)} \leq M(\xi) < \max(4^{t(t+1)/(\kappa-2t)}, M). \tag{2.11}$$

Assuming that $M \geq 4^{t(t+1)/(\kappa-2t)}$, we infer from part (i) of Lemma 2 that this number is at most

$$t \cdot \left(1 + \frac{\log\left(2\log M/\frac{t(t+1)}{\kappa-2t}\log 4\right)}{\log \theta}\right) \leq t \cdot \left(1 + \frac{\log(\theta \log M)}{\log \theta}\right)$$

$$\leq t \cdot \left(2 + \frac{\log\log 4M}{\log \theta}\right). \tag{2.12}$$

This is clearly also true if $M < 4^{t(t+1)/(\kappa-2t)}$.

We now estimate the number of solutions ξ of (1.5) with

$$M(\xi) < 4^{t(t+1)/(\kappa-2t)}.$$

From part (ii) of Lemma 2 with $A = 1$, $B = 4^{t(t+1)/(\kappa-2t)}$ it follows that this number is at most

$$t \cdot 2^{t^2+\kappa+1} \cdot \left(1 + \frac{\log(1 + \frac{2t(t+1)}{\kappa-2t})}{\log \theta}\right)$$

$$\leq t \cdot 2^{t^2+\kappa+1} \cdot 3\log\left(2t(t+1)\right) \cdot \left(1 + \frac{\log(2 + \frac{1}{\kappa-2t})}{\log \theta}\right).$$

Together with (2.12) this implies that the total number of solutions of (1.5) with

$$M(\xi) < \max(4^{t(t+1)/(\kappa-2t)}, M)$$

is at most

$$2^{t^2+t+\kappa+4}\left(1 + \frac{\log(2 + \frac{1}{\kappa-2t})}{\log\theta}\right) + t \cdot \frac{\log\log 4M}{\log\theta}$$

which is precisely the upper bound in part (ii) of Theorem 2.

3. Construction of the auxiliary polynomial

For an algebraic number ξ we put

$$\|\xi\| := \max(|\xi^{(1)}|, \ldots, |\xi^{(r)}|),$$

where $\xi^{(1)}, \ldots, \xi^{(r)}$ are the conjugates of ξ over $\mathbb{Q}$. More generally, for a vector $\mathbf{x} := (\xi_1, \ldots, \xi_R)$ with algebraic coordinates we put

$$\|\mathbf{x}\| := \max(\|\xi_1\|, \ldots, \|\xi_R\|) \,.$$

The ring of integers of an algebraic number field K (assumed to be contained in $\mathbb{C}$) is denoted by O_K. We need the following consequence of Siegel's lemma:

Lemma 3 *Let K be an algebraic number field of degree r. Further, let R, S be rational integers with*

$$0 < S \le R, \quad rS > (r-1)R, \tag{3.1}$$

let A be a positive real and let $\mathbf{a}_1, \ldots, \mathbf{a}_S \in K^R$ be K-linearly independent vectors for which there are rational integers $q_1, \ldots, q_S$ with

$$0 < q_i \le A, \quad q_i\mathbf{a}_i \in O_K^R, \quad \|q_i\mathbf{a}_i\| \le A \text{ for } i = 1, \ldots, S. \tag{3.2}$$

Then there are $\beta_1, \ldots, \beta_S \in O_K$ such that

$$\mathbf{x} := \sum_{i=1}^{S} \beta_i\mathbf{a}_i \in \mathbb{Z}^R \backslash \{\mathbf{0}\}, \tag{3.3}$$

$$\|\mathbf{x}\| \le \{C(K) \cdot SA\}^{\frac{rS}{rS-(r-1)R}}, \tag{3.4}$$

$$|\beta_i| \le \{C(K) \cdot SA\}^{\frac{rS}{rS-(r-1)R}} \text{ for } i = 1, \ldots, S, \tag{3.5}$$

where $C(K)$ is a constant depending only on K.

Proof Lemma 3 may be proved by applying a sophisticated version of Siegel's lemma of Bombieri-Vaaler type, but then some extra work must be done to get a good upper bound for the numbers $|\beta_i|$. Instead, we give a direct proof of

Lemma 3, following Wirsing [24]. $C_1(K), C_2(K), \ldots$ denote constants depending only on K.

Put $\mathbf{a}'_i := q_i \mathbf{a}_i$ for $i = 1, \ldots, S$. We search for $\beta'_1, \ldots, \beta'_S \in O_K$ such that

$$\mathbf{x} := \sum_{i=1}^{S} \beta'_i \mathbf{a}'_i \in \mathbb{Z}^R \backslash \{\mathbf{0}\} . \qquad (3.6)$$

Then (3.3) holds with

$$\beta_i = q_i \beta'_i \text{ for } i = 1, \ldots, S. \qquad (3.7)$$

Let $\{\omega_1, \ldots, \omega_r\}$ be a $\mathbb{Z}$-basis of O_K with $\omega_1 = 1$. We can express $\alpha \in O_K$ as

$$\alpha = \sum_{i=1}^{r} x_i \omega_i \text{ with } x_i \in \mathbb{Z}, \ |x_i| \leq C_1(K) \|\alpha\| \text{ for } i = 1, \ldots, r; \qquad (3.8)$$

the upper bounds for $|x_i|$ follow by taking conjugates and solving $x_1, \ldots, x_r$ from the system of linear equations $\alpha^{(j)} = \sum_{i=1}^{r} x_i \omega_i^{(j)}$ $(j = 1, \ldots, r)$, using Cramer's rule. Now we have

$$\mathbf{a}'_i = \sum_{j=1}^{r} \omega_j \mathbf{b}_{ij} \text{ with } \mathbf{b}_{ij} \in \mathbb{Z}^R \text{ for } i = 1, \ldots, S, \ j = 1, \ldots, r, \qquad (3.9)$$

$$\beta'_i = \sum_{k=1}^{r} \omega_k z_{ik} \text{ with } z_{ik} \in \mathbb{Z} \text{ for } i = 1, \ldots, S, \ k = 1, \ldots, r. \qquad (3.10)$$

Define the rational integers u_{jkl} by

$$\omega_j \omega_k = \sum_{l=1}^{r} u_{jkl} \omega_l \text{ for } j, k \in \{1, \ldots, r\} .$$

Then we obtain

$$\sum_{i=1}^{S} \beta'_i \mathbf{a}'_i = \sum_{i=1}^{S} \sum_{j=1}^{r} \sum_{k=1}^{r} \omega_j \omega_k z_{ik} \mathbf{b}_{ij}$$

$$= \sum_{l=1}^{r} \omega_l \Big\{ \sum_{i=1}^{S} \sum_{k=1}^{r} z_{ik} \mathbf{c}_{ikl} \Big\} \text{ with } \mathbf{c}_{ikl} := \sum_{j=1}^{r} u_{jkl} \mathbf{b}_{ij} \in \mathbb{Z}^R. \qquad (3.11)$$

By (3.8), (3.2) we have

$$\| \mathbf{b}_{ij} \| \leq C_2(K) \| \mathbf{a}'_i \| \leq C_2(K) A \text{ for } i = 1, \ldots, S, \ j = 1, \ldots, r, \qquad (3.12)$$

so

$$\|\, \mathbf{c}_{ikl}\,\| \leq C_3(K)A \text{ for } i = 1,\ldots,S,\ k = 1,\ldots,r,\ l = 1,\ldots,r. \qquad (3.13)$$

Recalling that $\omega_1 = 1$, we infer that $\sum_{i=1}^{S} \beta'_i \mathbf{a}'_i \in \mathbb{Z}^R$ if and only if the coefficients of $\omega_2,\ldots,\omega_r$ in (3.11) are 0, i.e.,

$$\sum_{i=1}^{S}\sum_{k=1}^{r} z_{ik}\mathbf{c}_{ikl} = 0 \text{ for } l = 2,\ldots,r. \qquad (3.14)$$

Since the vectors $\mathbf{c}_{ikl}$ have R coordinates, (3.14) is a system of $R(r-1)$ equations in Sr unknowns. Since $Sr > R(r-1)$, we have by the most basic form of Siegel's lemma (cf. [19], p. 127), that system (3.14) has a non-trivial solution in integers z_{ik} with

$$\max_{i,k}|z_{ik}| \leq \left\{ rS \cdot \max_{i,k,l}\|\,\mathbf{c}_{ikl}\,\| \right\}^{\frac{R(r-1)}{Sr-R(r-1)}} \qquad (3.15)$$

$$\leq \left\{ C_4(K)\cdot SA \right\}^{\frac{R(r-1)}{Sr-R(r-1)}} \quad \text{by (3.13)}.$$

By (3.14) we have that $\mathbf{x} := \sum_{i=1}^{S}\beta'_i\mathbf{a}'_i$ is equal to the coefficient of $\omega_1 = 1$ in (3.11), i.e.,

$$\mathbf{x} = \sum_{i=1}^{S}\sum_{k=1}^{r} z_{ik}\mathbf{c}_{1kl}\,.$$

Together with (3.15) and (3.13) this implies

$$\|\,\mathbf{x}\,\| \leq Sr \cdot \left(\max_{i,k}|z_{ik}| \right) \left(\max_{k,l}\|\,\mathbf{c}_{1kl}\,\| \right)$$

$$\leq \left(C_5(K)\cdot SA \right)^{1+\frac{R(r-1)}{Sr-R(r-1)}} = \left(C_5(K)\cdot SA \right)^{\frac{Sr}{Sr-R(r-1)}}.$$

Moreover, (3.10) and (3.15) imply

$$|\beta'_i| \leq C_6(K)\left\{ C_4(K)\cdot SA \right\}^{\frac{R(r-1)}{Sr-R(r-1)}}$$

and so, by (3.7),

$$|\beta_i| = |q_i||\beta'_i| \leq A|\beta'_i| \leq \left(C_7(K)\cdot SA \right)^{\frac{Sr}{Sr-R(r-1)}} \text{ for } i = 1,\ldots,S\,.$$

This completes the proof of Lemma 3.

Let $\alpha_1, \ldots, \alpha_t$ be the algebraic numbers from Theorem 2 and put $K := \mathbb{Q}(\alpha_1, \ldots, \alpha_t)$. As in (1.9), let $M = \max_{i=1,\ldots,r} M(\alpha_i)$, $r = [K : \mathbb{Q}]$. Let $\gamma_1, \ldots, \gamma_t$ be non-negative real numbers with $\gamma_1 + \cdots + \gamma_t = 1$. For $i = 0, 1, 2, \ldots$ we define the polynomial of degree i,

$$p_i(X) := (X - \alpha_1)^{j_1(i)} \cdots (X - \alpha_t)^{j_t(i)},$$

with $j_l(i) = [\gamma_l \cdot i]$ for $l = 1, \ldots, t-1$, and $j_t(i) = i - \sum_{l=1}^{t-1}[\gamma_l \cdot i]$. (3.16)

Let $k, d_1, \ldots, d_k$ be positive integers and put

$$\mathcal{I}_k = \{0, \ldots, d_1\} \times \cdots \times \{0, \ldots, d_k\} .$$

By **i** we denote a tuple $(i_1, \ldots, i_k) \in \mathcal{I}_k$. For a polynomial P with integer coefficients, we denote by $\| P \|$ the maximum of the absolute values of its coefficients. The next lemma gives our auxiliary polynomial:

Lemma 4 *Assume that*

$$\frac{2^{d_1 + \cdots + d_k}}{(d_1 + 1) \cdots (d_k + 1)} \geq C(K) \tag{3.17}$$

where $C(K)$ is the constant from Lemma 3. Let $\mathcal{I}$ be a subset of $\mathcal{I}_k$ with

$$\#\mathcal{I} \leq \frac{1}{2r}(d_1 + 1) \cdots (d_k + 1) . \tag{3.18}$$

Then there are $\beta_{\mathbf{i}} \in O_K$ for $\mathbf{i} \in \mathcal{I}_k \backslash \mathcal{I}$ such that

$$P(X_1, \ldots, X_k) := \sum_{\mathbf{i} \in \mathcal{I}_k \backslash \mathcal{I}} \beta_{\mathbf{i}} p_{i_1}(X_1) \cdots p_{i_k}(X_k)$$

$$\in \mathbb{Z}[X_1, \ldots, X_k] \backslash \{0\} , \tag{3.19}$$

$$\| P \| \leq (4M)^{2r(d_1 + \cdots + d_k)} , \tag{3.20}$$

$$|\beta_{\mathbf{i}}| \leq (4M)^{2r(d_1 + \cdots + d_k)} \text{ for } \mathbf{i} \in \mathcal{I}_k \backslash \mathcal{I}. \tag{3.21}$$

Proof Let $i > 0$. Then

$$p_i(X) = (X - \alpha_{i1}) \cdots (X - \alpha_{ii}) \text{ with } \alpha_{i1}, \ldots, \alpha_{ii} \in \{\alpha_1, \ldots, \alpha_t\}.$$

Let $q_{ij} \in \mathbb{Z}_{>0}$ be the leading coefficient of the minimal polynomial of α_{ij}. Clearly, by (1.9) we have

$$q_{ij} \leq M(\alpha_{ij}) \leq M, \quad q_{ij}\| \alpha_{ij} \| \leq M(\alpha_{ij}) \leq M . \tag{3.22}$$

The coefficient of $X_1^{j_1} \cdots X_k^{j_k}$ in $p_{i_1}(X_1) \cdots p_{i_k}(X_k)$ is equal to

$$\alpha(\mathbf{i},\mathbf{j}) = \pm \prod_{h=1}^{k} \sum_{S_h} \prod_{l_h \in S_h} \alpha_{i_h,l_h} , \qquad (3.23)$$

where for $h = 1, \ldots, k$, the sum is taken over all subsets S_h of $\{1, \ldots, i_h\}$ of cardinality $i_h - j_h$. Define the rational integer

$$q(\mathbf{i},\mathbf{j}) := \prod_{h=1}^{k} \prod_{j=1}^{i_h} q_{i_h,j} .$$

By (3.23) we have $q(\mathbf{i},\mathbf{j})\alpha(\mathbf{i},\mathbf{j}) \in O_K$ and by (3.23), (3.22) we have

$$q(\mathbf{i},\mathbf{j}) \leq M^{d_1 + \cdots + d_k}, \qquad (3.24)$$

$$\| q(\mathbf{i},\mathbf{j})\alpha(\mathbf{i},\mathbf{j}) \| \leq \prod_{h=1}^{k} \binom{i_h}{i_h - j_h} M^{i_1 + \cdots + i_k} \leq (2M)^{d_1 + \cdots + d_k} .$$

We apply Lemma 3 with $R = (d_1 + 1) \cdots (d_k + 1), S = R - \#\mathcal{I}$ and

$$\{\mathbf{a}_1, \ldots, \mathbf{a}_S\} = \{p_{i_1}(X_1) \cdots p_{i_k}(X_k) : \mathbf{i} \in \mathcal{I}_k \backslash \mathcal{I}\}.$$

The assertion (3.18) implies condition (3.1); and (3.24) implies (3.2) with $A = (2M)^{d_1 + \cdots + d_k}$. Note that by (3.17) we have

$$C(K) \cdot SA \leq C(K)(d_1 + 1) \cdots (d_k + 1)A \leq (4M)^{d_1 + \cdots + d_k}$$

and that by (3.18) we have $Sr \geq (r - \frac{1}{2})R$, whence

$$\frac{Sr}{Sr - R(r-1)} \leq \frac{(r - \frac{1}{2})R}{\frac{1}{2}R} \leq 2r .$$

Together with Lemma 3 this implies at once that there are $\beta_{\mathbf{i}}$ ($\mathbf{i} \in \mathcal{I}_k \backslash \mathcal{I}$) with (3.19)–(3.21).

4. Combinatorial lemmas We will have to estimate the values of the auxiliary polynomial constructed in §3 in certain points and for this purpose we need some combinatorial lemmas. We use the arguments from elementary probability theory introduced by Wirsing [24], except that we obtain better estimates by using the following lemma instead of Chebyshev's inequality:

Lemma 5 *Let $X_1, \ldots, X_k$ be mutually independent random variables on some probability space with probability measure P, such that for $i = 1, \ldots, k$, X_i has* *expectation* μ_i *and* $P(X_i \in [0,1]) = 1$. *Let* $\mu := \mu_1 + \cdots + \mu_k$ *and let* ϵ *be a real with* $0 < \epsilon < 2/3$. *Then*

$$P(|X_1 + \cdots + X_k - \mu| \geq \epsilon k) \leq 2e^{-\epsilon^2 k/3} \quad (e = 2.7182\ldots) \,. \quad (4.1)$$

Proof Clearly, (4.1) follows from

$$P(X_1 + \cdots + X_k - \mu \geq \epsilon k) \leq e^{-\epsilon^2 k/3} \,, \quad (4.2)$$

$$P(X_1 + \cdots + X_k - \mu \leq -\epsilon k) \leq e^{-\epsilon^2 k/3} \,, \quad (4.3)$$

and (4.3) follows from (4.2) by replacing X_i by $1 - X_i$, μ_i by $1 - \mu_i$ in (4.2) for $i = 1, \ldots, k$. So it suffices to prove (4.2).

For $i = 1, \ldots, k$, denote by σ_i^2 the variance of X_i, i.e., the expectation of $(X_i - \mu_i)^2$; since $P(X_i \in [0,1]) = 1$ this variance exists and is ≤ 1. Put $s^2 := \sum_{i=1}^{k} \sigma_i^2$. We may assume that $s^2 > 0$ since otherwise $P(X_i = \mu_i) = 1$ for $i = 1, \ldots, t$ and we are done. By the inequality at the bottom of p. 267, Section 19 of Loève [11] we have

$$P\left(\frac{X_1 + \cdots + X_k - \mu}{s} \geq \epsilon'\right) \leq \exp\left(-t\epsilon' + \frac{t^2}{2}\left(1 + \frac{tc}{2}\right)\right) \quad \text{for } \epsilon' > 0 \,, \quad (4.4)$$

where c is such that $P(|(X_i - \mu_i)/s| \leq c) = 1$ for $i = 1, \ldots, k$ and t is any real with $0 < t \leq c^{-1}$. (Loève uses the notation S' for $(X_1 + \cdots + X_k - \mu)/s$). We apply (4.4) with $\epsilon' = k\epsilon/s$, $c = 1/s$ and $t = \epsilon s$. Then the right-hand side of (4.4) becomes

$$\exp\left(-\epsilon^2 k + \frac{\epsilon^2 s^2}{2}\left(1 + \frac{\epsilon}{2}\right)\right) \leq \exp\left(-\epsilon^2 k/3\right)$$

since $s^2 \leq k$, $0 < \epsilon < 2/3$. This implies (4.2).

Let ϵ be a real and let $k, t, d_1, \ldots, d_k$ be positive integers with

$$0 < \epsilon < \frac{1}{6t} \,, \quad (4.5)$$

$$d_h > \frac{10^4}{\epsilon} \quad \text{for } h = 1, \ldots, k \,. \quad (4.6)$$

Define the sets

$$\mathcal{I}_k = \{0, \ldots, d_1\} \times \cdots \times \{0, \ldots, d_k\} \,,$$

$$\mathcal{C}_k = \{1, \ldots, t\}^k \,.$$

We will use **i** to denote a tuple $(i_1, \ldots, i_k) \in \mathcal{I}_k$ and **c** to denote a tuple $(c_1, \ldots, c_k) \in \mathcal{C}_k$.

Lemma 6   *There is a subset $\mathcal{I}$ of $\mathcal{I}_k$ with*

$$\#\mathcal{I} \le 24\epsilon^{-1}e^{-\epsilon^2 k/4}(d_1+1)\cdots(d_k+1)$$

such that for all $\mathbf{i} \in \mathcal{I}_k\backslash\mathcal{I}$ and all $x \in [0,1]$ we have

$$\left|\#\left\{h \in \{1,\ldots,k\} : \frac{i_h}{d_h} \le x\right\} - kx\right| \le \epsilon k \ .$$

Proof   For $x \in [0,1]$, $\mathbf{i} \in \mathcal{I}_k$ we put $s(\mathbf{i},x) := \#\{h \in \{1,\ldots,k\} : i_h/d_h \le x\}$. We endow $\mathcal{I}_k$ with the probability measure P such that each tuple $\mathbf{i} = (i_1,\ldots,i_k) \in \mathcal{I}_k$ has probability $1/\#\mathcal{I}_k = 1/(d_1+1)\cdots(d_k+1)$. Fix $x \in [0,1]$. For $h = 1,\ldots,k$, define the random variable $X_h = X_h(\mathbf{i})$ on $\mathcal{I}_k$ by $X_h = 1$ if $0 \le i_h/d_h \le x$ and $X_h = 0$ if $x < i_h/d_h \le 1$. Thus, $X_1,\ldots,X_k$ are mutually independent and X_h has expectation $\mu_h = P(X_h = 1) = ([xd_h]+1)/(d_h+1)$ for $h = 1,\ldots,k$. By Lemma 5 with $(0.9 - 10^{-4})\epsilon$ replacing ϵ we have

$$P\left(|X_1 + \cdots + X_k - (\mu_1 + \cdots + \mu_k)| > (0.9 - 10^{-4})\epsilon k\right) \le 2e^{-\epsilon^2 k/4} \ .$$

By (4.6) we have

$$|\mu_h - x| = \frac{|[xd_h]+1-xd_h-x|}{d_h+1} \le \frac{1}{d_h+1} < 10^{-4}\epsilon \ \text{ for } h = 1,\ldots,k \ .$$

Hence

$$P(|X_1 + \cdots + X_k - kx| > 0.9\epsilon k) \le 2e^{-\epsilon^2 k/4} \ .$$

This implies that for each fixed $x \in [0,1]$ there exists a subset $\mathcal{I}(x)$ of $\mathcal{I}_k$ with

$$|s(\mathbf{i},x) - kx| \le 0.9\epsilon k \ \text{ for } \mathbf{i} \in \mathcal{I}_k\backslash\mathcal{I}(x), \ \#\mathcal{I}(x) \le 2e^{-\epsilon^2 k/4} \ .$$

Now let $n = [10/\epsilon] + 1$ and take

$$\mathcal{I} := \bigcup_{m=0}^{n} \mathcal{I}(\frac{m}{n}) \ .$$

Then

$$\#\mathcal{I} \le 2(n+1)e^{-\epsilon^2 k/4} \le 24\epsilon^{-1}e^{-\epsilon^2 k/4}.$$

Let $x \in [0,1]$ and choose $m \in \{0,\ldots,n-1\}$ with $m/n \le x \le (m+1)/n$. Then for $\mathbf{i} \in \mathcal{I}_k\backslash\mathcal{I} = \cap_{m=0}^{n}(\mathcal{I}_k\backslash\mathcal{I}(\frac{m}{n}))$ we have

$$s(\mathbf{i},x) \le s(\mathbf{i},\frac{m+1}{n}) \le k(\frac{m+1}{n} + 0.9\epsilon) \le k(x + \frac{1}{n} + 0.9\epsilon) \le k(x + \epsilon),$$

$$s(\mathbf{i},x) \ge s(\mathbf{i},\frac{m}{n}) \ge k(\frac{m}{n} - 0.9\epsilon) \ge k(x - \frac{1}{n} - 0.9\epsilon) \ge k(x - \epsilon),$$

which is what we wanted to prove.

Lemma 7 *Let $\mathcal{I}$ be the set from Lemma 6. Then for $\mathbf{i} \in \mathcal{I}_k \backslash \mathcal{I}$, $h = 1, \ldots, k$ we have*

$$\left| \frac{i_{\pi(h)}}{d_{\pi(h)}} - \frac{h}{k} \right| \leq \epsilon \; ,$$

where π is the permutation of $(1, \ldots, k)$ such that

$$\frac{i_{\pi(1)}}{d_{\pi(1)}} \leq \cdots \leq \frac{i_{\pi(k)}}{d_{\pi(k)}} \; .$$

Proof Fix $\mathbf{i} \in \mathcal{I}_k \backslash \mathcal{I}$, $h \in \{1, \ldots, k\}$ and put $x := i_{\pi(h)}/d_{\pi(h)}$. By definition, the number of integers j with $j \in \{1, \ldots, k\}$, $i_j/d_j \leq x$ is equal to h. Lemma 6 implies that $|h - kx| \leq \epsilon k$. This implies Lemma 7.

Lemma 8 *There is a subset $\mathcal{C}$ of $\mathcal{C}_k$ with*

$$\#\mathcal{C} \leq 2t e^{-\epsilon^2 k/3} \cdot t^k \; ,$$

such that for each $\mathbf{c} \in \mathcal{C}_k \backslash \mathcal{C}$, $c \in \{1, \ldots, t\}$ we have

$$\left| \#\{h \in \{1, \ldots, k\} : c_k = c\} - \frac{k}{t} \right| \leq \epsilon k \; . \tag{4.7}$$

Proof Lemma 8 follows once we have proved that for each $c \in \{1, \ldots, t\}$ there is a subset $\mathcal{C}^{(c)}$ of $\mathcal{C}_k$ with $\#\mathcal{C}^{(c)} \leq 2e^{-\epsilon^2 k/3} t^k$ such that for each $\mathbf{c} \in \mathcal{C}_k \backslash \mathcal{C}^{(c)}$ we have (4.7). We endow $\mathcal{C}_k$ with the probability measure P such that each $\mathbf{c} = (c_1, \ldots, c_k) \in \mathcal{C}_k$ has probability $1/\#\mathcal{C}_k = 1/t^k$. Fix $c \in \{1, \ldots, t\}$. For $h = 1, \ldots, k$, define the random variable $X_h = X_h(\mathbf{c})$ on $\mathcal{C}_k$ by $X_h = 1$ if $c_h = c$ and $X_h = 0$ if $c_h \neq c$. Then $X_1, \ldots, X_k$ are mutually independent and X_h has expectation $1/t$ for $h = 1, \ldots, k$. Now by Lemma 5 we have

$$\#\left\{ \mathbf{c} \in \mathcal{C}_k : \left| \#\{h \in \{1, \ldots, k\} : c_k = c\} - \frac{k}{t} \right| \geq \epsilon k \right\} \cdot t^{-k}$$

$$= P(|X_1 + \cdots + X_k - \frac{k}{t}| \geq \epsilon k) \leq 2e^{-\epsilon^2 k/3}$$

which is what we wanted to prove.

The next lemma is the main result of this section:

Lemma 9 *Let $\varphi_1, \ldots, \varphi_t$ be non-negative reals satisfying (1.10), and let ϵ be a real and $k, t, d_1, \ldots, d_k$ integers satisfying (4.5), (4.6). Then there are subsets $\mathcal{I}$ of $\mathcal{I}_k = \{0, \ldots, d_1\} \times \cdots \times \{0, \ldots, d_k\}$ and $\mathcal{C}$ of $\mathcal{C}_k = \{1, \ldots, t\}^k$ with*

$$\#\mathcal{I} \leq 24\epsilon^{-1} e^{-\epsilon^2 k/4} \cdot (d_1 + 1) \cdots (d_k + 1) \; , \tag{4.8}$$

$$\#\mathcal{C} \leq 2t e^{-\epsilon^2 k/3} \cdot t^k \; , \tag{4.9}$$

and non-negative reals $\gamma_1, \ldots, \gamma_t$ *with* $\gamma_1 + \cdots + \gamma_t = 1$, *such that for all tuples* $\mathbf{i} \in \mathcal{I}_k \backslash \mathcal{I}$, $\mathbf{c} \in \mathcal{C}_k \backslash \mathcal{C}$ *we have*

$$\sum_{h=1}^{k} \frac{i_h}{d_h} \gamma_{c_h} \varphi_{c_h} \geq \left(\frac{k}{2t^2} - \frac{3\epsilon k}{t} \right) \cdot (2t + \delta) . \tag{4.10}$$

Remark The lower bound of (4.10) cannot be improved by another choice of $\gamma_1, \ldots, \gamma_t$.

Proof We prove Lemma 9 with the sets $\mathcal{I}$ from Lemmas 6 and and $\mathcal{C}$ from Lemma 8. These sets satisfy (4.8), (4.9), respectively. By (1.10), there is a subset I of $\{1, \ldots, t\}$ such that $(\#I)^2 \left(\sum_{j \in I} \varphi_j^{-1} \right)^{-1} \geq 2t + \delta$. Choose

$$\gamma_i := 0 \ \text{ for } i \in \{1, \ldots, t\} \backslash I, \quad \gamma_i := \varphi_i^{-1} / (\sum_{j \in I} \varphi_j^{-1})^{-1} \text{ for } i \in I.$$

Then (4.10) follows once we have proved that for every $\mathbf{i} = (i_1, \ldots, i_k) \in \mathcal{I}_k \backslash \mathcal{I}$, $\mathbf{c} = (c_1, \ldots, c_k) \in \mathcal{C}_k \backslash \mathcal{C}$,

$$\sum_{h:\ c_h \in I} \frac{i_h}{d_h} \geq \left(\frac{k}{2t^2} - \frac{3\epsilon k}{t} \right) (\#I)^2. \tag{4.11}$$

Fix $\mathbf{i} \in \mathcal{I}_k \backslash \mathcal{I}$, $\mathbf{c} \in \mathcal{C}_k \backslash \mathcal{C}$. Let $T := \#\{h \in \{1, \ldots, k\} : c_h \in I\}$ and let π be a permutation of $(1, \ldots, k)$ such that $i_{\pi(1)}/d_{\pi(1)} \leq \cdots \leq i_{\pi(k)}/d_{\pi(k)}$. By Lemma 8 and Lemma 7, respectively, we have

$$(\#I)k(\frac{1}{t} - \epsilon) \leq T \leq (\#I)k(\frac{1}{t} + \epsilon) , \quad \frac{i_{\pi(h)}}{d_{\pi(h)}} \geq \frac{h}{k} - \epsilon \text{ for } h = 1, \ldots, k.$$

Hence

$$\sum_{h:\ c_h \in I} \frac{i_h}{d_h} \geq \sum_{h=1}^{T} \frac{i_{\pi(h)}}{d_{\pi(h)}} \geq \sum_{h=1}^{T} (\frac{h}{k} - \epsilon) \geq \frac{1}{2k} T^2 - \epsilon T$$

$$\geq \frac{1}{2k} k^2 (\#I)^2 (\frac{1}{t} - \epsilon)^2 - \epsilon k (\frac{1}{t} + \epsilon) \#I \geq \left(\frac{k}{2t^2} - \frac{3\epsilon k}{t} \right) (\#I)^2$$

where we used that $\epsilon < \frac{1}{t}$ by (4.5) and $\#I \leq (\#I)^2$. This proves (4.11).

5. Estimation of certain values of the auxiliary polynomial Let $\alpha_1, \ldots, \alpha_t$ be the algebraic numbers and $\varphi_1, \ldots, \varphi_t$ the reals from Theorem 2. Thus, $\max_I (\#I)^2 \left(\sum_{i \in I} \varphi_i^{-1} \right)^{-1} \geq 2 + \delta$ with $0 < \delta < 1$. Define

$$\epsilon = \frac{\delta}{34t^2} , \tag{5.1}$$

$$k = [3.5 \times 10^4 \cdot t^4 \delta^{-2} (1 + \tfrac{1}{2} \log t)(1 + \tfrac{1}{2} \log \delta^{-1}) \log 4r], \tag{5.2}$$

and let $d_1, \ldots, d_k$ be integers satisfying

$$d_1 \geq d_2 \geq \cdots \geq d_k \geq \max\left(\frac{10^4 t}{\epsilon}, C(K)\right) \qquad (5.3)$$

where $C(K)$ is the constant from Lemma 3. Thus, (4.5) and (4.6) are satisfied and Lemma 9 is applicable. Let $\mathcal{I}$ and $\mathcal{C}$ be the sets, and $\gamma_1, \ldots, \gamma_t$ the reals from Lemma 9. Then

$$\#\mathcal{I} \leq \frac{1}{2r}(d_1 + 1) \cdots (d_t + 1) , \qquad (5.4)$$

$$\#\mathcal{C} \leq t\epsilon \cdot t^k . \qquad (5.5)$$

That is, (5.4) and (5.5) follow from (4.8), (4.9) and the inequalities

$$24\epsilon^{-1} e^{-k\epsilon^2/4} \leq \frac{1}{2r}, \quad 2te^{-k\epsilon^2/3} \leq t\epsilon,$$

and these inequalities hold true since by (5.1), (5.2) we have

$$\max\left(4\epsilon^{-2} \log \frac{48r}{\epsilon}, 3\epsilon^{-2} \log \frac{2}{\epsilon}\right)$$
$$= 4624 t^4 \delta^{-2}\left(\log 1632 + 2\log t + \log \delta^{-1} + \log r\right)$$
$$< 3.5 \times 10^4 \cdot t^4 \delta^{-2}\left(1 + \tfrac{1}{2}\log t\right)\left(1 + \tfrac{1}{2}\log \delta^{-1}\right) \log 4r - 1 < k .$$

We apply Lemma 4 with these $k, d_1, \ldots, d_k$, $\mathcal{I}$ and $\gamma_1, \ldots, \gamma_t$; this is possible since (5.3) and (5.4) imply the conditions (3.17) and (3.18) of Lemma 4. Let P be the auxiliary polynomial from Lemma 4, i.e.,

$$P(X_1, \ldots, X_k) = \sum_{i \in \mathcal{I}_k \backslash \mathcal{I}} \beta_i p_{i_1}(X_1) \cdots p_{i_k}(X_k) , \qquad (5.6)$$

where for $i = 1, 2, \ldots$ $p_i(X)$ is given by (3.16). Further, let $\xi_1, \ldots, \xi_k$ be solutions of (1.5) with

$$M(\xi_1) \geq (6M)^{68t^2(2r+t)/\delta} , \qquad (5.7)$$

$$M(\xi_1)^{d_1} \leq M(\xi_h)^{d_h} \leq M(\xi_1)^{d_1(1+\epsilon^2)} \quad \text{for } h = 1, \ldots, k. \qquad (5.8)$$

For a polynomial in k variables $X_1, \ldots, X_k$ and a tuple of non-negative integers $\mathbf{j} = (j_1, \ldots, j_k)$ define the differential operator

$$D^{\mathbf{j}} = \frac{1}{j_1! \cdots j_k!} \frac{\partial^{j_1 + \cdots + j_k}}{\partial X_1^{j_1} \cdots \partial X_k^{j_k}} ;$$

note that $D^{\mathbf{j}}$ maps polynomials with coefficients in $\mathbb{Z}$ to polynomials with coefficients in $\mathbb{Z}$. We need the following estimate:

Lemma 10   (i). *For each tuple* $\mathbf{c} = (c_1, \ldots, c_k) \in \mathcal{C}_k \backslash \mathcal{C}$ *and for each tuple of non-negative integers* $\mathbf{j} = (j_1, \ldots, j_k)$ *with*

$$\sum_{h=1}^{k} \frac{j_h}{d_h} < \frac{\epsilon k}{t} \tag{5.9}$$

we have

$$|D^{\mathbf{j}} P(\xi_1^{(c_1)}, \ldots, \xi_k^{(c_k)})| \leq (6M)^{(2r+t)(d_1 + \cdots + d_k)} \left(\prod_{h=1}^{k} |1, \xi_h^{(c_h)}|^{d_h} \right) \cdot$$

$$\cdot \left(M(\xi_1)^{d_1} \right)^{-(2t+\delta)(\{k/2t^2\} - 5k\epsilon/t)} . \tag{5.10}$$

(ii). For each tuple $\mathbf{c} \in \mathcal{C}_k$ *and each tuple of non-negative integers* $\mathbf{j}$ *we have*

$$|D^{\mathbf{j}} P(\xi_1^{(c_1)}, \ldots, \xi_k^{(c_k)})| \leq (6M)^{(2r+t)(d_1 + \cdots + d_k)} \left(\prod_{h=1}^{k} |1, \xi_h^{(c_h)}|^{d_h} \right) . \tag{5.11}$$

Proof   In addition to the hypotheses made above we assume that

$$0 \leq \varphi_l \leq 2t + \delta \text{ for } l = 1, \ldots, t. \tag{5.12}$$

This is no loss of generality. That is, suppose that for instance $\varphi_1 > 2t + \delta$. Then $\xi_1, \ldots, \xi_t$ satisfy (1.5) with $\varphi_1 = 2t + \delta$ and $\varphi_l = 0$ for $l = 2, \ldots, t$. Then these new φ_l satisfy (1.10) and we can prove Lemma 10 with these new φ_l.

For every non-negative integer j we define the differential operator for polynomials in one variable X, $D^j = (1/j!)d^j/dX^j$. Then for each $i \geq 0$, $j \geq 0$ we have

$$D^j p_i(X) = D^j \left(\prod_{l=1}^{t} (X - \alpha_l)^{j_l(i)} \right) = \sum_{\substack{0 \leq j_l \leq j_l(i) \\ j_1 + \cdots + j_t = j}} \prod_{l=1}^{t} \binom{j_l(i)}{j_l} (X - \alpha_l)^{j_l(i) - j_l} .$$

For $h \in \{1, \ldots, k\}$, $c \in \{1, \ldots, t\}$ we have by (1.5),

$$|\alpha_c - \xi_h^{(c)}| \leq 2|1, \alpha_c| \cdot |1, \xi_h^{(c)}| \cdot M(\xi_h)^{-\varphi_c} \tag{5.13}$$

and, trivially,

$$|\alpha_l - \xi_h^{(c)}| \leq 2|1, \alpha_l| \cdot |1, \xi_h^{(c)}| \text{ for } l = 1, \ldots, t. \tag{5.14}$$

Further, by (3.16) we have that $p_i(X) = \prod_{l=1}^{t} (X - \alpha_l)^{j_l(i)}$ where the $j_l(i)$ are non-negative integers with

$$\sum_{l=1}^{t} j_l(i) = i, \quad j_l(i) \geq \gamma_l i - 1 \text{ for } l = 1, \ldots, t .$$

Together with (5.8) and (5.12) these imply

$$|D^j p_i(\xi_h^{(c)})|$$

$$\leq \sum_{\substack{0 \leq j_l \leq j_l(i) \\ j_1 + \cdots + j_l = j}} \prod_{l=1}^{t} \binom{j_l(i)}{j_l} (2|1, \alpha_l| \cdot |1, \xi_h^{(c)}|)^{j_l(i) - j_l} M(\xi_h)^{-\varphi_c(j_c(i) - j_c)}$$

$$\leq (4M^t |1, \xi_h^{(c)}|)^i \cdot M(\xi_h)^{-\gamma_c \varphi_c i + \varphi_c(j+1)}$$

$$\leq (4M^t |1, \xi_h^{(c)}|)^i \cdot (M(\xi_1)^{d_1})^{-\gamma_c \varphi_c(i/d_h) + (1+\epsilon^2)(2t+\delta)(j+1)/d_h} . \qquad (5.15)$$

We now use Lemma 9. Let $\mathbf{c} = (c_1, \ldots, c_k) \in \mathcal{C}_k \backslash \mathcal{C}$, $\mathbf{i} = (i_1, \ldots, i_k) \in \mathcal{I}_k \backslash \mathcal{I}$, and $\mathbf{j} = (j_1, \ldots, j_k)$ a tuple of non-negative integers satisfying (5.9). Then by (5.15) we have

$$|D^{\mathbf{j}} \prod_{h=1}^{k} p_{i_h}(\xi_h^{(c_h)})| \leq AB (M(\xi_1)^{d_1})^{-C}$$

with $A = (4M^t)^{d_1 + \cdots + d_t}$, $B = \prod_{h=1}^{k} |1, \xi_h^{(c_h)}|^{d_h}$ and

$$C = \sum_{h=1}^{k} \frac{i_h}{d_h} \gamma_{c_h} \varphi_{c_h} - (1+\epsilon^2)(2t+\delta) \Big(\sum_{h=1}^{k} \frac{j_h}{d_h} + \sum_{h=1}^{k} \frac{1}{d_h} \Big) .$$

We have

$$\sum_{h=1}^{k} \frac{i_h}{d_h} \gamma_{c_h} \varphi_{c_h} \geq (2t+\delta) \Big(\frac{k}{2t^2} - \frac{3\epsilon k}{t} \Big) \text{ by Lemma 9,}$$

$$(1+\epsilon^2) \Big(\sum_{h=1}^{k} \frac{j_h}{d_h} + \sum_{h=1}^{k} \frac{1}{d_h} \Big) < \frac{2\epsilon k}{t} \text{ by (5.9), (5.1), (5.3);}$$

hence

$$C \geq (2t+\delta) \Big(\frac{k}{2t^2} - \frac{5\epsilon k}{t} \Big) .$$

Now (5.6) and the estimates for β_i in Lemma 4 give

$$|D^{\mathbf{j}} P(\xi_1^{(c_1)}, \ldots, \xi_k^{(c_k)})| \leq A' B (M(\xi_1)^{d_1})^{-C}$$

with

$$A' = A\Big(\sum_{i \in \mathcal{I}_k \setminus \mathcal{I}} |\beta_i| \Big)$$
$$\leq (4M^t)^{d_1 + \cdots + d_k} \cdot 2^{d_1 + \cdots + d_k} (4M)^{2r(d_1 + \cdots + d_k)}$$
$$\leq (6M)^{(2r+t)(d_1 + \cdots + d_k)} .$$

This proves part (i) of Lemma 10. We obtain part (ii) by observing that, as a consequence of (5.14), we can replace (5.15) by the trivial estimate $|D^j p_i(\xi_h^{(c)})|$ $\leq (4M^t|1, \xi_h^{(c)}|)^i$ and so all estimates made above remain valid if we replace the exponent $-C$ on $M(\xi_1)^{d_1}$ by 0.

Lemma 11 *Suppose that $\epsilon, k, d_1, \ldots, d_k$ satisfy (5.1)–(5.3) and that $\xi_1, \ldots, \xi_k$ are solutions of (1.5) satisfying (5.7) and (5.8). Let P be the polynomial from Lemma 4, with the set $\mathcal{I}$ and the reals $\gamma_1, \ldots, \gamma_t$ from Lemma 9. Then there is a tuple $\mathbf{c} = (c_1, \ldots, c_k) \in \mathcal{C}_k = \{1, \ldots, t\}^k$ such that for each tuple $\mathbf{j} = (j_1, \ldots, j_k)$ of non-negative integers with $\sum_{h=1}^k j_h/d_h < \epsilon k/t$ we have*

$$D^{\mathbf{j}} P(\xi_1^{(c_1)}, \ldots, \xi_t^{(c_t)}) = 0 . \tag{5.16}$$

Proof We assume the contrary. Let L be the normal extension of $\mathbb{Q}$ generated by the numbers $\xi_h^{(c)}$ ($h = 1, \ldots, k$, $c = 1, \ldots, t$). We call two tuples $\mathbf{c} = (c_1, \ldots, c_k)$, $\mathbf{c}' = (c'_1, \ldots, c'_k) \in \mathcal{C}_k$ *conjugate* if there is an $\mathbb{Q}$-automorphism of L mapping the ordered tuple $(\xi_1^{(c_1)}, \ldots, \xi_t^{(c_t)})$ to $(\xi_1^{(c'_1)}, \ldots, \xi_t^{(c'_t)})$. From our assumption, it follows that for every $\mathbf{c} \in \mathcal{C}_k$ there is a tuple $\mathbf{j_c}$ with (5.9) such that $D^{\mathbf{j_c}} P(\xi_1^{(c_1)}, \ldots, \xi_t^{(c_t)}) \neq 0$. Since P has its coefficients in $\mathbb{Z}$, there is no loss of generality in assuming that $\mathbf{j_c} = \mathbf{j_{c'}}$ whenever $\mathbf{c}$ and $\mathbf{c}'$ are conjugate. For $h = 1, \ldots, k$, let $q_h \in \mathbb{Z}_{>0}$ denote the leading coefficient of the minimal polynomial of ξ_h. Define the number

$$Z := \big(q_1^{d_1} \cdots q_k^{d_k} \big)^{t^{k-1}} \prod_{\mathbf{c} \in \mathcal{C}_k} D^{\mathbf{j_c}} P(\xi_1^{(c_1)}, \ldots, \xi_t^{(c_t)}) .$$

Then $Z \neq 0$. We will obtain a contradiction by showing that $Z \in \mathbb{Z}$ and $|Z| < 1$.

We first show that $Z \in \mathbb{Z}$. Since for conjugate tuples $\mathbf{c}, \mathbf{c}'$ we have $\mathbf{j_c} = \mathbf{j_{c'}}$, the number Z is invariant under automorphisms of L, i.e. $Z \in \mathbb{Q}$. Denote the fractional ideal with respect to the ring of integers of L generated by $\mu_1, \ldots, \mu_m \in L$ by $(\mu_1, \ldots, \mu_m)$. For $\mathbf{c} \in \mathcal{C}_k$ we have

$$D^{\mathbf{j_c}} P(\xi_1^{(c_1)}, \ldots, \xi_t^{(c_t)}) \in (1, \xi_1^{(c_1)})^{d_1} \cdots (1, \xi_k^{(c_k)})^{d_k} \tag{5.17}$$

since the polynomial $D^{\mathbf{j_c}} P$ has its coefficients in $\mathbb{Z}$ and has degree $\leq d_h$ in X_h. The minimal polynomial of ξ_h is $q_h \prod_{c=1}^t (X - \xi_h^{(c)})$. The coefficients of this polynomial are integers with gcd 1. On the other hand, by Gauss' lemma for fractional ideals in number fields, the ideal generated by the coefficients of this polynomial is equal to $q_h \prod_{c=1}^{t}(1, \xi_h^{(c)})$; therefore, $q_h \prod_{c=1}^{t}(1, \xi_h^{(c)}) = (1)$. Together with (5.17) this implies

$$Z \in (q_1^{d_1} \cdots q_k^{d_k})^{t^{k-1}} \Big(\prod_{c \in \mathcal{C}_k} (1, \xi_1^{(c_1)})^{d_1} \cdots (1, \xi_k^{(c_k)})^{d_k} \Big)$$

$$= \Big(\prod_{h=1}^{k} \{q_h(1, \xi_h^{(1)}) \cdots (1, \xi_h^{(t)})\}^{d_h} \Big)^{t^{k-1}} = (1) \, .$$

Hence $Z \in \mathbb{Z}$.

We now show that $|Z| < 1$. Lemma 10 gives

$$|Z| \le A_1 B_1 \big(M(\xi_1)^{d_1} \big)^{-C_1} \, ,$$

with

$$A_1 = (6M)^{(2r+t)(d_1 + \cdots + d_k)t^k} \, ,$$

$$B_1 = (q_1^{d_1} \cdots q_k^{d_k})^{t^{k-1}} \prod_{c \in \mathcal{C}_k} \Big(\prod_{h=1}^{k} |1, \xi_h^{(c_h)}|^{d_h} \Big) = \Big(\prod_{h=1}^{k} M(\xi_h)^{d_h} \Big)^{t^{k-1}} \, ,$$

$$C_1 = \big(\#\mathcal{C}_k \backslash \mathcal{C} \big) \cdot (2t + \delta) \{ \frac{k}{2t^2} - \frac{5\epsilon k}{t} \} \, .$$

Further,

$$A_1 \le (6M)^{(2r+t)kt^k d_1} \quad \text{by (5.3)},$$

$$B_1 \le M(\xi_1)^{kt^{k-1}d_1(1+\epsilon^2)} \quad \text{by (5.8)},$$

$$C_1 \ge (1 - \epsilon t)t^k (2t + \delta) \{ \frac{k}{2t^2} - \frac{5\epsilon k}{t} \}$$

$$= kt^{k-1} \cdot (1 - \epsilon t) \big(\frac{1}{2t} - 5\epsilon \big)(2t + \delta) \quad \text{by (5.5)}.$$

Therefore,

$$|Z| \le \Big(A_2 M(\xi_1)^{-C_2} \Big)^{kt^{k-1}d_1} \, ,$$

with

$$A_2 = (6M)^{t(2r+t)} \, ,$$

$$C_2 = -(1 + \epsilon^2) + (1 - \epsilon t) \big(\frac{1}{2t} - 5\epsilon \big)(2t + \delta)$$

$$= \frac{\delta}{2t} - 11\epsilon t - \frac{11}{2}\epsilon\delta + 10\epsilon^2 t^2 + 5\epsilon^2 t\delta - \epsilon^2 > \frac{\delta}{2t} - \frac{33}{2}\epsilon t \quad \text{since } \delta < t$$

$$\ge \frac{\delta}{68t} \quad \text{by (5.1)} \, .$$

Together with (5.7) this implies that $|Z| < 1$.

6. Completion of the proof of part (i) of Theorem 2   We apply Lemma 12 below, which is the sharpening of Roth's lemma from [5]. We mention that this sharpening was proved by making explicit the arguments in Faltings' proof of his Product theorem [7]. A result slightly weaker than Lemma 12 follows from Ferretti's work [8]. For further information on Faltings' Product theorem we refer to [13]. We recall that for a polynomial P with coefficients in $\mathbb{Z}$, $\| P \|$ denotes the maximum of the absolute values of its coefficients.

Lemma 12   *Let σ be a real and $k, d_1, \ldots, d_k$ integers such that $k \geq 2$, $0 < \sigma \leq k + 1$ and*

$$\frac{d_h}{d_{h+1}} \geq \omega_1 := \frac{2k^3}{\sigma} \quad \text{for } h = 1, \ldots, k - 1. \tag{6.1}$$

Further, let P be a non-zero polynomial in $\mathbb{Z}[X_1, \ldots, X_k]$ of degree at most d_h in X_h for $h = 1, \ldots, k$ and $\xi_1, \ldots, \xi_k$ non-zero algebraic numbers such that

$$M(\xi_h)^{d_h / \deg \xi_h} \geq \left(4^{d_1 + \cdots + d_k} \| P \| \right)^{\omega_2} \quad \text{for } h = 1, \ldots, k \tag{6.2}$$

with $\omega_2 := \left(3k^3 / \sigma \right)^k$. Then there is a tuple $\mathbf{j} = (j_1, \ldots, j_k)$ of non-negative integers with

$$\sum_{h=1}^{k} \frac{j_h}{d_h} < \sigma, \quad D^{\mathbf{j}} P(\xi_1, \ldots, \xi_k) \neq 0.$$

Proof   This follows from Theorem 3 and the Remark on pp. 221–222 of [5]. We mention that Theorem 3 of [5] has instead of (6.2) the assumption $H(\xi_h)^{d_h} > \{ e^{d_1 + \cdots + d_k} H(P) \}^{\omega_2}$, with heights $H(\xi_h)$, $H(P)$ defined in [5]. This is implied by (6.2) since $H(\xi_h) \geq M(\xi_h)^{1 / \deg \xi_h}$ and since for polynomials $P \in \mathbb{Z}[X_1, \ldots, X_k]$, $H(P)$ is equal to the Euclidean norm of the vector of coefficients of P so $H(P) \leq \{ (d_1 + 1) \cdots (d_k + 1) \}^{1/2} \| P \|$.

Let $\varphi_1, \ldots, \varphi_t$ be non-negative reals satisfying (1.10). Let ϵ and k be given by (5.1), (5.2), respectively. Put

$$\sigma := \frac{\epsilon k}{t}. \tag{6.3}$$

Thus, the quantities ω_1, ω_2 in Lemma 12 are equal to

$$\omega_1 = \frac{2k^2 t}{\epsilon}, \quad \omega_2 = \left(\frac{3k^2 t}{\epsilon} \right)^k = \left(\frac{3\omega_1}{2} \right)^k. \tag{6.4}$$

We prove the following:

Lemma 13 *The inequality* (1.5) *has no solutions* $\xi_1, \ldots, \xi_k$ *with*

$$M(\xi_1) \geq (4M)^{3rk\omega_2} , \tag{6.5}$$

$$M(\xi_{h+1}) \geq M(\xi_h)^{3\omega_1/2} \quad \text{for } h = 1, \ldots, k-1 . \tag{6.6}$$

Proof We assume the contrary and obtain a contradiction by applying Lemmas 11 and 12. We choose integers $d_1, \ldots, d_k$ as follows: take

$$d_k \geq \max\left(\frac{10^4 t}{\epsilon^2}, C(K)\right) \tag{6.7}$$

and let $d_1, \ldots, d_{k-1}$ be the integers defined by

$$d_k \log M(\xi_k) - \log M(\xi_1) < d_1 \log M(\xi_1) \leq d_k \log M(\xi_k),$$
$$d_1 \log M(\xi_1) \leq d_h \log M(\xi_h) < d_1 \log M(\xi_1) + \log M(\xi_h)$$
$$\text{for } h = 2, \ldots, k-1 .$$

The inequality (6.7) implies that

$$\frac{\log M(\xi_1)}{d_k \log M(\xi_k)} < 10^{-4}\epsilon^2 ,$$

$$\frac{\log M(\xi_h)}{d_1 \log M(\xi_1)} \leq \left(1 - 10^{-4}\epsilon^2\right)^{-1} \frac{\log M(\xi_h)}{d_k \log M(\xi_k)} \leq \epsilon^2 \quad \text{for } h = 2, \ldots, k-1,$$

so

$$M(\xi_1)^{d_1} \leq M(\xi_h)^{d_h} \leq M(\xi_1)^{d_1(1+\epsilon^2)} \quad \text{for } h = 1, \ldots, k. \tag{6.8}$$

Further, (6.8) and (6.6) imply that

$$\frac{d_h}{d_{h+1}} \geq (1+\epsilon^2)^{-1} \cdot \frac{\log M(\xi_{h+1})}{\log M(\xi_h)} \geq (1+\epsilon^2)^{-1} \frac{3\omega_1}{2} > \omega_1 . \tag{6.9}$$

We apply Lemma 11. Let P be the polynomial from Lemma 4. We assumed (5.1) and (5.2); and (5.3) is a consequence of (6.7) and (6.9). Further, (5.7) follows from (6.5), (5.1), (5.2), and (6.4), while (5.8) follows from (6.8). So by Lemma 11 we have that there is a tuple $\mathbf{c} \in \mathcal{C}_k$ such that for each tuple of non-negative integers $\mathbf{j} = (j_1, \ldots, j_k)$ satisfying (5.9) we have $D^{\mathbf{j}} P(\xi_1^{(c_1)}, \ldots, \xi_t^{(c_t)}) = 0$.

We now apply Lemma 12 with $\sigma = \epsilon k / t$ and with $\xi_h^{(c_h)}$ replacing ξ_h for $h = 1, \ldots, k$. From (6.9) we know already that (6.1) holds. Further, we have for $h = 1, \ldots, k$,

$$M(\xi_h)^{d_h} \geq M(\xi_1)^{d_1} \geq (4M)^{3rkd_1\omega_2} \quad \text{by (6.8), (6.5)}$$
$$\geq (4M)^{3r(d_1+\cdots+d_k)\omega_2} \quad \text{by (6.9)}$$
$$\geq \left(4^{d_1+\cdots+d_k} \| P \|\right)^{\omega_2} \quad \text{by (3.20)}.$$

Hence (6.2) is also satisfied. It follows that there is a tuple $\mathbf{j}$ with (5.9) for which $D^{\mathbf{j}}P(\xi_1^{(c_1)}, \ldots, \xi_t^{(c_t)}) \neq 0$. This is contrary to what we proved above. Thus, our assumption that Lemma 13 is false leads to a contradiction.

We now complete the proof of part (i) of Theorem 2. Define a sequence of solutions $\xi_1, \xi_2, \ldots$ of (1.5) as follows: ξ_1 is a solution ξ of (1.5) such that $M(\xi) \geq (4M)^{3rk\omega_2}$ and $M(\xi)$ is minimal; and for $h = 1, 2, \ldots, \xi_{h+1}$ is a solution ξ of (1.5) such that $M(\xi) \geq M(\xi_h)^{3\omega_1/2}$ and $M(\xi)$ is minimal. From Lemma 13 it follows, that this sequence has at most $k - 1$ elements.

Let $A := \max(4^{t(t+1)/(\kappa-2t)}, M)$ be the lower bound in part (i) of Theorem 2. Put $\theta := 1 + (\kappa - 2t)/t$. By assumption, the solutions of (1.5) lie in the union of the intervals $I_0 = [A, (4M)^{3rk\omega_2}]$ and $I_h = [M(\xi_h), M(\xi_h)^{3\omega_1/2}]$ $(h = 1, 2, \ldots)$. By part (i) of Lemma 2 and $4M \leq A^{\theta}$ we have that the number of solutions ξ in I_0 is at most

$$t\Big(1 + \frac{\log(2\log\{(4M)^{3rk\omega_2}\}/\log A)}{\log \theta}\Big) \leq t\Big(1 + \frac{\log\{6rk\omega_2\theta\}}{\log \theta}\Big)$$

$$\leq t\Big(2 + \frac{\log 6rk}{\log \theta} + k\frac{\log 3\omega_1/2}{\log \theta}\Big)$$

$$\leq t\Big(2 + k\frac{\log 3\omega_1}{\log \theta}\Big) \quad \text{by (5.1), (5.2), (6.4).}$$

Moreover, by part (i) of Lemma 2 we have for $h = 1, 2, \ldots,$ that the number of solutions in I_h is at most

$$t\Big(1 + \frac{\log(2\log\{M(\xi_h)^{3\omega_1/2}\}/\log M(\xi_h))}{\log \theta}\Big) \leq t\Big(1 + \frac{\log 3\omega_1}{\log \theta}\Big).$$

Since we have at most $k - 1$ intervals I_h $(h \geq 1)$, it follows that (1.5) has at most

$$N := t\Big(k + 1 + (2k - 1)\frac{\log 3\omega_1}{\log \theta}\Big)$$

solutions with $M(\xi) \geq A$. We estimate this from above. From (1.7) it follows that $\kappa = \sum_{l=1}^{t} \varphi_l \geq 2t + \delta$ so

$$\log \theta \geq \log(1 + \frac{\delta}{t}) \geq \frac{\delta}{2t}.$$

Further,

$$\log 3\omega_1 = \log \frac{6k^2 t}{\epsilon} \quad \text{by (6.4)}$$

$$\leq \log \Big(2.5 \times 10^{11} \cdot t^{11}\delta^{-5}\big(1 + \tfrac{1}{2}\log t\big)^2\big(1 + \tfrac{1}{2}\log \delta^{-1}\big)^2(\log 4r)^2\Big)$$

$$\text{by (5.1), (5.2)}$$

$$< 27 + 12\log t + 6\log \delta^{-1} + 2\log\log 4r \quad \text{using } \big(1 + \tfrac{1}{2}\log x\big)^2 \leq x \text{ for } x \geq 1$$

$$< 85\big(1 + \tfrac{1}{2}\log t\big)\big(1 + \tfrac{1}{2}\log \delta^{-1}\big) \cdot \log\log 4r \quad \text{using } \log\log 4r \geq \log\log 4.$$

Together with (5.2) this implies

$$N \le kt \cdot (1 + \frac{4t}{\delta} \log 3\omega_1) < 5kt^2 \delta^{-1} \log 3\omega_1$$

$$< 5 \times 3.5 \times 10^4 \times 85 \cdot t^6 (1 + \tfrac{1}{2} \log t)^2 \cdot \delta^{-3} (1 + \tfrac{1}{2} \log \delta^{-1})^2 \log 4r \log\log 4r$$

$$< 2 \times 10^7 \cdot t^7 \delta^{-4} \log 4r \cdot \log\log 4r \ .$$

This completes the proof of part (i) of Theorem 2.

7. Proof of Theorem 3 We need the following combinatorial lemma:

Lemma 14 *Let θ be a real with $0 < \theta < 1$ and t an integer ≥ 1. There exists a set P, consisting of tuples $\underline{\rho} = (\rho_1, \ldots, \rho_t)$ with $\rho_1 \ge \rho_2 \ge \cdots \ge \rho_t \ge 0$ and $1 - \theta \le \sum_{i=1}^{t} \rho_i \le 1$, such that $\#P \le 4\{e^2(\tfrac{1}{2} + \frac{1+\theta^{-1}}{t})\}^{t-1}$ and such that for all reals $F_1, \ldots, F_t, \Lambda$ with*

$$0 < F_1 \le F_2 \le \cdots \le F_t \le 1, \quad F_1 \cdots F_t \le \Lambda$$

there is a tuple $\underline{\rho} \in P$ with $F_i \le \Lambda^{\rho_i}$ for $i = 1, \ldots, t$.

Proof We assume without loss of generality that $F_1 \cdots F_t = \Lambda$ and that $t \ge 2$ (otherwise we may take $\rho_1 = 1$). Define c_i by $F_i = \Lambda^{c_i}$ for $i = 1, \ldots, t$; thus, $c_1 \ge \cdots \ge c_t \ge 0$ and $c_1 + \cdots + c_t = 1$. Put

$$g := [\theta^{-1}(t-1)] + 1, \quad f_i = [c_i g], \quad \rho_i = f_i/g \text{ for } i = 1, \ldots, t \ .$$

Then clearly, $F_i \le \Lambda^{\rho_i}$ for $i = 1, \ldots, t$. Since $c_i g - 1 < f_i \le c_i g$, we have $g - t < \sum_{i=1}^{t} f_i \le g$ and therefore, $g - t + 1 \le \sum_{i=1}^{t} f_i \le g$ since the f_i are integers. It follows that $1 - \theta \le \sum_{i=1}^{t} \rho_i \le 1$. Further, the tuple $\underline{\rho} = (\rho_1, \ldots, \rho_t)$ belongs to the set

$$P := \left\{ \left(\frac{f_1}{g}, \ldots, \frac{f_t}{g}\right) : f_1, \ldots, f_t \in \mathbb{Z}, \ f_1 \ge \cdots \ge f_t \ge 0, \ g - t + 1 \le \sum_{i=1}^{t} f_i \le g \right\} .$$

The map $(f_1/g, \ldots, f_t/g) \mapsto (f_1 + t - 1, f_2 + t - 2, \ldots, f_t)$ maps P bijectively onto

$$P' := \left\{ (h_1, \ldots, h_t) \in \mathbb{Z}^t : h_1 > h_2 > \cdots > h_t \ge 0, \ g' - t + 1 \le \sum_{i=1}^{t} h_i \le g' \right\} ,$$

with $g' = g + \tfrac{1}{2}t(t-1) = [\theta^{-1}(t-1)] + \tfrac{1}{2}t(t-1) + 1$. Clearly, the cardinality of P' is at most $1/t!$ times the cardinality of the set of all (not necessarily decreasing) tuples of non-negative integers $(h_1, \ldots, h_t)$ with $g'-t+1 \leq \sum_{i=1}^{t} h_i \leq g'$. Using that

$$\binom{x+y}{y} \leq \frac{(x+y)^{x+y}}{x^x y^y} = (1 + \frac{y}{x})^x (1 + \frac{x}{y})^y \leq (e(1 + \frac{x}{y}))^y \text{ for } x, y \geq 1$$

we infer

$$\#P = \#P' \leq \frac{1}{t!} \sum_{h=g'-t+1}^{g'} \binom{h+t-1}{t-1} \leq \frac{1}{t!} \cdot t \cdot \binom{g'+t-1}{t-1}$$

$$\leq \frac{e^t}{t^{t-1}} \cdot \left(e(1 + \theta^{-1} + \frac{t}{2} + \frac{1}{t-1})\right)^{t-1}$$

$$\leq \frac{e^t}{t^{t-1}} \cdot \left(e(1 + \theta^{-1} + \frac{t}{2})\right)^{t-1} \cdot \left(1 + \frac{1}{3(t-1)}\right)^{t-1} \text{ since } t \geq 2, \theta < 1$$

$$\leq 4 \cdot \left(e^2(\frac{1}{2} + \frac{1+\theta^{-1}}{t})\right)^{t-1}.$$

This ends the proof of the lemma.

Let f be the polynomial from Theorem 3, i.e.

$$f(X) = a_0(X - \alpha_1) \cdots (X - \alpha_r)$$

where the coefficients of f are rational integers, f is primitive, and $\alpha_1, \ldots, \alpha_r$ are distinct. Further, let g be a primitive, irreducible polynomial in $\mathbb{Z}[X]$ of degree t satisfying the conditions (1.17) and (1.18) with κ satisfying (1.16). Then

$$g(X) = b_0(X - \xi^{(1)}) \cdots (X - \xi^{(t)})$$

where $\xi^{(1)}, \ldots, \xi^{(t)}$ are the conjugates of an algebraic number ξ of degree t and $b_0 \in \mathbb{Z}$. We order $\xi^{(1)}, \ldots, \xi^{(t)}$ in such a way that

$$\min_{j=1,\ldots,r} \frac{|\alpha_j - \xi^{(1)}|}{|1, \alpha_j|} \leq \cdots \leq \min_{j=1,\ldots,r} \frac{|\alpha_j - \xi^{(t)}|}{|1, \alpha_j|}. \tag{7.1}$$

We show that ξ satisfies one from a finite collection of systems (1.5) to which Theorem 2 is applicable. From (1.12) it follows that

$$\frac{|R(f,g)|}{M(f)^t M(g)^r} = \prod_{i=1}^{t} \prod_{j=1}^{r} \frac{|\alpha_j - \xi^{(i)}|}{|1, \alpha_j| \cdot |1, \xi^{(i)}|}. \tag{7.2}$$

For $i = 1, \ldots, t$, let α_{j_i} be the zero of f for which

$$\frac{|\alpha_{j_i} - \xi^{(i)}|}{|1, \alpha_{j_i}|} = \min_{j=1,\ldots,r} \frac{|\alpha_j - \xi^{(i)}|}{|1, \alpha_j|}.$$

From triangle inequality (2.7) it follows that for $j \neq j_i$,

$$\frac{|\alpha_j - \xi^{(i)}|}{|1, \alpha_j| \cdot |1, \xi^{(i)}|} \geq \frac{1}{2} \left(\frac{|\alpha_j - \xi^{(i)}|}{|1, \alpha_j| \cdot |1, \xi^{(i)}|} + \frac{|\alpha_{j_i} - \xi^{(i)}|}{|1, \alpha_{j_i}| \cdot |1, \xi^{(i)}|} \right) \geq \frac{|\alpha_{j_i} - \alpha_j|}{2|1, \alpha_{j_i}| \cdot |1, \alpha_j|} .$$

Further, using that the discriminant $D(f) = a_0^{2r-2} \prod_{1 \leq p < q \leq r} (\alpha_p - \alpha_q)^2$ is a non-zero rational integer,

$$\prod_{j \neq j_i} \frac{|\alpha_{j_i} - \alpha_j|}{2|1, \alpha_{j_i}| \cdot |1, \alpha_j|} \geq \prod_{1 \leq p < q \leq r} \frac{|\alpha_p - \alpha_q|}{2|1, \alpha_p| \cdot |1, \alpha_q|} = \frac{|D(f)|^{1/2}}{2^{r(r-1)/2} M(f)^{r-1}}$$

$$\geq \frac{1}{2^{r(r-1)/2} M(f)^{r-1}} .$$

Together with (7.2) this implies that

$$\frac{|R(f, g)|}{M(f)^t M(g)^r} \geq C^{-1} \prod_{i=1}^{t} \frac{|\alpha_{j_i} - \xi^{(i)}|}{2|1, \alpha_{j_i}| \cdot |1, \xi^{(i)}|} , \tag{7.3}$$

$$\text{with } C = \left(2^{1+r(r-1)/2} M(f)^{r-1} \right)^t .$$

Put $\kappa' := (1 + \frac{1}{3} + \cdots + \frac{1}{2t-1})(2t + \frac{3}{4}\delta)$. From (1.18) it follows that $M(g) \geq C^{(\kappa-\kappa')^{-1}}$. Combining this with (7.3) and (1.17), and using that $M(g) = M(\xi)$, we get

$$\prod_{i=1}^{t} \frac{|\alpha_{j_i} - \xi^{(i)}|}{2|1, \alpha_{j_i}| \cdot |1, \xi^{(i)}|} \leq C \cdot M(g)^{-\kappa} \leq M(\xi)^{-\kappa'} .$$

We now apply Lemma 14 to $F_j := |\alpha_{j_i} - \xi^{(i)}| / (2 \cdot |1, \alpha_{j_i}| \cdot |1, \xi^{(i)}|)$ for $j = 1, \ldots, t$ and $\Lambda = M(\xi)^{-\kappa'}$. It is trivial that $F_t \leq 1$ and together with (7.1) this gives $0 < F_1 \leq \cdots \leq F_t \leq 1$. Put

$$\kappa'' := (1 + \frac{1}{3} + \cdots + \frac{1}{2t-1})(2t + \frac{1}{2}\delta), \quad \theta := 1 - \kappa''/\kappa' = \delta/(8t + 3\delta) . \tag{7.4}$$

Letting P be the set from Lemma 14, we infer that there is a tuple $\rho = (\rho_1, \ldots, \rho_t) \in P$ such that

$$\frac{|\alpha_{j_i} - \xi^{(i)}|}{2|1, \alpha_{j_i}| \cdot |1, \xi^{(i)}|} \leq M(\xi)^{-\rho_i \kappa'} = M(\xi)^{-\varphi_i} \text{ for } i = 1, \ldots, t, \tag{7.5}$$

where $\varphi_i := \rho_i \kappa'$. Note that $\sum_{i=1}^{t} \varphi_i \geq \kappa''$. Together with (1.7) this implies

$$\max_I (\#I)^2 \left(\sum_{i \in I} \varphi_i^{-1} \right)^{-1} \geq \left(\sum_{j=1}^{t} \frac{1}{2j-1} \right)^{-1} \cdot \kappa'' = 2t + \frac{\delta}{2} .$$

Further, we have

$$M(f) \geq \max_{i=1,\ldots,r} M(\alpha_i), \quad [\mathbb{Q}(\alpha_{j_1},\ldots,\alpha_{j_t}) : \mathbb{Q}] \leq r^t,$$

$$M(\xi) = M(g) \geq \max(4^{t(t+1)/(\kappa''-2t)}, M(f)) \quad \text{by (1.18)}.$$

Hence from part (i) of Theorem 2 with $\delta/2$ replacing δ it follows that each system (7.5) has at most $3.2 \times 10^8 t^7 \delta^{-4} \log 4r^t \log\log 4r^t$ solutions ξ coming from an irreducible polynomial g satisfying (1.17), (1.18).

By (7.4) we have

$$\#P \leq 4\left(e^2\left(\frac{1}{2} + \frac{1+\theta^{-1}}{t}\right)\right)^{t-1} = 4\left(e^2\left(\frac{1}{2} + \frac{8}{\delta} + \frac{4}{t}\right)\right)^{t-1}$$

$$\leq 4\left(e^2\left(\frac{1}{2} + \frac{8}{\delta}\right)\right)^{t-1}\left(1 + \frac{1}{2t}\right)^{t-1} \leq 7(63\delta^{-1})^{t-1}.$$

Further, for the tuple $(j_1,\ldots,j_t)$ we have at most r^t possibilities. Therefore, we have at most $7r^t(63\delta^{-1})^{t-1}$ possibilities for the system (7.5). We conclude that the total number of primitive, irreducible polynomials g satisfying (1.17), (1.18) is at most

$$7r^t(63\delta^{-1})^{t-1} \cdot 3.2 \times 10^8 t^7 \delta^{-4} \log 4r^t \log\log 4r^t$$

$$\leq 10^{15}(\delta^{-1})^{t+3} \cdot (100r)^t \log 4r \log\log 4r.$$

This completes the proof of Theorem 3.

References

[1] E. Bombieri and A.J. van der Poorten. Some quantitative results related to Roth's theorem. *J. Austral. Math. Soc.* (A), **45** (1988), 233–248; Corrigenda. ibid., **48** (1990), 154–155.

[2] P. Corvaja. Approximation diophantienne sur la droite. Thèse de Doctorat de Mathématiques, Univ. Paris VI, 1995.

[3] H. Davenport and K.F. Roth. Rational approximations to algebraic numbers. *Mathematika*, **2** (1955), 160–167.

[4] H. Esnault and E. Viehweg. Dyson's lemma for polynomials in several variables (and the theorem of Roth). *Invent. math.*, **78** (1984), 445–490.

[5] J.-H. Evertse. An explicit version of Faltings' Product theorem and an improvement of Roth's lemma. *Acta Arith.*, **73** (1995), 215–248.

[6] —. An improvement of the quantitative Subspace theorem. *Compos. Math.*, **101** (1996), 225–311.

[7] G. Faltings. Diophantine approximation on abelian varieties. *Ann. Math.*, **133** (1991), 549–576.

[8] R. Ferretti. An effective version of Faltings' Product Theorem. *Forum Math.*, **8** (1996), 401–427.

[9] S. Lang. *Algebra* (2nd ed). Addison–Wesley, Redwood City, 1984.

[10] W.J. Leveque. *Topics in Number Theory*, vol. II. Addison–Wesley, Redwood City, 1956.

[11] M. Loève. *Probability Theory* (4th ed), vol. I. Springer-Verlag, Berlin, 1977.

[12] M. Mignotte. Quelques remarques sur l'approximation rationelle des nombres algébriques. *J. reine angew. Math.*, **268/269** (1974), 341–347.

[13] M. van der Put. The Product theorem. *Diophantine Approximation and Abelian Varieties, Proc. Conf. Soesterberg, The Netherlands, 1992* (B. Edixhoven and J.-H. Evertse, Eds.), Lect. Notes in Math., **1566**, Springer-Verlag, Berlin, 1993, pp. 77–82.

[14] K. Ramachandra. Approximation of algebraic numbers. *Nachr. Akad. Wiss. Göttingen, Math.-Phys. Kl. II*, (1966), 45–52.

[15] K.F. Roth. Rational approximations to algebraic numbers. *Mathematika*, **2** (1955), 1–20.

[16] Min Ru and P.M. Wong. Integral points of $\mathbf{P}^n \backslash \{2n + 1$ hyperplanes in general position$\}$. *Invent. math.*, **106** (1991), 195–216.

[17] W.M. Schmidt. Simultaneous approximation to algebraic numbers by rationals. *Acta Math.*, **125** (1970), 189–201.

[18] —. Inequalities for resultants and for decomposable forms. *Proc. Conf. Diophantine Approximation and its Applications, Washington D.C., 1972* (C.F. Osgood, Ed.), Academic Press, New York, 1973, pp. 235–253.

[19] —. *Diophantine Approximation*. Lect. Notes in Math., **785**, Springer-Verlag, Berlin, 1980.

[20] —. The subspace theorem in diophantine approximations. *Compos. Math.*, **69** (1989), 121–173.

[21] —. The number of exceptional approximations in Roth's theorem. *J. Austral. Math. Soc.*, (A) **59** (1995), 375–383.

[22] C.L. Siegel. Approximation algebraischer Zahlen. *Math. Z.*, **10** (1921), 173–213.

[23] —. Über Näherungswerte algebraischer Zahlen. *Math. Ann.*, **84** (1921), 80–99.

[24] E. Wirsing. On approximations of algebraic numbers by algebraic numbers of bounded degree. *Proc. Symp. Pure Math.*, **20**, A.M.S., Providence, 1971, pp. 213–248.

Jan-Hendrik Evertse
Mathematical Institute, University of Leiden
P.O. Box 9512, 2300 RA Leiden, The Netherlands

6

The Brun–Titchmarsh Theorem

JOHN FRIEDLANDER and HENRYK IWANIEC

1. Introduction The distribution of primes in arithmetic progressions is a central issue in analytic number theory. For $(a, q) = 1$ we let $\pi(x; q, a)$ denote the number of primes $p \leq x$, $p \equiv a \pmod q$. The asymptotic formula

$$\pi(x; q, a) \sim \frac{x}{\varphi(q) \log x} \tag{1.1}$$

holds for each given progression as $x \to \infty$ and one is interested in estimates uniform with q as large as possible. The Generalized Riemann Hypothesis allows q as large as $x^{1/2-\varepsilon}$ in (1.1) but all that is known unconditionally is the much smaller range $q < (\log x)^A$ for arbitrary constant A, known as the Siegel-Walfisz theorem and now about sixty years old. On the other hand, one expects (1.1) to hold with q as large as $x^{1-\varepsilon}$ although it is known to fail in the range $q < x(\log x)^{-A}$ for every A (see [1]).

Given the incompleteness of the above results it has proved to be of great utility that, almost from the birth of the Brun sieve, it was possible to give upper bounds

$$\pi(x; q, a) < \frac{cx}{\varphi(q) \log x/q}$$

with c an absolute constant, which yield the correct order of magnitude throughout the range $q < x^{1-\varepsilon}$. This is known as the Brun–Titchmarsh theorem due to the work [6]. Progress in the sieve led to this result with $c = 2 + \varepsilon$, which can now be obtained in several ways (Selberg sieve, combinatorial sieve, large sieve). This is the limit of the method in several respects. An improvement in the constant c for small q would have striking consequences for the problem of exceptional zeros of L-functions, hence for class numbers, etc. Moreover, it is known that there are sequences of integers satisfying the same standard sieve axioms (consider the set of those integers in the progression composed of an odd number of prime factors) for which the corresponding bound cannot be improved.

In view of this limitation it was a great achievement when Y. Motohashi [5], by using non-trivial information about the nature of arithmetic progressions, was able to improve the constant $c = 2$ for certain larger ranges of q. This work has had a significant impact on subsequent developments in general sieve theory. Throughout, we shall write $q = x^\theta$. Motohashi's original results gave improvements in the range $0 < \theta < \frac{2}{5}$. The largest range in which such improvements are known till now is $0 < \theta < \frac{2}{3}$ given in [2]. The limit $\frac{2}{3}$ is set by an application of Weil's estimate for Kloosterman sums and, despite many recent developments in the latter topic, has resisted further widening. In this paper we make a small improvement throughout the full range $0 < \theta < 1$. Of course, in view of the earlier results, we may assume $\theta \geq \frac{2}{3}$.

Theorem 1 *Let* $\frac{6}{11} < \theta < 1$ *and* $c = 2 - (\frac{1-\theta}{4})^6$. *Then we have*

$$\pi(x; q, a) < \frac{cx}{\varphi(q) \log x/q} \tag{1.2}$$

for all x sufficiently large in terms of θ.

Remark The exponent 6 can certainly be improved to some extent but we have not tried to find the best result.

As in the Motohashi argument our proof is based on a non-trivial treatment of the remainder term in the linear sieve. Whereas Motohashi used multiplicative characters we employ a Fourier series expansion of the remainder term which leads to incomplete sums of Kloosterman type. The case of large q corresponds to very short sums and if one assumes conjectures about these, such as the Hooley R^*-conjecture, then one knows [2, Theorem 9] that the bound of Theorem 1 holds with $c = \frac{5}{3} + \varepsilon$.

Very recently A.A. Karatsuba [4] has established non-trivial estimates for certain remarkably short two-variable exponential sums of Kloosterman type. His method works only if the variables are in certain isolated ranges. By arranging the sieve machinery to meet these ranges we are able to exploit his estimates to give Theorem 1.

There are a number of ways of applying the sieve to $\pi(x; q, a)$. The method we use here actually gives the upper bound of Theorem 1 for the number $\pi(x, z; q, a)$ of integers $n \leq x$, $n \equiv a \pmod{q}$ which have no prime factors less than $z = (x/q)^{1/3}$. This is clearly a larger number.

For completeness we present the proof of a case of Karatsuba's theorem which suffices for our purpose. Specifically we consider a bilinear form

$$S = \sum_{\substack{(mn,q)=1}}\sum \alpha_m \beta_n e_q(a\overline{mn}) \quad \text{with } (a,q) = 1, \tag{1.3}$$

where α_m, β_n are arbitrary complex numbers with $|\alpha_m| \leq 1$, $|\beta_n| \leq 1$ supported on the dyadic intervals $M < m \leq 2M$, $N < n \leq 2N$. As usual

$$e(t) = e^{2\pi i t}, \quad e_q(t) = e(t/q), \quad d\overline{d} \equiv 1 \pmod{q}.$$

We also assume that α_m, β_n are supported on primes. This gives cleaner, stronger estimates and in any case is what occurs in our application.

Theorem 2 *Let k, ℓ be positive integers with $k < N$, $\ell < M$. Suppose*

$$k(2N)^{2k-1} < q, \quad \ell(2M)^{2\ell-1} < q .\tag{1.4}$$

Then we have

$$|S| \leq k\ell MN \left(qM^{-\ell}N^{-k}\right)^{\frac{1}{2k\ell}} .\tag{1.5}$$

The proof of Theorem 2 is completely elementary, no use of anything of the depth of Weil's bound. The argument is simpler than but reminiscent of Vinogradov's treatment of exponential sums. The latter yields a corresponding Brun–Titchmarsh theorem for primes in short intervals valid in the whole range (see [2], Theorem 14).

Acknowledgements We thank Karatsuba for preprints of his then-unpublished work. J. F. was supported in part by NSERC grant A5123 and Macquarie University. H. I. was supported in part by NSF grant DMS-9500797. We began this work while enjoying the hospitality of IAS Princeton.

2. Proof of Theorem 1 Throughout we keep the standard notation and appeal to results familiar from the linear sieve theory. Thus, given $(a, q) = 1$ we consider the sifting sequence

$$\mathcal{A} = \{n \leq x : n \equiv a \,(\mathrm{mod}\, q)\}$$

and the sifting function

$$S(\mathcal{A}, z) = |\{n \in \mathcal{A} : (n, P(z)) = 1\}|$$

where $P(z)$ denotes the product of all primes $p < z$, $p \nmid q$.

Let $w < y < z$. Then, by an iteration of Buchstab's identity

$$S(\mathcal{A}, z) = S(\mathcal{A}, w) - \sum_{y \leq p < z} S\left(\mathcal{A}_p, p\right) - \sum_{w \leq p < y} S\left(\mathcal{A}_p, w\right)$$

$$+ \sum_{w \leq p_2 < p_1 < y} \sum S\left(\mathcal{A}_{p_1 p_2}, p_2\right),\tag{2.1}$$

where $\mathcal{A}_d = \{n \in \mathcal{A} : n \equiv 0 \,(\mathrm{mod}\, d)\}$. We choose $z = (x/q)^{\frac{1}{3}}$. If we apply classical estimates of the linear sieve [3] for all of the sifting functions on the right we would get the same bound as applying such estimates for the left-hand side because it is in the nature of these bounds that they are, with the above choice of z, stable under transformations via the Buchstab identity. Applying these to either side of (2.1) we obtain

$$S(\mathcal{A}, z) < \frac{2x}{\varphi(q)\log D} \quad \text{with } D = \frac{x^{1-\varepsilon}}{q}\tag{2.2}$$

for any $\varepsilon > 0$ provided x is sufficiently large in terms of ε and θ. Thus, to improve (2.2) it suffices to improve any part on the right-hand side, and Theorem 2 allows us to do this for the double sum if y and w are chosen suitably.

The classical bound for each sifting function in this double sum on the right side of (2.1) is, for $w > D^{\frac{1}{5}}$,

$$S\left(\mathcal{A}_{p_1 p_2}, p_2\right) < \frac{2x}{\varphi(q)p_1\, p_2 \log D_{12}} \quad \text{with } D_{12} = \frac{D}{p_1\, p_2} \qquad (2.3)$$

because the sequence $\mathcal{A}_{p_1 p_2}$ has level of distribution D_{12} and $p_2^3 > D_{12}$; the latter follows from the stronger condition $p_2^3 > D'_{12}$ below. In Section 4 we shall show that on average over p_1, p_2 this sequence has a larger level of distribution. Specifically, let ℓ be an integer such that

$$1 + \frac{3\theta}{1 - \theta} < 2\ell < \frac{5\theta}{1 - \theta}\ ; \qquad (2.4)$$

this exists and is ≥ 3 because $\frac{6}{11} < \theta < 1$. We choose

$$w = q^{\frac{1}{2\ell}+\varepsilon} \quad \text{and} \quad y = q^{\frac{1-\delta}{2\ell-1}-\varepsilon} \qquad (2.5)$$

where

$$\delta = \frac{1}{2\ell(2\ell^2 - \ell + 1)}\ .$$

In this range the improved level of distribution is

$$D'_{12} = \left(p_1\, p_2\, q^{-1/\ell}\right)^{\frac{1}{2\ell}} D_{12}\ . \qquad (2.6)$$

Therefore, on average

$$S\left(\mathcal{A}_{p_1 p_2}, p_2\right) < \frac{2x}{\varphi(q)p_1\, p_2 \log D'_{12}} \qquad (2.7)$$

provided $p_2^3 > D'_{12}$. It suffices to verify this last condition in the worst scenario $p_1 = p_2 = w$ in which case it is just the upper bound in (2.4). The lower bound in (2.4) suffices to imply the requirement $y < z$ for our choice of y and z.

The new bound (2.7) is smaller than the classical one (2.3) by the amount

$$\frac{2x}{\varphi(q)p_1\, p_2} \left(\frac{1}{\log D_{12}} - \frac{1}{\log D'_{12}}\right) > \frac{2x}{\varphi(q)p_1\, p_2}\, \frac{\log\left(p_1\, p_2\, q^{-1/\ell}\right)}{2\ell \log^2\left(xq^{-1-1/\ell}\right)}$$

because $D'_{12} < xq^{-1-1/\ell}$. Summing over p_1, p_2 we get

$$\sum_{w \le p_2 < p_1 < y}\sum (p_1 p_2)^{-1} \log\left(p_1 p_2 q^{-1/\ell}\right) \sim c(\ell) \log q$$

where

$$c(\ell) = \iint_{\frac{1}{2\ell}<\alpha<\beta<\frac{1-\delta}{2\ell-1}} \left(\alpha + \beta - \frac{1}{\ell}\right) d\alpha\, d\beta$$

$$= \frac{1}{2}\left(\frac{1-2\delta\ell}{2\ell(2\ell-1)}\right)^3 = \frac{1}{16(2\ell^2-\ell+1)^3}.$$

Therefore our improvement is

$$\frac{c(\ell)x \log q}{\ell\varphi(q)\log^2(xq^{-1-1/\ell})} = \frac{c(\theta,\ell)x}{\varphi(q)\log x/q}$$

where

$$c(\theta,\ell) = \frac{\ell c(\ell)\theta(1-\theta)}{((1-\theta)\ell-\theta)^2}.$$

Choose the smallest ℓ from (2.4); this is less than $\frac{3}{2(1-\theta)}$. The above constant satisfies

$$c(\theta,\ell) > c\left(\theta, \frac{3}{2(1-\theta)}\right) > \left(\frac{1-\theta}{4}\right)^6,$$

which proves Theorem 1, subject to the verification of the level of distribution (2.6).

3. The Karatsuba theorem

Here we prove Theorem 2. By Hölder's inequality we obtain

$$|S|^k \le M^{k-1}\sum_m \left|\sum_n \beta_n e_q(a\overline{mn})\right|^k$$

$$\le M^{k-1}\sum_{n_1}\cdots\sum_{n_k}\left|\sum_m \varepsilon_m e_q\left(a\overline{m}(\overline{n}_1+\cdots+\overline{n}_k)\right)\right|$$

$$= M^{k-1}\sum_{b\,(\mathrm{mod}\,q)} \nu(b)\left|\sum_m \varepsilon_m e_q(ab\overline{m})\right|$$

with some ε_m having $|\varepsilon_m| = 1$ and where $\nu(b)$ denotes the number of k-tuples $(n_1,\ldots,n_k)$ such that

$$\overline{n}_1 + \cdots + \overline{n}_k \equiv b\,(\mathrm{mod}\,q). \tag{3.1}$$

Again by Hölder's inequality

$$|S|^{2k\ell} \leq M^{2\ell(k-1)} \left(\sum_b \nu(b) \right)^{2\ell-2} \left(\sum_b \nu^2(b) \right) \sum_b \Big| \sum_m \varepsilon_m \, e_q(ab\overline{m}) \Big|^{2\ell} .$$

Here the first sum is $\leq N^k$ and the second sum is $\leq \mu_k(N)$, the number of $2k$-tuples $(n_1, \dots, n_{2k})$ such that

$$\overline{n}_1 + \cdots + \overline{n}_k \equiv \overline{n}_{k+1} + \cdots + \overline{n}_{2k} \pmod{q}. \tag{3.2}$$

Also for the third sum we have

$$\sum_{b \,(\mathrm{mod}\, q)} \Big| \sum_m \varepsilon_m \, e_q(ab\overline{m}) \Big|^{2\ell} \leq \mu_\ell(M) q .$$

From these estimates we obtain

$$|S|^{2k\ell} \leq \mu_\ell(M)\mu_k(N) M^{2\ell(k-1)} N^{2k(\ell-1)} q . \tag{3.3}$$

We require a bound for $\mu_k(N)$. To this end we note that from the assumption (1.4) it follows that every solution of the congruence (3.2) is also a solution of the equation

$$\frac{1}{n_1} + \cdots + \frac{1}{n_k} = \frac{1}{n_{k+1}} + \cdots + \frac{1}{n_{2k}} \tag{3.4}$$

and hence it suffices to estimate the number of solutions of (3.4). We shall also make the (weak) assumption that $k < N$ since under this condition we can say that $n_1^{-1} + \cdots + n_k^{-1}$ may be written as a sum of fractions an^{-1} having distinct n and $1 \leq a \leq k$ and hence $(a, n) = 1$. Recall that the n_j are primes. Thus

$$\frac{1}{n_1} + \cdots + \frac{1}{n_k} = \frac{b}{[n_1, \dots, n_k]}$$

with $(b, n_1 \dots n_k) = 1$ where $[n_1, \dots, n_k]$ denotes the least common multiple, and a similar expression holds for the right-hand size of (3.4). Hence, for any solution of (3.4) we have $n_j = n_1$ for some $k < j \leq 2k$. By induction the solutions of (3.4) are diagonal up to permutation and hence

$$\mu_k(N) \leq k! \, N^k , \tag{3.5}$$

with a similar bound for $\mu_\ell(M)$. Inserting these into (3.3) we conclude that

$$|S|^{2k\ell} \leq k! \, \ell! \, M^{2k\ell-\ell} N^{2k\ell-k} q . \tag{3.6}$$

This yields (1.5).

4. The level of distribution for $\mathcal{A}_{p_1 p_2}$ Now we are ready to prove that the subsequence $\mathcal{A}_{p_1 p_2}$ has level of distribution

$$D'_{12} = \left(\frac{P_1 P_2}{q^{1/\ell}} \right)^{\frac{1}{2\ell}} \frac{x^{1-3\varepsilon}}{q\, P_1 P_2} \tag{4.1}$$

on average over p_1, p_2 in the dyadic box $P_1 < p_1 \le 2P_1$, $P_2 < p_2 \le 2P_2$ for any P_1, P_2 with $w \le P_1, P_2 < y$. This was required to establish (2.7) with this D'_{12}.

We shall estimate the sum of error terms

$$R_d = \sum_{p_1} \sum_{p_2} \left(\left| \mathcal{A}_{d\,p_1\,p_2} \right| - \frac{x}{qdp_1\,p_2} \right) \tag{4.2}$$

over p_1, p_2 in the above box for every individual d with $(d, q) = 1$. We shall prove that

$$R_d \ll \frac{x^{1-\varepsilon}}{qd} \quad \text{if } d < D'_{12} . \tag{4.3}$$

We write

$$\left| \mathcal{A}_{dp_1\,p_2} \right| = \sum_{\substack{n\equiv 0\,(\mathrm{mod}\,dp_1\,p_2) \\ n\equiv a\,(\mathrm{mod}\,q)}} f(n) + O\left(\sideset{}{'}\sum_{\substack{n\equiv 0\,(\mathrm{mod}\,dp_1\,p_2) \\ n\equiv a\,(\mathrm{mod}\,q)}} 1 \right)$$

where $f(t)$ is a smooth non-negative function supported on $x^{1-\varepsilon} < t < x + x^{1-\varepsilon}$ such that $t^j f^{(j)}(t) \ll x^\varepsilon$ for $j \ge 0$ and $\hat{f}(0) = x$, while $\sum'$ restricts the summation to n in either of the two short intervals $x < n < x + x^{1-\varepsilon}$ or $0 < n < x^{1-\varepsilon}$. By Poisson summation

$$\sum_n f(n) = \frac{1}{qdp_1\,p_2} \sum_h \hat{f}\left(\frac{h}{qdp_1\,p_2} \right) e_q\left(-ah\overline{dp_1 p_2} \right) .$$

The zero frequency $h = 0$ gives the main term x/qdp_1p_2 whereas the contribution from $|h| \ge H$ with

$$H = qdP_1 P_2 x^{2\varepsilon - 1}$$

is negligible as is seen by partial integration. Therefore we have

$$R_d = \sum_{0<|h|<H} \sum_{p_1} \sum_{p_2} (qdp_1p_2)^{-1} \hat{f}\left(\frac{h}{qdp_1p_2} \right) e_q(-ah\overline{dp_1p_2}) + O\left(\frac{x^{1-\varepsilon}}{qd} \right) .$$

To separate the variables p_1, p_2 we write the Fourier transform as

$$\hat{f}\left(\frac{h}{qdp_1p_2}\right) = p_2 \int f(tp_2)e\left(\frac{ht}{qdp_1}\right) dt \ .$$

Hence

$$R_d \ll \frac{x}{qdP_1P_2} \sum_{0<|h|<H} \Big| \sum_{p_1}\sum_{p_2} \alpha_{p_1}\beta_{p_2}e_q(ah\overline{dp_1p_2}) \Big| + \frac{x^{1-\varepsilon}}{qd}$$

for some $|\alpha_{p_1}| \leq 1$ and $|\beta_{p_2}| \leq 1$. To the double sum over p_1, p_2 we apply Theorem 2 with $k = \ell$ and the modulus $q/(q,h) \geq q/H$. The condition (1.4) requires

$$q > \ell(2P)^{2\ell-1}H \tag{4.4}$$

for both $P = P_1$ and P_2. Employing $d < D'_{12}$ given by (4.1) and $P_1, P_2 < y$ given by (2.5) one verifies (4.4). Here the presence of δ in the choice of y corresponds to the presence of H in the condition (4.4) and the latter emerges because it is necessary to put the fraction h/q into lowest terms. For progressions of prime modulus q this would be unnecessary and the result would be simpler and slightly stronger. Now, since (4.4) is verified, we derive by Theorem 2 that

$$R_d \ll \frac{xH}{qd}\left(\frac{q^{1/\ell}}{P_1P_2}\right)^{\frac{1}{2\ell}} + \frac{x^{1-\varepsilon}}{qd}$$

which gives (4.3).

Note that

$$S\left(\mathcal{A}_{p_1p_2}, p_2\right) \leq S\left(\mathcal{A}_{p_1p_2}, P_2\right) \ .$$

Applying the linear sieve to the larger sifting function we derive by (4.3) the estimate (2.7) on average over p_1, p_2 in every relevant dyadic box. This completes the proof of Theorem 1.

References

[1] J. Friedlander and A. Granville. Limitations to the equi-distribution of primes, I. *Ann. Math.*, **129** (1989), 363–382.

[2] H. Iwaniec. On the Brun–Titchmarsh theorem. *J. Math. Soc. Japan*, **34** (1982), 95–123.

[3] W.B. Jurkat and H.-E. Richert. An improvement of Selberg's sieve method, I. *Acta Arith.*, **11** (1965), 217–240.

[4] A.A. Karatsuba. The distribution of inverses in a residue ring modulo a given modulus. *Russian Acad. Sci. Dokl. Math.*, **48** (1994), 452–454.

[5] Y. Motohashi. On some improvements of the Brun–Titchmarsh theorem. *J. Math. Soc. Japan*, **27** (1975), 444–453.

[6] E.C. Titchmarsh. A divisor problem. *Rend. Circ. Mat. Palermo*, **54** (1930), 414–429.

John Friedlander
Scarborough College, University of Toronto
Scarborough, Ontario M1C 1A4, Canada

Henryk Iwaniec
Mathematics Department, Rutgers University
New Brunswick, New Jersey 08903, U.S.A.

7

A Decomposition of Riemann's Zeta-Function

ANDREW GRANVILLE

1. Introduction It is currently very much in vogue to study sums of the form

$$\zeta(p_1, p_2, \ldots, p_g) := \sum_{a_1 > a_2 > \cdots > a_g \geq 1} \frac{1}{a_1^{p_1}} \frac{1}{a_2^{p_2}} \cdots \frac{1}{a_g^{p_g}}$$

where all the a_i's and p_i's are positive integers, with $p_1 \geq 2$. Note that it is necessary that $p_1 \geq 2$ else $\zeta(\mathbf{p})$ diverges. These sums are related to polylogarithm functions (see [1], [2], and [3]) as well as to zeta functions (the Riemann zeta function is of course the case $g = 1$). In this note we prove an identity that was conjectured by Moen [5] and Markett [7]:

Proposition *If g and N are positive integers with $N \geq g + 1$ then*

$$\zeta(N) = \sum_{\substack{p_1 + p_2 + \cdots + p_g = N \\ \text{Each } p_j \geq 1, \text{ and } p_1 \geq 2}} \zeta(p_1, p_2, \ldots, p_g). \tag{1}$$

This identity was proved for $g = 2$ by Euler, and for $g = 3$ by Hoffman and Moen [6]. The above proposition has been proved independently by Zagier [9], who writes of his proof, 'Although this proof is not very long, it seems too complicated compared with the elegance of the statement. It would be nice to find a more natural proof': Unfortunately much the same can be said of the very different proof that I have presented here.

Markett [7] and J. Borwein and Girgensohn [3] were able to evaluate

$$\zeta(p_1, p_2, p_3)$$

in terms of values of $\zeta(p)$ whenever $p_1 + p_2 + p_3 \leq 6$, and in terms of $\zeta(p)$ and $\zeta(a, b)$ whenever $p_1 + p_2 + p_3 \leq 10$ — it would be interesting to know whether such 'descents' are always possible or, as most researchers seem to believe, that there is only a small class of such sums that can be so evaluated.

Proof of (1) We may re-write the sum on the right side of (1) as

$$\sum_{a_1 > a_2 > \cdots > a_g \geq 1} \sum_{\substack{p_1 + p_2 + \cdots + p_g = N \\ \text{Each } p_j \geq 1, \text{ and } p_1 \geq 2}} \frac{1}{a_1^{p_1}} \frac{1}{a_2^{p_2}} \cdots \frac{1}{a_g^{p_g}}.$$

The author is a Presidential Faculty Fellow, supported in part by the National Science Foundation.

The second sum here is the coefficient of x^n in the power series

$$\sum_{p_1 \geq 2} \left(\frac{x}{a_1}\right)^{p_1} \prod_{j=2}^{g} \sum_{p_j \geq 1} \left(\frac{x}{a_j}\right)^{p_j} = \frac{x^2/a_1^2}{(1-x/a_1)} \prod_{j=2}^{g} \frac{x/a_j}{(1-x/a_j)}$$

$$= \frac{x^{g+1}}{a_1} \prod_{j=1}^{g} \frac{1}{(a_j - x)} = \frac{x^{g+1}}{a_1} \sum_{j=1}^{g} \frac{1}{(a_j - x)} \prod_{\substack{i=1 \\ i \neq j}}^{g} \frac{1}{(a_i - a_j)}.$$

Therefore the sum above is

$$\sum_{a_1 > a_2 > \cdots > a_g \geq 1} \frac{1}{a_1} \sum_{j=1}^{g} \frac{1}{a_j^{N-g}} \prod_{\substack{i=1 \\ i \neq j}}^{g} \frac{1}{(a_i - a_j)}$$

$$= \sum_{m \geq 1} \frac{1}{m^{N-g}} \sum_{j=1}^{g} A(m, j-1)(-1)^{g-j} B(m, g-j) \qquad (2)$$

where we take each $a_j = m$ in turn, with

$$A(m, j-1) := \sum_{a_1 > a_2 > \cdots > a_{j-1} > m} \frac{1}{a_1} \prod_{i=1}^{j-1} \frac{1}{(a_i - m)}$$

$$= \sum_{b_1 > b_2 > \cdots > b_{j-1} \geq 1} \frac{1}{(b_1 + m)b_1 b_2 \ldots b_{j-1}}$$

taking each $b_i = a_i - m$, and

$$B(m, g-j) := \sum_{m > a_{j+1} > a_{j+2} > \cdots > a_g \geq 1} \prod_{i=j+1}^{g} \frac{1}{(m - a_i)}$$

$$= \sum_{0 < b_{j+1} < b_{j+2} < \cdots < b_g < m} \frac{1}{b_{j+1} b_{j+2} \ldots b_g},$$

now taking each $b_i = m - a_i$. Note that the generating function for B is

$$\sum_{i \geq 0} B(m, i) x^i = \prod_{b=1}^{m-1} \left(1 + \frac{x}{b}\right). \qquad (3)$$

Dealing with A is somewhat more difficult. We start by noting that

$$\sum_{b_1 > b_2} \frac{1}{(b_1 + m)b_1} = \frac{1}{m} \sum_{b_1 > b_2} \left(\frac{1}{b_1} - \frac{1}{b_1 + m}\right)$$

$$= \frac{1}{m} \sum_{c_2 = 1}^{m} \frac{1}{b_2 + c_2},$$

as this is a telescoping sum. Substituting this back into the definition for A we next have to deal with

$$\sum_{b_2>b_3}\sum_{c_2=1}^{m}\frac{1}{(b_2+c_2)b_2}=\sum_{c_2=1}^{m}\frac{1}{c_2}\sum_{b_2>b_3}\left(\frac{1}{b_2}-\frac{1}{(b_2+c_2)}\right)$$

$$=\sum_{c_2=1}^{m}\frac{1}{c_2}\sum_{c_3=1}^{c_2}\frac{1}{b_3+c_3},$$

for the same reasons. Putting this back into the definition we have to do the same calculation again, now with the indices moved up one. Iterating this procedure we end up with

$$A(m,j-1):=\frac{1}{m}\sum_{m\geq c_2\geq c_3\geq\cdots\geq c_j\geq 1}\frac{1}{c_2c_3\ldots c_j},$$

which has generating function

$$\sum_{i\geq 0}A(m,i)x^i=\frac{1}{m}\prod_{c=1}^{m}\left(1+\frac{x}{c}+\left(\frac{x}{c}\right)^2+\left(\frac{x}{c}\right)^3+\ldots\right)$$

$$=\frac{1}{m}\prod_{c=1}^{m}\left(1-\frac{x}{c}\right)^{-1}. \tag{4}$$

Therefore, by (2), (3), and (4), the sum on the right side of (1) is

$$\sum_{m\geq 1}\frac{1}{m^{N-g+1}}$$

times the coefficient of x^{g-1} in the power series

$$\prod_{b=1}^{m-1}\left(1-\frac{x}{b}\right)\prod_{c=1}^{m}\left(1-\frac{x}{c}\right)^{-1}=\left(1-\frac{x}{m}\right)^{-1}=\sum_{i\geq 0}\left(\frac{x}{m}\right)^i.$$

We thus get $\sum_{m\geq 1}1/m^N=\zeta(N)$, giving (1).

2. Evaluations of $\zeta(r,s)$ Euler demonstrated that if $N=r+s$ is odd with s even then

$$\zeta(r,s)=-\frac{1}{2}\left\{\binom{N}{r}+1\right\}\zeta(N)+\sum_{\substack{a+b=N\\a,b\geq 2\\a\text{ odd}}}\left\{\binom{a-1}{s-1}+\binom{a-1}{t-1}\right\}\zeta(a)\zeta(b). \tag{5}$$

One can then obtain the value of $\zeta(s,r)$, provided $r > 1$, from the trivial identity

$$\zeta(r,s) + \zeta(s,r) = \zeta(r)\zeta(s) - \zeta(r+s). \tag{6}$$

(To prove this just write out the zeta-functions on both sides and compare terms). In the special case $r = 1$ he proved, for any $N \geq 3$ that

$$\zeta(N-1,1) = \frac{N-1}{2}\zeta(N) - \frac{1}{2}\sum_{\substack{a+b=N \\ a,b\geq 2}} \zeta(a)\zeta(b). \tag{7}$$

Pages 47–49 of [8], Equation (2) of [2], and Theorem 4.1 of [7] are all equivalent to, for $N \geq 4$,

$$\zeta(N-2,1,1) = \frac{(N-1)(N-2)}{6}\zeta(N) + \frac{1}{2}\zeta(2)\zeta(N-2)$$
$$- \frac{N-2}{4}\sum_{\substack{a+b=N \\ a,b\geq 2}} \zeta(a)\zeta(b) + \frac{1}{6}\sum_{\substack{a+b+c=N \\ a,b,c\geq 2}} \zeta(a)\zeta(b)\zeta(c). \tag{8}$$

Proof of (7) We evaluate the last sum in (7), using (6):

$$\sum_{\substack{a+b=N \\ a,b\geq 2}} \zeta(a)\zeta(b) = \sum_{\substack{a+b=N \\ a,b\geq 2}} (\zeta(a,b) + \zeta(b,a) + \zeta(N))$$

$$= 2\sum_{\substack{a+b=N \\ a,b\geq 2}} \zeta(a,b) + (N-3)\zeta(N)$$

$$= 2(\zeta(N) - \zeta(N-1,1)) + (N-3)\zeta(N)$$

using (1) with $g = 2$, and the result follows after some re-arrangement.

Proof of (8) We begin by proving

$$\sum_{\substack{p+q=N-1 \\ p\geq 2,\, q\geq 1}} \zeta(p,1,q) = \zeta(2,N-2) + \zeta(N-1,1). \tag{9}$$

Now the sum here equals

$$\sum_{a>b>c\geq 1} \frac{1}{b} \sum_{\substack{p+q=N-1 \\ p\geq 2,\, q\geq 1}} \frac{1}{a^p c^q} = \sum_{a>b>c\geq 1} \frac{1}{ab(a-c)}\left(\frac{1}{c^{N-3}} - \frac{1}{a^{N-3}}\right)$$

The first term is

$$\sum_{c\geq1}\frac{1}{c^{N-2}}\sum_{b>c}\frac{1}{b}\sum_{a>b}\left(\frac{1}{a-c}-\frac{1}{a}\right)=\sum_{c\geq1}\frac{1}{c^{N-2}}\sum_{b>c}\frac{1}{b}\sum_{i=0}^{c-1}\frac{1}{b-i}$$

$$=\zeta(2,N-2)+\sum_{c>i\geq1}\frac{1}{c^{N-2}}\frac{1}{i}\sum_{b>c}\left(\frac{1}{b-i}-\frac{1}{b}\right)$$

$$=\zeta(2,N-2)+\zeta(N-1,1)+\sum_{c>i\geq1}\frac{1}{c^{N-2}}\frac{1}{i}\sum_{j=1}^{i-1}\frac{1}{c-j}$$

But the final sum in both of the last two displays are identical (after the change of variables $(a,b,c)\to(c,i,j)$), and so we have proved (9). We next prove that

$$\sum_{\substack{p+q=N-1\\p\geq2,\ q\geq1}}\zeta(p,q,1)=\zeta(2)\zeta(N-2)-\zeta(2,N-2)+\zeta(N-1,1)-\zeta(N).\quad(10)$$

Using (7) and then (1) we have

$$(N-1)\zeta(N)-2\zeta(N-1,1)-\zeta(2)\zeta(N-2)=\sum_{N-3\geq a\geq2}\zeta(a)\zeta(N-a)$$

$$=\sum_{\substack{a+b+c=N\\a,b\geq2,\ c\geq1}}\zeta(a)\zeta(b,c)$$

Just as in the proof of (6) we may determine such a product in terms of zeta-functions by considering each term. We thus get $\zeta(a)\zeta(b,c)=\zeta(a,b,c)+\zeta(b,a,c)+\zeta(b,c,a)+\zeta(a+b,c)+\zeta(b,a+c)$. Summing up over all possibilities with $a+b+c=N$ and $a,b\geq2$ we get three times the sum over all $\zeta(A,B,C)$ in the sum (1) other than a few terms corresponding to when $a=1$ or $b=1$, and some multiples of $\zeta(A,B)$. Precisely we get:

$$3\sum_{\substack{a+b+c=N\\a,b\geq2,\ c\geq1}}\zeta(a,b,c)-\sum_{\substack{a+c=N-1\\a\geq2,\ c\geq1}}(2\zeta(a,1,c)+\zeta(a,c,1))$$

$$+\sum_{\substack{d+c=N\\d\geq3,\ c\geq1}}(d-3)\zeta(d,c)+\sum_{\substack{f+b=N\\b\geq2,\ f\geq2}}(f-2)\zeta(b,f)$$

$$=3\zeta(N)-\sum_{\substack{a+c=N-1\\a\geq2,\ c\geq1}}\zeta(a,c,1)$$

$$+\sum_{\substack{a+b=N\\a\geq2,\ b\geq1}}(a+b-5)\zeta(a,b)-\zeta(2,N-2)-\zeta(N-1,1)$$

$$=(N-2)\zeta(N)-\sum_{\substack{a+c=N-1\\a\geq2,\ c\geq1}}\zeta(a,c,1)-\zeta(2,N-2)-\zeta(N-1,1)$$

using (1) and (9). Combining the last two displays gives (10).
We now try to evaluate the last sum in (8):

$$\sum_{\substack{a+b+c=N \\ a,b,c\geq 2}} \zeta(a)\zeta(b)\zeta(c) = \sum_{\substack{a+b+c=N \\ a,b,c\geq 2}} \sum_{x,y,z\geq 1} \frac{1}{x^a y^b z^c}.$$

By analogy with the proof of (6), with a, b, c fixed we break up this sum according to how x, y, z are ordered by size. For example, when $x > y > z$ we get precisely $\zeta(a, b, c)$. We thus get the sum of $\zeta(u, v, w)$ as u, v, w ranges over all six orderings of a, b, c; plus the sum of $\zeta(u, N - u) + \zeta(N - u, u)$ for each $u \in \{a, b, c\}$; plus $\zeta(N)$. Thus, using (6), we have

$$6 \sum_{\substack{a+b+c=N \\ a,b,c\geq 2}} \zeta(a,b,c) + 3 \sum_{\substack{a+b+c=N \\ a,b,c\geq 2}} (\zeta(c)\zeta(a+b) - \zeta(N)) + \sum_{\substack{a+b+c=N \\ a,b,c\geq 2}} \zeta(N).$$

Using (1) with $g = 3$, we thus have

$$6\left(\zeta(N) + \zeta(N-2,1,1) - \sum_{\substack{a+d+1=N \\ a\geq 2,\ d\geq 1}} (\zeta(a,d,1) + \zeta(a,1,d))\right)$$

$$+ 3 \sum_{\substack{c+d=N \\ d\geq 4,\ c\geq 2}} (d-3)\zeta(c)\zeta(d) - 2\binom{N-4}{2}\zeta(N)$$

$$= 6\zeta(N-2,1,1) - (N-1)(N-8)\zeta(N) - 3\zeta(N-2)\zeta(2) - 12\zeta(N-1,1)$$

$$- \frac{3}{2} \sum_{\substack{c+d=N \\ d\geq 2,\ c\geq 2}} (N-6)\zeta(c)\zeta(d)$$

using (9) and (10), and combining the (c, d) and (d, c) terms in the final sum. Using (7) to remove the $\zeta(N - 1, 1)$ terms, we obtain (8).

Acknowledgements   Thanks are due to Roland Girgensohn for supplying the reference [5], to Don Zagier and Michael Hoffman for their useful email correspondence, and to the authors of [1], [2], [4], and [6] for making available their preprints.

References

[1] D.H. Bailey, J.M. Borwein and R. Girgensohn. Experimental evaluation of Euler sums. *Experimental Math.*, **3** (1994), 17–30.

[2] D. Borwein, J.M. Borwein and R. Girgensohn. Explicit evaluation of Euler sums. *Proc. Edin. Math. Soc.*, **38** (1995), 273–294

[3] J.M. Borwein and R. Girgensohn. Evaluating triple Euler sums. *Electronic J. Combinatorics*, **3** (1996). 27pp.

[4] J.P. Buhler and R.E. Crandall. On the evaluation of Euler Sums. *Experimental Math.*,**3** (1994), 275–285.

[5] M.E. Hoffman. Multiple harmonic series. *Pacific J. Math.*, **152** (1992), 275–290.

[6] M.E. Hoffman and C. Moen. Sums of triple harmonic series. *J. Number Th.*, **60** (1996), 329–331.

[7] C. Markett. Triple sums and the Riemann zeta function. *J. Number Th.*, **48** (1994), 113–132.

[8] N. Nielsen. *Handbuch der Theorie der Gammafunktion*. Chelsea, New York, 1965.

[9] D. Zagier. Multiple zeta values. In preparation.

Andrew Granville
Department of Mathematics, The University of Georgia
Athens, Georgia 30602, U.S.A.

8

Multiplicative Properties of Consecutive Integers

ADOLF J. HILDEBRAND

1. Introduction A variety of conjectures in number theory are based on the heuristic that the multiplicative structures of consecutive integers are independent. Thus, one would normally expect the probability that two consecutive integers n and $n+1$ both possess a given multiplicative property to be approximately the square of the probability that an individual integer n possesses this property. Of course, there may be obstructions preventing this (as in the case of the property "divisible by 2"), but these obstructions are usually related to congruence conditions, and the problem is easily modified to take into account such conditions.

While it is easy to formulate conjectures based on such independence assumptions, these conjectures often turn out to be extremely difficult to prove and in many cases seem intractable. The most famous problem of this type is the twin prime conjecture according to which n and $n+2$ are simultaneously prime infinitely often. A quantitative form of this conjecture asserts that the probability that n and $n+2$ are both prime is proportional to $(\log n)^{-2}$, the square of the probability that an individual integer n is prime.

We consider here more general problems such as that of proving, under suitable conditions on the multiplicative structure of a set A of positive integers, the existence of infinitely many integers n (or a positive proportion of integers n) such that both n and $n+1$ belong A. Over the past decade, there has been some progress on problems of this type, but many open questions remain. The purpose of this paper is to give a survey of results that have been obtained and to discuss some open problems that arise in this connection.

We begin by stating four concrete problems which motivated much of the work that we will discuss here.

Problem A (Erdös [6]) *For $n \geq 2$ let $P(n)$ denote the largest prime factor of n. Show that for every $\epsilon > 0$ there exist infinitely many integers $n \geq 2$ such that $P(n)$ and $P(n+1)$ both exceed $n^{1-\epsilon}$.*

Problem B (Chowla [4]) *Let k be an integer ≥ 2. Show that there exist constants $c_0(k)$ and $p_0(k)$ such that for every prime $p \geq p_0(k)$ there exists a positive integer $n \leq c_0(k)$ such that n and $n+1$ are both kth power residues modulo p.*

Problem C (Graham and Hensley [8]) *Let $\lambda(n)$ be the Liouville function,* *defined by $\lambda(n) = 1$ if the total number of prime factors of n is even, and $\lambda(n) = -1$ otherwise. Show that the set of positive integers n for which $\lambda(n) = \lambda(n+1)$ has positive lower density.*

Problem D (Chowla [5]) *Show that any finite pattern $\underline{\epsilon} = (\epsilon_1, \ldots, \epsilon_k)$ of values $\epsilon_i = \pm 1$ occurs infinitely often in the sequence $\{\lambda(n)\}_{n \geq 1}$.*

Problem A had been a long-standing conjecture of Erdös. In an attempt to attack this problem, Balog [1] in 1982 proposed a conjecture giving a simple general condition on the multiplicative structure and the density of a set A of positive integers that would imply that A contains infinitely many pairs $(n, n+1)$ of consecutive integers. Problem A is a particular case of this conjecture. Balog's conjecture, and hence the assertion of Problem A, was established in [12].

In investigating Problem B, it is convenient to use the following reformulation which can be shown to be (essentially) equivalent to Problem B; see, e.g., Mills [22]. Let F_k denote the set of all completely multiplicative functions whose values are kth roots of unity, i.e.,

$$ F_k = \{f : \mathbb{N} \to \mathbb{C} : f^k \equiv 1, \quad f(n_1 n_2) = f(n_1)f(n_2) \quad (n_1, n_2 \in \mathbb{N})\}. $$

Problem B* *Let k be an integer ≥ 2. Show that there exists a constant $c_0(k)$ such that for all $f \in F_k$ there exists a positive integer $n \leq c_0(k)$ satisfying $f(n) = f(n+1) = 1$.*

For small values of k, the assertion of the latter problem can be verified by considering all possible assignments of values $f(p)$ for primes p below a suitable bound $c_0(k)$ and showing that each such assignment will result in a pair of consecutive integers below $c_0(k)$ on which f takes on the value 1. For example, if $k = 2$, it is easy to check that this will be the case with $c_0(2) = 10$. Using similar case-by-case verifications, and in part with the aid of computers, the assertion of Problems B and B* had been established in a series of papers in the 1960s for all values $k \leq 7$; see Problems 11-15 in Chap. 8 of [5] and the references in [13].

The general case of Problems B and B*, however, remained open for many years; it was settled in [13] for prime values of k, and in [18] for general k. The proof of these results uses the ideas behind the proof of Balog's conjecture along with some additional arguments.

Problem C appeared in the problems section of the *American Mathematical Monthly* and is easily solved by considering the values of the function $h(n) = \lambda(n)\lambda(n+1)$, which is equal to 1 if $\lambda(n) = \lambda(n+1)$, and -1 if $\lambda(n) = -\lambda(n+1)$. The complete multiplicativity of $\lambda(n)$ implies that whenever $h(n) = -1$ then either $h(2n) = 1$ or $h(2n+1) = 1$. Setting

$$ N(x) = \#\{n \leq x : \lambda(n) = \lambda(n+1)\} = \#\{n \leq x : h(n) = 1\}, $$

it therefore follows that, for all $x \geq 1$, $[x] - N(x) \leq N(2x + 1)$. Combining this inequality with the trivial bound $N(2x + 1) \geq N(x)$ gives

$$N(2x + 1) \geq \max(N(x), [x] - N(x)) \geq [x]/2 \quad (x \geq 1),$$

which implies that $\liminf_{x \to \infty} N(x)/x \geq 1/4$. Hence $\lambda(n) = \lambda(n + 1)$ holds on a set of lower density $\geq 1/4$.

In fact, the above argument shows that the same conclusion remains valid for any completely multiplicative function f with values ± 1 in place of $\lambda(n)$.

While Problem C, as it stands, is not difficult, the question of whether the opposite relation $\lambda(n) = -\lambda(n + 1)$ holds on a set of positive density remains open, although some partial results in this direction are known.

The conjecture of Problem D was stated by Chowla as Problem 56 in Chapter 8 of his book [5]. Chowla commented that "for $k \geq 3$ this seems an extremely hard conjecture." The case $k = 3$ of this conjecture was established in [15], but very little is known beyond this case.

The plan for the remainder of this paper is as follows. In Section 2 we describe Balog's conjecture and some of its applications. In Section 3 we present quantitative versions of these results. In Sections 4 and 5 we consider generalizations to pairs of linear forms $(an + b, cn + d)$ and to strings of k consecutive integers in place of a pair of consecutive integers. In Section 6 we discuss problems on consecutive values of multiplicative functions that arise in connection with Problems B* and C. In the final section we consider Problem D and, more generally, the occurrence of patterns $\underline{\epsilon} = (\epsilon_1, \ldots, \epsilon_k)$, $\epsilon_i = \pm 1$, among the values of a completely multiplicative function with values ± 1. We will formulate a general conjecture that gives a condition on f under which every such pattern should occur with its expected frequency.

2. Balog's conjecture Motivated by Problem A above, A. Balog formulated a general conjecture from which the asserted result would follow. To this end he introduced the concept of a "stable set", defined as follows.

Definition (Balog [1]) *Let d be a positive integer. A set $A \subset \mathbb{N}$ is called d-stable if the implication*

$$n \in A \iff dn \in A$$

holds for all positive integers n, with the possible exception of a set of density zero. The set A is called stable if it is p-stable for every prime p (and hence also d-stable for every positive integer d).

A stable set is thus a set that is invariant, modulo sets of density zero, with respect to multiplication or division by a fixed integer. It is easily seen that the sets $\{n \in \mathbb{N} : P(n) > n^{1-\epsilon}\}$ arising in Problem A, and more generally any set of the type

$$Q_{\alpha,\beta} = \{n \in \mathbb{N} : n^\alpha < P(n) \leq n^\beta\}, \tag{1}$$

where $0 \leq \alpha < \beta \leq 1$, are stable.

Balog observed that if A is 2-stable and has density greater than $1/3$, then A contains infinitely many pairs $(n, n+1)$ of consecutive integers, and he showed that the same conclusion holds if A is p-stable for $p = 2$ and $p = 3$ and $d(A) > 3/10$. These and similar results led him to conjecture that the conclusion holds for any set A that is p-stable for every prime p and which has positive density.

Balog's conjecture was proved in [12], in the following slightly stronger and more general form. Here, and in the sequel, we denote by $\underline{d}(A)$ (resp. $\overline{d}(A)$) the lower (resp. upper) asymptotic density of A.

Theorem 1 ([12])   *For every $\epsilon > 0$ there exist constants $\delta(\epsilon) > 0$ and $p_0(\epsilon)$ such that if $A \subset \mathbb{N}$ is p-stable for every prime $p \leq p_0(\epsilon)$ and $\underline{d}(A) \geq \epsilon$ then $\underline{d}(A \cap (A+1)) > \delta(\epsilon)$. Moreover, the same result holds with the upper density $\overline{d}$ in place of the lower density $\underline{d}$.*

A key element in the proof of this result is played by sets $S = \{d_1 < \cdots < d_r\}$ of positive integers having the property

$$d_j - d_i = (d_i, d_j) \quad (1 \leq i < j \leq r). \tag{2}$$

Following Heath-Brown, we will call such sets S "special sets". An example of a special set is the set $\{12, 15, 16, 18\}$. The proof of Theorem 1 depends on the existence of special sets of arbitrarily large cardinality, a fact that is not obvious and which was first established by Heath-Brown [10]. A simple inductive construction of special sets S_r of cardinality r, also due to Heath-Brown [11], goes as follows: For $r = 2$, we can take $S_2 = \{1, 2\}$. If $r \geq 2$ and a special set $S_r = \{d_i\}_{i=1}^r$ has already been constructed, then setting

$$D = \prod_{i=1}^{r} d_i, \quad S_{r+1} = \{D - d_i : 1 \leq i \leq r\} \cup \{D\}$$

gives a special set with $r + 1$ elements.

Applying Theorem 1 to the sets $Q_{\alpha,\beta}$ defined in (1) (which, as noted above, are stable and which also have positive density), we obtain as an immediate corollary a result that contains the assertion of Problem A as a particular case:

Corollary 1 ([12])   *Let $0 \leq \alpha < \beta \leq 1$. Then the set of positive integers n for which $n^\alpha < P(n) \leq n^\beta$ and $(n+1)^\alpha < P(n+1) \leq (n+1)^\beta$ both hold has positive lower density.*

Another application concerns the least pair of consecutive quadratic non-residues modulo a prime p. Let $n_1(p)$ denote the least quadratic non-residue modulo p, and $n_2(p)$ the least positive integer n such that n and $n + 1$ are both quadratic non-residues modulo p. The best known bound for $n_1(p)$ is the

Burgess-Vinogradov estimate $n_1(p) \ll p^{\theta+\epsilon}$, where $\theta = 1/4\sqrt{e}$ and ϵ is any fixed positive number. The following result shows that the same bound holds for $n_2(p)$.

Corollary 2 ([16]) *For any fixed $\epsilon > 0$ and all sufficiently large primes p we have $n_2(p) \ll p^{\theta+\epsilon}$.*

Since trivially $n_2(p) \geq n_1(p)$, this estimate is best-possible in the sense that any further improvement would result in an improvement of the bound for $n_1(p)$.

While not a direct consequence of Theorem 1, Corollary 2 can be deduced from an appropriate finite version of Theorem 1 as follows. Given a large prime p, let A be the set of quadratic non-residues modulo p, and set $x_0 = p^{\theta+\epsilon}$. The method of Burgess-Vinogradov then shows that $|A \cap [0, x]| \gg_\epsilon x$ for $x \geq x_0$. Suppose first that $(*)$ $n_1(p) > c_0$ with a sufficiently large constant $c_0 = c_0(\epsilon)$. Then all positive integers $d \leq c_0$ are quadratic residues modulo p, and multiplying or dividing an arbitrary integer n by some integer $d \leq c_0$ does not change the quadratic residue character of n. Hence A is d-stable (in a rather strong sense) for all $d \leq c_0$. If $c_0 = c_0(\epsilon)$ is sufficiently large, then a suitable quantitative version of Theorem 1 will imply that, for $x \geq x_0$, $|A \cap (A-1) \cap [1, x]| \gg_\epsilon x$. Hence $A \cap (A-1)$ contains a positive integer $n \leq x_0$, i.e., we have $n_2(p) \leq x_0 = p^{\theta+\epsilon}$ as claimed.

A similar, but simpler argument can be used in the case when $(*)$ is not satisfied.

3. Quantitative results In trying to obtain good quantitative versions of results such as Corollary 1, it is essential to have special sets S available that are, in a certain sense, not too large. The key parameter turns out to be the quantity

$$\tau(S) = \text{l.c.m.}\{d_j - d_i : 1 \leq i < j \leq r\},$$

which one would like to be as small as possible as a function of r. The special sets S_r constructed above are far from optimal in this respect. The problem of constructing special sets S with small values of $\tau(S)$ was considered by Heath-Brown who obtained the following result.

Theorem 2 (Heath-Brown [11]) *For any $r \geq 2$ there exists a special set S of cardinality r for which $\log \tau(S) \ll r^3 \log r$. Moreover, any such set satisfies $\log \tau(S) \gg r \log r$.*

Using this result, Heath-Brown obtained the following quantitative version of Problem A.

Theorem 3 (Heath-Brown [11]) *Let $\epsilon(n) = c(\log \log n / \log n)^{1/4}$, where c is a sufficiently large constant. Then there exist infinitely many positive integers n such that $P(n) > n^{1-\epsilon(n)}$ and $P(n+1) > n^{1-\epsilon(n)}$ both hold.*

Another application of Theorem 2 was given by Balog, Erdös, and Tenenbaum:

Theorem 4 (Balog, Erdös, and Tenenbaum [2])   *The estimate*

$$\#\{n \le x : P(n(n+1)) \le y\} \gg xu^{-u^{7u}}, \quad u = \frac{\log x}{\log y}$$

holds uniformly in the range

$$x \ge 3, \quad \max\left(2, \exp\left\{\frac{8 \log x \log_3 x}{\log_2 x}\right\}\right) \le y \le x,$$

where $\log_k x$ *denotes the k times iterated logarithm.*

Taking y to be of the form $y = \exp(c \log x \log_3 x / \log_2 x)$ yields the following corollary, which is analogous to Theorem 3 above.

Corollary   *Let* $\epsilon(n) = c \log_3 n / \log_2 n$, *where c is a sufficiently large constant. Then there exist infinitely many positive integers n such that* $P(n(n+1)) \le n^{\epsilon(n)}$.

4. Generalization to two linear forms   As a natural generalization of Theorem 1 one might try to obtain a similar result with $(n, n+1)$ replaced by a pair $(an+b, cn+d)$ of linear forms in n. Such a generalization was recently given by Balog and Ruzsa [3].

Theorem 5 (Balog and Ruzsa [3])   *Let* $a > 0$, $b > 0$ *and* $c \ne 0$ *be integers satisfying* $(a,b)|c$. *If A is a stable set and* $\overline{d}(A) > 0$ *then* $\overline{d}(bA \cap (aA + c)) > 0$. *Moreover, the same result holds with* $\underline{d}$ *in place of* $\overline{d}$.

The proof of Theorem 5 hinges again on the construction of special sets S that satisfy (2) as well as some additional conditions.

By specializing A to be of the form (1), one obtains the following corollary, which generalizes the corresponding corollary to Theorem 1.

Corollary 1 (Balog and Ruzsa [3])   *Let* $0 \le \alpha < \beta \le 1$, *and let* $a > 0$ *and* $c \ne 0$ *be given integers. Then the set of positive integers n satisfying*

$$n^\alpha < P(n) \le n^\beta, \quad (an+c)^\alpha < P(an+c) \le (an+c)^\beta$$

has positive lower density.

This result has (and, in fact, was motivated by) the following surprising application due to Fouvry and Mauduit [7]. Let $s(n)$ denote the sum of the binary digits of n and set

$$A^+ = \{n \in \mathbb{N} : s(n) \equiv 0 \bmod 2\}, \quad A^- = \{n \in \mathbb{N} : s(n) \equiv 1 \bmod 2\}.$$

Thus, A^+ (resp. A^-) is the set of positive integers that have an even (resp. odd) number of 1's in their binary expansion. A difficult unsolved problem is to show that each of these two sets contains infinitely many primes. This led Fouvry and Mauduit to consider the corresponding problem with the set of primes replaced by other sets such as $\{n \in \mathbb{N} : P(n) > n^{1-\epsilon}\}$, $\{n \in \mathbb{N} : P(n) \leq n^{\epsilon}\}$, and more generally, any set $Q_{\alpha,\beta}$ with $0 \leq \alpha < \beta \leq 1$.

Corollary 2 (Fouvry and Mauduit [7])   *Let* $0 \leq \alpha < \beta \leq 1$. *Then the sets* $Q_{\alpha,\beta} \cap A^+$ *and* $Q_{\alpha,\beta} \cap A^-$ *both have positive lower density.*

To obtain this result, it suffices to observe that $n \in A^+$ holds if and only if $2n + 1 \in A^-$ and to apply Corollary 1 with the linear form $an + c = 2n + 1$.

While Theorem 5 covers a large class of pairs of linear forms $(an + b, cn + d)$, it does not apply to pairs of forms like $(n, -n + c)$, which arise in connection with Goldbach type problems. The method of Balog and Ruzsa seems to break down in this case, and the question whether a Goldbach type analog of Theorem 1 holds remains open. It seems plausible that such an analog holds under the same conditions as those in Theorem 1:

Conjecture 1   *If* A *is a stable set with* $\underline{d}(A) > 0$, *then every sufficiently large integer* N *can be represented as a sum of two elements of* A. *Moreover, the number of such representations is* $\gg N$.

By taking A to be of the form $Q_{\alpha,\beta}$ with $(\alpha, \beta) = (1 - \epsilon, 1)$, this conjecture would imply that, for any $\epsilon > 0$, every sufficiently large integer N can be written as $N = a + b$ with $P(a)$ and $P(b)$ both exceeding $> N^{1-\epsilon}$. Similarly, taking $(\alpha, \beta) = (0, \epsilon)$ leads to such representations in which $P(a)$ and $P(b)$ are both bounded from above by N^{ϵ}. Currently, results of this type are only known for values ϵ of the order of 0.3.

5. Generalization to strings of three or more consecutive integers
Another natural extension of Problem A is to show, for given $\epsilon > 0$ and $k \geq 3$, the existence of infinitely many integers n such that the greatest prime factors of $n, n+1, \ldots, n+k-1$ all exceed $n^{1-\epsilon}$. To approach this general problem in a manner analogous to Theorem 1, one might try to find, for any given integer $k \geq 3$, conditions on a set A that imply the existence of infinitely many k-tuples of consecutive integers in A, or even the stronger relation

$$\underline{d}(A \cap (A+1) \cap \cdots \cap (A+k-1)) > 0. \tag{3}$$

In the case $k = 2$, Theorem 1 shows that (3) holds for any set A that is stable and has positive lower density. A priori, there seems to be no reason why the same conditions should not imply (3) for any $k \geq 3$. We are therefore led to the following conjecture.

Conjecture 2 *If A is a stable set with $\underline{d}(A) > 0$, then (3) holds for any $k \geq 2$. In particular, A contains arbitrarily long strings of consecutive integers.*

This conjecture appears to be quite deep, and very little is known when $k \geq 3$, even in special cases like the generalization of Problem A mentioned above. A modest step towards the above conjecture is contained in the following result:

Theorem 6 ([17]) *Let $k \geq 2$ be given. If A is a stable set with $\underline{d}(A) > (k-2)/(k-1)$ then (3) holds.*

It is easy to see that (3) holds for any set A with $\underline{d}(A) > 1 - 1/k$. Without additional conditions on the set A, this bound on $\underline{d}(A)$ is best possible, as can be seen by taking $A = \{n \in \mathbb{N} : k \nmid n\}$. Theorem 6 shows that if one assumes that A is stable then the density condition can be relaxed to $\underline{d}(A) > 1 - 1/(k-1)$. Note that, in the case $k = 2$, this condition reduces to $\underline{d}(A) > 0$, i.e., the condition of Theorem 1.

6. Consecutive values of multiplicative functions   The solution to Problems B and B* is given by the following results.

Theorem 7 ([18]) *Let k be an integer ≥ 2. There exist constants $c_0(k)$ and $p_0(k)$ such that for every prime $p \geq p_0(k)$ there exists a positive integer $n \leq c_0(k)$ for which n and $n+1$ are both kth power residues modulo p.*

Theorem 7* ([18]) *Let k be a positive integer. There exists a constant $c_0(k)$ such that for any function $f \in F_k$ there exists a positive integer $n \leq c_0(k)$ satisfying (*) $f(n) = f(n+1) = 1$.*

The argument of [18] in fact yields the existence of infinitely many integers n satisfying (*), and it can probably be adapted to show that the set of such integers has positive lower density.

Theorem 7 follows easily from Theorem 7* by taking f to be a kth power residue character. (In fact, as already noted, the two results are essentially equivalent.) The proof of Theorem 7* has its roots in the same ideas as that of Theorem 1, but it requires additional arguments and is significantly more complicated.

The similarity between Theorems 1 and 7* becomes apparent if we set $A = \{n \in \mathbb{N} : f(n) = 1\}$ in Theorem 7* and note that the conclusion of that theorem amounts to the existence an element of $A \cap (A - 1)$ below $c_0(k)$. While this set A is not stable in the sense of Balog's definition, it nonetheless has the following closely related property: if $d \in A$ then $dA \subset A$ and $A/d \cap \mathbb{N} \subset A$, while if $d \in A^c$ then $dA \subset A^c$ and $A/d \cap \mathbb{N} \subset A^c$. Thus it is not surprising that arguments similar to those used to prove Theorem 1 can be applied to the proof of Theorem 7*.

The main task in the proof of Theorem 7* is the construction of a special set $S = \{d_1 < \cdots < d_r\}$ of cardinality $r \geq 2$ whose elements are bounded in terms of k and which satisfies, in addition to (2), the condition

$$f(d_i) = \omega \quad (1 \leq i \leq r), \quad f((d_i, d_j))) = \omega \quad (1 \leq i < j \leq r). \tag{4}$$

for some kth root of unity ω. Given such a set S, the conclusion (*) of Theorem 7* is easily obtained: From (2), (4), and the complete multiplicativity of f it follows that for $1 \leq i < j \leq r$

$$f\left(\frac{d_i}{(d_i, d_j)}\right) = \frac{f(d_i)}{f((d_i, d_j))} = 1,$$
$$f\left(\frac{d_i}{(d_i, d_j)} + 1\right) = f\left(\frac{d_i + (d_i, d_j)}{(d_i, d_j)}\right) = f\left(\frac{d_j}{(d_i, d_j)}\right) = 1,$$

whence (*) holds with $n = d_i/(d_i, d_j)$.

The construction of special sets satisfying (4) is quite delicate and requires a variety of combinatorial and number theoretical tools, including sieve estimates, estimates for sums of multiplicative functions, and Ramsey's theorem.

One can generalize Problems B* and C to strings of $k \geq 3$ consecutive integers. For example, if f is a completely multiplicative function f with values ± 1, do there exist infinitely many (or a positive proportion of) strings of k consecutive integers on which f takes on the value $+1$? One would certainly expect this to be the case when f is the Liouville function. However, there exist functions f that do not have this property for $k = 3$. Examples are the completely multiplicative functions $f_\pm$ defined by

$$f_\pm(p) = \begin{cases} \pm 1 & \text{if } p = 3, \\ 1 & \text{if } p \equiv 1 \bmod 3, \\ -1 & \text{if } p \equiv -1 \bmod 3. \end{cases}$$

On integers coprime to 3, these functions coincide with the non-principal character modulo 3. The functions $f_\pm$ therefore take on both values ± 1 in every interval of length 3 and thus cannot be equal to 1 on three consecutive integers. However, I. and G. Schur proved that the functions $f_\pm$ are the only such exceptions.

Theorem 8 (I. and G. Schur [23]) *For any completely multiplicative function $f \not\equiv f_\pm$ with values ± 1, there exists a positive integer n for which $f(n) = f(n+1) = f(n+2) = 1$.*

This result is proved by a fairly complex case-by-case analysis of possible assignments of the values $f(p)$ for small primes p. The proof, in fact, shows that an integer n with the desired property $f(n) = f(n+1) = f(n+2) = 1$ must occur below some absolute bound n_0. However, it does not seem to yield the existence of a positive proportion of integers with this property.

Theorem 8 suggests two natural generalizations. The first is to prove an analogous result for functions f whose values are roots of unity of a given order. This problem remains wide open. The second possible generalization is to extend the result to sequences of $k \geq 4$ consecutive integers. Of course, the set of "exceptional" functions will increase with k. In the case $k = 4$, Hudson [19] found, in addition to the functions $f_\pm$ above, 13 completely multiplicative functions with values ± 1 that do not have four consecutive values of 1, and he conjectured that these 15 functions are the only ones with this property. Again, very little is known in this direction.

7. General patterns of values   As a natural generalization of Problems C and D, and of results such as Theorem 8, one might try to investigate the occurrence of general patterns $\underline{\epsilon} = (\epsilon_1, \ldots, \epsilon_k)$ of values ± 1 in the sequence of values of a completely multiplicative function f with values ± 1. This appears to be a difficult, and in its full generality probably intractable, problem. Very few results are known, even in particular cases such as the Liouville function $\lambda(n)$. In this section we will survey some of these results and formulate a general conjecture.

We let

$$M = \{f : \mathbb{N} \to \{\pm 1\} : f(n_1 n_2) = f(n_1)f(n_2) \quad (n_1, n_2 \in \mathbb{N})\}$$

be the set of all completely multiplicative functions with values ± 1. Given a function $f \in M$ and a pattern $\underline{\epsilon} = (\epsilon_1, \ldots, \epsilon_k)$ of values ± 1, we set

$$N(\underline{\epsilon}; x) = N_f(\underline{\epsilon}; x) = \#\{n \leq x : f(n+i-1) = \epsilon_i \quad (1 \leq i \leq k)\};$$

i.e., $N(\underline{\epsilon}; x)$ counts the number of occurrences of the pattern $\underline{\epsilon}$ among the first $[x]$ terms of the sequence $\{f(n)\}_{n \geq 1}$. The ultimate goal is to show that, as $x \to \infty$, $N(\underline{\epsilon}; x)/x$ tends to a limit, and to evaluate that limit. However, we are very from proving such a result for general patterns $\underline{\epsilon}$, and in most cases it is not even known whether $N(\underline{\epsilon}; x)$ tends to infinity with x.

The case of patterns $\underline{\epsilon}$ of length 1 is the only case in which the asymptotic behavior of $N_f(\underline{\epsilon}; x)$ is known for general functions $f \in M$, and even this case is highly non-trivial. In view of the relations

$$N((1); x) + N((-1); x) = [x], \quad N((1); x) - N((-1); x) = \sum_{n \leq x} f(n),$$

the convergence of $N_f(\underline{\epsilon}; x)/x$ for $\underline{\epsilon} = (1)$ or $\underline{\epsilon} = (-1)$ is equivalent to the existence of the mean value $\lim_{x \to \infty} (1/x) \sum_{n \leq x} f(n)$ of f. In the case $f = \lambda$, this mean value exists and is zero, a result that is equivalent to the prime number theorem. For more general classes of multiplicative functions f, Wirsing [26] proved a mean value theorem which in the case $f \in M$ may be stated as follows.

Theorem 9 (Wirsing [26]) *Let $f \in M$. If*

$$\sum_{f(p) \neq 1} \frac{1}{p} = \infty, \tag{5}$$

then for $\underline{\epsilon} = (1)$ and $\underline{\epsilon} = (-1)$ we have

$$\lim_{x \to \infty} \frac{N_f(\underline{\epsilon}; x)}{x} = \frac{1}{2}. \tag{6}$$

If the series in (5) converges, then the two limits (6) exist, but are different from 1/2.

We denote by M_0 the set of functions $f \in M$ that satisfy (5). By Wirsing's theorem these are exactly the functions in M that have mean value 0; in particular, the Liouville function λ belongs to M_0. We will, for the most part, restrict ourselves here to functions $f \in M_0$. Only for this subclass of M is the general problem described above non-routine and, in most cases, still unsolved. For functions $f \in M \setminus M_0$, i.e., for functions f for which the series (5) converges, the asymptotic behavior of $N_f(\underline{\epsilon}; x)$ can be completely determined by standard convolution arguments.

We now turn to the case of patterns of length 2. For the two patterns $\underline{\epsilon} = (1, 1)$ and $\underline{\epsilon} = (-1, -1)$ the simple argument given in Section 1 shows that

$$\liminf_{x \to \infty} \frac{1}{x}(N(1, 1); x) + N((-1, -1); x))$$

$$= \liminf_{x \to \infty} \frac{1}{x} \#\{n \leq x : f(n) = f(n+1)\} \geq \frac{1}{4}.$$

Since for any function $f \in M$ with mean value 0,

$$N((1, 1); x) = N((-1, -1); x) + o(x),$$

it follows that for $f \in M_0$ and for each of the patterns $\underline{\epsilon} = (1, 1)$ and $\underline{\epsilon} = (-1, -1)$,

$$\liminf_{x \to \infty} \frac{N_f(\underline{\epsilon}; x)}{x} > 0. \tag{7}$$

The remaining two patterns of length 2, $\underline{\epsilon} = (1, -1)$ and $\underline{\epsilon} = (-1, 1)$, by contrast, lead to difficult problems. In particular, the question whether (7) holds for these patterns is still open. The best known result in this direction is the following theorem.

Theorem 10 ([14])   *For any function $f \in M_0$ and for $\underline{\epsilon} = (-1, 1)$ and $\underline{\epsilon} = (1, -1)$ we have*

$$\limsup_{x \to \infty} \frac{N_f(\underline{\epsilon}; x)}{x (\log \log x)^{-4}} > 0.$$

Even in the special case $f = \lambda$, a better bound for $N(\underline{\epsilon}; x)$ than that of Theorem 10 is not known. In particular, we cannot exclude the possibility that $N_\lambda(\underline{\epsilon}; x) = o(x)$ holds for each of the patterns $\underline{\epsilon} = (1, -1)$ and $\underline{\epsilon} = (-1, 1)$, or equivalently, that the number of sign changes up to x in the sequence $\{\lambda(n)\}_{n \geq 1}$ is of order $o(x)$. In this case, the sequence $\{\lambda(n)\}_{n \geq 1}$ would consist of alternating blocks of 1's and (-1)'s, with the average block length tending to infinity, a behavior that seems highly unlikely.

For patterns $\underline{\epsilon}$ of length 3 or longer, only a few isolated results are known. In the case $f = \lambda$, the conjecture of Chowla stated in Problem D predicts that every pattern $\underline{\epsilon}$ occurs infinitely often. In fact, assuming the sequence $\{\lambda(n)\}_{n \geq 1}$ behaves like a random sequence, one would expect that much more is true: namely, that a pattern $\underline{\epsilon}$ occurs with its "proper" frequency 2^{-k}, where k is the length of the pattern.

The following result establishes Chowla's original conjecture for all patterns of length $k = 3$.

Theorem 11 ([15])   *For any pattern $\underline{\epsilon} = (\epsilon_1, \epsilon_2, \epsilon_3)$ of values $\epsilon_i = \pm 1$ there exist infinitely many positive integers n such that $\lambda(n + i) = \epsilon_i$ for $i = 1, 2, 3$.*

Beyond this result, very little is known. In particular, it is not known whether (7) holds for any pattern of length $k = 3$; the argument of [15] gives only a much weaker lower bound for the counting function $N_\lambda(\underline{\epsilon}; x)$. The case $k \geq 4$ of Problem D is completely open; indeed, not a single pattern of length $k \geq 4$ is known for which Chowla's conjecture holds.

It seems plausible that the result of Theorem 11 holds for all but finitely many functions $f \in M_0$. The proof of Theorem 11 depends only mildly on particular properties of the function $\lambda(n)$. It immediately generalizes to functions $f \in M$ that have the same values as the function λ at the primes 2, 3, 5, 7, 29, and 31; other classes of functions f can be covered by similar arguments.

For general functions $f \in M_0$, the full analog of Theorem 11 is still open. However, a number of partial results are known. Some of these results establish analogs of Theorem 11 for particular patterns $\underline{\epsilon}$ of length 3 or for functions f satisfying some additional hypotheses. An example is Theorem 8 of the previous section, which shows that for all except two functions $f \in M$ the pattern $\underline{\epsilon} = (1, 1, 1)$ occurs at least once in the sequence $\{f(n)\}_{n \geq 1}$. Another result of this type is the following theorem.

Theorem 12 (Sudo [24])   *Let $f \in M$ and suppose that $f(p) = -1$ for at least* *two primes p. Then there exists a positive integer n such that*

$$(f(n), f(n+1), f(n+2)) = (-1, 1, -1).$$

Moreover, if $f(2) = 1$, there exist infinitely many such integers.

Several authors have investigated the occurrence of general patterns of the form (*) $f(n + a_i) = 1$, $1 \leq i \leq k$, where $a_1 < \ldots < a_k$ are given integers. In particular, the question which tuples $(a_1, \ldots, a_k)$ have the property that the equation (*) has a solution $n \leq c_0$ with a suitable constant $c_0 = c_0(a_1, \ldots, a_k)$ for all $f \in M$ has received some attention because of the following (easily proved) connection with patterns of quadratic residues: The tuples $(a_1, \ldots, a_k)$ with this property are exactly those for which, for every sufficiently large prime p, there exists an integer $n \leq c_0$ such that each of the numbers $n + a_i$, $1 \leq i \leq k$, is a quadratic residue modulo p. This problem was initiated by E. Lehmer [21] who discussed a number of special cases and formulated several conjectures; see also Section F6 in [9]. We quote two typical results concerning this problem.

Theorem 13 (Hudson [20]) *Let $f \in M$, and suppose that $f(2) = -1$ and that, for some q with $(q, 5) = 1$, $f(q) \neq \left(\dfrac{q}{5}\right)$. Then there exists a positive integer $n \leq 12q$ such that $f(n) = f(n+2) = f(n+3) = 1$.*

Theorem 14 (Walum [25]) *Let $S \geq 1$ be given. Then there exists a constant $B = B(S)$ such that for any function $f \in M$ there exists a positive integer $n \in [S, S + B]$ with $f(n) = f(n+3) = f(n+4) = 1$.*

The above results suggest that, for a given value of k, all but finitely many functions $f \in M$ take on any pattern of length k infinitely often. Moreover, in results such as Theorem 8 in which a complete determination of all exceptional functions f was obtained, the exceptional functions turned out to be functions that are very close to a quadratic character. We are thus led to conjecture that if a function $f \in M$ is, in a suitable sense, not too close to a character, then the sequence $\{f(n)\}_{n \geq 1}$ behaves like a random sequence of values ± 1 in the sense that each of the patterns $\underline{\epsilon}$ occurs in this sequence with the expected frequency. To measure the "closeness" to a character, we introduce the following definition.

Definition *A function $f \in M$ is called characterlike, if, for some Dirichlet character χ,*

$$\sum_{f(p) \neq \chi(p)} \frac{1}{p} < \infty \qquad (*)$$

holds.

Note that in the case χ is a principal character, condition (*) is equivalent to $\sum_{f(p) \neq 1} 1/p < \infty$, which by Wirsing's theorem characterizes those functions

$f \in M$ that do not have mean value 0. Thus, a function $f \in M$ that is not characterlike in the sense of the above definition belongs to the class M_0 of functions $f \in M$ with mean value 0.

For characterlike functions f, the asymptotic behavior of the frequencies $N(\underline{\epsilon}; x)$ can easily be determined by representing f as a convolution product $f = \chi * g$, where χ is the character associated with f. The interesting case therefore is that of functions that are not characterlike. For this case we propose the following conjecture.

Conjecture 3 *Let $f \in M$, and assume that f is not characterlike. Then, for each pattern $\underline{\epsilon} = (\epsilon_1, \dots, \epsilon_k)$ of values $\epsilon_i = \pm 1$, we have $\lim_{x \to \infty} N_f(\underline{\epsilon}; x)/x = 2^{-k}$.*

References

[1] A. Balog. Problem in Tagungsbericht 41 (1982). Math. Forschungsinstitut Oberwolfach, p. 29.

[2] A. Balog, P. Erdös and G. Tenenbaum. On arithmetic functions involving consecutive divisors. *Analytic Number Theory* (Allerton Park, IL 1989), Progress in Math., **85**, Birkhäuser, Boston, 1990, pp. 77–90.

[3] A. Balog and I.Z. Ruzsa. On an additive property of stable sets. Preprint, 1995.

[4] P. Chowla and S. Chowla. On kth power residues. *J. Number Th.*, **10** (1978), 351–353.

[5] S. Chowla. *The Riemann Hypothesis and Hilbert's Tenth Problem*. Gordon and Breach, New York, 1965.

[6] P. Erdös. Problems and results on number theoretic properties of consecutive integers and related questions. *Congressus Numeratium XVI, Utilitas Math.*, Winnipeg, 1976, pp. 25–44.

[7] E. Fouvry and C. Mauduit. Sommes des chiffres et nombres presque premiers. *Math. Ann.*, **305** (1996), 571–599.

[8] S. Graham and D. Hensley. Problem E 3025. *Amer. Math. Soc. Monthly*, **90** (1983), 707.

[9] R. Guy. *Unsolved Problems in Number Theory* (2nd ed). Springer-Verlag, Berlin, 1994.

[10] D.R. Heath-Brown. The divisor function at consecutive integers. *Mathematika*, **31** (1984), 141–149.

[11] —. Consecutive almost-primes. *J. Indian Math. Soc.*, **52** (1987), 39–49; Correction and foot note to 'Consecutive almost-primes', to appear.

[12] A. Hildebrand. On a conjecture of Balog. *Proc. Amer. Math. Soc.*, **95** (1985), 517–523.

[13] —. On consecutive kth power residues. *Monatsh. Math.*, **102** (1986), 103–114.

[14] —. Multiplicative functions at consecutive integers. *Math. Proc. Cambridge Philos. Soc.*, **100** (1986), 229–236.

[15] —. On consecutive values of the Liouville function. *L'Enseignement Math.*, **32** (1986), 219–226.

[16] —. On the least pair of consecutive quadratic non-residues. *Michigan Math. J.*, **34** (1987), 57–62.

[17] —. On integer sets containing strings of consecutive integers. *Mathematika*, **36** (1989), 60–70.

[18] —. On consecutive kth power residues, II. *Michigan Math. J.*, **38** (1991), 241–253.

[19] R. Hudson. Totally multiplicative sequences with values ±1 which exclude four consecutive values of 1. *J. reine angew. Math.*, **271** (1974), 218–220.

[20] —. On the first occurrence of certain patterns of quadratic residues and nonresidues. *Israel J. Math.*, **44** (1983), 23–32.

[21] E. Lehmer. Patterns of power residues. *J. Number Th.*, **17** (1983), 37–46.

[22] W.H. Mills. Bounded consecutive residues and related problems. *Proc. Sympos. Pure Math.*, AMS, Providence, R.I., 1965, pp. 170–174.

[23] I. Schur and G. Schur. Multiplikativ signierte Folgen positiver ganzer Zahlen. *Gesammelte Abhandlungen von Issai Schur*, vol. 3, Springer-Verlag, Berlin, 1973, pp. 392–399.

[24] M. Sudo. On totally multiplicative signatures of natural numbers. *Proc. Japan Acad. Ser. A*, **60** (1984), 273–275.

[25] H. Walum. A recurrent pattern in the list of quadratic residues mod a prime and in the values of the Liouville λ function. *J. Number Th.*, **12** (1980), 53–56.

[26] E. Wirsing. Das asymptotische Verhalten von Summen über multiplikative Funktionen. II. *Acta Math. Acad. Sci. Hung.*, **18** (1967), 411–467.

Adolf J. Hildebrand
Department of Mathematics, University of Illinois
Urbana, Illinois 61801, U.S.A.

9

On the Equation $(x^m - 1)/(x - 1) = y^q$ with x Power

Noriko Hirata-Kohno and Tarlok N. Shorey

1. Introduction   We consider the equation

$$\frac{x^m - 1}{x - 1} = y^q \text{ in integers } x > 1,\ y > 1,\ m > 2,\ q \geq 2. \tag{1}$$

Ljunggren [5] proved that equation (1) with $q = 2$ has no solution other than $x = 3$, $y = 11$, $m = 5$ and $x = 7$, $y = 20$, $m = 4$. Thus there is no loss of generality in assuming that q is an odd prime number. Recently, Saradha and Shorey [7] proved that equation (1) has finitely many solutions whenever x is a square. In this paper, we consider analogous question if x is a cube or a higher power. Thus we shall consider equation (1) with $x = z^\mu$ where $z > 1$ and $\mu \geq 3$ are integers. There is no loss of generality in assuming that μ is an odd prime number. We shall follow the above notation and assumptions throughout the paper.

Theorem   *Let $z > 1$ be an integer and $\mu \geq 3$ be a prime number. Assume that*

$$q > 2(\mu - 1)(2\mu - 3). \tag{2}$$

Then equation (1) with $x = z^\mu$ implies that $\max(x, y, m, q)$ is bounded by an effectively computable number c depending only on μ.

If $\mu = q$, Shorey [9] showed that the assertion of the Theorem is valid with c replaced by an absolute constant. In fact, Maohua Le [4] proved that equation (1) has no solution whenever x is a q-th power. The case $q = 3$ of the preceding result is due to Inkeri [3, Lemma 4]. Consequently, we derive from the Theorem that equation (1) with $x = z^3$ and $q \neq 5, 7, 11$ implies that $\max(x, y, m, q)$ is bounded by an effectively computable absolute constant.

2. Lemmas   This section consists of lemmas for the proof of the Theorem. We start with the following result of Shorey and Tijdeman [11] on equation (1).

Lemma 1   *Equation (1) has only finitely many solutions if either x is fixed or m has a fixed prime divisor. Furthermore, the assertion is effective.*

Now we apply Lemma 1 to secure the following factorisation for equation (1) with $x = z^\mu$.

Lemma 2　*Equation* (1) *with* $x = z^\mu$ *implies that either* $\max(x, y, m, q)$ *is bounded by an effectively computable number depending only on* μ *or*

$$\frac{z^m - 1}{z - 1} = y_1^q, \quad \frac{z^{m(\mu-1)} + z^{m(\mu-2)} + \cdots + 1}{z^{\mu-1} + z^{\mu-2} + \cdots + 1} = y_2^q \tag{3}$$

where $y_1 > 1$ *and* $y_2 > 1$ *are relatively prime integers such that* $y_1 y_2 = y$.

Proof　If either $\gcd(\mu, m) > 1$ or $\gcd(\mu - 1, m) > 1$, then m has a fixed prime divisor and we apply Lemma 1 to conclude that $\max(x, y, m, q)$ is bounded by an effectively computable number depending only on μ. Thus we may suppose that $\gcd(\mu, m) = \gcd(\mu - 1, m) = 1$. We write equation (1) as

$$ABC^{-1} = y^q$$

where

$$A = \frac{z^m - 1}{z - 1}, \quad B = \frac{z^{\mu m} - 1}{z^m - 1}, \quad C = \frac{z^\mu - 1}{z - 1}.$$

Let p be a prime number dividing A and B and let ν be the least positive integer such that $z^\nu \equiv 1 \pmod{p}$. Then $p = \mu$, $\nu | m$ and $\nu | (p - 1)$ which imply that $\nu = 1$ since $\gcd(\mu - 1, m) = 1$. Now we see that $\mu | m$ contradicting $\gcd(\mu, m) = 1$. Thus $\gcd(A, B) = 1$. Similarly $\gcd(A, C) = 1$. Consequently $A = y_1^q$ where $y_1 > 1$ is an integer and (3) follows.

The next result, due to the first author, is a slight refinement of Lemma 1 of [10], which is proved by the method of Baker's article [2] on the approximations of certain algebraic numbers by rationals using Padé approximations.

Lemma 3　*Let* A, B, K *and* n *be positive integers such that* $A > B$, $K < n$, $n \geq 3$ *and* $\omega = (B/A)^{1/n}$ *is not a rational number. For* $0 < \phi < 1$, *put*

$$\delta = 1 + \frac{2 - \phi}{K}, \quad s = \frac{\delta}{1 - \phi},$$

$$u_1 = \left(3^{2K+1} \cdot 2^{s(4K+2+3n(K+1)) + (1 + (3n)/2)(K+1)}\right)^{1/(Ks-1)},$$

$$u_2^{-1} = 3^{2K+1} K^2 \left(1 + 2^{-29}\right)^{K-1} n^{2K} 2^{K+s+2+3n(K+1)}.$$

Assume that

$$A(A - B)^{-\delta} u_1^{-1} > 1. \tag{4}$$

Then

$$\left| \omega - \frac{p}{q} \right| > \frac{u_2}{Aq^{K(s+1)}}$$

for all integers p *and* q *with* $q > 0$.

Proof   We put

$$l(r) = n^{2K}(r+1)^{2K+1}r^{2K+1}A^r 2^{3n(r+1)(K+1)/2+r(K+1)},$$

$$\lambda_1 = 3^{2K+1}A \cdot 2^{(K+1)((3n)/2+1)},$$

$$\lambda_2 = 2^{4K+2+3n(K+1)}(A-B)^{K+1}A^{-K},$$

$$c = 2^{3n(K+1)/2}n^{2K},$$

and

$$\Lambda = \frac{\log \lambda_1}{\log \lambda_2}.$$

By (4) and $0 < \phi < 1$, we observe that $0 < \lambda_2 < 1$, $s > 1$ and $0 < -\Lambda \leq s$. We follow the proof of Lemma 4 and Lemma 5 both in Baker [2] with $m_j = j$ for $0 \leq j \leq K$ to conclude that for integers r, p and q with $r > 0$, $q > 0$ and $p \neq q$, there exists a polynomial $P_r(X) \in \mathbb{Z}[X]$ satisfying the four conditions: (i) $\deg P_r \leq K$, (ii) $H(P_r) \leq l(r)$, (iii) $P_r(p/q) \neq 0$, and (iv) $|P_r(\omega)| \leq \lambda_2^r$. Here $H(P_r)$ denotes the maximum of the absolute values of the coefficients of P_r. We remark that for $p = q$, the lemma follows immediately. We may assume that $|\omega - p/q| < 2^{-29}$ and we define r as the smallest integer such that

$$\lambda_2^r \leq \frac{1}{2q^K}.$$

We see $l(r) \leq c\lambda_1^r$. We suppose that $r \geq 2$. As

$$\lambda_2^r > \frac{\lambda_2}{2q^K},$$

we obtain

$$l(r) \leq c\lambda_1^r = c\lambda_2^{r\Lambda} \leq c\left(\frac{\lambda_2}{2q^K}\right)^{\Lambda} = c\lambda_1 2^{-\Lambda}q^{-K\Lambda} \leq c\lambda_1 2^s q^{Ks}.$$

When $r = 1$, also we have $l(r) \leq c\lambda_1 2^s q^{Ks}$. Further, we observe that

$$\frac{1}{q^K} \leq \left|P_r\left(\frac{p}{q}\right)\right| \leq \left|P_r\left(\frac{p}{q}\right) - P_r(\omega)\right| + |P_r(\omega)|$$

$$\leq \left|P_r\left(\frac{p}{q}\right) - P_r(\omega)\right| + \frac{1}{2q^K}.$$

Thus

$$\left|P_r\left(\frac{p}{q}\right) - P_r(\omega)\right| \geq \frac{1}{2q^K}.$$

On the other hand, we have

$$\left| P_r\left(\frac{p}{q}\right) - P_r(\omega) \right| = \left| \int_{p/q}^{\omega} P_r'(X)dX \right|$$

$$\leq K^2\left(1 + 2^{-29}\right)^{K-1}l(r)\left|\omega - \frac{p}{q}\right|.$$

Consequently

$$\left|\omega - \frac{p}{q}\right| > \frac{u_2}{Aq^{K(s+1)}}$$

by definition of u_2.

In an earlier paper, Baker [1] proved a result of the above form under the validity of a more restrictive assumption, namely (4) with $\delta = 2$ which is not satisfied in the proof of the Theorem. Now we apply the above lemmas to prove the Theorem whenever q is fixed.

Lemma 4   *Equation (1) with $x = z^\mu$ and (2) implies that $\max(x, y, m)$ is bounded by an effectively computable number depending only on q and μ.*

Proof   Let equation (1) with $x = z^\mu$ and (2) be satisfied. Then we may suppose (3) and we apply Lemma 1 to assume that $\min(m, z)$ exceeds a sufficiently large effectively computable number c_1 depending only on q and μ. Further, we re-write (3) as

$$(z-1)y_1^q = z^m - 1, \quad (z^{\mu-1} + \cdots + 1)y_2^q = z^{m(\mu-1)} + \cdots + 1. \qquad (5)$$

Then

$$0 < (z^{\mu-1} + \cdots + 1)y_2^q - (z-1)^{\mu-1}y_1^{q(\mu-1)} \leq \mu z^{m(\mu-2)} \qquad (6)$$

which implies that

$$0 < \left|\omega - \frac{y_2}{y_1^{\mu-1}}\right| < \frac{2\mu z^{m(\mu-2)}}{z^{\mu-1}y_1^{q(\mu-1)}} \qquad (7)$$

where

$$\omega = \left(\frac{(z-1)^{\mu-1}}{z^{\mu-1} + \cdots + 1}\right)^{1/q}. \qquad (8)$$

For applying Lemma 3, we put $A = z^{\mu-1} + \cdots + 1$, $B = (z-1)^{\mu-1}$, $n = q \geq 3$, $K = 2(\mu - 2) < q$ by (2), $\phi = \mu^{-4}$, $\delta = (\mu - 1 - (\phi/2))/(\mu - 2)$, $s = \delta/(1 - \phi)$, $p = y_2$ and $q = y_1^{\mu-1}$. Then we utilise (2) to estimate

$$K(s+1)(\mu-1)/q = \frac{(2 - \phi)(K+1)(\mu-1)}{(1-\phi)q} < 1 - \mu^{-4}. \qquad (9)$$

Further, the assumption (4) of Lemma 3 is satisfied if

$$z^{\mu-1}(\mu z^{\mu-2})^{-\delta} u_1^{-1} > 1$$

which is the case since $z \geq c_1$ with c_1 sufficiently large. Hence we apply Lemma 3 to conclude that

$$\left| \omega - \frac{y_2}{y_1^{\mu-1}} \right| > \frac{u_2}{2 z^{\mu-1} y_1^{K(s+1)(\mu-1)}}. \tag{10}$$

Here we observe that (10) follows immediately if ω is rational and therefore it involves no loss of generality in assuming that ω is irrational while applying Lemma 3. Now we derive from (7), (10), and $z < y_1^{q/(m-1)}$ by (5) that

$$y_1^{q(\mu-1)-K(s+1)(\mu-1)} \leq c_2 y_1^{(\mu-2)qm/(m-1)},$$

where c_2 is an effectively computable number depending only on q and μ. Therefore, since $m \geq c_1$, $z \geq c_1$, and c_1 is sufficiently large, we observe that

$$\frac{1}{\mu-1} \leq \frac{K(s+1)}{q} + \frac{1}{\mu^5}$$

which contradicts (9).

In view of the above lemma, it remains to show that equation (1) with $x = z^\mu$ implies that q is bounded. The proof depends on Baker's theory of linear forms in logarithms. The following refinement of an estimate of Shorey [8, Lemma 4] is a consequence of a result of Philippon and Waldschmidt [6, Theorem 2.2] on linear forms in logarithms.

Lemma 5   *Let $n > 1$ be an integer and $\tau_1 \geq 1$, $\tau_2 \geq 1$ be real numbers. Let $\alpha_1, \cdots, \alpha_{n-1}$ and α_n be positive rational numbers of heights not exceeding A_1 and A, respectively, where $A_1 \geq 3$, $A \geq 3$ and*

$$(\log A)(\log A_1)^{-1} \geq \tau_1^{-1}.$$

Further, assume that $|\log \alpha_i| \leq A_1^{-1/\tau_2}$ for $1 \leq i \leq n$. Let $b_1, \cdots, b_n$ be rational integers of absolute values not exceeding $B \geq 2$ such that

$$\wedge = b_1 \log \alpha_1 + \cdots + b_n \log \alpha_n \neq 0.$$

There exists an effectively computable number c_3 depending only on n, τ_1, and τ_2 such that

$$|\wedge| \geq \exp\left(-c_3 \left(1 + \frac{\log B}{\log A_1} \right) \log A \right).$$

Here the height of a non-zero rational number a/b with $\gcd(a,b) = 1$ is defined as $\max(|a|, |b|)$.

3. Proof of the Theorem

We denote by $c_4, \cdots, c_7$ effectively computable positive numbers depending only on μ. Suppose that equation (1) with $x = z^\mu$ is satisfied. By Lemma 4, it suffices to show that $q \leq c_4$. Now we refer to Lemma 1 to suppose that $\min(q, m, z) \geq c_5$ with c_5 sufficiently large. Then we derive (3) which, as in the proof of Lemma 4, implies (6). Consequently

$$0 < |\log \alpha_1 + q \log \alpha_2| < 8\mu z^{-m} \tag{11}$$

where $\alpha_1 = \omega^q$, $\alpha_2 = y_1^{\mu-1}/y_2$ and ω is given by (8). We observe that

$$|\log \alpha_1| < \log\left(1 + \frac{\mu z^{\mu-2}}{(z-1)^{\mu-1}}\right) < \frac{2\mu}{z} < z^{-1/2}$$

which, together with (11), implies that

$$0 < |\log \alpha_2| < z^{-1/2}.$$

Further, we observe from (5) that the heights of α_1 and α_2 do not exceed $2z^{\mu-1}$ and $2z^{(\mu-1)(m-1)/q}$ by (5) and (2), respectively. Consequently

$$\frac{1}{2z^{(\mu-1)(m-1)/q}} \leq \frac{|y_1^{\mu-1} - y_2|}{\max(y_1^{\mu-1}, y_2)} \leq 2z^{-1/2}$$

which implies that

$$q \leq 4(\mu - 1)(m - 1)$$

and

$$\frac{\log(2z^{(\mu-1)(m-1)/q})}{\log(2z^{\mu-1})} \geq \frac{m-1}{2q} \geq \frac{1}{8(\mu-1)}.$$

Now we apply Lemma 5 with $n = 2$, $A_1 = 2z^{\mu-1}$, $A = 2z^{(\mu-1)(m-1)/q}$, $\tau_1 = \tau_2 = 8(\mu - 1)$, and $B = q$ to conclude that

$$|\log \alpha_1 + q \log \alpha_2| \geq \exp(-c_6 m q^{-1} \log(qz)). \tag{12}$$

Finally, we combine (11) and (12) to conclude that $q \leq c_7$.

References

[1] A. Baker. Rational approximations to $\sqrt[3]{2}$ and other algebraic numbers. *Quart. J. Math. Oxford*, **15** (1964), 375–383.

[2] —. Simultaneous rational approximations to certain algebraic numbers. *Proc. Cambridge Philos. Soc.*, **63** (1967), 693–702.

[3] K. Inkeri. On the diophantine equation $a(x^n-1)/(x-1) = y^m$. *Acta Arith.*, **21** (1972), 299–311.

[4] M.H. Le. A note on the diophantine equation $(x^m-1)/(x-1) = y^n + 1$ *Math. Proc. Cambridge Philos. Soc.*, **116** (1994), 385–389.

[5] W. Ljunggren. Noen setninger om ubestemte likninger av formen $\frac{x^n-1}{x-1} = y^q$. *Norsk. Mat. Tidsskr.*, **25** (1943), 17–20.

[6] P. Philippon and M. Waldschmidt. Lower bounds for linear forms in logarithms. *New Advances in Transcendence Theory* (A. Baker, Ed.), Cambridge Univ. Press, London, 1988, pp. 280–312

[7] N. Saradha and T.N. Shorey. The equation $\frac{x^n-1}{x-1} = y^q$ with x square. To appear.

[8] T.N. Shorey. Perfect powers in values of certain polynomials at integer points. *Math. Proc. Cambridge Philos. Soc.*, **99** (1986), 195–207

[9] —. On the equation $z^q = (x^n-1)/(x-1)$. *Indag. Math.*, **89** (1986), 345–351.

[10] T.N. Shorey and Yu.V. Nesterenko. Perfect powers in products of integers from a block of consecutive integers. II. *Acta Arith.*, **76** (1996), 191–198.

[11] T.N. Shorey and R. Tijdeman. New applications of Diophantine approximation to Diophantine equations. *Math. Scand.*, **39** (1976), 5–18.

Noriko Hirata-Kohno
Department of Mathematics, College of Science and Technology, Nihon University
Surugadai, Tokyo-101, Japan

Tarlok N. Shorey
School of Mathematics, Tata Institute of Fundamental Research
Homi Bhabha Road, Bombay 400 005, India

10

Congruence Families of Exponential Sums

MARTIN N. HUXLEY and NIGEL WATT

1. Introduction In the Bombieri–Iwaniec method for single exponential sums [1], as set out in [5], the exponential sum with a primitive Dirichlet character χ mod k,

$$S_\chi = \sum_{M}^{M_2} \chi(m)\mathrm{e}(f(m)), \tag{1.1}$$

where $M_2 \leq 2M$, has to be estimated in terms of a type 1 congruence family of sums

$$S_\ell = \sum_{M}^{M_2} \mathrm{e}\left(f(m) + \frac{\ell m}{k}\right), \qquad \ell = 0, ..., k-1 \tag{1.2}$$

or a type 2 congruence family

$$S_\ell = \sum_{M}^{M_2} \mathrm{e}\left(f\left(m + \frac{\ell}{k}\right)\right), \qquad \ell = 0, ..., k-1. \tag{1.3}$$

In Jutila's sums with modular form coefficients [8], twisting by a character makes no essential difference. The Iwaniec–Mozzochi double exponential sum can be twisted by a character in two different ways [7]; the corresponding congruence families were considered in [6].

The main idea of this paper is that resonance curves for coincidences between different pairs of sums of the congruence family are related by a translation in the plane. The curves are parametrised by matrices of the modular group. As in [7], the appropriate congruence subgroups play a special role.

Theorem 1 *Let $F(x)$ be a function four times continuously differentiable on the interval $1 \leq x \leq 2$, whose derivatives satisfy the following conditions:*

$$|F^{(r)}(x)| \leq C_1 \tag{1.4}$$

for $r = 3, 4$,

$$|F^{(r)}(x)| \geq 1/C_1 \tag{1.5}$$

for $r = 3$, where C_1 is a positive constant. Let the congruence family S_ℓ be defined by (1.2) (type 1) or by (1.3) (type 2), where $f(x) = TF(x/M)$. Suppose that for some $C_2 \geq 1$ either case 1 or case 2 holds.

Case 1. $M \leq C_2\sqrt{T}$, *and* (1.5) *holds for* $r = 4$ *also;*
Case 2. $M \geq C_2^{-1}\sqrt{T}$ *and* (1.4) *and* (1.5) *hold for* $r = 2$ *also, and*

$$|F''(x)F^{(4)}(x) - 3(F^{(3)}(x))^2| \geq C_3$$

for some positive constant C_3. *Let* $\epsilon > 0$ *be arbitrary. If* M *is sufficiently large in terms of* C_1, *then we have*

$$\sum_{0}^{k-1} |S_\ell|^5 \ll \sqrt{d(k)k} M^{5/2} T^{89/114+\epsilon} \tag{1.6}$$

for any $\epsilon > 0$, *when*

$$k \ll T^{2/57} \quad and \quad T^{49} \ll M^{114} \ll T^{65}.$$

The T^ϵ *in* (1.6) *may be replaced by* $(\log T)^5$ *for type 1 sums in Case 1 and for type 2 sums in Case 2.*
   Secondly we have in Case 1

$$\sum_{0}^{k-1} |S_\ell|^5 \ll \sqrt{d(k)k} \left(M^{13/4} T^{11/24} + M^{7/4} T^{13/12} \right) \log^5 T \tag{1.7}$$

when

$$T^{1/3+\epsilon} \ll M \ll T^{1/2}$$

and

$$\frac{k}{d(k)} \ll \min \left(\left(\frac{T}{M^2} \right)^{3/8}, \left(\frac{M^3}{T} \right)^{1/4} \right)$$

for type 1 families, or

$$k \ll \min \left(\left(\frac{T}{M^2} \right)^{1/4}, \max \left(\left(\frac{M^3}{T} \right)^{1/12}, \left(\frac{M^{42}}{T^{17}} \right)^{1/24} \right) \right)$$

for type 2 families.
   Thirdly, we have in Case 2

$$\sum_{0}^{k-1} |S_\ell|^5 \ll \sqrt{d(k)k} \left(M^{7/4} T^{29/24} + M^{13/4} T^{1/3} \right) \log^5 T \tag{1.8}$$

when

$$T^{1/2} \ll M \ll T^{2/3-\epsilon}$$

and

$$k \ll \min\left(\left(\frac{M^2}{T}\right)^{1/4}, \left(\frac{T^2}{M^3}\right)^{1/6}\right)$$

for type 1 families, or

$$\frac{k}{d(k)} \ll \min\left(\left(\frac{M^2}{T}\right)^{3/8}, \left(\frac{T^2}{M^3}\right)^{1/4}\right)$$

for type 2 families.

The implied constants are constructed from C_1, C_2, C_3, from the implied constants in the various ranges for M and k, and, where appropriate, from ϵ.

Theorem 1 can be stated more generally with bounded weights $g(m/M)$, where $g(x)$ has bounded variation. The upper bounds for k can be relaxed, at the cost of extra terms in (1.6), (1.7), and (1.8). Theorem 1 can be extended to a family of congruence families of sums indexed by a parameter y; the number of cases rises from eight in Theorem 17.2.2 of [5] to about eighteen.

Theorem 2 *Let $\epsilon > 0$ be given. There is a constant $C(\epsilon)$ with*

$$\left|L\left(\tfrac{1}{2} + it, \chi\right)\right| \le C(\epsilon)(d(k))^{1/10}k^{2/5}|t|^{89/570+\epsilon}$$

for each primitive character χ mod k, when $|t| \ge k^{57/2}$. For prime k the factor $|t|^\epsilon$ may be replaced by a bounded power of $\log|t|$.

The exponent $89/570$ of $|t|$ in Theorem 2 is the same as for the Riemann zeta function in [4]; Watt [10] had $89/560$. Theorem 17.3.1 of [5] had $89/570$, but a larger power of k. The exponent $2/5$ of k is much larger than the exponent $3/16$ for Dirichlet L-functions at a fixed height t found by Burgess [2]. In applications $\log|t|$ and $\log k$ are often comparable. Then a result of Motohashi [9] gives useful estimates, stronger than ours.

2. Resonance curves The sum S is divided into short intervals of length N. On each interval we pick a rational value a/q of $f''(x)/2$ to label the interval as a Farey arc. We form an approximating polynomial as in chapter 7 of [5]; its terms in x^2 and x^3 do not involve ℓ. We introduce extra notation: for type 1 sums $q\ell = h(\ell)k + j(\ell)$, for type 2 sums $2a\ell = h(\ell)k + j(\ell)$, with $h(\ell)$ and $j(\ell)$ integers chosen so that

$$|\kappa k + j(\ell)| \le k/2.$$

The exponential sum over a Farey arc is transformed by Poisson summation. The large sieve shows that the transformed sums are small in mean square, provided that there are not too many coincidences between transformed sums from different minor arcs (which may come from different sums of the family). These coincidences, or *resonances*, are associated with *magic matrices* with integer entries and determinant one.

We group the Farey arcs into blocks corresponding to values of a/q in an interval between two consecutive Farey fractions e/r and f/s. When we fix the magic matrix, the arc labelled a/q can only resonate with the arc a_1/q_1 obtained from a/q by Möbius action. Fix ℓ and ℓ_1, and suppose that a/q labels an arc of S_ℓ, a_1/q_1 labels an arc of S_{ℓ_1}. Resonances occur when there is an integer point (c, d) close to a *resonance curve* $R(\ell, \ell_1)$, $z = K(y)$, defined implicitly in terms of a/q by

$$\frac{a}{q} = \frac{ex + f}{rx + s},$$

$$y = \alpha(\ell, \ell_1) - g'(x), \quad z = \beta(\ell, \ell_1) + xg'(x) - g(x)$$

as in Lemmas 15.2.1 and 15.3.1 of [5]. The *Fortean* (coincidence-detecting) function $g(x)$ does not depend on ℓ or ℓ_1, but

$$\alpha(\ell, \ell_1) = \alpha + (j(\ell) - j_1(\ell_1))/k,$$

$$\beta(\ell, \ell_1) = \beta + \frac{h(\ell)s}{r} + \frac{j(\ell)s}{kr} - \frac{h_1(\ell_1)s_1}{r_1} - \frac{j_1(\ell_1)s_1}{kr_1},$$

where α, β denote $\alpha(0,0)$ and $\beta(0,0)$. For the relevant values of x, y lies in an interval of length $O(R^2U/sN)$, and z lies in an interval of length $O(R^2U/rN)$, where U is the number of minor arcs in the block, and R is a parameter related to N by $NR^2 \asymp M^3/T$. In [4] the shorter interval (that for z, in the case $r > s$) had length less than one. In the case $r > s$, the integer d close to a value of z in was unique. For the short arc of the resonance curve with z close to d, there was at most one integer c close to the values of y. In this paper we show that, for many pairs of values ℓ and ℓ_1, there is no integer c, and so no coincident pair of minor arcs.

For type 1 families

$$\alpha(\ell, \ell_1) \equiv \alpha + \frac{r\ell - r_1\ell_1}{k} \pmod 1, \quad \beta(\ell, \ell_1) = \beta + \frac{s\ell - s_1\ell_1}{k},$$

and for type 2 families

$$\alpha(\ell, \ell_1) \equiv \alpha + \frac{2e\ell - 2e_1\ell_1}{k} \pmod 1,$$

$$\beta(\ell, \ell_1) = \beta + \frac{2es\ell}{kr} - \frac{2e_1s_1\ell_1}{kr_1}$$

$$= \beta + \frac{2f\ell - 2f_1\ell_1}{k} \mp \frac{\ell}{kr} \pm \frac{\ell_1}{kr_1}$$

$$= \beta + \frac{2f\ell - 2f_1\ell_1}{k} + O\left(\frac{1}{r}\right).$$

To restore the symmetry between type 1 and type 2 families, we put

$$\beta(\ell, \ell_1) = \beta + \frac{2f\ell - 2f_1\ell_1}{k}$$

for type 2 sums, and we add $O(1/r)$ to the approximation error.

If the point (y, z) lies on $R(\ell, \ell_1)$, then the point

$$(y - \alpha(\ell, \ell_1) + \alpha, \ z - \beta(\ell, \ell_1) + \beta)$$

lies on $R(0,0)$. Thus $R(\ell, \ell_1)$ is $R(0,0)$ shifted by a vector $(\alpha(\ell, \ell_1) - \alpha, \beta(\ell, \ell_1) - \beta)$, which is $1/k$ times an integer vector. Any integer point (A, B) close to $R(\ell, \ell_1)$ corresponds to a rational point $(a/k, b/k)$ close to $R(0,0)$, with

$$a = Ak - r\ell + r_1\ell_1, \qquad b = Bk - s\ell + s_1\ell_1$$
$$a = Ak - 2e\ell + 2e_1\ell_1, \qquad b = Bk - 2f\ell + 2f_1\ell_1$$

for type 1 or type 2 families respectively.

If $R(\ell, \ell_1)$ and $R(\ell_2, \ell_3)$ are shifted by vectors whose difference is an integer vector, then there is an integer point close to the curve $R(\ell, \ell_1)$ if and only if there is an integer point close to the curve $R(\ell_2, \ell_3)$. We must count how many shift vectors are congruent modulo one.

For type 1 families we consider the simultaneous congruences

$$r\ell - r_1\ell_1 \equiv c \ (\text{mod} \ k), \qquad s\ell - s_1\ell_1 \equiv d \ (\text{mod} \ k).$$

The fractions are related by the magic matrix:

$$\begin{pmatrix} f_1 & e_1 \\ s_1 & r_1 \end{pmatrix} = \begin{pmatrix} A & B \\ C & D \end{pmatrix} \begin{pmatrix} f & e \\ s & r \end{pmatrix},$$

where all the matrices have determinant one, so

$$rs_1 - r_1 s = r(Cf + Ds) - s(Ce + Dr) = C.$$

Solving the congruences, we get

$$C\ell \equiv cs_1 - dr_1 \ (\text{mod} \ k), \qquad C\ell_1 \equiv cs - dr \ (\text{mod} \ k),$$

so only shifts modulo one with $cs - dr \equiv 0 \ (\text{mod} \ (C, k))$ occur, and these occur (C, k) times. Similarly for type 2 families only shifts modulo one with $cf \equiv de$ (mod $(2B, k)$) occur, and these occur $(2B, k)$ times. As in [7], magic matrices in congruence subgroups play a special role.

We consider together all the minor arcs between the arcs labelled by the reference fractions e/r and f/s. For type 2 sums the number of curves $R(\ell, \ell_1)$

close to an integer point is at most $(2B, k)$ times the number of integer points (a, b) close to the enlarged curve $kR(0,0)$ with a congruence condition

$$af - be \equiv 0 \; (\text{mod } (2B, k)).$$

The approximation is an expansion about the centre of the arc labelled e/r (in the case $r > s$). The U consecutive minor arcs start with the arc adjacent to the arc labelled e/r. The arc labelled f/s is not one of the U consecutive minor arcs, and it may be distant from them. If $s > r$, then we take our U consecutive arcs next to the arc labelled f/s; the calculations are analogous.

We count values of b first, then values of a. The integer b is a multiple of $(2B, f, k)$ and lies in an interval of length $O(kR^2U/rN)$, so there are

$$O\left(1 + \frac{kR^2U}{(2B, f, k)rN}\right) \tag{2.1}$$

possible values of b. When b is chosen, then the point kz lies within an interval of length

$$O\left(\frac{kR^2}{rG(x)} + \frac{k}{r}\right), \tag{2.2}$$

by Lemma 15.3.1 of [5], where

$$G(x) = \frac{1}{3\mu r(rx + s)}$$

in the notation of Lemma 15.2.1 of [5]. Since $dz/dy = -x$, the point ky lies within an interval of length

$$O\left(\frac{kR^2}{rxG(x)} + \frac{k}{rx}\right).$$

Again by Lemma 15.3.1 of [5],

$$|ky - a| \ll krG(x)/N^2.$$

Since a is a multiple of $(2B, e, k)$, there are

$$O\left(1 + \frac{kR^2}{(2B, e, k)rxG(x)} + \frac{k}{(2B, e, k)rx} + \frac{krG(x)}{(2B, e, k)N^2}\right)$$

possible integers a. We have $G(x) \ll NU$, and for $x = u/t$, $q = ru + st$,

$$xG(x) = \frac{x}{3\mu r(rx + s)} = \frac{u}{3\mu r(ru + st)}.$$

The minimum order of magnitude of $x = u/t$ occurs at the minor arc furthest from e/r, with

$$\frac{U}{R^2} \asymp \frac{eu + ft}{ru + st} - \frac{e}{r} = \frac{t}{r(ru + st)}. \tag{2.3}$$

Since the arc labelled f/s is outside the block of U arcs,

$$\frac{1}{R^2} \asymp \frac{f}{s} - \frac{eu + ft}{ru + st} = \frac{u}{r(ru + st)}. \tag{2.4}$$

We see from (2.3) and (2.4) that

$$\frac{s}{rx} = \frac{st}{ru} \ll U, \tag{2.5}$$

and so

$$\frac{1}{xG(x)} \ll \frac{r^2 U}{N R^2}.$$

To estimate k/rx, we recall that minor arcs were grouped by the size of q into ranges $Q \le q < 2Q$, so $ru + st \asymp Q$. If $ru \ge st$, then $u \asymp Q/r$, so

$$\frac{k}{rx} = \frac{kt}{ru} \asymp \frac{krU}{R^2} \ll \frac{kQU}{R^2},$$

If $ru \le st$, then $t \asymp Q/s$, and in (2.3) $U \asymp R^2/rs$, so

$$\frac{k}{rx} = \frac{kt}{ru} \asymp \frac{kQ}{rsu} \ll \frac{kQU}{R^2}.$$

The number of possible integers a has now been estimated as

$$O\left(1 + \frac{krU}{(2B, e, k)N} + \frac{kQU}{(2B, e, k)R^2}\right). \tag{2.6}$$

Multiplying the estimates (2.6) and (2.1), we see that the number of possible integer pairs (a, b) is

$$O\Bigg(1 + \frac{kR^2 U}{(2B, f, k)rN} + \frac{krU}{(2B, e, k)N}$$

$$+ \frac{kQU}{(2B, e, k)R^2} + \frac{k^2 R^2 U^2}{(2B, ef, k)N^2} + \frac{k^2 QU^2}{(2B, ef, k)rN}\Bigg)$$

Each pair of integers (a,b) modulo k corresponds to $(2B,k)$ coincident pairs, so the number of possible coincidences is

$$O\left((2B,k)\left(1+\frac{kR^2U}{rN}+\frac{krU}{N}+\frac{kQU}{R^2}+\frac{k^2R^2U^2}{N^2}+\frac{k^2QU^2}{rN}\right)\right) \qquad (2.7)$$

The choice of reference fractions depends on Q, with

$$U \asymp (N/Q)^{2/3},$$

For any two reference fractions e/r and f/s (not necessarily consecutive) with $f/s - e/r = \pm 1/rs$,

$$R/\sqrt{U} \ll \max(r,s) \ll Q.$$

We estimate (2.7) as

$$O\left((2B,k)\left(1+\frac{kR}{Q}+k\left(\frac{Q}{N}\right)^{1/3}\right.\right.$$
$$\left.\left.+k\left(\frac{N^2Q}{R^6}\right)^{1/3}+k^2\left(\frac{R^3}{Q^2N}\right)^{2/3}+k^2\left(\frac{N^2}{Q^2R^3}\right)^{1/3}\right)\right)$$
$$\ll (2B,k)k\left(1+k\left(\frac{R^2}{N^2}+\frac{N^2}{R^5}\right)^{1/3}\right). \qquad (2.8)$$

We have excluded coincidences between the arcs containing reference fractions. From [9] and [5], there are at most

$$k+O(k^2Q/N) \qquad (2.9)$$

coincidences between sums of the congruence family on any given pair of minor arcs.

Similarly the number of coincidences for type 1 families is

$$O\left((C,k)k\left(1+k\left(\frac{R}{N}\right)^{2/3}+\frac{kQ}{N}\right)\right), \qquad (2.10)$$

with one less term, because there is no term k/r in (2.2).

For lower triangular matrices (2.8) is trivial. The matrix acts by

$$\begin{pmatrix} f_1 & e_1 \\ s_1 & r_1 \end{pmatrix} = \begin{pmatrix} 1 & 0 \\ C & 1 \end{pmatrix}\begin{pmatrix} f & e \\ s & r \end{pmatrix},$$

so $e = e_1$, $f = f_1$, and

$$\alpha(\ell, \ell_1) \equiv \alpha + 2e(\ell - \ell_1)/k \pmod 1,$$
$$\beta(\ell, \ell_1) = \beta + 2f(\ell - \ell_1)/k.$$

If there is a coincidence, then for some point (y, z) on the resonance curve $R(0, 0)$, corresponding to some $x = u/t$, the rational $2e(\ell - \ell_1)/k$ modulo one lies in an interval of length $O(rG(x)/N^2)$, and $2f(\ell - \ell_1)/k$ modulo one lies in an interval of length

$$O\left(\frac{R^2}{rG(x)} + \frac{1}{r}\right).$$

The values of $2e(\ell - \ell_1)/k$ modulo one are spaced $(2e, k)/k$ apart, and they are each taken $(2e, k)$ times, so there are

$$O\left((2e, k)\left(1 + \frac{krG(x)}{(2e, k)N^2}\right)\right) \ll (2e, k) + \frac{krG(x)}{N^2}$$

possible residue classes for $\ell - \ell_1 \bmod k$ which give coincidences. Similarly, considering $\beta(\ell, \ell_1)$, we see that there are

$$O\left((2f, k) + \frac{kR^2}{rG(x)} + \frac{k}{r}\right)$$

possible residue classes for $\ell - \ell_1$ modulo k.

The minimum of these two bounds is

$$O\left((2e, k) + (2f, k) + \sqrt{\frac{krG(x)}{N^2}\left(\frac{kR^2}{rG(x)} + \frac{k}{r}\right)}\right)$$

$$\ll (2e, k) + (2f, k) + \frac{kR}{N} + \frac{k\sqrt{G(x)}}{N}.$$

Including the endpoint terms from (2.9), we find that the number of coincidences from all the pairs S_ℓ, S_{ℓ_1} of type 2 sums in the block of minor arcs between two consecutive reference fractions given by a particular lower triangular matrix is

$$O\left((2e, k)k + (2f, k)k + \frac{k^2 Q}{N} + \frac{k^2\sqrt{G(x)}}{N}\right).$$

The corresponding bound for upper triangular matrices and type 1 sums is

$$O\left((r, k)k + (s, k)k + \frac{k^2 Q}{N}\right).$$

3. Completing the proofs   First we count coincidences. Magic matrices are classified as follows. Type 1 consists of the identity, and a bounded number of other matrices with small entries. Type 2 consists of upper and lower triangular matrices. All other magic matrices are type 3, with

$$A/C \asymp |D/C| \asymp T/M^2.$$

In Lemma 5.2 of the sister paper [7], with the help of a theorem of Wolke [11] on small divisors of integers, we show that the proportion of type 3 magic matrices with $d|C$ (or, by symmetry, with $d|B$) is $O(N^\epsilon/d)$ in all cases; $O(1/d)$ if $d = 1$, or if $A/C \gg 1$ in the case $d|C$, or if $A/C \ll 1$ in the case $d|B$; and $O((\log N)^\beta/d)$ for some β if $A/C \gg 1/d$ and $C \gg d^2$ in the case $d|C$, or if $A/C \ll d$ and $B \gg d^2$ in the case $d|B$. We use this for each factor d of k to see that the average of $(2B, k)$ or of (C, k) over type 3 magic matrices is $O(d(k)N^\epsilon)$. An easier calculation gives the average of $(2B, k)$ over upper triangular matrices as $O(d(k))$, and similarly for the average of (C, k) over lower triangular matrices. For type 1 magic matrices we use (2.9) on each minor arc.

A triangular matrix can give a coincidence for any arc of the sum. For the lower triangular matrices acting on type 2 families we want the average of (e, k) over the reference fractions e/r. We note that $e/r \asymp P/Q$. There is some choice in the construction of reference fractions. The number of minor arcs between consecutive reference fractions should lie between bounded multiples of $W = (N/Q)^{2/3}$. Fractions a/q with

$$q < B_1 R/\sqrt{W}$$

are chosen first (B_1 and B_2 below are some positive constants). If these choices leave a long gap between consecutive fractions e/r and f/s with $r > s$, then we fill it with fractions

$$\left|\frac{a_i}{q_i}\right| = \frac{e + ft_i}{r + st_i}, \qquad \left|\frac{a_{i+1}}{q_{i+1}} - \frac{a_i}{q_i}\right| \asymp \frac{W}{R^2}.$$

We can make the choices in such a way that the average of (e, k) over reference fractions is no more than a constant multiple of the corresponding average over all Farey fractions with denominator $O(R/\sqrt{W})$ in an interval $0 < a/q < B_2 P/Q$. This average is again $O(d(k))$. Similarly for type 1 congruence families we can choose the reference fractions so that the average of (r, k) is $O(d(k))$.

Hence the number of coincidences between minor arcs with $Q \le q < 2Q$ in a congruence family of sums is

$$O\left(d(k)k(PQ)^\epsilon + k^2\left(\left(\frac{R}{N}\right)^{2/3} + \left(\frac{N^2}{R^5}\right)^{1/3} + \frac{Q}{N}\right)\right)$$

times the estimate for coincidences between minor arcs of a single sum. The ranges with $Q \asymp R$ dominate as in [5], and the total number of coincidences is

$$O\left(d(k)kM^{\epsilon} + k^2\left(\left(\frac{R}{N}\right)^{2/3} + \left(\frac{N^2}{R^5}\right)^{1/3}\right)\right) \qquad (3.1)$$

times the estimate for the number of coincidences between minor arcs of a single sum. The factor M^{ϵ} can be omitted in some cases. The term in N^2/R^5 can be omitted for type 1 families. In ranges where all magic matrices are lower triangular for type 1 families, the factor (3.1) can be replaced by

$$O\left(d(k)k + \frac{k^2R}{N}\right). \qquad (3.2)$$

In ranges where all magic matrices are upper triangular for type 2 families, the factor (3.1) can be replaced by

$$O\left(d(k)k + \frac{k^2R}{N} + \frac{k^2}{(NR^2)^{1/6}}\right). \qquad (3.3)$$

The cases of Theorem 1 now follow like Theorems 17.1.4 and 17.3.1 of [5], with the choices of the parameters N and R in Lemma 17.1.2.

We begin Theorem 2 with the approximate functional equation for $L(s, \chi)$ and partial summation. We must estimate sums S_{χ} and $S_{\bar{\chi}}$ with

$$f(m) = T \log(m/M), \qquad M \ll \sqrt{kT}.$$

This gives type 1 congruence families with $T/M^2 \gg 1/k$. Since d runs through all the factors of k, there is always a factor d of k for which the factor $O(N^{\epsilon})$ occurs in the bound for type 3 matrices, except when k is prime. The three cases of Theorem 1 are strong enough when

$$T^{23/57} \ll M \ll T^{34/57}. \qquad (3.4)$$

Outside this range we use the van der Corput iteration [3]. For the differencing step we choose complex numbers η_{ℓ} of unit modulus so that $\eta_{\ell}S_{\ell}$ is real and positive, the *Halász multipliers*. We apply the differencing step to $\sum \eta_{\ell}S_{\ell}$, and group terms after the Cauchy inequality to form congruence families again. A power of k enters the differencing parameter. For example, using the classical exponent pair $(2/7, 4/7)$ after the differencing step, we get

$$\Sigma|S_{\ell}| \ll k^{8/9}T^{1/9}M^{11/18}$$

for $M \gg k^2$. The exponent of k is less than $9/10$, and this bound is strong enough when

$$k^2 \ll M \ll T^{77/190},$$

overlapping the range (3.4). For $M \ll k^2$ we use the trivial estimate

$$S_\chi \ll M \ll T^{4/57}.$$

References

[1] E. Bombieri and H. Iwaniec. On the order of $\zeta\left(\frac{1}{2} + it\right)$. *Ann. Scuola Norm. Sup. Pisa Cl. Sci.*, (4) **13** (1986), 449–72.

[2] D.A. Burgess. On character sums and L-series. II. *Proc. London Math. Soc.*, (3) **13** (1963), 524–36.

[3] S.W. Graham and G. Kolesnik. *Van der Corput's Method of Exponential Sums*. Cambridge Univ. Press, London, 1991.

[4] M.N. Huxley. Exponential sums and the Riemann zeta function. IV. *Proc. London Math. Soc.*, (3) **66** (1993), 1–40.

[5] —. *Area, Lattice Points, and Exponential Sums*. Oxford Univ. Press, Oxford, 1996.

[6] M.N. Huxley and N. Watt. The number of ideals in a quadratic field. *Proc. Indian Acad. Sci. (Math. Sci.)*, **104** (1994), 157–165.

[7] —. The number of ideals in a quadratic field II. To appear.

[8] M. Jutila. *A Method in the Theory of Exponential Sums*. Lect. Math. Phy., **80**, Tata Inst. Fund. Res.–Springer-Verlag, Bombay, 1987.

[9] Y. Motohashi. A note on the mean value of the zeta and L-functions I, II, III. *Proc. Japan Acad., Ser. A*, **61** (1985), 222–4, 313–16, **62** (1986), 152–154.

[10] N. Watt. A hybrid bound for Dirichlet L-functions on the critical line. *Proc. Amalfi Conf. Analytic Number Theory* (E. Bombieri, A. Perelli, S. Salerno and U. Zannier, Eds.), Univ. di Salerno, Salerno, 1992, pp. 387–92.

[11] D. Wolke. A new proof of a theorem of van der Corput. *J. London Math. Soc.*, (2) **5** (1972), 609–612.

Martin N. Huxley
School of Mathematics, University of Wales Cardiff
Cardiff CF2 4YH, Wales, U.K.

Nigel Watt
Department of Mathematics, University of Nottingham
Nottingham, U.K.

On Some Results Concerning the Riemann Hypothesis

Aleksandar Ivić

1. Introduction A central place in Analytic Number Theory is occupied by the Riemann zeta-function $\zeta(s)$, defined for $\mathrm{Re}\, s > 1$ by

$$\zeta(s) = \sum_{n=1}^{\infty} n^{-s} = \prod_{p\,:\,\mathrm{prime}} (1 - p^{-s})^{-1}, \qquad (1.1)$$

and otherwise by analytic continuation. It admits meromorphic continuation to the whole complex plane, its only singularity being the simple pole $s = 1$ with residue 1. For general information on $\zeta(s)$ the reader is referred to the monographs [7], [15], and [58]. From the functional equation

$$\zeta(s) = \chi(s)\zeta(1-s), \quad \chi(s) = 2^s \pi^{s-1} \sin\left(\frac{\pi s}{2}\right)\Gamma(1-s), \qquad (1.2)$$

which is valid for any complex s, it follows that $\zeta(s)$ has zeros at $s = -2, -4, \ldots$. These zeros are traditionally called the "trivial" zeros of $\zeta(s)$, to distinguish them from the complex zeros of $\zeta(s)$, of which the smallest ones (in absolute value) are $\frac{1}{2} \pm 14.134725\ldots i$. It is well-known that all complex zeros of $\zeta(s)$ lie in the so-called "critical strip" $0 < \sigma = \mathrm{Re}\, s < 1$, and if $N(T)$ denotes the number of zeros $\rho = \beta + i\gamma$ (β, γ real) of $\zeta(s)$ for which $0 < \gamma \leq T$, then

$$N(T) = \frac{T}{2\pi} \log\left(\frac{T}{2\pi}\right) - \frac{T}{2\pi} + \frac{7}{8} + S(T) + O\left(\frac{1}{T}\right) \qquad (1.3)$$

with

$$S(T) = \frac{1}{\pi} \arg \zeta(\tfrac{1}{2} + iT) = O(\log T). \qquad (1.4)$$

Here $S(T)$ is obtained by continuous variation along the straight lines joining $2, 2 + iT, \frac{1}{2} + iT$, starting with the value 0; if T is the ordinate of a zero, let $S(T) = S(T + 0)$. This is the so-called Riemann–von Mangoldt formula. *The Riemann hypothesis* (henceforth RH for short) is the conjecture, stated by B. Riemann in his epoch-making memoir [52], that *very likely – sehr wahrschein-lich – all complex zeros of $\zeta(s)$ have real parts equal to $\frac{1}{2}$*. For this reason the line $\sigma = \frac{1}{2}$ is called the "critical line" in the theory of $\zeta(s)$. The RH is undoubtedly one of the most celebrated and difficult open problems in whole

Mathematics. Its proof (or disproof) would have very important consequences in multiplicative number theory, especially in problems involving the distribution of primes. It would also very likely lead to generalizations to many other zeta-functions (Dirichlet series) sharing similar properties with $\zeta(s)$.

The RH can be put into many equivalent forms. One of the classical is

$$\pi(x) = \operatorname{li} x + O(\sqrt{x} \log x), \tag{1.5}$$

where $\pi(x)$ is the number of primes not exceeding x (≥ 2) and

$$\operatorname{li} x = \int_0^x \frac{dt}{\log t} = \lim_{\varepsilon \to 0+} \left(\int_0^{1-\varepsilon} + \int_{1+\varepsilon}^x \right) \frac{dt}{\log t}$$

$$= \sum_{n=1}^N (n-1)! \frac{x}{\log^n x} + O\left(\frac{x}{\log^{N+1} x} \right) \tag{1.6}$$

for any fixed integer $N \geq 1$. One can give a purely arithmetic equivalent of the RH without mentioning primes: We can define recursively the Möbius function $\mu(n)$ as

$$\mu(1) = 1, \ \mu(n) = - \sum_{d \mid n, d < n} \mu(d) \qquad (n > 1).$$

Then the RH is equivalent to the assertion that for any given integer $k \geq 1$ there exists an integer $N_0 = N_0(k)$ such that, for integers $N \geq N_0$, one has

$$\left(\sum_{n=1}^N \mu(n) \right)^{2k} \leq N^{k+1}. \tag{1.7}$$

The above definition of $\mu(n)$ is elementary and avoids explicit mention of primes. A non-elementary definition of $\mu(n)$ is through the series representation

$$\sum_{n=1}^\infty \mu(n) n^{-s} = \frac{1}{\zeta(s)} \qquad (\operatorname{Re} s > 1), \tag{1.8}$$

and an equivalent form of the RH is that (1.8) holds for $\sigma > 1/2$. The inequality (1.7) is in fact the bound

$$\sum_{n \leq x} \mu(n) \ll_\varepsilon x^{\frac{1}{2} + \varepsilon} \tag{1.9}$$

in disguise, where ε corresponds to $1/(2k)$, x to N, and the $2k$-th power avoids absolute values. The bound (1.9) (see [15] and [58]) is one of the classical equivalents of the RH. The sharper bound

$$\left| \sum_{n \leq x} \mu(n) \right| < \sqrt{x} \qquad (x > 1)$$

was proposed in 1897 by Mertens on the basis of numerical evidence, and later became known in the literature as *the Mertens conjecture*. It was disproved in 1985 by A.M. Odlyzko and H.J.J. te Riele [47].

Instead of working with the complex zeros of $\zeta(s)$ on the critical line, it is convenient to introduce the function

$$Z(t) = \chi^{-1/2}(\tfrac{1}{2} + it)\zeta(\tfrac{1}{2} + it), \qquad (1.10)$$

where $\chi(s)$ is given by (1.2). Since $\chi(s)\chi(1-s) = 1$ and $\overline{\Gamma(s)} = \Gamma(\overline{s})$, it follows that $|Z(t)| = |\zeta(\tfrac{1}{2} + it)|$, $Z(t)$ is even, and $\overline{Z(t)} = Z(t)$. Hence $Z(t)$ is real if t is real, and the zeros of $Z(t)$ correspond to the zeros of $\zeta(s)$ on the critical line. Let us denote by $0 < \gamma_1 \leq \gamma_2 \leq \ldots$ the positive zeros of $Z(t)$ with multiplicities counted (all known zeros are simple). If the RH is true, then it is known (see [58]) that

$$S(T) = O\Big(\frac{\log T}{\log\log T}\Big), \qquad (1.11)$$

and this seemingly small improvement over (1.4) is significant: If (1.11) holds, then from (1.3) one infers that $N(T + H) - N(T) > 0$ for $H = C/\log\log T$ with a suitable $C > 0$ and $T \geq T_0$. Consequently we have, assuming the RH, the bound

$$\gamma_{n+1} - \gamma_n \ll \frac{1}{\log\log \gamma_n} \qquad (1.12)$$

for the gap between consecutive zeros on the critical line. For some unconditional results on $\gamma_{n+1} - \gamma_n$, see [16], [17], and [25].

It turned out that already Riemann had computed several zeros of the zeta-function and had a deep understanding of its analytic behaviour. Siegel provided rigorous proof of a formula that had its genesis in Riemann's work. It came to be known later as *the Riemann–Siegel formula* (see [15], [38], [56], [58], and (8.9)) and, in a weakened form, it says that

$$Z(t) = 2 \sum_{n \leq (t/2\pi)^{1/2}} n^{-1/2} \cos\Big(t \log \frac{\sqrt{t/2\pi}}{n} - \frac{t}{2} - \frac{\pi}{8}\Big) + O(t^{-1/4}), \qquad (1.13)$$

where the O-term is actually best possible, namely it is $\Omega_\pm(t^{-1/4})$. The Riemann–Siegel formula is an indispensable tool in the theory of $\zeta(s)$.

There exists a large and rich literature on numerical calculations involving $\zeta(s)$ and its zeros (see [36], [44], [45], [46], and [51], which contain references to further works). This literature reflects the development of Mathematics in general, and of Numerical Analysis and Analytic Number Theory in particular. Suffice to say that it is known that the first 1.5 billion complex zeros of $\zeta(s)$ in the upper half-plane are simple and do have real parts equal to 1/2, as predicted by the RH. Moreover, many large blocks of zeros of much greater height have been thoroughly investigated, and all known zeros satisfy the RH. However, one should be very careful in relying on numerical evidence in Analytic Number Theory. A classical example for this is the inequality $\pi(x) < \operatorname{li} x$ (see (1.5) and (1.6)), noticed already by Gauss, which is known to be true for all x for which the functions in question have been actually computed. But the inequality $\pi(x) < \operatorname{li} x$ is false; not only does $\pi(x) - \operatorname{li} x$ assume positive values for some arbitrarily large values of x, but J.E. Littlewood [35] proved that

$$\pi(x) = \operatorname{li} x + \Omega_\pm \left(\sqrt{x}\, \frac{\log\log\log x}{\log x} \right).$$

By extending the methods of R. Sherman Lehman [55], H.J.J. te Riele [50] showed that $\pi(x) < \operatorname{li} x$ fails for some (unspecified) $x < 6.69 \times 10^{370}$. For values of t which are this large we may hope that $Z(t)$ will also show its true asymptotic behaviour. Nevertheless, we cannot compute by today's methods the values of $Z(t)$ for t this large, actually even $t = 10^{100}$ seems out of reach at present. To assess why the values of t where $Z(t)$ will "really" exhibit its true behaviour must be "very large", it suffices to compare (1.4) and (1.11) and note that the corresponding bounds differ by a factor of $\log\log T$, which is a very slowly varying function.

The aim of this text is to discuss some topics and recent results connected, one way or another, with the RH. The topics will include the Lehmer phenomenon, the Davenport–Heilbronn zeta-function, mean and large values on the critical line, and zeros of a class of convolution functions connected with $\zeta(s)$. This choice is to a large extent motivated by the author's own research in recent years, but anyway all important aspects of recent research on the RH certainly cannot be covered in one paper.

Acknowledgement   I want to thank Professors M. Jutila, K. Matsumoto, Y. Motohashi and A.M. Odlyzko for valuable remarks on an earlier version of this text.

2. Lehmer's phenomenon   The function $Z(t)$, defined by (1.10), has a negative local maximum $-0.52625\ldots$ at $t = 2.47575\ldots$. This is the only known occurrence of a negative local maximum, while no positive local minimum is known. *Lehmer's phenomenon* (named after D.H. Lehmer, who in his works [33], [34] made significant contributions to the subject) is the fact (see [46] for a thorough discussion) that the graph of $Z(t)$ sometimes barely crosses the t-axis. This means that the absolute value of the maximum or minimum of $Z(t)$ between its two consecutive zeros is small. For instance, A.M. Odlyzko found 1976 values of n such that $|Z(\frac{1}{2}(\gamma_n + \gamma_{n+1}))| < 0.0005$ in the block that he investigated (in the version of [46] available to the author, but Odlyzko kindly informed him that many more examples occur in the computations that are going on now). Several extreme examples are also given by van de Lune et al. in [36]. The Lehmer phenomenon shows the delicacy of the RH. For should it happen that, for $t \geq t_0$, $Z(t)$ attains a negative local maximum or a positive local minimum, then the RH would be disproved. This assertion follows (see [7]) from the following

Proposition *If the RH is true, then the graph of $Z'(t)/Z(t)$ is monotonically decreasing between the zeros of $Z(t)$ for $t \geq t_0$.*

Namely suppose that $Z(t)$ has a negative local maximum or a positive local minimum between its two consecutive zeros γ_n and γ_{n+1}. Then $Z'(t)$ would have at least two distinct zeros x_1 and x_2 ($x_1 < x_2$) in (γ_n, γ_{n+1}), and hence so would $Z'(t)/Z(t)$. But we have

$$\frac{Z'(x_1)}{Z(x_1)} > \frac{Z'(x_2)}{Z(x_2)},$$

which is a contradiction, since $Z'(x_1) = Z'(x_2) = 0$.

To prove the Proposition, consider the function

$$\xi(s) := \frac{1}{2}s(s-1)\pi^{-s/2}\Gamma(\frac{s}{2})\zeta(s),$$

so that $\xi(s)$ is an entire function of order one (see Chapter 1 of [15]), and one has unconditionally

$$\frac{\xi'(s)}{\xi(s)} = B + \sum_{\rho}(\frac{1}{s-\rho} + \frac{1}{\rho}), \quad B = \log 2 + \frac{1}{2}\log \pi - 1 - \frac{1}{2}C_0, \qquad (2.1)$$

where ρ denotes complex zeros of $\zeta(s)$ and C_0 is Euler's constant. By (1.2)

$$Z(t) = \chi^{-1/2}(\frac{1}{2}+it)\zeta(\frac{1}{2}+it) = \frac{\pi^{-it/2}\Gamma(\frac{1}{4}+\frac{1}{2}it)\zeta(\frac{1}{2}+it)}{|\Gamma(\frac{1}{4}+\frac{1}{2}it)|},$$

so that we may write

$$\xi(\tfrac{1}{2}+it) = -f(t)Z(t), \quad f(t) := \frac{1}{2}\pi^{-1/4}(t^2+\tfrac{1}{4})|\Gamma(\tfrac{1}{4}+\tfrac{1}{2}it)|.$$

Consequently logarithmic differentiation gives

$$\frac{Z'(t)}{Z(t)} = -\frac{f'(t)}{f(t)} + i\frac{\xi'(\frac{1}{2}+it)}{\xi(\frac{1}{2}+it)}. \qquad (2.2)$$

Assume now that the RH is true. Then by using (2.1) with $\rho = \frac{1}{2} + i\gamma$, $s = \frac{1}{2} + it$ we obtain, if $t \neq \gamma$,

$$\left(\frac{i\xi'(\frac{1}{2}+it)}{\xi(\frac{1}{2}+it)}\right)' = -\sum_{\gamma}\frac{1}{(t-\gamma)^2} < -C(\log\log t)^2 \quad (C > 0)$$

for $t \geq t_0$, since (1.12) holds. On the other hand, by using Stirling's formula for the gamma-function and $\log|z| = \operatorname{Re}\log z$ it is readily found that

$$\frac{d\big(f'(t)/f(t)\big)}{dt} \ll \frac{1}{t},$$

so that from (2.2) it follows that $(Z'(t)/Z(t))' < 0$ if $t \geq t_0$, which implies the Proposition. Actually the value of t_0 may be easily effectively determined and seen not to exceed 1000. Since $Z(t)$ has no positive local minimum or negative local maximum for $3 \leq t \leq 1000$, it follows that the RH is false if we find (numerically) the occurrence of a single negative local maximum (besides the one at $t = 2.47575\ldots$) or a positive local minimum of $Z(t)$.

3. The Davenport–Heilbronn zeta-function This is a zeta-function (Dirichlet series) which satisfies a functional equation similar to the classical functional equation (1.2) for $\zeta(s)$, but for this zeta-function the analogue of the RH does not hold. This function was introduced by H. Davenport and H. Heilbronn [6] as

$$f(s) = 5^{-s}\big(\zeta(s, \tfrac{1}{5}) + \tan\theta\,\zeta(s, \tfrac{2}{5}) - \tan\theta\,\zeta(s, \tfrac{3}{5}) - \zeta(s, \tfrac{4}{5})\big), \qquad (3.1)$$

where $\theta = \arctan((\sqrt{10 - 2\sqrt{5}} - 2)/(\sqrt{5} - 1))$ and, for $\operatorname{Re} s > 1$,

$$\zeta(s, a) = \sum_{n=0}^{\infty}(n + a)^{-s} \qquad (0 < a \leq 1)$$

is the familiar *Hurwitz zeta-function*, defined for $\operatorname{Re} s \leq 1$ by analytic continuation. With the above choice of θ (see [6], [31], or [58]) it can be shown that $f(s)$ satisfies the functional equation

$$f(s) = X(s)f(1 - s), \qquad X(s) = \frac{2\Gamma(1 - s)\cos(\frac{\pi s}{2})}{5^{s-\frac{1}{2}}(2\pi)^{1-s}}, \qquad (3.2)$$

whose analogy with the functional equation (1.2) for $\zeta(s)$ is evident. Let $1/2 < \sigma_1 < \sigma_2 < 1$. Then it can be shown (see Chapter 6 of [31]) that $f(s)$ has infinitely many zeros in the strip $\sigma_1 < \sigma = \operatorname{Re} s < \sigma_2$, and it also has (see Chapter 10 of [58]) an infinity of zeros in the half-plane $\sigma > 1$, while from the product representation in (1.1) it follows that $\zeta(s) \neq 0$ for $\sigma > 1$, so that in the half-plane $\sigma > 1$ the behaviour of zeros of $\zeta(s)$ and $f(s)$ is different. Actually the number of zeros of $f(s)$ for which $\sigma > 1$ and $0 < t = \text{Im } s \leq T$ is $\gg T$, and similarly each rectangle $0 < t \leq T, 1/2 < \sigma_1 < \sigma \leq \sigma_2 \leq 1$ contains at least $c(\sigma_1, \sigma_2)T$ zeros of $f(s)$. R. Spira [57] found that $0.808517 + 85.699348i$ (the values are approximate) is a zero of $f(s)$ lying in the critical strip $0 < \sigma < 1$, but not on the critical line $\sigma = 1/2$. On the other hand, A.A. Karatsuba [30] proved that the number of zeros $\frac{1}{2} + i\gamma$ of $f(s)$ for which $0 < \gamma \leq T$ is at least $T(\log T)^{1/2-\varepsilon}$ for any given $\varepsilon > 0$ and $T \geq T_0(\varepsilon)$. This bound is weaker than A. Selberg's classical result [53] that there are $\gg T \log T$ zeros $\frac{1}{2} + i\gamma$ of $\zeta(s)$ for which $0 < \gamma \leq T$. From the Riemann–von Mangoldt formula (1.3) it follows that, up to the value of the $\ll$–constant, Selberg's result on $\zeta(s)$ is best possible. There are certainly $\ll T \log T$ zeros $\frac{1}{2} + i\gamma$ of $f(s)$ for which $0 < \gamma \leq T$ and it may be that almost all of them lie on the critical line $\sigma = 1/2$, although this has not been proved yet. The Davenport–Heilbronn zeta-function is not the only example of a zeta-function that exhibits the phenomena described above, and many so-called *Epstein zeta-functions* (see Bombieri and Hejhal [5]) also have complex zeros off their respective critical lines.

What is the most important difference between $\zeta(s)$ and $f(s)$ which is accountable for the difference of distribution of zeros of the two functions, which occurs at least in the region $\sigma > 1$? It is most likely that the answer is the lack of the Euler product for $f(s)$, similar to the one in (1.1) for $\zeta(s)$. But $f(s)$ can be written as a linear combination of two L-functions which have Euler products (with a common factor) and this fact plays the crucial rôle in Karatsuba's proof of the lower bound result for the number of zeros of $f(s)$. In any case the example of $f(s)$ shows that it is not clear whether the influence of the Euler product for $\zeta(s)$ will extend all the way to the line $\sigma = 1/2$ and produce the zero-free region predicted by the RH.

4. Mean value formulas on the critical line   Mean values of $|\zeta(\frac{1}{2} + it)|$ were subject of extensive research in recent years, thanks largely to the application of poweful methods from spectral theory (see Y. Motohashi's fundamental monograph [43]). The connection with the RH is only indirect, as will be seen a little later. For $k \geq 1$ a fixed integer let

$$\int_0^T |\zeta(\tfrac{1}{2} + it)|^{2k} \, dt = T P_{k^2}(\log T) + E_k(T), \tag{4.1}$$

where for some suitable coefficients $a_{j,k}$ one has

$$P_{k^2}(y) = \sum_{j=0}^{k^2} a_{j,k} y^j. \tag{4.2}$$

An extensive literature exists on $E_k(T)$, especially on $E_1(T) \equiv E(T)$ (see F.V. Atkinson's classical paper [2]), and the reader is referred to [19] for a comprehensive account. It is known that

$$P_1(y) = y + 2C_0 - 1 - \log(2\pi),$$

and $P_4(y)$ is a quartic polynomial whose leading coefficient equals $1/(2\pi^2)$ (see [21] for an explicit evaluation of its coefficients). One hopes that

$$E_k(T) = o(T) \qquad (T \to \infty) \tag{4.3}$$

will hold for each fixed integer $k \geq 1$, but so far this is known to be true only in the cases $k = 1$ and $k = 2$, when $E_k(T)$ is a true error term in the asymptotic formula (4.1). In fact heretofore it has not been clear how to define properly (even on heuristic grounds) the values of $a_{j,k}$ in (4.2) for $k \geq 3$ (see [23] for an extensive discussion concerning the case $k = 3$). The connection between $E_k(T)$ and the RH is indirect, namely there is a connection with the *Lindelöf hypothesis* (LH for short). The LH is also a famous unsettled problem, and it states that

$$\zeta(\tfrac{1}{2} + it) \ll_\varepsilon t^\varepsilon \tag{4.4}$$

for any given $\varepsilon > 0$ and $t \geq t_0 > 0$ (since $\overline{\zeta(\tfrac{1}{2} + it)} = \zeta(\tfrac{1}{2} - it)$, t may be assumed to be positive). It is well-known (see [58] for a proof) that the RH implies

$$\zeta(\tfrac{1}{2} + it) \ll \exp\left(\frac{A \log t}{\log \log t}\right) \qquad (A > 0, t \geq t_0), \tag{4.5}$$

so that obviously the RH implies the LH. In the other direction it is unknown whether the LH (or (4.5)) implies the RH. However, it is known that the LH has considerable influence on the distribution of zeros of $\zeta(s)$. If $N(\sigma, T)$ denotes the number of zeros $\rho = \beta + i\gamma$ of $\zeta(s)$ for which $\sigma \leq \beta$ and $|\gamma| \leq T$, then it is known (see Chapter 11 of [15]), that the LH implies that $N(\sigma, T) \ll T^{2-2\sigma+\varepsilon}$ for $1/2 \leq \sigma \leq 1$ (this is a form of *the density hypothesis*) and $N(\tfrac{3}{4}+\delta, T) \ll T^\varepsilon$, where $\varepsilon = \varepsilon(\delta)$ may be arbitrarily small for any $0 < \delta < 1/4$.

The best unconditional bound for the order of $\zeta(s)$ on the critical line, known at the time of the writing of this text, is

$$\zeta(\tfrac{1}{2} + it) \ll_\varepsilon t^{c+\varepsilon} \tag{4.6}$$

with $c = 89/570 = 0.15614\ldots$. This is due to M.N. Huxley [12], and represents the last in a long series of improvements over the past 80 years. The result is obtained by intricate estimates of exponential sums of the type $\sum_{N < n \leq 2N} n^{it}$ $(N \ll \sqrt{t})$, and the value $c = 0.15$ appears to be the limit of the method.

Estimates for $E_k(T)$ in (4.1) (both pointwise and in the mean sense) have many applications. From the knowledge about the order of $E_k(T)$ one can deduce a bound for $\zeta(\frac{1}{2} + iT)$ via the estimate

$$\zeta(\tfrac{1}{2} + iT) \ll (\log T)^{(k^2+1)/(2k)} + \Big(\log T \max_{t\in[T-1,T+1]} |E_k(t)|\Big)^{1/(2k)}, \qquad (4.7)$$

which is Lemma 4.2 of [19]. Thus the best known upper bound

$$E(T) \equiv E_1(T) \ll T^{72/227}(\log T)^{679/227} \qquad (4.8)$$

of M.N. Huxley [13] yields (4.6) with $c = 36/227 = 0.15859\ldots$. Similarly the sharpest known bound

$$E_2(T) \ll T^{2/3}\log^C T \qquad (C > 0) \qquad (4.9)$$

of Y. Motohashi and the author (see [19], [26], and [28]) yields (4.6) with the classical value $c = 1/6$ of Hardy and Littlewood. Since the difficulties in evaluating the left-hand side of (4.1) greatly increase as k increases, it is reasonable to expect that the best estimate for $\zeta(\frac{1}{2} + iT)$ that one can get from (4.7) will be when $k = 1$.

The LH is equivalent to the bound

$$\int_0^T |\zeta(\tfrac{1}{2} + it)|^{2k}\, dt \ll_{k,\varepsilon} T^{1+\varepsilon} \qquad (4.10)$$

for any $k \geq 1$ and any $\varepsilon > 0$, which in turn is the same as

$$E_k(T) \ll_{k,\varepsilon} T^{1+\varepsilon}. \qquad (4.11)$$

The enormous difficulty in settling the truth of the LH, and so *a fortiori* of the RH, is best reflected in the relatively modest upper bounds for the integrals in (4.10) (see Chapter 8 of [15] for sharpest known results). On the other hand, we have Ω-results in the case $k = 1, 2$, which show that $E_1(T)$ and $E_2(T)$ cannot be always small. Thus J.L. Hafner and the author [10], [11] proved that

$$E_1(T) = \Omega_+\Big((T\log T)^{\frac{1}{4}}(\log\log T)^{\frac{3+\log 4}{4}} e^{-C\sqrt{\log\log\log T}}\Big) \qquad (4.12)$$

and

$$E_1(T) = \Omega_-\Big(T^{\frac{1}{4}}\exp\Big(\frac{D(\log\log T)^{\frac{1}{4}}}{(\log\log\log T)^{\frac{3}{4}}}\Big)\Big) \qquad (4.13)$$

for some absolute constants $C, D > 0$. Moreover, the author has proved in [18] that there exist constants $A, B > 0$ such that, for $T \geq T_0$, every interval $[T, T + B\sqrt{T}]$ contains points t_1, t_2 for which

$$E_1(t_1) > At_1^{1/4}, \quad E_1(t_2) < -At_2^{1/4}.$$

Numerical investigations concerning $E_1(T)$ and its zeros were carried out by H.J.J. te Riele and the author [29].

The Ω–result

$$E_2(T) = \Omega(\sqrt{T}) \tag{4.14}$$

was proved by Y. Motohashi and the author (see [26], [28], and Chapter 5 of [19]). The method of proof involved differences of values of the functions $E_2(T)$, so that (4.14) was the limit of the method. The basis of this, as well of other recent investigations involving $E_2(T)$, is Y. Motohashi's fundamental explicit formula for

$$(\Delta\sqrt{\pi})^{-1} \int_{-\infty}^{\infty} |\zeta(\tfrac{1}{2} + it + iT)|^4 \, e^{-(t/\Delta)^2} \, dt \quad (\Delta > 0), \tag{4.15}$$

obtained by deep methods involving spectral theory of the non-Euclidean Laplacian (see [39], [40], [42], [43], and Chapter 5 of [19]). On p. 310 of [19] it was pointed out that a stronger result than (4.14), namely

$$\limsup_{T\to\infty} |E_2(T)| T^{-1/2} = +\infty$$

follows if certain quantities connected with the discrete spectrum of the non-Euclidean Laplacian are linearly independent over the integers. Y. Motohashi [41] (see also [43]) unconditionally improved (4.14) by showing that

$$E_2(T) = \Omega_\pm(\sqrt{T}). \tag{4.16}$$

The author [24] recently reproved (4.16) and showed that there is a sign change of $E_2(t)t^{-1/2}$ in every interval of the form $[T, AT]$ for a suitable constant $A > 1$ and $T \geq T_0$. The key step in proving (4.16) is to show that the function

$$Z_2(\xi) := \int_1^{\infty} |\zeta(\tfrac{1}{2} + it)|^4 \, t^{-\xi} \, dt,$$

defined initially as a function of the complex variable ξ for $\mathrm{Re}\,\xi > 1$, is meromorphic over the whole complex plane. In the half-plane $\mathrm{Re}\,\xi > 0$ it has a pole of order five at $\xi = 1$, infinitely many simple poles of the form $\tfrac{1}{2} \pm \kappa i$, while the remaining poles for $\mathrm{Re}\,\xi > 0$ are of the form $\rho/2$, $\zeta(\rho) = 0$. Here $\kappa^2 + \tfrac{1}{4}$ is in the discrete spectrum of the non-Euclidean Laplacian over the full modular group (see [41] for the details). By using (4.1) and integration by parts it follows that

$$Z_2(\xi) = C + \xi \int_1^\infty P_4(\log t) t^{-\xi} \, dt + \xi \int_1^\infty E_2(t) t^{-\xi - 1} \, dt \qquad (4.17)$$

with a suitable constant C, where the integrals are certainly absolutely convergent for $\operatorname{Re} \xi > 1$ (actually the second for $\operatorname{Re} \xi > 1/2$ in view of (4.20)). Now (4.16) is an immediate consequence of (4.17) and a classical oscillation result of E. Landau (see [1] for a proof).

It may be asked then how do the Ω-results for $E_1(T)$ and $E_2(T)$ affect the LH, and thus indirectly the RH? One may conjecture that these Ω-results lie fairly close to the truth, in other words that

$$E_k(T) = O_{k,\varepsilon}(T^{\frac{1}{4}k + \varepsilon}) \qquad (4.18)$$

holds for $k = 1, 2$. This view is suggested by estimates in the mean for the functions in question. Namely the author [14] proved that

$$\int_1^T |E_1(t)|^A \, dt \ll_\varepsilon T^{1 + \frac{1}{4}A + \varepsilon} \qquad (0 \le A \le \frac{35}{4}), \qquad (4.19)$$

and the range for A for which (4.19) holds can be slightly increased by using the best known estimate (4.6) in the course of the proof. Also Y. Motohashi and the author [27], [28] proved that

$$\int_0^T E_2(t) \, dt \ll T^{3/2}, \qquad \int_0^T E_2^2(t) \, dt \ll T^2 \log^C T \quad (C > 0). \qquad (4.20)$$

The bounds (4.19) and (4.20) show indeed that, in the mean sense, the bound (4.18) does hold when $k = 1, 2$. Curiously enough, it does not seem possible to show that the RH implies (4.18) for $k \le 3$. If (4.18) holds for any k, then in view of (4.7) we would obtain (4.6) with the hitherto sharpest bound $c \le 1/8$. What can one expect about the order of magnitude of $E_k(T)$ for $k \ge 3$? It was already mentioned that the structure of $E_k(T)$ becomes increasingly complex as k increases. Thus perhaps we should not expect a smaller exponent than $k/4$ in (4.18) for $k \ge 3$, as it would by (4.7) yield a result of the type $\mu(1/2) < 1/8$, which in view of the Ω-results is not obtainable from (4.18) when $k = 1, 2$. Should this be true, then by analogy with the cases $k = 1, 2$ one would be led to conjecture that

$$E_k(T) = \Omega(T^{k/4}) \qquad (4.21)$$

holds for any fixed $k \ge 1$. But already for $k = 5$ the omega–result (4.21) yields, in view of (4.1),

$$\int_0^T |\zeta(\tfrac{1}{2} + it)|^{10} \, dt = \Omega_+(T^{5/4}), \qquad (4.22)$$

which contradicts (4.10) and disproves *both* the LH and the RH. It would be of great interest to obtain more detailed information on $E_k(T)$ in the cases when $k = 3$ and especially when $k = 4$, as the latter probably represents a turning point in the asymptotic behaviour of mean values of $|\zeta(\frac{1}{2}+it)|$. In [23] the author proved that $E_3(T) \ll_\varepsilon T^{1+\varepsilon}$ conditionally, that is, provided that a certain conjecture involving the ternary additive divisor problem holds, which concerns the sum

$$\sum_{n\leq N} d_3(n)d_3(n+f) \quad (f \geq 1),$$

where $d_3(n)$ is the divisor function generated by $\zeta^3(s)$. Y. Motohashi ([39] p. 339, [41], and [43]) proposes, on heuristic grounds based on analogy with explicit formulas known in the cases $k = 1, 2$, a formula for the analogue of (4.15) for the sixth moment, and also conjectures (4.21) for $k = 3$.

5. Large values on the critical line

Another topic of extensive research in recent years are large values of $|\zeta(\frac{1}{2}+it)|$. R. Balasubramanian and K. Ramachandra (see [3], [4], [48], and [49]) proved unconditionally that

$$\max_{T\leq t\leq T+H} |\zeta(\tfrac{1}{2}+it)| > \exp\Big(\frac{3}{4}\Big(\frac{\log H}{\log\log H}\Big)^{1/2}\Big) \tag{5.1}$$

for $T \geq T_0$ and $\log\log T \ll H \leq T$, and probably on the RH this can be further improved (but no results seem to exist yet). Anyway (5.1) shows that $|Z(t)|$ assumes large values relatively often. On the other hand, on the RH one expects that the bound in (1.11) can be also further reduced, very likely (see [46]) to

$$S(T) \ll_\varepsilon (\log T)^{\frac{1}{2}+\varepsilon}. \tag{5.2}$$

That is, H.L. Montgomery [37] proved, assuming the RH, that

$$S(T) = \Omega_\pm\Big(\Big(\frac{\log T}{\log\log T}\Big)^{1/2}\Big),$$

which is in accord with (5.2). K.-M. Tsang [60], improving a classical result of A. Selberg [53], has shown that one has unconditionally

$$S(T) = \Omega_\pm\Big(\Big(\frac{\log T}{\log\log T}\Big)^{1/3}\Big).$$

K.-M. Tsang [60] also proved that unconditionally

$$\Big(\sup_{T\leq t\leq 2T} \log|\zeta(\tfrac{1}{2}+it)|\Big)\Big(\sup_{T\leq t\leq 2T} \pm S(t)\Big) \gg \frac{\log T}{\log\log T},$$

which shows that either $|\zeta(\frac{1}{2}+it)|$ or $|S(t)|$ must assume large values in $[T, 2T]$. It may be pointed out that the calculations relating to the values of $S(T)$ (see, e.g., [45], [46]) show that all known values of $S(T)$ are relatively small. In other words they are not anywhere near the values predicted by the above Ω–results, which supports the view that the values of s for which $\zeta(s)$ will exhibit its true asymptotic behaviour must be very large.

If (5.2) is true on the RH, then clearly (1.12) can be improved to

$$\gamma_{n+1} - \gamma_n \ll_\varepsilon (\log \gamma_n)^{\varepsilon-1/2}. \tag{5.3}$$

This means that, as $n \to \infty$, the gap between the consecutive zeros of $Z(t)$ tends to zero not so slowly. Now take $H = T$ in (5.1), and let t_0 be the point in $[T, 2T]$ where the maximum in (5.1) is attained. This point falls into an interval of length $\ll (\log T)^{\varepsilon-1/2}$ between two consecutive zeros, so that in the vicinity of t_0 the function $Z(t)$ must have very large oscillations, which will be carried over to $Z'(t), Z''(t), \ldots$ etc. For example, for $T = 10^{5000}$ we shall have

$$|Z(t_0)| > 2.68 \times 10^{11}, \tag{5.4}$$

while $(\log T)^{-1/2} = 0.00932\ldots$, which shows how large the oscillations of $Z(t)$ near t_0 will be. It seems that $\zeta(s)$ is the only zeta-function for which we know sharp results on the critical line of the type discussed above.

6. A class of convolution functions It does not appear easy to put the discussion of Section 5 into a quantitative form, and to see what will be the consequence of large oscillations of $Z(t)$ for the distribution of its zeros. The basic idea, used in [20] and [22], is to connect the order of $Z(t)$ with the distribution of its zeros and the order of its derivatives (see (7.4)). A.A. Lavrik [32] proved the useful result that, uniformly for $0 \le k \le \frac{1}{2} \log t$, one has

$$Z^{(k)}(t) = 2 \sum_{n \le (t/2\pi)^{1/2}} n^{-1/2} \Big(\log \frac{(t/2\pi)^{1/2}}{n}\Big)^k$$

$$\times \cos\Big(t \log \frac{(t/2\pi)^{1/2}}{n} - \frac{t}{2} - \frac{\pi}{8} + \frac{\pi k}{2}\Big) + O\Big(t^{-1/4}(\tfrac{3}{2} \log t)^{k+1}\Big). \tag{6.1}$$

The range for which (6.1) holds is large, but it is difficult to obtain good uniform bounds for $Z^{(k)}(t)$ from (6.1). To overcome this obstacle the author introduced in [20] the class of convolution functions

$$M_{Z,f}(t) := \int_{-\infty}^{\infty} Z(t+x)f(\frac{x}{G}) \, dx, \tag{6.2}$$

where $G > 0$, and $f(x)$ (≥ 0) is an even function belonging to the class of C^∞-functions $f(x)$ called S_α^β by Gel'fand and Shilov [9]. The functions $f(x)$ satisfy for any real x the inequalities

$$|x^k f^{(q)}(x)| \leq CA^k B^q k^{k\alpha} q^{q\beta} \qquad (k, q = 0, 1, 2, \ldots) \qquad (6.3)$$

with suitable constants $A, B, C > 0$ depending on f alone. For $\alpha = 0$ it follows that $f(x)$ is of bounded support, namely it vanishes for $|x| \geq A$. For $\alpha > 0$ the condition (6.3) is equivalent (see [9]) to the condition

$$|f^{(q)}(x)| \leq CB^q q^{q\beta} \exp(-a|x|^{1/\alpha}) \qquad (a = \alpha/(eA^{1/\alpha})) \qquad (6.4)$$

for all x and $q \geq 0$. Let us denote by E_α^β the subclass of S_α^β with $\alpha > 0$ consisting of even functions $f(x)$ such that $f(x)$ is not the zero-function. It is shown in [9] that S_α^β is non-empty if $\beta \geq 0$ and $\alpha + \beta \geq 1$. If these conditions hold then E_α^β is also non-empty, since $f(-x) \in S_\alpha^\beta$ if $f(x) \in S_\alpha^\beta$, and $f(x) + f(-x)$ is always even.

One of the main properties of the convolution function $M_{Z,f}(t)$, which follows by k-fold integration by parts from (6.2), is that for any integer $k \geq 0$

$$M_{Z,f}^{(k)}(t) = M_{Z^{(k)},f}(t) = (-1/G)^k \int_{-\infty}^{\infty} Z(t+x) f^{(k)}\left(\frac{x}{G}\right) dx. \qquad (6.5)$$

This relation shows that the order of $M^{(k)}$ depends only on the orders of Z and $f^{(k)}$, and the latter is by (6.4) of exponential decay, which is very useful in dealing with convergence problems etc. The salient point of our approach is that the difficulties inherent in the distribution of zeros of $Z(t)$ are transposed to the distribution of zeros of $M_{Z,f}(t)$, and for the latter function (6.5) provides good uniform control of its derivatives.

Several analogies between $Z(t)$ and $M_{Z,f}(t)$ are established in [20], especially in connection with mean values and the distribution of their respective zeros. We shall retain here the notation introduced in [20], so that $N_M(T)$ denotes the number of zeros of $M_{Z,f}(t)$ in $(0, T]$, with multiplicities counted. If $f(x) \in E_\alpha^\beta$, $f(x) \geq 0$, and $G = \delta/\log(T/(2\pi))$ with suitable $\delta > 0$, then Theorem 4 of [20] says that

$$N_M(T+V) - N_M(T-V) \gg \frac{V}{\log T}, \qquad V = T^{c+\varepsilon}, \quad c = 0.329021\ldots, \qquad (6.6)$$

for any given $\varepsilon > 0$. The nonnegativity of $f(x)$ was needed in the proof of this result. For the function $Z(t)$ the analogous result is that

$$N_0(T+V) - N_0(T-V) \gg V \log T, \qquad V = T^{c+\varepsilon}, \quad c = 0.329021\ldots, \qquad (6.7)$$

where as usual $N_0(T)$ denotes the number of zeros of $Z(t)$ (or of $\zeta(\frac{1}{2} + it)$) in $(0, T]$, with multiplicities counted. Thus the fundamental problem in the theory of $\zeta(s)$ is to estimate $N(T) - N_0(T)$, and the RH may be reformulated as $N(T) = N_0(T)$ for $T > 0$. The bound (6.7) was proved by A.A. Karatsuba (see [31] for a detailed account). As explained in [20], the bound (6.6) probably falls short (by a factor of $\log^2 T$) from the expected (true) order of magnitude for the number of zeros of $M_{Z,f}(t)$ in $[T - V, T + V]$. This is due to the method of proof of (6.6), which is not as strong as the classical method of A. Selberg [53] (see also Chapter 10 of [58]).

In what follows we shall need the following technical result, which is proved similarly to Lemma 3 in [20]. We state it here as

Lemma 1 *If $L = (\log T)^{\frac{1}{2}+\varepsilon}$, $P = \sqrt{T/(2\pi)}$, $0 < G < 1, L \ll V \leq T^{\frac{1}{3}}$, and $f(x) \in E_\alpha^\beta$, then*

$$\int_{T-VL}^{T+VL} |M_{Z,f}(t)|e^{-(T-t)^2 V^{-2}}\, dt$$

$$\geq GV \left\{ |\widehat{f}(\frac{G}{2\pi} \log P)| + O(T^{-1/4} + V^2 T^{-3/4} L^2) \right\}. \tag{6.8}$$

Here and in the sequel $\widehat{f}(x)$ is the Fourier transform of $f(x)$, namely

$$\widehat{f}(x) = \int_{-\infty}^{\infty} f(u)e^{2\pi i x u}\, du.$$

7. Technical preparation In this section we shall lay the groundwork for the investigation of the distribution of zeros of $Z(t)$ via the convolution functions $M_{Z,f}(t)$. To do this we shall first briefly outline a method based on a generalized form of the mean value theorem from the differential calculus. This can be conveniently obtained from the expression for the n-th divided difference

$$[x, x_1, x_2, \cdots, x_n] = \frac{F(x)}{(x - x_1)(x - x_2) \cdots (x - x_n)}$$

$$+ \frac{F(x_1)}{(x_1 - x)(x_1 - x_2) \cdots (x_1 - x_n)} + \cdots + \frac{F(x_n)}{(x_n - x)(x_n - x_1) \cdots (x_n - x_{n-1})}$$

where $x_i \neq x_j$ if $i \neq j$, and $F(t)$ is a real-valued function of the real variable t. We have the representation, with $x_{n+1} = x$,

$$[x, x_1, x_2, \cdots, x_n]$$

$$= \int_0^1 \int_0^{t_1} \cdots \int_0^{t_{n-1}} F^{(n)}\Big(x_1 + \sum_{\nu=1}^{n}(x_{\nu+1} - x_\nu)t_\nu\Big)\, dt_n \cdots dt_1$$

$$= \frac{F^{(n)}(\xi)}{n!} \tag{7.1}$$

if $F(t) \in C^n[I]$, $\xi = \xi(x, x_1, \cdots, x_n)$, and I is the smallest interval containing all the points $x, x_1, \cdots, x_n$. If we suppose additionally that $F(x_j) = 0$ for $j = 1, \cdots, n$, then on comparing the two expressions for $[x, x_1, x_2, \cdots, x_n]$ it follows that

$$F(x) = (x - x_1)(x - x_2) \cdots (x - x_n) \frac{F^{(n)}(\xi)}{n!}, \qquad (7.2)$$

where $\xi = \xi(x)$ if we consider $x_1, \cdots, x_n$ as fixed and x as a variable. The underlying idea is that, if the (distinct) zeros x_j of $F(x)$ are sufficiently close to one another, then (7.2) may lead to a contradiction if $F(x)$ is assumed to be large and one has good bounds for its derivatives.

To obtain the analogue of (7.2) when the points x_j are not necessarily distinct, note that if $F(z)$ is a regular function of the complex variable z in a region which contains the distinct points $x, x_1, \cdots, x_n$, then for a suitable closed contour $\mathcal{C}$ containing these points one obtains by the residue theorem

$$[x, x_1, x_2, \cdots, x_n] = \frac{1}{2\pi i} \int_{\mathcal{C}} \frac{F(z)}{(z - x)(z - x_1) \cdots (z - x_n)} \, dz. \qquad (7.3)$$

By comparing (7.1) and (7.3) and using analytic continuation we obtain

$$|F(x)| \leq \prod_{k=1}^{n} |x - x_k| \frac{|F^{(n)}(\xi)|}{n!} \qquad (\xi = \xi(x)). \qquad (7.4)$$

Now we shall apply (7.4) to $F(t) = M_{Z,f}(t)$, $f(x) \in E_\alpha^\beta$, with n replaced by k, to obtain

$$|M_{Z,f}(t)| \leq \prod_{t-H \leq \gamma \leq t+H} |\gamma - t| \frac{|M_{Z,f}^{(k)}(\tau)|}{k!}, \qquad (7.5)$$

where γ runs over the zeros of $M_{Z,f}(t)$ in $[t - H, t + H]$, $\tau = \tau(t, H) \in [t - H, t + H]$, $|t - T| \leq T^{1/2+\varepsilon}$, and $k = k(t, H)$ is the number of zeros of $M_{Z,f}(t)$ in $[t - H, t + H]$. We shall choose

$$H = \frac{A \log_3 T}{\log_2 T} \qquad (\log_r T = \log(\log_{r-1} T)) \qquad (7.6)$$

for a sufficiently large $A > 0$. One intuitively feels that, with a suitable choice of G and f (see (8.1) and (8.2)), the functions $N(T)$ and $N_M(T)$ will not differ by much. Thus we shall suppose that the analogues of (1.3) and (1.11) hold for $N_M(T)$, namely that

$$N_M(T) = \frac{T}{2\pi} \log\left(\frac{T}{2\pi}\right) - \frac{T}{2\pi} + S_M(T) + O(1) \qquad (7.7)$$

with a continuous function $S_M(T)$ satisfying

$$S_M(T) = O\left(\frac{\log T}{\log\log T}\right), \tag{7.8}$$

although it is hard to imagine what should be the appropriate analogue for $S_M(T)$ of the defining relation $S(T) = \pi^{-1} \arg\zeta(\frac{1}{2} + iT)$ in (1.4). We also suppose that

$$\int_T^{T+U} (S_M(t+H) - S_M(t-H))^{2m}\, dt \ll U(\log(2 + H\log T))^m \tag{7.9}$$

holds for any fixed integer $m \geq 1$, $T^a < U \leq T$, $\frac{1}{2} < a \leq 1$, $0 < H < 1$. Such a result holds unconditionally (even in the form of an asymptotic formula) if $S_M(T)$ is replaced by $S(T)$, as shown in the works of A. Fujii [8] and K.-M. Tsang [59]. Thus it seems plausible that (7.9) will also hold. It has already been mentioned that it is reasonable to expect that $S(T)$ and $S_M(T)$ will be close to one another. The author feels that this "closeness" should hold also in the mean sense, and that instead of (7.9) one could impose a condition which links directly $S_M(T)$ and $S(T)$, such as that for any fixed integer $m \geq 1$ one has

$$\int_T^{T+U} (S_M(t) - S(t))^{2m}\, dt \ll U(\log\log T)^m, \tag{7.10}$$

where $T^a < U \leq T$, $\frac{1}{2} < a \leq 1$. If (7.7) holds, then

$$
\begin{aligned}
k &= N_M(t+H) - N_M(t-H) + O(1) \\
&= \frac{H}{\pi}\log\left(\frac{T}{2\pi}\right) + S_M(t+H) - S_M(t-H) + O(1).
\end{aligned} \tag{7.11}
$$

To bound from above the product in (7.5) we proceed as follows. First we have trivially

$$\prod_{|\gamma - t| \leq 1/\log_2 T} |\gamma - t| \leq 1.$$

The remaining portions of the product with $t - H \leq \gamma < t - 1/\log_2 T$ and $t + 1/\log_2 T < \gamma \leq t + H$ are treated analogously, so we shall consider in detail only the latter. We have

$$
\begin{aligned}
\log\left(\prod_{t+1/\log_2 T < \gamma \leq t+H} |\gamma - t| \right) &= \sum_{t+1/\log_2 T < \gamma \leq t+H} \log(\gamma - t) \\
&= \int_{t+1/\log_2 T+0}^{t+H} \log(u - t)\, dN_M(u).
\end{aligned}
$$

By using integration by parts and (7.8) we shall obtain then

Lemma 2 *Suppose that (7.7) and (7.8) hold. If γ denotes zeros of $M_{Z,f}(t)$, H is given by (7.6) and $|T - t| \leq T^{1/2+\varepsilon}$, then*

$$\prod_{t-H \leq \gamma \leq t+H} |\gamma - t|$$

$$\leq \exp\left\{\frac{1}{\pi} \log(\frac{T}{2\pi}) \cdot (H \log H - H + O(\frac{\log_3 T}{\log_2 T}))\right\}. \tag{7.12}$$

8. The asymptotic formula for the convolution function

In this section we shall prove a sharp asymptotic formula for $M_{Z,f}(t)$, which is given by Theorem 1. This will hold if $f(x)$ belongs to a specific subclass of functions from E_α^0 ($\alpha > 1$ is fixed), and for such $M_{Z,f}(t)$ we may hope that (7.7)–(7.10) will hold. To construct this subclass of functions first of all let $\varphi(x) \geq 0$ (but $\varphi(x) \not\equiv 0$) belong to E_0^α. Such a choice is possible, since it is readily checked that $f^2(x) \in S_\alpha^\beta$ if $f(x) \in S_\alpha^\beta$, and trivially $f^2(x) \geq 0$. Thus $\varphi(x)$ is of bounded support, so that $\varphi(x) = 0$ for $|x| \geq a$ for some $a > 0$. We normalize $\varphi(x)$ so that $\int_{-\infty}^\infty \varphi(x)\, dx = 1$, and for an arbitrary constant $b > \max(1, a)$ we put

$$\Phi(x) := \int_{x-b}^{x+b} \varphi(t)\, dt.$$

Then $0 \leq \Phi(x) \leq 1$, $\Phi(x)$ is even (because $\varphi(x)$ is even) and nonincreasing for $x \geq 0$, and

$$\Phi(x) = \begin{cases} 0 & \text{if } |x| \geq b + a, \\ 1 & \text{if } |x| \leq b - a. \end{cases}$$

One can also check that $\varphi(x) \in S_0^\alpha$ implies that $\Phi(x) \in S_0^\alpha$. Namely $|x^k \Phi(x)| \leq (b+a)^k$, and for $q \geq 1$ one uses (6.3) (with $k = 0$, $f^{(q)}$ replaced by $\varphi^{(q-1)}$ and $(\alpha, \beta) = (0, \alpha)$) to obtain

$$|x^k \Phi^{(q)}(x)| \leq (b+a)^k |\Phi^{(q)}(x)|$$

$$\leq (b+a)^k \left(|\varphi^{(q-1)}(x+b)| + |\varphi^{(q-1)}(x-b)|\right)$$

$$\leq (b+a)^k 2CB^{q-1}(q-1)^{(q-1)\alpha} \leq \frac{2C}{B}(b+a)^k B^q q^{q\alpha}.$$

Hence (6.3) will hold for Φ in place of f, with $A = b + a$ and a suitable C. Let

$$f(x) := \int_{-\infty}^\infty \Phi(u) e^{-2\pi i x u}\, du = \int_{-\infty}^\infty \Phi(u) \cos(2\pi x u)\, du.$$

A fundamental property of the class S_α^β (see [9]) is that $\widehat{S_\alpha^\beta} = S_\beta^\alpha$, where in general $\widehat{U} = \{\widehat{f}(x) : f(x) \in U\}$, and $\widehat{f}(x)$ is the Fourier transform of $f(x)$. Thus $f(x) \in S_\alpha^0$, $f(x)$ is even (because $\Phi(x)$ is even), and by the inverse Fourier transform we have $\widehat{f}(x) = \Phi(x)$. The function $f(x)$ is not necessarily nonnegative, but this property is not needed in the sequel.

Henceforth let

$$G = \frac{\delta}{\log(T/(2\pi))} \qquad (\delta > 0). \tag{8.1}$$

In view of (1.3) it is seen that, on the RH, G is of the order of the average spacing between the zeros of $Z(t)$. If $f(x)$ is as above, then we have

Theorem 1 *For $|t - T| \le VL$, $L = \log^{\frac{1}{2}+\varepsilon} T$, $\log^\varepsilon T \le V \le T^{\frac{1}{4}}/\log T$, $0 < \delta < 2\pi(b-a)$ and any fixed $N \ge 1$ we have*

$$M_{Z,f}(t) = G\big(Z(t) + O(T^{-N})\big). \tag{8.2}$$

Proof. Before we give the proof of (8.2) it may be remarked that the hypotheses on t in the formulation of the theorem may be relaxed.

In order to prove (8.2) it will be convenient to work with the real-valued function $\theta(t)$, defined by

$$Z(t) = e^{i\theta(t)}\zeta(\tfrac{1}{2}+it) = \chi^{-1/2}(\tfrac{1}{2}+it)\zeta(\tfrac{1}{2}+it), \tag{8.3}$$

and one has from the functional equation (1.2)

$$\theta(t) = \operatorname{Im}\log\Gamma(\tfrac{1}{4}+\tfrac{1}{2}it) - \frac{t}{2}\log\pi. \tag{8.4}$$

We have the explicit representation (see Chapter 3 of [31])

$$\theta(t) = \frac{t}{2}\log\frac{t}{2\pi} - \frac{t}{2} - \frac{\pi}{8} + \Delta(t) \tag{8.5}$$

with $(\psi(x) = x - [x] - 1/2)$

$$\Delta(t) := \frac{t}{4}\log(1+\frac{1}{4t^2}) + \frac{1}{4}\arctan\frac{1}{2t} + \frac{t}{2}\int_0^\infty \frac{\psi(u)\,du}{(u+\frac{1}{4})^2+t^2}. \tag{8.6}$$

This formula is very useful, since it allows one to evaluate explicitly all the derivatives of $\theta(t)$. For $t \to \infty$ it is seen that $\Delta(t)$ admits an asymptotic expansion in terms of negative powers of t, and from (8.4) and Stirling's formula for the gamma-function it is found that (B_k is the k-th Bernoulli number)

$$\Delta(t) \sim \sum_{n=1}^\infty \frac{(2^{2n}-1)|B_{2n}|}{2^{2n}(2n-1)2nt^{2n-1}}. \tag{8.7}$$

The meaning of (8.7) is that, for an arbitrary integer $N \geq 1$, $\Delta(t)$ equals the sum of the first N terms of the series in (8.7), plus the error term which is $O_N(t^{-2N-1})$. In general we shall have, for $k \geq 0$ and suitable constants $c_{k,n}$,

$$\Delta^{(k)}(t) \sim \sum_{n=1}^{\infty} c_{k,n} t^{1-2n-k}. \tag{8.8}$$

For complex s not equal to the poles of the gamma-factors we have the Riemann–Siegel formula (this is equation (56) of C.L. Siegel [56])

$$\pi^{-s/2}\Gamma(\frac{s}{2})\zeta(s) = \pi^{-s/2}\Gamma(\frac{s}{2}) \int_{0\nearrow 1} \frac{e^{i\pi x^2} x^{-s}}{e^{i\pi x} - e^{-i\pi x}} \, dx$$
$$+ \pi^{(s-1)/2}\Gamma(\frac{1-s}{2}) \int_{0\searrow 1} \frac{e^{-i\pi x^2} x^{s-1}}{e^{i\pi x} - e^{-i\pi x}} \, dx. \tag{8.9}$$

Here $0 \nearrow 1$ (resp. $0 \searrow 1$) denotes a straight line which starts from infinity in the upper complex half-plane, has slope equal to 1 (resp. to -1), and cuts the real axis between 0 and 1. Setting in (8.9) $s = \frac{1}{2} + it$ and using (8.4), we have that

$$Z(t) = \text{Im}\left(e^{-i\theta(t)} \int_{0\searrow 1} \Xi(z,t) dz\right), \tag{8.10}$$

where

$$\Xi(z,t) = \frac{e^{-i\pi z^2} z^{-1/2+it}}{\sin(\pi z)}.$$

The contribution of the portion of the integral in (8.10) for which $|z| \geq \log t$ is $\ll \exp(-\log^2 t)$, hence we obtain

$$Z(t) = \text{Im}\left(e^{-i\theta(t)} \int_{0\searrow 1, |z| < \log t} \Xi(z,t) dz\right) + O(\exp(-\log^2 t)). \tag{8.11}$$

From the decay property (6.4) it follows that

$$M_{Z,f}(t) = \int_{-\log^{2\alpha-1} t}^{\log^{2\alpha-1} t} Z(t+x) f\left(\frac{x}{G}\right) dx + O(\exp(-c\log^2 t)), \tag{8.12}$$

where c denotes positive, absolute constants which may not be the same ones at each occurrence. Thus from (8.10) and (8.12) we obtain that

$$M_{Z,f}(t) = \text{Im}\left(\int_{0\searrow 1, |z| < \log t} \Xi(z,t) \int_{-\log^{2\alpha-1} t}^{\log^{2\alpha-1} t} e^{-i\theta(t+x)} z^{ix} f\left(\frac{x}{G}\right) dx \, dz\right)$$
$$+ O(\exp(-c\log^2 t)). \tag{8.13}$$

By using Taylor's formula we have

$$\theta(t+x) = \theta(t) + \frac{x}{2}\log\frac{t}{2\pi} + x\Delta'(t) + R(t,x) \qquad (8.14)$$

with $\Delta'(t) \ll t^{-2}$ and

$$R(t,x) = \sum_{n=2}^{\infty}\left(\frac{(-1)^n}{2n(n-1)t^{n-1}} + \frac{\Delta^{(n)}(t)}{n!}\right)x^n.$$

Now we put

$$e^{-iR(t,x)} = 1 + S(t,x), \qquad (8.15)$$

say, and use (8.5), (8.6), (8.8), and (8.14). We obtain

$$S(t,x) = \sum_{k=1}^{\infty}\frac{(-i)^k R^k(t,x)}{k!} = \sum_{n=2}^{\infty}g_n(t)x^n, \qquad (8.16)$$

where each $g_n(t) \in C^{\infty}(0,\infty)$ has an asymptotic expansion of the form

$$g_n(t) \sim \sum_{k=0}^{\infty}d_{n,k}t^{-k-[(n+1)/2]} \quad (t\to\infty) \qquad (8.17)$$

with suitable constants $d_{n,k}$. From (8.13)–(8.15) we have

$$M_{Z,f}(t) = \operatorname{Im}\left(I_1 + I_2\right) + O(e^{-c\log^2 t}), \qquad (8.18)$$

where

$$I_1 := \int_{0\searrow 1,|z|<\log t} e^{-i\theta(t)}\Xi(z,t)$$
$$\times \int_{-\log^{2\alpha-1}t}^{\log^{2\alpha-1}t}\exp\left(-x\arg z + 2i\pi x\frac{\Lambda}{G}\right)f\left(\frac{x}{G}\right)dx\,dz \qquad (8.19)$$

with

$$\Lambda := \frac{G}{2\pi}\log\frac{|z|e^{-\Delta'(t)}}{\sqrt{t/2\pi}},$$

and I_2 is the same as I_1, only it has the extra factor $S(t,x)$ in the inner integral. We have

$$I_1 = \int_{0\searrow 1,|z|<\log t} e^{-i\theta(t)}\Xi(z,t)h(z)\,dz + O(e^{-c\log^2 t}),$$

where

$$h(z) = G \int_{-\infty}^{\infty} e^{-Gy \arg z} f(y) \exp\left(2i\pi y \Lambda\right) dy$$

$$= G \sum_{n=0}^{\infty} \frac{(-G \arg z)^n}{n!} \int_{-\infty}^{\infty} y^n f(y) \exp\left(2i\pi y \Lambda\right) dy,$$

where change of summation and integration is justified by absolute convergence. But

$$\int_{-\infty}^{\infty} f(y) \exp\left(2i\pi y \Lambda\right) dy = \widehat{f}(\Lambda) = 1$$

for $\delta < 2\pi(b - a)$, since

$$|\Lambda| = \frac{\delta}{2\pi \log(T/(2\pi))} \left(\log\sqrt{\frac{t}{2\pi}} - \log|z| + \Delta'(t)\right)$$

$$= \left(\frac{\delta}{4\pi} + o(1)\right) < \frac{\delta}{2\pi} < b - a,$$

and $\widehat{f}(x) = 1$ for $|x| < b - a$. Moreover, for $n \geq 1$ and $|x| < b - a$ we have

$$\widehat{f}^{(n)}(x) = (2\pi i)^n \int_{-\infty}^{\infty} y^n e^{2\pi i x y} f(y) dy = 0.$$

Hence we obtain

$$I_1 = G \int_{0 \searrow 1} e^{-i\theta(t)} \Xi(z, t) dz + O(e^{-c \log^2 t}). \tag{8.20}$$

Similarly from (8.16) we have

$$I_2 = \sum_{n=1}^{N} e^{-i\theta(t)} t^{-n} \int_{0 \searrow 1, |z| < \log t} \Xi(z, t)$$

$$\times \int_{-\log^{2\alpha-1} t}^{\log^{2\alpha-1} t} P_n(x) \exp\left(-x \arg z + 2i\pi x \frac{\Lambda}{G}\right) f\left(\frac{x}{G}\right) dx dz$$

$$+ O\left(\frac{1}{t^{N+1}} \int_{-\log^{2\alpha-1} t}^{\log^{2\alpha-1} t} (1 + x^{N+2}) |f\left(\frac{x}{G}\right)| \left| \int_{0 \searrow 1} \Xi(z, t) z^{ix} dz \right| dx\right)$$

$$+ O(e^{-c \log^2 t}),$$

where each $P_n(x)$ is a polynomial in x of degree $n \geq 2$. The integral over z in the error term is similar to the one in (8.10). Hence by the residue theorem we have, for $Q = [\sqrt{t/2\pi}]$,

$$\int_{0 \searrow 1} \Xi(z, t) z^{ix} dz = 2\pi i \sum_{n=1}^{Q} \operatorname*{Res}_{z=n} + \int_{Q \searrow Q+1} \cdots dz,$$

similarly as in the derivation of the Riemann-Siegel formula. It follows that the left-hand side is $\ll t^{1/4}$. Thus analogously as in the case of I_1 we find that, for $n \geq 1$,

$$\int_{-\log^{2\alpha-1} t}^{\log^{2\alpha-1} t} P_n(x) \exp\left(-x \arg z + 2i\pi x \frac{\Lambda}{G}\right) f\left(\frac{x}{G}\right) dx$$

$$= \int_{-\infty}^{\infty} P_n(x) \cdots dx + O(e^{-c\log^2 t}) = O(e^{-c\log^2 t}).$$

Hence it follows that, for any fixed integer $N \geq 1$,

$$I_2 \ll_N T^{-N}. \tag{8.21}$$

Theorem 1 now follows from (8.10) and (8.18)–(8.21), since clearly it suffices to assume that N is an integer. One can generalize Theorem 1 to derivatives of $M_{Z,f}(t)$.

Theorem 1 shows that $Z(t)$ and $M_{Z,f}(t)/G$ differ only by $O(T^{-N})$, for any fixed $N \geq 1$, which is a very small quantity. This certainly supports the belief that, for this particular subclass of functions $f(x)$, the assertions (7.7)–(7.10) will be true, but *proving* it may be very hard. On the other hand, nothing precludes the possibility that the error term in Theorem 1, although it is quite small, represents a function possessing many small "spikes" (like $t^{-N} \sin(t^{N+2})$, say). These spikes could introduce many new zeros, thus violating (7.7)–(7.10). Therefore it remains an open question to investigate the distribution of zeros of $M_{Z,f}(t)$ of Theorem 1.

9. Convolution functions and the RH In this section we shall discuss the possibility to use convolution functions to disprove the RH, of course in the case should it be false. Let us denote by T_α^β the subclass of S_α^β with $\alpha > 1$ consisting of functions $f(x)$, which are not identically zero, and for which $\int_{-\infty}^{\infty} f(x)\,dx > 0$. It is clear that T_α^β is non-empty. Our choice for G will be the same one as in (8.1), so that for suitable δ we shall have

$$\widehat{f}\left(\frac{G}{4\pi}\log\left(\frac{T}{2\pi}\right)\right) = \widehat{f}\left(\frac{\delta}{4\pi}\right) \gg 1. \tag{9.1}$$

In fact by continuity (9.1) will hold for $|\delta| \leq C_1$, where $C_1 > 0$ is a suitable constant depending only on f, since if $f(x) \in T_\alpha^\beta$, then we have $\widehat{f}(0) = \int_{-\infty}^{\infty} f(x)\,dx > 0$. Moreover, if $f(x) \in S_\alpha^0$, then $\widehat{f}(x) \in S_0^\alpha$ and thus it is of bounded support, and consequently $G \ll 1/\log T$ must hold if the bound in (9.1) is to be satisfied. This choice of $f(x)$ turns out to be better suited for our purposes than the choice made in Section 8, which perhaps would seem more natural in view of Theorem 1.

Now observe that if we replace $f(x)$ by $f_1(x) := f(Dx)$ for a given $D > 0$, then obviously $f_1(x) \in S_\alpha^0$, and moreover uniformly for $q \geq 0$ we have

$$f_1^{(q)}(x) = D^q f^{(q)}(Dx) \ll (BD)^q \exp(-aD^{1/\alpha}|x|^{1/\alpha}). \qquad (9.2)$$

In other words the constant B in (6.3) or (6.4) is replaced by BD. Take now $D = \eta/B$, where $\eta > 0$ is an arbitrary, but fixed number, and write f for Df_1. If the RH holds, then from (4.5), (6.4), (6.5), and (9.2) we have, for k given by (7.11),

$$M_{Z,f}^{(k)}(t) \ll \left(\frac{\eta}{G}\right)^k \exp\left(\frac{B_1 \log t}{\log \log t}\right) \qquad (9.3)$$

with a suitable constant $B_1 > 0$.

We shall assume now that the RH holds and that (7.7), (7.10) hold for some $f(x) \in T_\alpha^0$ (for which (9.3) holds, which is implied by the RH), and we shall obtain a contradiction. This is similar to the method of [22], so we shall be brief. Take $U := T^{1/2+\varepsilon}$, so that we may apply (7.9) or (7.10), and let $V = T^{1/4}/\log T$, $L = \log^{1/2+\varepsilon} T$. We shall consider the mean value of $|M_{Z,f}(t)|$ over $[T - U, T + U]$ in order to show that, on the average, $|M_{Z,f}(t)|$ is not too small. We have first

$$\int_{T-U}^{T+U} |M_{Z,f}(t)|\, dt \gg GUL^{-1}. \qquad (9.4)$$

We have assumed that (7.10) holds, but this implies that (7.9) holds also. Namely it holds unconditionally with $S(t)$ in place of $S_M(t)$. Thus for any fixed integer $m \geq 1$ we have

$$\int_T^{T+U} \left(S_M(t+H) - S_M(t-H)\right)^{2m} dt$$

$$\ll \int_{T+H}^{T+H+U} \left(S_M(t) - S(t)\right)^{2m} dt + \int_T^{T+U} \left(S(t+H) - S(t-H)\right)^{2m} dt$$

$$+ \int_{T-H}^{T-H+U} \left(S(t) - S_M(t)\right)^{2m} dt \ll U(\log \log T)^m,$$

where $T^a < U \leq T$, $\frac{1}{2} < a \leq 1$. Let $\mathcal{D}$ be the subset of $[T - U, T + U]$ where

$$|S_M(t+H) - S_M(t-H)| \leq \log^{1/2} T \qquad (9.5)$$

fails. The bound (7.9) implies that

$$m(\mathcal{D}) \ll U \log^{-C} T \qquad (9.6)$$

for any fixed $C > 0$. If we take $C = 10$ in (9.6) and use the Cauchy-Schwarz inequality for integrals we shall have

$$\int_{\mathcal{D}} |M_{Z,f}(t)|\, dt \le (m(\mathcal{D}))^{1/2} \Big(\int_{T-U}^{T+U} M_{Z,f}^2(t)\, dt \Big)^{1/2} \ll GU \log^{-4} T. \qquad (9.7)$$

Therefore (9.4) and (9.7) yield

$$GUL^{-1} \ll \int_{\mathcal{D}'} |M_{Z,f}(t)|\, dt, \qquad (9.8)$$

where $\mathcal{D}' = [T - U, T + U] \setminus \mathcal{D}$, hence in (9.8) integration is over t for which (9.5) holds. If t is in $\mathcal{D}'$ and γ denotes the zeros of $M_{Z,f}(t)$, then from (7.6) and (7.11) we obtain

$$\log k = \log H - \log \pi + \log_2(\frac{T}{2\pi}) + O((\log T)^{\varepsilon - 1/2}) \qquad (9.9)$$

for any given $\varepsilon > 0$, where $\log_r t = \log(\log_{r-1} t)$. To bound $M_{Z,f}(t)$ we use (7.5), with k given by (9.9), $\tau = \tau(t, k)$, (9.3) and

$$k! = \exp(k \log k - k + O(\log k)).$$

We obtain, denoting by B_j positive absolute constants,

$$GUL^{-1} \ll \int_{\mathcal{D}'} \prod_{|\gamma - t| \le H} |\gamma - t| \frac{|M_{Z,f}^{(k)}(\tau)|}{k!}\, dt \ll \exp\Big(\frac{B_3 \log T}{\log_2 T} \Big)$$

$$\times \exp\Big(\frac{H}{\pi} \log(\frac{T}{2\pi}) \big(\log \frac{\eta}{\delta} + \log_2(\frac{T}{2\pi}) - \log H + \log \pi - \log_2(\frac{T}{2\pi}) + 1 \big) \Big)$$

$$\times \int_{\mathcal{D}'} \prod_{|\gamma - t| \le H} |\gamma - t|\, dt. \qquad (9.10)$$

It was in evaluating $k \log k$ that we needed (9.9), since the bound (7.8) would not suffice. If the product under the last integral is bounded by (7.12), we obtain from (9.10)

$$GUL^{-1} \ll U \exp\Big(\frac{H}{\pi} \log(\frac{T}{2\pi}) \cdot (\log \frac{\eta}{\delta} + B_4) \Big),$$

and thus for $T \ge T_0$

$$1 \le \exp\Big(\frac{H}{\pi} \log(\frac{T}{2\pi}) \cdot (\log \frac{\eta}{\delta} + B_5) \Big). \qquad (9.11)$$

Now we choose e.g. $\eta = \delta^2$, $\delta = \min(C_1, e^{-2B_5})$, where C_1 is the constant for which (9.1) holds if $|\delta| \leq C_1$, so that (9.11) gives

$$1 \leq \exp\left(\frac{-B_5 H}{\pi} \log(\frac{T}{2\pi})\right),$$

which is a contradiction for $T \geq T_1$. Thus we have proved the following

Theorem 2 *If (7.7) and (7.10) hold for suitable $f(x) \in T_\alpha^0$ with G given by (8.1), then the Riemann hypothesis is false.*

Theorem 2 is similar to the result proved also in [22]. Perhaps it should be mentioned that (7.10) is not the only condition which would lead to the disproof of the RH. It would be enough to assume, under the RH, that one had (7.7)–(7.9) for a suitable $f(x)$, or

$$N_M(t) = N(t) + O\left(\frac{\log T}{(\log \log T)^2}\right) \tag{9.18}$$

for $t \in [T - U, T + U]$ with a suitable $U(= T^{1/2+\varepsilon}$, but smaller values are possible), to derive a contradiction. The main drawback of this approach is the necessity to impose conditions like (7.7)–(7.9) which can be, for all we know, equally difficult to settle as the assertions which we originally set out to prove (or disprove). For this reason our results can only be conditional.

References

[1] R.J. Anderson and H.M. Stark. Oscillation theorems. *Lect. Notes in Math.*, **899**, Springer-Verlag, Berlin, 1981, pp. 79–106.

[2] F.V. Atkinson. The mean value of the Riemann zeta-function. *Acta Math.*, **81** (1949), 353–376.

[3] R. Balasubramanian. On the frequency of Titchmarsh's phenomenon for $\zeta(s)$. IV. *Hardy–Ramanujan J.*, **9** (1986), 1–10.

[4] R. Balasubramanian and K. Ramachandra. On the frequency of Titch-marsh's phenomenon for $\zeta(s)$. III. *Proc. Indian Acad. Sci. Section A*, **86** (1977), 341–351.

[5] E. Bombieri and D. Hejhal. Sur les zéros des fonctions zeta d'Epstein. *Comptes Rendus Acad. Sci. Paris*, **304** (1987), 213–217.

[6] H. Davenport and H. Heilbronn. On the zeros of certain Dirichlet series. I, II. *J. London Math. Soc.*, **11** (1936), 181–185, 307–312.

[7] H.M. Edwards. *Riemann's Zeta-Function*. Academic Press, New York–London, 1974.

[8] A. Fujii. On the distribution of the zeros of the Riemann zeta-function in short intervals. *Bull. Amer. Math. Soc.*, **8** (1975), 139–142.

[9] I.M. Gel'fand and G.E. Shilov. *Generalized Functions.* vol. 2. Academic Press, New York–London, 1968.

[10] J.L. Hafner and A. Ivić. On the mean square of the Riemann zeta-function on the critical line. *J. Number Th.*, **32** (1989), 151–191.

[11] —. On some mean value results for the Riemann zeta-function. *Proc. International Number Theory Conf. Québec 1987*, Walter de Gruyter and Co., Berlin–New York, 348–358.

[12] M.N. Huxley. Exponential sums and the Riemann zeta-function. IV. *Proc. London Math. Soc.*, (3) **66** (1993), 1–40.

[13] —. A note on exponential sums with a difference. *Bull. London Math. Soc.*, **29** (1994), 325–327.

[14] A. Ivić. Large values of the error term in the divisor problem. *Invent. math.*, **71** (1983), 513–520.

[15] —. *The Riemann Zeta-Function.* John Wiley and Sons, New York, 1985.

[16] —. On consecutive zeros of the Riemann zeta-function on the critical line. *Séminaire de Théorie des Nombres*, Université de Bordeaux 1986/87, Exposé no. **29**, 14 pp.

[17] —. On a problem connected with zeros of $\zeta(s)$ on the critical line. *Monatshefte Math.*, **104** (1987), 17–27.

[18] —. Large values of certain number-theoretic error terms. *Acta Arith.*, **56** (1990), 135–159.

[19] —. *Mean values of the Riemann Zeta-Function.* Lect. Math. Phy., **82**, Tata Inst. Fund. Res.–Springer-Verlag, Bombay, 1991.

[20] —. On a class of convolution functions connected with $\zeta(s)$. *Bull. CIX Acad. Serbe des Sciences et des Arts, Sci. Math.*, **20** (1995), 29–50.

[21] —. On the fourth moment of the Riemann zeta-function. *Publs. Inst. Math.* (Belgrade), **57(71)** (1995), 101–110.

[22] —. On the distribution of zeros of a class of convolution functions. *Bull. CXI Acad. Serbe des Sciences et des Arts, Sci. Math.*, **21** (1996), 61–71.

[23] —. On the ternary additive divisor problem and the sixth moment of the zeta-function. *Sieve Methods, Exponential Sums, and their Applications in Number Theory* (G.R.H. Greaves, G. Harman and M.N. Huxley, Eds.), Cambridge Univ. Press, London, 1997, pp. 205–243.

[24] —. On the Mellin transform and the Riemann zeta-function. *Proc. Conf. Elementary and Analytic Number Theory, Vienna, July 1996* (W.G. Nowak and J. Schoißengeier, Eds.), Univ. Wien and Univ. für Bodenkultur, Vienna, 1997, pp. 112–127.

[25] A. Ivić and M. Jutila. Gaps between consecutive zeros of the Riemann zeta-function. *Monatshefte Math.*, **105** (1988), 59–73.

[26] A. Ivić and Y. Motohashi. A note on the mean value of the zeta and *L*-functions. VII. *Proc. Japan Acad. Ser. A*, **66** (1990), 150–152.

[27] —. The mean square of the error term for the fourth moment of the zeta-function. *Proc. London Math. Soc.*, (3)**66** (1994), 309–329.

[28] —. The fourth moment of the Riemann zeta-function. *J. Number Theory*, **51** (1995), 16–45.

[29] A. Ivić and H.J.J. te Riele. On the zeros of the error term for the mean square of $|\zeta(\frac{1}{2} + it)|$. *Math. Comp.*, **56** No **193** (1991), 303–328.

[30] A.A. Karatsuba. On the zeros of the Davenport–Heilbronn function lying on the critical line. *Izv. Akad. Nauk SSSR ser. mat.*, **54** (1990), 303–315. (Russian)

[31] A.A. Karatsuba and S.M. Voronin. *The Riemann Zeta-Function*. Walter de Gruyter, Berlin–New York, 1992.

[32] A.A. Lavrik. Uniform approximations and zeros of derivatives of Hardy's Z-function in short intervals. *Analysis Mathem.*, **17** (1991), 257–259. (Russian)

[33] D.H. Lehmer. On the roots of the Riemann zeta function. *Acta Math.*, **95** (1956), 291–298.

[34] —. Extended computation of the Riemann zeta-function. *Mathematika*, **3** (1956), 102–108.

[35] J.E. Littlewood. Sur la distribution des nombres premiers. *Comptes rendus Acad. Sci. (Paris)*, **158** (1914), 1869–1872.

[36] J. van de Lune, H.J.J. te Riele and D.T. Winter. On the zeros of the Riemann zeta-function in the critical strip. IV. *Math. Comp.*, **46** (1987), 273–308.

[37] H.L. Montgomery. Extreme values of the Riemann zeta-function. *Comment. Math. Helv.*, **52** (1977), 511–518.

[38] Y. Motohashi. *Riemann–Siegel Formula*. Ulam Chair Lectures, Colorado Univ., Boulder, 1987.

[39] —. The fourth power mean of the Riemann zeta-function. *Proc. Amalfi Conf. Analytic Number Theory 1989* (E. Bombieri, A. Perelli, S. Salerno and U. Zannier, Eds.), Univ. di Salerno, Salerno, 1992, pp. 325–344.

[40] —. An explicit formula for the fourth power mean of the Riemann zeta-function. *Acta Math.*, **170** (1993), 181–220.

[41] —. A relation between the Riemann zeta-function and the hyperbolic Laplacian. *Ann. Sc. Norm. Sup. Pisa, Cl. Sci. IV ser.*, **22** (1995), 299–313.

[42] —. The Riemann zeta-function and the non-Euclidean Laplacian. *AMS Sugaku Expositions*, **8** (1995), 59–87.

[43] —. *Spectral theory of the Riemann zeta-function*. Cambridge Univ. Press, in press.

[44] A.M. Odlyzko. On the distribution of spacings between the zeros of the zeta-function. *Math. Comp.*, **48** (1987), 273–308.

[45] —. Analytic computations in number theory. *Proc. Symp. Applied Math.*, **48** (1994), 451–463.

[46] —. The 10^{20}-th zero of the Riemann zeta-function and 175 million of its neighbors. To appear.

[47] A.M. Odlyzko and H.J.J. te Riele. Disproof of the Mertens conjecture. *J. reine angew. Math.*, **357** (1985), 138–160.

[48] K. Ramachandra. Progress towards a conjecture on the mean value of Titchmarsh series. *Recent Progress in Analytic Number Theory, Symp. Durham, 1979* (H. Halberstam and C. Hooley, Eds.), vol. 1, Academic Press, London, 1981, pp. 303–318.

[49] —. *On the Mean-Value and Omega-Theorems for the Riemann Zeta-Function.* Lect. Math. Phys., **85**, Tata Inst. Fund. Res.–Springer-Verlag, Bombay, 1995.

[50] H.J.J. te Riele. On the sign of the difference of $\pi(x) - \mathrm{li}\, x$. *Math. Comp.*, **48** (1987), 323–328.

[51] H.J.J. te Riele and J. van de Lune. Computational number theory at CWI in 1970–1994, CWI Quarterly, **7**(4) (1994), 285–335.

[52] B. Riemann. Über die Anzahl der Primzahlen unter einer gegebenen Grösse. *Monats. Preuss. Akad. Wiss.*, (1859–1860), 671–680.

[53] A. Selberg. On the zeros of Riemann's zeta-function. *Skr. Norske Vid. Akad. Oslo*, **10** (1942), 1–59.

[54] —. Contributions to the theory of the Riemann zeta-function. *Arch. Math. Naturvid.*, **48** (1946), 89–155.

[55] R. Sherman Lehman. On the difference $\pi(x) - \mathrm{li}\, x$. *Acta Arith.*, **11** (1966), 397–410.

[56] C.L. Siegel. Über Riemanns Nachlaß zur analytischen Zahlentheorie. *Quell. Stud. Gesch. Mat. Astr. Physik*, **2** (1932), 45–80.

[57] R. Spira. Some zeros of the Titchmarsh counterexample. *Math. Comp.*, **63** (1994), 747–748.

[58] E.C. Titchmarsh. *The Theory of the Riemann Zeta-Function.* Clarendon Press, Oxford, 1951.

[59] K.-M. Tsang. Some Ω-theorems for the Riemann zeta-function. *Acta Arith.*, **46** (1986), 369–395.

[60] —. The large values of the Riemann zeta-function. *Mathematica*, **40** (1993), 203–214.

Aleksandar Ivić
Katedra Matematike RGF-a, Universiteta u Beogradu
Djušina 7, 11000 Beograd, Serbia (Yugoslavia)

12

Mean Values of Dirichlet Series via Laplace Transforms

MATTI JUTILA

1. Introduction Given a continuous function g of at most exponential growth on the real interval $[0, \infty)$ with the Laplace transform

$$L(p) = \int_0^\infty g(t)e^{-pt}\,dt,$$

its integral function may be written as the following Laplace inversion integral:

$$\int_0^T g(t)\,dt = \frac{1}{2\pi i}\int_{(a)} L(p)p^{-1}e^{pT}\,dp. \tag{1.1}$$

Here the notation means that the integral is taken over the line $\mathrm{Re}\,p = a$, and a is a sufficiently large constant. While apparently resembling Perron's formula for discrete mean values, the device (1.1) nevertheless fails to enjoy the same status as a standard tool. But in principle both formulae are of comparable scope, and we wish to illustrate and popularize the Laplace transform approach to mean values by concrete applications to Dirichlet series.

True, Laplace transforms are by no means any novelty in analytic number theory. For instance, the classical Tauberian theory draws conclusions about the original function from the behaviour of its Laplace transform on the positive real axis near the origin. A sample of results of this kind is the following: if $g(t)$ is a non-negative function such that $L(\delta) \sim \delta^{-1}$ for $\delta \to 0+$, then its integral function (1.1) is asymptotically $\sim T$ (see [32], §7.12). However, this argument fails to give any error term; such a sharpening would require information about the Laplace transform *off* the real axis.

The present work is methodically closely related to our recent papers [17], [18]. The former of these was concerned with weighted mean values of $|\zeta(\frac{1}{2} + it)|^4$ with an attempt to find a new approach via Laplace transforms and spectral theory to a very deep theorem of Motohashi [25] from a more general point of view. In the latter paper, we reproved a celebrated formula of Atkinson [3] on the mean square of Riemann's zeta-function, again by use of Laplace transforms, in a new way applicable even to automorphic L-functions. The last mentioned aspect of generality is actually a typical feature of the method. The reason is that the argument relies largely on certain *properties* of functions –

in the first place functional equations – rather than on their diverse individual definitions.

More specifically, we are going to deal with functions of three kind, namely $|\zeta(\frac{1}{2} + it)|^4$ and squares of L-functions attached to holomorphic or non-holomorphic cusp forms for the full modular group (good introductions to the theory of automorphic functions are, e.g., [1], [11], [29]; the last two references cover also their spectral theory).

Given a holomorphic cusp form of weight k, represented by its Fourier series

$$\sum_{n=1}^{\infty} a(n)e(nz),$$

the corresponding L-function is

$$\sum_{n=1}^{\infty} a(n)n^{-s}.$$

The "critical line" for this series is $\sigma = k/2$ because its functional equation relates values at s and $k - s$ lying symmetrically to this line. However, to stress the analogy with the Riemann zeta-function, it is convenient to introduce the "normalized" coefficients $\tilde{a}(n) = a(n)n^{-(k-1)/2}$, for then the Riemannian critical line $\sigma = 1/2$ will play the same role even for the series

$$F(s) = \sum_{n=1}^{\infty} \tilde{a}(n)n^{-s}.$$

Turning to analogous L-series for non-holomorphic cusp forms, let $u(z) = u(x + yi)$ be such a form with the Fourier series

$$u(z) = y^{1/2} \sum_{\substack{n=-\infty \\ n\neq 0}}^{\infty} \rho(n)K_{i\kappa}(2\pi|n|y)e(nx).$$

By definition, $u(z)$ is an eigenfunction of the hyperbolic Laplacian for a certain eigenvalue $\lambda > 1/4$, and $\kappa = \sqrt{\lambda - 1/4} > 0$. Also, it is customary to assume that $u(x+yi)$ is either even or odd as a function of x, and accordingly the *parity symbol* ω is defined to be either 1 or -1. Another standard assumption is that our cusp forms (holomorphic or not) are eigenfunctions of all Hecke operators. Then the coefficients $a(n)$, normalized by $a(1) = 1$, and $t(n) = \rho(n)/\rho(1)$ (for $n \geq 1$) will be *real* as Hecke eigenvalues. We now define the L-function related to the cusp form $u(z)$ as the series

$$H(s) = \sum_{n=1}^{\infty} t(n)n^{-s}.$$

Our main topic will be the mean square of $\zeta^2(s)$, $F(s)$, and $H(s)$ over a segment of the critical line. The similarity of these functions goes back to the similarity of their functional equations, which read as follows:

$$\zeta^2(s) = 2^{2s-1}\pi^{2(s-1)}\Gamma^2(1-s)(1 - \cos(\pi s))\zeta^2(1-s), \qquad (1.2)$$

$$F(s) = (-1)^{k/2}(2\pi)^{2s-1}\frac{\Gamma(1-s+(k-1)/2)}{\Gamma(s+(k-1)/2)}F(1-s), \qquad (1.3)$$

$$H(s) = 2^{2s-1}\pi^{2(s-1)}\Gamma(1-s+i\kappa)\Gamma(1-s-i\kappa)$$
$$\times\ (\omega\cosh(\pi\kappa) - \cos(\pi s))H(1-s); \qquad (1.4)$$

in particular, the function $\zeta^2(s)$ behaves heuristically like a function $H(s)$ related to the (admittedly non-existing) eigenvalue $1/4$. The first two relations are standard results, and for (1.4) see, e.g., [23] or [29], Lemma 3.4.

To emphasize the analogy between our functions $\zeta^2(s)$, $F(s)$, and $H(s)$, we are going to adopt the common symbol $\varphi(s)$ for all of them, and their coefficients $d(n)$, $\tilde{a}(n)$, and $t(n)$ will be denoted by $c(n)$. Consider now the mean value equation (1.1) for $|\varphi(\frac{1}{2}+it)|^2$. With the purpose of analyzing the Laplace inversion integral on the right, we proceed through the following steps of the argument:

(1) An arithmetical formula is established for the Laplace transform of $|\varphi(\frac{1}{2}+it)|^2$ (Lemma 1 in §2).

(2) Evaluating the inversion integral approximately by use of the theorem of residues, we end up with an arithmetic formula for the mean value in question (Theorem 1 in §3). This expression is related to the "generalized additive divisor problem" concerning the summatory function of $c(n)c(n+f)$, where f is the "shift". At this stage, some smoothing must be made with respect to T in (1.1) in order to accelerate the convergence of the integral on the right. Therefore the mean value will be equipped with a smooth weight function.

(3) The error term for the arithmetic formula mentioned above is translated into the language of the spectral theory (Theorem 2 in §5).

(4) Mean value results (Theorem 3 and its corollary in §7) are deduced from the preceding formula by appealing to known facts from the spectral theory.

Our main goal in this paper is to put the fundamental work of Motohashi [25] on the fourth moment of the zeta-function into a more general context, admittedly at the cost of losing the remarkable accuracy of his main theorem as a price to be paid for the generality and relative simplicity of our argument. However, we try to avoid being too wasteful, for an attempt is made to keep error terms small enough (that is, well below the "barrier" $\sqrt{T}$) to give a basis for recovering the main results in the important joint work of Ivić and Motohashi (see [9], [10], or [7]) to be briefly surveyed below. As we pointed out above, an approach to Motohashi's theory along these lines was outlined already in [17], but the present version is more precise and detailed.

The asymptotic formula for the mean square of $|\varphi(\frac{1}{2} + it)|^2$ is of the general type

$$I_{2,\varphi}(T) = \int_0^T |\varphi(\tfrac{1}{2} + it)|^2\, dt = TP_\varphi(\log T) + E_{2,\varphi}(T), \qquad (1.5)$$

where P_φ is a polynomial (of degree 4 if $\varphi(s) = \zeta^2(s)$, and of degree 1 otherwise). In the zeta-function case, the error term $E_{2,\varphi}(T)$ is usually written as $E_2(T)$. The main results of Ivić and Motohashi on $E_2(T)$ are:

$$E_2(T) \ll T^{2/3} \log^C T, \qquad (1.6)$$

$$\int_0^T E_2^2(t)\, dt \ll T^2 \log^C T, \qquad (1.7)$$

$$E_2(T) = \Omega_\pm(\sqrt{T}); \qquad (1.8)$$

see [9] for (1.6) and (1.8) (the latter in the form $\Omega(\sqrt{T})$), and [10] for (1.7); the assertion (1.8) in its full force is due to Motohashi [28]. Another proof of the latter result was recently given by Ivić [8] by use of a spectral theoretic formula for the Laplace transform of $E_2(t)$. The estimate (1.6), in a slightly weaker form, was first established by Zavorotnyi [34].

Our principal result is Theorem 2 in §5, giving a spectral theoretic formula for a weighted variant of the error term $E_{2,\varphi}(T)$, and this can be used as a basis for new proofs of (1.6)–(1.8). In addition, (1.6) and (1.7) hold for $E_{2,F}(T)$ as well, and even for $E_{2,H}(T)$, at least if the factor $\log^C T$ is weakened to T^ε. The analogue of (1.8) for cusp form L-functions is still an open problem; Motohashi [27] reduces this question (in the case of holomorphic cusp forms) to a certain highly plausible non-vanishing conjecture.

The Ω-estimate (1.8) may be deduced from our considerations in two ways. The easier way (see [20]) is to derive the above mentioned formula of Ivić [8] for the Laplace transform of $E_2(t)$ from the Laplace transform of $|\zeta(\frac{1}{2} + it)|^4$ in Lemma 1. Alternatively, following Motohashi [28], we may consider the function

$$\int_1^\infty |\varphi(\tfrac{1}{2} + it)|^2 t^{-\xi}\, dt. \qquad (1.9)$$

Since the error terms in Theorem 2 are $O(T^{2/5+\varepsilon})$, this function can be analytically continued to a meromorphic function in the half-plane $\sigma > 2/5$ having a pole at $\xi = 1$, the other possible poles being situated among the points $\frac{1}{2} \pm i\kappa_j$ and $\rho/2$, where κ_j is the κ-parameter for the jth Maass wave form, and ρ runs over the complex zeros of the zeta-function. Then (1.8) may be deduced as in [28] by use of a lemma of Landau and Motohashi's non-vanishing theorem

([24], Theorem 3); the latter guarantees that at least one of the points $\frac{1}{2} \pm i\kappa_j$ is really a pole of the function (1.9) for $\varphi(s) = \zeta^2(s)$. If the same property would be shared by the Hecke L-functions, then (1.8) could be generalized to $E_{2,\varphi}(T) = \Omega_\pm(\sqrt{T})$. However, this remains still an open problem because an appropriate analogue of Motohashi's non-vanishing theorem is not available.

Notation Generally C stands for a positive numerical constant, and ε for a small fixed positive number; the meaning of these symbols is not necessarily the same at each occurrence. Also, we write $A \asymp B$ to mean that $A \ll B \ll A$, and $A \sim B$ to mean that $B \leq A \leq 2B$. The constants implied by the notations $O(\cdots)$ etc. are either absolute, or may depend on parameters involved; in particular, since the cusp forms related to the series $F(s)$ and $H(s)$ will be fixed, some constants may depend on κ or k. Otherwise the notation is standard, as to for instance Bessel or hypergeometric functions, or it will be explained in the text.

2. Laplace transforms To get started with our unified approach to the mean values $I_{2,\varphi}(T)$ in (1.5), we need appropriately uniform expressions for the Laplace transforms of the functions $|\varphi(\frac{1}{2} + it)|^2$ to be averaged. In the classical case of $|\zeta(\frac{1}{2} + it)|^4$, two formulae for its Laplace transform are available: one, due to Titchmarsh [31], involving the ordinary divisor function $d(n)$, and the other, due to Atkinson [2], involving the divisor function $d_4(n)$. As in [17], we prefer to choose the former alternative. The argument of Titchmarsh can be immediately extended to cusp form L-functions.

Let $\varphi(s)$ be any one of the functions $\zeta^2(s)$, $F(s)$, or $H(s)$, and define

$$\phi(z) = \frac{1}{2\pi i} \int_{(a)} \varphi(s)\Gamma(s)z^{-s}\, ds \qquad (0 < a < 1), \tag{2.1}$$

for Re $z > 0$. Then, moving the integration to the line Re $s = 2$, we see by Mellin's formula and the theorem of residues that $\phi(z)$ equals

$$\sum_{n=1}^{\infty} d(n)e^{-nz} - (\gamma - \log z)/z, \tag{2.2}$$

$$\sum_{n=1}^{\infty} \tilde{a}(n)e^{-nz}, \tag{2.3}$$

$$\sum_{n=1}^{\infty} t(n)e^{-nz}, \tag{2.4}$$

as the case may be; here γ denotes Euler's constant. The argument of Titchmarsh then proceeds as follows: by (2.1), the Mellin transform of the function

$\phi(ixe^{-i\delta})$ with $0 < \delta < \pi/2$ is $\varphi(s)\Gamma(s)e^{-i(\pi/2-\delta)s}$, so Parseval's formula for Mellin transforms gives

$$\frac{1}{2\pi} \int_{-\infty}^{\infty} |\Gamma(\tfrac{1}{2} + it)\varphi(\tfrac{1}{2} + it)|^2 e^{(\pi - 2\delta)t} \, dt$$

$$= \int_0^{\infty} |\phi(ixe^{-i\delta})|^2 \, dx = \int_0^{\infty} \phi(ixe^{-i\delta})\phi(-ixe^{i\delta}) \, dx.$$

This may be analytically continued to the strip $0 < \text{Re } \delta < \pi/2$. Putting p in place of 2δ, we find that the Laplace transform of the function

$$\frac{e^{\pi t}}{2 \cosh \pi t} |\varphi(\tfrac{1}{2} + it)|^2 \tag{2.5}$$

equals

$$\int_0^{\infty} \phi(ixe^{-ip/2})\phi(-ixe^{ip/2}) \, dx, \tag{2.6}$$

up to a certain function representing the contribution of the negative values of t. This "correction" function is holomorphic even in the strip $|\text{Re } p| < \pi$. Another similar correction function arises if the factor in front of $|\varphi(\tfrac{1}{2} + it)|^2$ in (2.5) is omitted. In this way, we end up with a formula for the Laplace transform of $|\varphi(\tfrac{1}{2} + it)|^2$. In the next lemma, its expression is simplified to a more explicit form suitable for applications. Recall that $c(n)$ stands for the coefficients of $\varphi(s)$.

Lemma 1 *The Laplace transform of the function $|\varphi(\tfrac{1}{2} + it)|^2$ is, for $0 < \text{Re } p < \pi$,*

$$L(p) = 2\pi \int_0^{\infty} \phi(2\pi ixe^{-ip/2})\phi(-2\pi ixe^{ip/2}) \, dx + \lambda(p), \tag{2.7}$$

where the function $\lambda(p)$ is holomorphic even in the strip $|\text{Re } p| < \pi$, and it is bounded in the strip $|\text{Re } p| \le \theta$ for any fixed $\theta < \pi$. Also, for $0 < \text{Re } p \le \theta < \pi$, $|\text{Im } p| \ll 1$, we have

$$L(p) = 2i \sum_{m,n=1}^{\infty} c(m)c(n)L_{m,n}(p) + L_0(p), \tag{2.8}$$

where

$$L_{m,n}(p) = \frac{e(-me^{-ip/2} + ne^{ip/2})}{-me^{-ip/2} + ne^{ip/2}}, \tag{2.9}$$

and

$$L_0(p) \ll \log^3(1 + (\text{Re } p)^{-1}). \tag{2.10}$$

Proof The formula (2.7) and the properties of the function $\lambda(p)$ follow from (2.6) and the above discussion. It remains to analyze the integral on the right.

The key result here is an approximate functional equation connecting $\phi(1/z)$ and $\phi(4\pi^2 z)$ for Re $z > 0$. A relation like this is given in [32], eq. (7.16.2), for $\varphi(s) = \zeta^2(s)$. We need a generalized and refined version of this formula, where the error term is kept explicit. Following Titchmarsh, write the equation (2.1) for $\phi(1/z)$, change the variable s to $1 - s$, and apply the functional equation $\varphi(s) = \chi_\varphi(s)\varphi(1 - s)$. Then, changing $1 - a$ back to a, we obtain

$$\phi(1/z) = \frac{z}{2\pi i} \int_{(a)} \Gamma(1 - s)\chi_\varphi^{-1}(s)\varphi(s)z^{-s}\,ds. \tag{2.11}$$

By the functional equations (1.2)–(1.4) and Stirling's formula, it is easy to verify that for $\delta = \pm 1$ we have in any fixed vertical strip

$$\frac{\Gamma(1 - s)(2\pi)^{2s-1}}{\Gamma(s)\chi_\varphi(s)} = -\delta i + \eta_{\varphi,\delta}(s), \tag{2.12}$$

where

$$\eta_{\varphi,\delta}(s) \ll \begin{cases} (|t| + 1)^{-1}, & \text{if } \operatorname{sgn} t = \delta, \\ 1, & \text{otherwise.} \end{cases} \tag{2.13}$$

Recall that by Stirling's formula

$$\log\Gamma(s) = \frac{1}{2}\log 2\pi + (\sigma - \tfrac{1}{2})\log t$$
$$- \frac{\pi}{2}t + i\left((\sigma - \tfrac{1}{2})\frac{\pi}{2} + t\log t - t\right) + O(t^{-1}) \tag{2.14}$$

if σ is bounded and $t \to \infty$. Moreover, (2.13) may be sharpened to $\eta_{\varphi,\delta}(s) \ll e^{-\pi|t|}$ in the case $\varphi(s) = \zeta^2(s)$ for $\operatorname{sgn} t = \delta$. Let now δ be the sign of Im z. Then, applying (2.12) and (2.1), we may rewrite (2.11) as follows:

$$\phi(1/z) = -2\pi i\delta z\phi(4\pi^2 z) - iz\int_{(a)} \eta_{\varphi,\delta}(s)\Gamma(s)\varphi(s)(4\pi^2 z)^{-s}\,ds.$$

In particular, for $z = \delta(2\pi)^{-1}ixe^{-\delta ip/2}$ with variable $x > 0$, this gives

$$\phi\left(\frac{2\pi}{\delta ixe^{-\delta ip/2}}\right) = xe^{-\delta ip/2}\phi(2\pi\delta ixe^{-\delta ip/2}) + \rho_{\varphi,\delta}(x), \tag{2.15}$$

where

$$\rho_{\varphi,\delta}(x) = \frac{1}{2\pi i}\int_{(a)} \eta_{\varphi,\delta}(s)\Gamma(s)\varphi(s)(2\pi)^{-s}\left(\delta ie^{-\delta ip/2}\right)^{1-s}x^{1-s}\,ds. \tag{2.16}$$

Returning now to (2.7), we calculate by use of (2.15)

$$\int_0^1 \phi(2\pi i x e^{-ip/2})\phi(-2\pi i x e^{ip/2})\, dx = \int_1^\infty \phi\left(\frac{2\pi}{-ixe^{ip/2}}\right)\phi\left(\frac{2\pi}{ixe^{-ip/2}}\right)\frac{dx}{x^2}$$

$$= \int_1^\infty \phi(-2\pi i x e^{ip/2})\phi(2\pi i x e^{-ip/2})\, dx + \int_1^\infty \left\{ x^{-1}e^{ip/2}\phi(-2\pi i x e^{ip/2})\rho_{\varphi,1}(x) \right.$$

$$\left. + x^{-1}e^{-ip/2}\phi(2\pi i x e^{-ip/2})\rho_{\varphi,-1}(x) + x^{-2}\rho_{\varphi,-1}(x)\rho_{\varphi,1}(x) \right\}\, dx.$$

The leading integral may be combined with the tail of the integral in (2.7) to give the main term

$$4\pi \int_1^\infty \phi(2\pi i x e^{-ip/2})\phi(-2\pi i x e^{ip/2})\, dx \tag{2.17}$$

for $L(p)$. In the remaining three integrals, we substitute $\rho_{\varphi,\delta}(x)$ from (2.16) and the factors $\phi(\cdots)$ from (2.1), choosing $a = 1/2 + \min(1/4, \operatorname{Re} p)$ in both cases. Then the x-integration may be performed under the respective double complex integrals, and the s-integrals may be estimated by use of (2.16) and the standard mean value estimate

$$\int_0^T |\varphi(a+it)|^2\, dt \ll T \log^C T,$$

where $C = 4$ for $\varphi(s) = \zeta^2(s)$, and $C = 1$ in the cusp form cases. The first mentioned bigger value of C is compensated by the exponential decay of $\eta_{\varphi,\delta}(s)$ mentioned above. Then, if p is restricted as assumed in the latter part of the lemma, the estimate (2.10) may be verified by straightforward estimations of the three terms under consideration. Finally, the integral (2.17) gives immediately the double series in (2.8) in the cusp form cases. In the zeta-function case, there are still certain cross terms to be taken into account. However, these may be readily absorbed into the function $L_0(p)$.

3. An arithmetic formula for the mean value The mean value $I_{2,\varphi}(T)$ defined in (1.5) may be written by (1.1) (with p replaced by $2p$ for convenience) as the complex integral

$$I_{2,\varphi}(T) = \frac{1}{2\pi i}\int_{(a)} L(2p)p^{-1}e^{2pT}\, dp, \tag{3.1}$$

where $a > 0$ and $L(2p)$ is given in Lemma 1. We choose $a = 1/T$. Wanting to truncate this integral to a very short interval lying symmetrically to the real axis, we accelerate its convergence by a smoothing device with respect to T. To this end, let

$$T^{2/5+\varepsilon} \leq \Delta \leq T^{2/3}, \tag{3.2}$$

and define

$$I_{2,\varphi}(T, \Delta) = \frac{1}{\sqrt{\pi}\Delta} \int_{-\Delta \log T}^{\Delta \log T} I_{2,\varphi}(T + \tau) e^{-(\tau/\Delta)^2} d\tau. \tag{3.3}$$

The following theorem is analogous to a lemma of Heath-Brown ([6], Lemma 3) and to Lemma 2 in [17], with three main differences: the result is not restricted to the zeta-function, the integral is taken over the whole interval $[0, T]$ instead of a segment $[T_1, T_2]$, and finally the parameter Δ is allowed to take values smaller than $\sqrt{T}$.

Theorem 1 *Under the assumption* (3.2) *on* Δ, *we have for large values of* T

$$I_{2,\varphi}(T, \Delta) = 2 \sum_{n \leq T/2\pi} c^2(n) n^{-1} (T - 2\pi n)$$

$$+ 4 \sum_{\substack{2n+f \leq T/\pi \\ f \geq 1}} c(n) c(n+f) (n(n+f))^{-1/2} (\log(1 + f/n))^{-1}$$

$$\times \sin(T \log(1 + f/n)) \exp\left(-\tfrac{1}{4}\Delta^2 \log^2(1 + f/n)\right)$$

$$+ O(\Delta \log^C T), \tag{3.4}$$

where C *is a numerical constant.*

Proof First we combine (3.1) and (3.3) using the familiar formula

$$\int_{-\infty}^{\infty} e^{Ax - Bx^2} dx = \sqrt{(\pi/B)} e^{A^2/4B} \qquad (\text{Re } B > 0)$$

to get

$$I_{2,\varphi}(T, \Delta) = \frac{1}{2\pi i} \int_{(a)} L(2p) p^{-1} e^{2pT + p^2 \Delta^2} dp + O(1).$$

Write here $p = a + ui$. Obviously the integral may be truncated to the interval $|u| \leq U$, where

$$U \asymp \Delta^{-1} \log T. \tag{3.5}$$

The contribution of the term $L_0(2p)$ from Lemma 1 to this integral is $\ll \log^4 T$. Therefore, by this lemma, we have

$$I_{2,\varphi}(T, \Delta) = \frac{1}{\pi} \sum_{m,n=1}^{\infty} c(m) c(n) \int_{a-Ui}^{a+Ui} p^{-1} L_{m,n}(2p) e^{2pT + p^2 \Delta^2} dp$$

$$+ O(\log^4 T). \tag{3.6}$$

Clearly, the sums over m and n may be truncated to finite sums over $m, n \ll T \log T$, for the double series converges exponentially.

For symmetry, we combine in (3.6) the contributions of the pairs (m, n) and (n, m) if $m \neq n$. Thus, writing

$$J_{m,n}(r) = \frac{1}{2\pi} \int_{r-iU}^{r+iU} M_{m,n}(p) \, dp \tag{3.7}$$

with

$$M_{m,n}(p) = p^{-1} \left(L_{m,n}(2p) + L_{n,m}(2p) \right) e^{2pT+p^2\Delta^2}, \tag{3.8}$$

we have

$$I_{2,\varphi}(T, \Delta) = \sum_{m,n \ll T \log T} c(m)c(n) J_{m,n}(a) + O(\log^4 T). \tag{3.9}$$

We are going to evaluate the integrals $J_{m,n}(a)$ by the theorem of residues on completing the segment of integration to a rectangular contour. To see how to do this, let us consider the integrand in the rectangle $|\operatorname{Re} p| \leq \Delta^{-1}$, $|\operatorname{Im} p| \leq U$. Then, with $f = m - n$, we have

$$e(-me^{-ip} + ne^{ip})e^{2pT+p^2\Delta^2} = \exp \left[-2p(\pi(m+n) - T) + \pi i f p^2 + p^2 \Delta^2 + O((m+n)|p|^3) \right], \tag{3.10}$$

$$-me^{-ip} + ne^{ip} = -f + (m+n)ip + O(|f||p|^2) + O((m+n)|p|^3). \tag{3.11}$$

In general, the expression (3.10) becomes smaller if $|\operatorname{Re} p|$ increases so that the sign of $\operatorname{Re} p$ coincides with that of $\pi(m+n) - T$. Therefore it is reasonable to complete the path in $J_{m,n}(a)$ leftwards for $m + n \leq T/\pi$ and rightwards for $m+n > T/\pi$. More precisely, we move the integration to the line $\operatorname{Re} p = -\Delta^{-1}$ if $m + n \leq T/\pi$, and to the line $\operatorname{Re} p = \Delta^{-1}$ if $m + n > T/\pi$, unless

$$|m + n - T/\pi| \leq P, \quad |f| > \Delta^2 \tag{3.12}$$

with

$$P = (T/\Delta) \log^2 T, \tag{3.13}$$

in which case the integration is taken to the imaginary axis.

The relevant singularities of the function $M_{m,n}(p)$ defined by (3.8) and (2.9) are the poles $\pm \frac{1}{2} i \log(n/m)$ on the imaginary axis. These are simple for $m \neq n$, while for $m = n$ they coincide to give a double pole at $p = 0$. The poles lie inside the contour if

$$m + n \leq T/\pi, \quad \frac{1}{2}|\log(m/n)| < U. \tag{3.14}$$

The residue of $M_{n,n}(p)$ at $p = 0$ is $2(in)^{-1}(T - 2\pi n)$. Further, for $m \neq n$, the sum of the residues of $M_{m,n}(p)$ at its poles is

$$-\frac{2i\sin(T\log(m/n))\exp(-(1/4)\Delta^2\log^2(m/n))}{\sqrt{mn}\log(m/n)}.$$

Now the explicit terms in (3.4) arise from the above mentioned residues for all pairs (m, n) with $m + n \leq T/\pi$, thus the second condition in (3.14) is ignored. However, the contribution of the extra terms in (3.4) is clearly negligible.

Consider next the integrals over the horizontal sides lying on the lines $\text{Im } p = \pm U$. To keep the poles $\frac{1}{2}i\log(n/m)$ away from the path, we choose U so that the points $\pm Ui$ lie half-way between two neighbouring poles. It is now easily seen, by (3.10), that the integrand is very small on the horizontal sides.

It remains to deal with the integrals $J_{m,n}(r)$ with $r = \pm\Delta^{-1}$ or $r = 0$ as specified above. The latter possibility may occur only for $\Delta \leq T^{1/2}$, which is thus a more delicate case than $\Delta > T^{1/2}$. Once the former case is discussed, the latter can be settled by similar but more straightforward arguments. Let us therefore suppose in the sequel that

$$T^{2/5+\varepsilon} \leq \Delta \leq T^{1/2}.$$

By (3.10), (3.11), (3.13), and our restriction for Δ, the integrals $J_{m,n}(r)$ are very small for $|m + n - T/\pi| > P$, and also for $|m + n - T/\pi| > \Delta\log^2 T$ if $r = \pm\Delta^{-1}$ and $|f| \leq \Delta^2$. Therefore we may suppose henceforth that $m + n$ lies, in any case, in the critical interval $|m + n - T/\pi| \leq P$, and that even $|m + n - T/\pi| \leq \Delta\log^2 T$ for $|f| \leq \Delta^2$.

To begin with our analysis of the integrals $J_{m,n}(r)$, we simplify these by use of (3.10)–(3.11). A little calculation shows that

$$J_{m,n}(r) = \frac{1}{\pi i}\int_{r-Ui}^{r+Ui}\exp(-2p(\pi(m+n)-T)+p^2\Delta^2)$$
$$\times\frac{f\sin(\pi fp^2)+(m+n)p\cos(\pi fp^2)}{p(f^2+(m+n)^2p^2)}\,dp$$
$$+O(\min(|f|^{-1}T\Delta^{-3},\Delta^{-2})\log^4 T) \tag{3.15}$$

for $r = \pm\Delta^{-1}$ and the corresponding pairs (m, n); note that the modulus of the expression (3.11) is $\gg T/\Delta$ for $f \ll (T/\Delta)\log T$, and it is $\asymp |f|$ otherwise.

Also,

$$J_{m,n}(0) = \frac{2}{\pi i f} \int_0^U \sin(\pi f u^2) \sin(2u(\pi(m+n) - T)) e^{-u^2 \Delta^2} u^{-1}\, du$$

$$+ O(|f|^{-1} T \Delta^{-3} \log^4 T) \tag{3.16}$$

in the case (3.12); now even the term $(m+n)ip$ in (3.11) was treated as an error term.

At this stage, we need information about the size of the coefficients $c(n)$, at least in mean. An ideal estimate would be $|c(n)| \leq d(n)$, which holds for $c(n) = \tilde{a}(n)$ by Deligne's theorem, but for $t(n)$ this is so far only a conjecture. However, the following unconditional estimate of the Rankin type is well-known:

$$\sum_{x \leq n \leq x+y} |t(n)|^2 \ll y + x^{3/5+\varepsilon} \qquad (1 \leq y \leq x), \tag{3.17}$$

and the same holds for $c(n)$ in general if y is replaced by $y \log^3 x$. This is crude for $y < x^{3/5}$, but then we may use the estimate (see [23] or [5])

$$\sum_{n \leq x} t(n) \ll x^{2/5}. \tag{3.18}$$

Consider now the contribution of the error terms in (3.15) and (3.16) to $I_{2,\varphi}(T, \Delta)$ for all pairs (m, n) such that $|m + n - T/\pi| \leq P$. Those pairs with m and n of the same parity are of the form $(N+h, N-h)$, where $|2N - T/\pi| \leq P$. The contribution of this set of pairs to (3.9) is

$$\ll \log^4 T \sum_N \sum_{0 \leq h \ll T} \min(h^{-1} T/\Delta^3, \Delta^{-2}) |c(N+h) c(N-h)|$$

$$\ll (T/\Delta^3)(P \log^3 T + T^{3/5+\varepsilon}) \log^5 T$$

by Cauchy's inequality (applied to the sum over N) and the above mentioned generalized variant of (3.17). This is $\ll \Delta$ by (3.13) and our assuption $\Delta \gg T^{2/5+\varepsilon}$. The pairs with m and n of different parity may be treated similarly.

Next we deal with the integrals (3.15) for $|m+n-T/\pi| \leq \Delta \log^2 T$, $|f| \leq \Delta^2$. These are of the order

$$\ll \frac{f^2 \Delta^{-2} + T \Delta^{-1}}{f^2 + (T\Delta^{-1})^2}.$$

Then, for $c(n) = d(n)$ or $\tilde{a}(n)$, the resulting contribution to $I_{2,\varphi}(T, \Delta)$ is $\ll \Delta \log^4 T$. In the case $c(n) = t(n)$, the argument must be modified a bit; the sum over m for given n is first treated by partial summation and (3.18) to get some saving, after which the sum over n is estimated by Cauchy's inequality and (3.17). The result is $\ll (T^{2/5} + T^{1/5}\Delta^{1/2})T^\varepsilon \ll \Delta$.

It remains to estimate the contribution of the integrals (3.16) for the pairs (m, n) satisfying (3.12). This amounts to integrals

$$\int_0^U \left| \sum_{m,n} c(m)c(n)w_1(m+n)w_2(f)f^{-1} \right.$$

$$\left. \times \sin(\pi f u^2)e(\pm(m+n)u) \right| u^{-1}e^{-u^2\Delta^2}\,du \qquad (3.19)$$

with $f = m - n$, where the weight functions w_1 and w_2 correspond to the conditions in (3.12); for instance, w_1 is the characteristic function of the interval $[T/\pi - P, T/\pi + P]$. However, we may average over the parameter P in (3.13) to make w_1 a smooth function with $w_1^{(j)}(x) \ll P^{-j}$ for $j = 1, 2$, and analogously for w_2. Then, by partial summation, we may reduce the double sum in (3.19) to separate "standard" exponential sums involving coefficients $c(n)$. The latter sums can be estimated by the following lemma.

Lemma 2   *Let $x \geq 2$, $0 < y \leq x$, and $0 < \alpha \leq 1$. Then*

$$\sum_{x-y \leq n \leq x} c(n)e(n\alpha) \ll x^{1/2}\log x + E\{\min(\alpha^{-1}, y) + y\alpha\}\log x,$$

where $E = 1$ if $c(n) = d(n)$, and $E = 0$ if $c(n) = \tilde{a}(n)$ or $t(n)$.

For the divisor function, this is an easy corollary of a transformation formula of Wilton [33] for exponential sums (for a different proof and generalization, see [13]). For the coefficients $t(n)$, this follows immediately from Theorem 8.1 in [11], and the case of holomorphic cusp forms is a well-known classical result.

Consider the case $c(n) = d(n)$ in (3.19); the others are easier because $E = 0$ in Lemma 2. Let us sum first over m by Lemma 2 and partial summation. Then the coefficient of $d(n)$ in the double sum will be $e(\pm nu)$ times a function of n and u of the order

$$\ll |n - T/2\pi|^{-1} \min(1, Tu^2)(\sqrt{T} + \min(P, u^{-1}))\log T.$$

Moreover, differentiating this function with respect to n we see that it is stationary as n runs over an interval of length at most P. Therefore partial summation is applicable even to the n-sum, decomposed into segments of length about P, and the integrand in (3.19) is seen to be

$$\ll P^{-1}\min(1, Tu^2)(\sqrt{T} + \min(P, u^{-1}))^2(\log^3 T)u^{-1}e^{-u^2\Delta^2}.$$

The integrals of this over the ranges $(0, P^{-1}]$, $(P^{-1}, T^{-1/2}]$, and $(T^{-1/2}, U]$ are each $\ll \Delta\log^2 T$. Thus the proof of Theorem 1 is complete in the case

$\Delta \leq T^{1/2}$, and as we pointed out above, the remaining case $\Delta > T^{1/2}$ is analogous but easier.

Remark As a preparation for an analytic treatment of the non-diagonal sum in (3.4), it is helpful to equip it with a smooth weight function. For this purpose, let $\nu(x)$ be a smooth function which equals 1 for $x \leq 1 - \Delta/T$, and vanishes for $x \geq 1 + \Delta/T$. The weight function $\nu((2n + f)(T/\pi)^{-1})$ then makes the sum smoother. Using (3.18) or estimates by absolute values as above to estimate the approximation error, we find that (3.4) remains valid with the same error term even with the weights inserted. Moreover, the f-sum may be truncated to the interval $[1, f_0]$ with $f_0 = (T/\Delta) \log T$. Thus

$$
\begin{aligned}
I_{2,\varphi}(T, \Delta) =& S_0(T) + \sum_{1 \leq f \leq f_0} S_f(T, \Delta) + O(\Delta \log^C T) \\
=& S_0(T) + S(T, \Delta) + O(\Delta \log^C T),
\end{aligned}
\tag{3.20}
$$

say, where

$$
S_0(T) = 2 \sum_{n \leq T/2\pi} c^2(n) n^{-1}(T - 2\pi n),
\tag{3.21}
$$

and

$$
S_f(T, \Delta) = \sum_{n=1}^{\infty} c(n) c(n + f) W_f(n/f)
\tag{3.22}
$$

with

$$
\begin{aligned}
W_f(x) =& 4f^{-1} \nu((\pi f/T)(2x + 1))(x(x + 1))^{-1/2} (\log(1 + 1/x))^{-1} \\
& \times \sin\left(T \log(1 + 1/x)\right) \exp\left(-\tfrac{1}{4}\Delta^2 \log^2(1 + 1/x)\right).
\end{aligned}
\tag{3.23}
$$

4. The main term for $I_{2,\varphi}(T)$

It is natural to try to single out a main term for $I_{2,\varphi}(T)$ from the formula (3.20) for its smoothed version $I_{2,\varphi}(T, \Delta)$ defined in (3.3). In any case, the leading sum $S_0(T)$ is no problem, for its main term and error term, say $S_{00}(T)$ and $S_{01}(T)$, are by standard arguments equal to

$$
S_{00}(T) = 4\pi \operatorname{Res}\left(\frac{C_0(s + 1)(T/2\pi)^{s+1}}{s(s + 1)}, 0\right)
\tag{4.1}
$$

and

$$
S_{01}(T) = -2i \int_{(a)} \frac{C_0(s + 1)(T/2\pi)^{s+1}}{s(s + 1)} \, ds \quad (-1/2 < a < 0),
\tag{4.2}
$$

where

$$
C_0(s) = \sum_{n=1}^{\infty} c^2(n) n^{-s}.
\tag{4.3}
$$

For $c(n) = d(n)$, this function is $\zeta^4(s)/\zeta(2s)$, and otherwise it is the Rankin zeta-function related to the respective cusp form. In the former case, the residue (4.1) comes from a pole of fifth order, and in the latter case from a double pole. Therefore the term $S_{00}(T)$ is of a suitable form to be included into the main term in (1.5), or even to be this main term itself in the cusp form cases, for then the sum $S(T, \Delta)$ in (3.20) is not likely to contribute any additional smooth function to the main term.

The case $c(n) = d(n)$ is of different nature since $d(n)$ is positive, so the only cancellation in the non-diagonal sums $S_f(T, \Delta)$ is due to the trigonometric factor. However, this factor is a slowly oscillating function of n if f is small, and the corresponding sum cannot be viewed as a genuine error term. Thus a contribution from the non-diagonal part should be included into the main term; this observation goes back to Atkinson [2].

To single out a main term from a non-diagonal sum for a given shift f, we use an asymptotic formula in the additive divisor problem concerning the sum

$$D(N; f) = \sum_{n \leq N} d(n)d(n + f) \qquad (f \geq 1).$$

Suppose that a smooth function $D_0(N; f)$ is an appropriate main term for $D(N; f)$. Then partial summation shows that a natural main term (if there is any) for a general sum of the type

$$\sum_{n=1}^{\infty} d(n)d(n + f)W(n/f), \tag{4.4}$$

with $W(x)$ of compact support in $(0, \infty)$, is given by the integral

$$\int_0^{\infty} W(x/f) \, dD_0(x; f) = f \int_0^{\infty} W(x)D_0'(fx; f) \, dx. \tag{4.5}$$

In particular, the main term for the off-diagonal sum $S(T, \Delta)$ in (3.20) is

$$\sum_{1 \leq f \leq f_0} f \int_0^{\infty} W_f(x)D_0'(fx; f) \, dx. \tag{4.6}$$

The next question now is: how to define the main term $D_0(N; f)$ in the additive divisor problem ? The classical and well-known answer to this problem is that a certain function of the type

$$D_1(N; f) = NQ_f(\log N), \tag{4.7}$$

where Q_f is a quadratic polynomial depending on f (see [6], Theorem 2), may be chosen to play the role of $D_0(N; f)$. However, this main term makes sense only if f is sufficiently small compared with N. The following more "uniform" main term has been given by Motohashi [26]:

$$D_2(N;f) = (6/\pi^2) \int_0^{N/f} m_2(x;f)\,dx, \qquad (4.8)$$

where

$$\begin{aligned} m_2(x;f) = {} & \sigma(f)\log x \log(x+1) \\ & + \left\{ \sigma(f)(2\gamma - 2\frac{\zeta'}{\zeta}(2) - \log f) + 2\sigma'(f) \right\} \log(x(x+1)) \\ & + \sigma(f)\left\{ \left(2\gamma - 2\frac{\zeta'}{\zeta}(2) - \log f\right)^2 - 4\left(\frac{\zeta'}{\zeta}\right)'(2) \right\} \\ & + 4\sigma'(f)(2\gamma - 2\frac{\zeta'}{\zeta}(2) - \log f) + 4\sigma''(f) \qquad (4.9) \end{aligned}$$

with $\sigma^{(\nu)}(f) = \sum_{d\mid f} d\log^\nu d$. Now $D_1(N;f)$ is close to $D_2(n;f)$, in a certain sense to be made more precise below, if f is small, but this connection breaks down if f approaches N, or even exceeds it. On the other hand, the function $D_2(N;f)$ serves as a main term in the very wide range $1 \le f \ll N^{10/7-\varepsilon}$. The construction of this function is based on the Kuznetsov–Motohashi identity (see [26], Theorem 3).

A third approach to the additive divisor problem, and indeed a very natural one at least in principle, is via the generating Dirichlet series

$$\sum_{n=1}^{\infty} d(n)d(n+f)n^{-s} \qquad (4.10)$$

and Perron's formula. This argument has been worked out by Tahtadjan and Vinogradov [30] (see also [14], [15]). The function (4.10) itself is not easily tractable, but there is a certain meromorphic function $\zeta_f^*(s)$ (more about this in the next section), defined in terms of non-holomorphic Eisenstein and Poincaré series, which approximates the function (4.10) relatively well and may be expressed explicitly in terms of the spectral resolution of the hyperbolic Laplacian. In particular, if the function (4.10) is analytically continued to a meromorphic function with the aid of $\zeta_f^*(s)$, then both functions have a triple pole at $s = 1$ with the same principal part of the Laurent expansion. Therefore, by Perron's formula, the function

$$D_3(N;f) = \operatorname{Res}\left(\zeta_f^*(s)N^s s^{-1}, 1\right),$$

may be expected to give the main term for $D(N;f)$, in other words it is another candidate for the function $D_0(N;f)$, and this was shown in [30]. Now, because

$D_3(N; f)$ is functionally of the same type as $D_1(N; f)$ in (4.7), these functions must be identical, because there cannot be two different main terms of such a simple form.

It is interesting to analyze the connection between $D_2(N; f)$ and $D_3(N; f)$. The latter function may be made explicit by the theory in [30], and a rather tedious calculation shows that

$$D_3(N; f) = (6/\pi^2) \int_0^{N/f} m_3(x; f)\, dx,$$

where the function $m_3(x; f)$ is defined like $m_2(x; f)$ in (4.9), except that $x + 1$ is replaced by x at the two places where it occurs. Thus, in practice, $\log(x+1)$ is replaced by $\log x$, which does not make a big difference if x is large, or if the range $[0, N/f]$ for x is long. However, the situation is quite different if f approaches N.

Now, depending on the choice of the function D_0, there are the two possibilities

$$(6/\pi^2) \sum_{1 \leq f \leq f_0} \int_0^\infty W_f(x) m_i(x; f)\, dx \quad (i = 2, 3) \tag{4.11}$$

for the main term (4.6) of the sum $S(T, \Delta)$. Since

$$m_2(x/f; f) - m_3(x/f; f) \ll \sigma(f) \log\left(1 + f/x\right) \left(|\log(x/f)| + \log f + 1\right),$$

and this difference is a smooth function, it is easy to see that the ambiguity of the main term in question is $\ll T^\varepsilon$, which may be neglected. Thus, when calculating the main term by the formula (4.6), we may apply the simpler function $D_1(x; f)$ (equal to $D_3(x; f)$) as a substitute for $D_0(x; f)$ in spite of its weaker accuracy compared with $D_2(x; f)$. Then the analysis of the main term amounts to that carried out by Heath-Brown [6]. To get rid of the parameter Δ, we may specify it to be a suitable function of T, say $\Delta = T^{2/5+\varepsilon}$.

5. Spectral formulae for error terms

Having extracted a smooth main term from the expression (3.20) for $I_{2,\varphi}(T, \Delta)$, we now turn to its finer oscillatory behaviour. The diagonal part $S_0(T)$ is controlled by the function $C_0(s)$ defined in (4.3), so our main concern will be the off-diagonal part $S(T, \Delta)$.

Aiming at a unified discussion of the sum $S(T, \Delta)$, we appeal to the Dirichlet series method for the sake of its universality, though the case $c(n) = d(n)$ could be dealt with by the Kuznetsov–Motohashi identity [26] as well. As the main tool, we are going to apply the Dirichlet series

$$C_f(s) = \sum_{n=1}^\infty c(n)c(n+f) \left(\frac{n}{n+f}\right)^\alpha (n+f)^{-s} \quad (\sigma > 1), \tag{5.1}$$

where $\alpha = (k-1)/2$ in the case of the holomorphic cusp forms, and $\alpha = 0$ otherwise. This series is analogous to (4.10); the present slightly different definition is adopted in order to admit natural interpretations in terms of inner products of certain automorphic functions [15]. To motivate the choice of α in the case $c(n) = \tilde{a}(n)$, note that then

$$C_f(s-k+1) = \sum_{n=1}^{\infty} a(n)a(n+f)(n+f)^{-s}.$$

By the Mellin inversion formula and (3.20), we have

$$S_f(T,\Delta) = \frac{1}{2\pi i} \int_{(a)} C_f(s)M_f(s)\,ds \quad (a > 1), \tag{5.2}$$

where

$$M_f(s) = f^s \int_0^{\infty} W_f(x)\,(1+1/x)^{\alpha}\,(x+1)^{s-1}\,dx \tag{5.3}$$

is the Mellin transform of the function $W_f((x-f)/f)(x/(x-f))^{\alpha}$ with the convention that $W(x) = 0$ for $x < 0$.

An important property of the function $C_f(s)$ is its analytic continuability to a meromorphic function, holomorphic in the half-plane $\sigma > 1/2$ up to a triple pole at $s = 1$ in the case $\varphi(s) = \zeta^2(s)$. Supposing the last mentioned property for a moment, we move the integration in (5.2) to a line with $1/2 < a < 1$, and the possible main term for $S_f(T,\Delta)$ is given by the residue at $s = 1$; this is another interpretation for the same main term discussed in the preceding section. Then, to summarize the decomposition of $I_{2,\varphi}(T,\Delta)$ into a main term and an error term, we rewrite the previous decomposition (3.20) as follows:

$$I_{2,\varphi}(T,\Delta) = S_{00}(T) + \sum_{1 \leq f \leq f_0} \text{Res}\,(C_f(s)M_f(s),1) + E_{2,\varphi}(T,\Delta), \tag{5.4}$$

where the error term is given by

$$E_{2,\varphi}(T,\Delta)$$
$$= S_{01}(T) + \sum_{1 \leq f \leq f_0} \frac{1}{2\pi i} \int_{(a)} C_f(s)M_f(s)\,ds \quad (1/2 < a < 1) \tag{5.5}$$

with $S_{01}(T)$ as in (4.2). Our main goal in this section is expressing $E_{2,\varphi}(T,\Delta)$ in a spectral theoretic form (see Theorem 2 below).

The argument now goes on as follows: the function $C_f(s)$ in (5.4)-(5.5) is replaced by a certain more easily tractable function $C_f^*(s)$ without affecting the residue in (5.4), the new integral involving $C_f^*(s)$ in (5.5) is evaluated, and the approximation error is finally estimated. As a preparation of technical nature for this procedure, we need an approximate formula for the Mellin transform $M_f(s)$. We ignore for a moment the smoothing function ν in (3.23), but its effect will be commented afterwards.

Lemma 3 *Let a and b_i $(1 \leq i \leq 4)$ be constants with $0 < b_1 \leq b_2 < 1$, $0 < b_3 \leq b_4 < 1$, T be a large number, $1 \ll X \ll T$, $T^{b_1} \ll \Delta \ll T^{b_2}$, and $s = \sigma + it$ with σ bounded, $t > 0$, and $T^{b_3} \ll t \ll T^{b_4}$. Then*

$$
\int_0^X x^a (x+1)^s \sin\left(T \log\left(1 + 1/x\right)\right) \left(\log\left(1 + 1/x\right)\right)^{-1}
$$
$$
\times \exp\left(-\tfrac{1}{4}\Delta^2 \log^2\left(1 + 1/x\right)\right) dx
$$
$$
= \sqrt{\pi/2}\, e\left(-\tfrac{1}{8}\right)(1 + t/T)^{iT}(1 + T/t)^{it}
$$
$$
\times t^{-a-3/2-\sigma}(T+t)^{\sigma+1/2} T^{a+1/2}
$$
$$
\times \left(\log\left(1 + t/T\right)\right)^{-1} \exp\left(-\tfrac{1}{4}\Delta^2 \log^2\left(1 + t/T\right)\right)
$$
$$
+ O\left(X^{3+a+\sigma}(\sqrt{t}X + |T - tX|)^{-1} \exp\left(-\tfrac{1}{5}\Delta^2/X^2\right)\right)
$$
$$
+ O\left(t^{-7/2-a-\sigma}T^{2+a+\sigma} \exp\left(-\tfrac{1}{5}\Delta^2 t^2/T^2\right)\right); \tag{5.6}
$$

the leading term and the second error term are to be omitted if $t < T/X$.

Proof It is more convenient to work with the variable $y = 1/x$. Then, the saddle point $y = y_0$ for our integral is the zero of the function

$$
\psi(y) = \frac{Ty - t}{y(y+1)},
$$

thus $y_0 = t/T$, which lies in the interval $[X^{-1}, \infty)$ if $t \geq T/X$. The saddle point method (in the form of [3], Lemma 1, for instance) gives the main term in (5.6). The integral over y may be restricted to a finite interval, say $[X^{-1}, b]$ for a small positive constant b. The functions F, μ, and Φ occurring in Atkinson's saddle point lemma may be chosen as follows:

$$
F(y) = Ty + t, \quad \mu(y) = cy, \quad \Phi(y) = y^{-a-\sigma-3} \exp(-\tfrac{1}{5}\Delta^2 y^2),
$$

where c is a small positive constant. Then the error terms of the lemma give those in (5.6). It is a minor complication that the condition $(0 <)\psi'(y) \gg F(y)y^{-2}$ of the lemma is not satisfied in the whole interval of integration. However, this does hold when $\psi(y)$ is small, say for $y \in [y_0/2, 2y_0]$, and otherwise the lower bound $|\psi(y)| \gg F(y)y^{-1}$ may be used to play the same rôle in the proof of the saddle point result.

Corollary *Let $M_f(s)$ be defined by (5.3), where α is a constant, σ is bounded, and $W_f(x)$ is as in (3.23). Then, putting $\tau = |t| + 1$, we have*

$$M_f(s) \ll f^{\sigma-1}\tau^{-1/2-\sigma}T^\sigma \exp(-\tfrac{1}{5}(\Delta t/T)^2)$$

$$+ \frac{T^{\sigma+\varepsilon}}{f(\tau+f)} \exp(-(\Delta f/T)^2), \tag{5.7}$$

where the first term may be omitted if $f \geq \tau/(2\pi)$.

Proof If the smoothing factor $\nu(\cdots)$ in (3.23) is ignored for a moment, and the lemma is applied with X close to $T/(2\pi f)$, then the assertion follows immediately from (5.6), even without the factor T^ε in the second term, if $|t| \gg T^\varepsilon$. The same result then plainly follows even if the function ν is taken into account. Finally, to deal with the case $t \ll T^\varepsilon$, note that integration by parts gives, in any case, the bound $\ll \tau f^{-2}T^\sigma \exp(-(\Delta f/T)^2)$, which completes the proof of (5.7).

Remark Suppose that the integral (5.6) is smoothed, for instance by taking a weighted average with respect to the parameter X over a certain interval $[X - X_0, X + X_0]$. If the saddle point T/t (now in terms of the variable x) does not lie in this interval, then a saving by a power T^{-A} for any fixed $A > 0$ may be obtained in the first error term in (5.6) if

$$cX \geq X_0 \gg X^2|T - tX|^{-1}T^\varepsilon \tag{5.8}$$

for some small positive constant c. To verify this, one may apply a smoothed variant of Atkinson's saddle point lemma ([12], Theorem 2.2) in the proof of Lemma 3.

Specified to the Mellin transform $M_f(s)$, where we now have $1 \leq f \leq f_0 = (T/\Delta)\log T$, $X \asymp T/f$, and $X_0 \asymp \Delta/f$ (the latter follows from our construction of the smoothing function ν in (3.23)), the condition (5.8) means that $M_f(s)$ is very small for $|t| \gg T^{1+\varepsilon}/\Delta$. Also, the condition in the end of the lemma means that the main term may occur only for $f \ll |t|$. These observations imply truncation conditions for the range of the integration in (5.2) and for the range of summation over f for given s, as far as the most significant contributions are concerned.

After this digression, we introduce the functions $C_f^*(s)$ following [15]. Let $E(z,s)$ be the non-holomorphic Eisenstein series, $P_f(z,s)$ the non-holomorphic Poincaré series, in the standard notation, and with $\xi(s) = \pi^{-s/2}\Gamma(s/2)\zeta(s)$, define $E^*(z,s) = \xi(2s)E(z,s)$ and $E^*(z) = E^*(z,1/2)$. Then the function

$$\zeta_f^*(s) = \frac{\pi^s\Gamma(s)}{\Gamma^4(s/2)}\langle P_f(z,s), |E^*(z)|^2\rangle, \tag{5.9}$$

where $\langle \cdot, \cdot \rangle$ denotes the Petersson inner product, is our approximation $C_f^*(s)$ to the the function $C_f(s)$ in the case $c(n) = d(n)$. Further, in the cases of holomorphic or nonholomorphic cusp forms, say $A(z)$ and $u(z)$, the respective function $C_f^*(s)$ is

$$\frac{(4\pi)^s}{\Gamma(s)} \langle P_f(z,s), y^k |A(z)|^2 \rangle, \tag{5.10}$$

$$\frac{4\pi^s \Gamma(s)}{|\rho(1)|^2 \Gamma^2(s/2)\Gamma(s/2 + i\kappa)\Gamma(s/2 - i\kappa)} \langle P_f(z,s), |u(z)|^2 \rangle. \tag{5.11}$$

As a matter of fact, the function (5.10) is *identical* with the corresponding function $C_f(s)$.

An important property of these functions, allowing their analytic continuation to meromorphic functions and giving a starting point for the calculation of the respective integrals (5.2) with $C_f(s)$ replaced by $C_f^*(s)$, is their representability in terms of the spectrum of the hyperbolic Laplacian. Let $u_j(z)$ be the jth Maass wave form, attached to the eigenvalue $1/4 + \kappa_j^2$, and write $z_j = 1/2 + i\kappa_j$, $\rho_j = \rho_j(1)$. Let $H_j(s)$ be the L-series $H(s)$ (in the notation of the introduction) related to the form u_j, and denote its coefficients by $t_j(n)$. Then (see [14], Lemma 2), for $\sigma > 1/2$, we have

$$\zeta_f^*(s) = Z_f(s) + \frac{f^{1/2-s}}{4^s \Gamma^4(s/2)} \sum_{j=1}^{\infty} |\rho_j|^2 t_j(f) H_j^2(\tfrac{1}{2}) |\Gamma(z_j/2)|^4 \Gamma(s - z_j)\Gamma(s - \overline{z}_j)$$

$$+ \frac{\pi f^{1/2-s}}{4^s \Gamma^4(s/2)} \int_{-\infty}^{\infty} \frac{\sigma_{2iu}(f) |\xi(\tfrac{1}{2} + iu)|^4 \Gamma(s - \tfrac{1}{2} + iu)\Gamma(s - \tfrac{1}{2} - iu)}{f^{iu} |\Gamma(\tfrac{1}{2} + iu)|^2 |\zeta(1 + 2iu)|^2} du, \tag{5.12}$$

where $Z_f(s)$ is a meromorphic function defined in terms of the gamma-function which has a triple pole at $s = 1$. The latter function is responsible for the main term $D_0(N; f)$ in the additive divisor problem, and the residue of the function $Z_f(s)M_f(s)$ at $s = 1$ gives the main term for the sum $S_f(T, \Delta)$. The contribution of $Z_f(s)$ to the error terms in these two problems is negligible.

To formulate the corresponding formulae in the cusp form cases, write

$$c(t) = \langle E(z, \tfrac{1}{2} + it), y^k |A(z)|^2 \rangle, \quad \tilde{c}(t) = \langle E(z, \tfrac{1}{2} + it, |u(z)|^2 \rangle,$$
$$c_j = \langle u_j, y^k |A(z)|^2 \rangle, \qquad\qquad \tilde{c}_j = \langle u_j, |u(z)|^2 \rangle.$$

The functions $C_f^*(s)$ for holomorphic and nonholomorphic cusp forms now read as follows (see [15], Theorem 1; as to (5.13) below, note that because of different scaling, the variable s in [15] is to be replaced by $s + k - 1$):

$$\frac{(4\pi)^k f^{1/2-s}}{2\Gamma(s)\Gamma(s + k - 1)} \left\{ \sum_{j=1}^{\infty} c_j \overline{\rho_j} t_j(f) \Gamma(s - z_j)\Gamma(s - \overline{z}_j) \right.$$

$$\left. + \frac{1}{2\sqrt{\pi}} \int_{-\infty}^{\infty} \frac{c(u)\sigma_{2iu}(f)\Gamma(s - \tfrac{1}{2} + iu)\Gamma(s - \tfrac{1}{2} - iu)}{(\pi f)^{iu} \Gamma(\tfrac{1}{2} - iu)\zeta(1 - 2iu)} du \right\} \tag{5.13}$$

and

$$\frac{8\pi f^{1/2-s}}{|\rho(1)|^2 4^s \Gamma^2(s/2)\Gamma(s/2+i\kappa)\Gamma(s/2-i\kappa)} \left\{ \sum_{j=1}^{\infty} \tilde{c}_j \overline{\rho}_j t_j(f)\Gamma(s-z_j)\Gamma(s-\overline{z}_j) \right.$$

$$\left. + \frac{1}{2\sqrt{\pi}} \int_{-\infty}^{\infty} \frac{\tilde{c}(u)\sigma_{2iu}(f)\Gamma(s-\frac{1}{2}+iu)\Gamma(s-\frac{1}{2}-iu)}{(\pi f)^{iu}\Gamma(\frac{1}{2}-iu)\zeta(1-2iu)} \, du \right\}. \quad (5.14)$$

In the above formulae, the series and integral represent the contribution of the discrete and continuous spectrum of the hyperbolic Laplacian, respectively. The functions $C_f^*(s)$ are holomorphic in the half-plane $\sigma > 1/2$, up to the possible pole at $s = 1$, and the points z_j, $\overline{z}_j$ are simple poles on the line $\sigma = 1/2$. The integrals are holomorphic on this line, and their poles in the half-plane $\sigma > 0$ are situated at the zeros of $\zeta(2s)$. To see this, the integrals should be analytically continued over the half-line. For this purpose, note that each one of these integrals is a holomorphic function in the strip $0 < \sigma < 1/2$, and if a suitable meromorphic function having poles at the zeros of $\zeta(2s)$ is added to it, the boundary values of the new function on the half-line coincide with those of the integral understood as a function in the half-plane $\sigma > 1/2$. In this way, a meromorphic continuation of the integral to the half-plane $\sigma > 0$ is established, and the same procedure may be repeated step by step to extend the meromorphic continuation to the whole plane.

In concrete applications, we have to cope with various spectral averages, for which purpose we quote a couple of inequalities. A prototype of a spectral mean value estimate is Iwaniec's spectral large sieve: putting $\alpha_j = |\rho_j|^2 / \cosh(\pi\kappa_j)$, we have

$$\sum_{K \leq \kappa_j \leq K+K_0} \alpha_j \left| \sum_{n \leq N} a_n t_j(n) \right|^2 \ll (KK_0 + N) \log^C(KN) \sum_{n \leq N} |a_n|^2 \quad (5.15)$$

for $1 \leq K_0 \leq K$ and any complex numbers a_n (for a simple proof, see [19] or [29], Theorem 3.3). An important application of this is the mean value estimate

$$\sum_{K \leq \kappa_j \leq 2K} H_j^4(\tfrac{1}{2}) \ll K^2 \log^C K; \quad (5.16)$$

for a proof, represent $H_j^2(\frac{1}{2})$ first by its approximate functional equation. Another proof, independent of the spectral large sieve, has been given by Motohashi [24].

For the inner products occurring in (5.13) and (5.14), we have

$$\sum_{\kappa_j \leq K} |c_j|^2 \exp(\pi\kappa_j) + \int_{-K}^{K} |c(u)|^2 \exp(\pi|u|) \, du \ll K^{2k}, \quad (5.17)$$

$$\sum_{\kappa_j \le K} |\tilde{c}_j|^2 \exp(\pi \kappa_j) + \int_{-K}^{K} |\tilde{c}(u)|^2 \exp(\pi|u|)\, du \ll K^\varepsilon. \qquad (5.18)$$

The former result is due to Good [4], and we proved the latter result recently in [15], [16] by a method applicable in both cases.

Returning to the integral $S_f(T, \Delta)$ in (5.2), its spectral part is given by

$$S_f^*(T, \Delta) = \frac{1}{2\pi i} \int_{(a)} C_f^*(s) M_f(s)\, ds \qquad (1/2 < a < 1). \qquad (5.19)$$

Here the contributions of the discrete and continuous spectra are formally analogous but the latter is less significant, so the former deserves more attention. The corresponding part of the above integral may be reduced to integrals of the type

$$\frac{1}{2\pi i} \int_{(a)} \frac{\Gamma(s - z_j)\Gamma(s - \overline{z}_j)}{4^s \Gamma^2(s/2)\Gamma(s/2 + i\kappa)\Gamma(s/2 - i\kappa)} X^s ds \qquad (\kappa \in \mathbb{R}), \qquad (5.20)$$

$$\frac{1}{2\pi i} \int_{(a)} \frac{\Gamma(s - z_j)\Gamma(s - \overline{z}_j)}{\Gamma(s)\Gamma(s + k - 1)} X^s ds \qquad (5.21)$$

if $M_f(s)$ is substituted from (5.3) and the order of the integrations is inverted.

To begin with our analysis of the integrals $S_f^*(T)$, let us briefly comment the case of the holomorphic cusp forms, which appears to be the easiest one. The integral (5.21) is of the Mellin-Barnes type and its value is

$$\frac{2}{\pi} \cosh(\pi \kappa_j) X^{1/2} \operatorname{Re}\left\{ X^{-i\kappa_j} \frac{\Gamma(\frac{1}{2} + i\kappa_j)\Gamma(-2i\kappa_j)}{\Gamma(k - \frac{1}{2} - i\kappa_j)} \right.$$
$$\left. \times F(\tfrac{1}{2} + i\kappa_j, \tfrac{3}{2} - k + i\kappa_j; 1 + 2i\kappa_j; 1/X) \right\}, \qquad (5.22)$$

as one may verify by a calculation similar to that in [21], sec. 3.6. The contribution of the continuous spectrum may can be treated analogously, after which the integration over s is done, and the remaining integral over x may be evaluated approximately by the saddle point method. The sum over f produces something like a factor $H_j(\frac{1}{2})$ to the discrete part of the formula, and a zeta-factor to its continuous part. The result of this procedure then appears to be of the same flavour as an identity due to Motohashi [27] for a local weighted mean square of a Hecke L-function for a holomorphic cusp form.

The remaining cases concerning the zeta-function and the Hecke L- series for a nonholomorphic cusp form may be treated simultaneously with the integral (5.20) serving as the common starting point; the case $\kappa = 0$ is related to the zeta-function, and otherwise κ will retain its previous meaning in terms of the

192   M. Jutila

eigenvalue corresponding to the cusp form. This integral is not as easy as (5.21) to write down explicitly, so we shall be content with approximate formulae. The approximation $C_f(s) \approx C_f^*(s)$ entails additional error terms. All these will be manageable essentially within the limit of a similar error term as in Theorem 1.

For a neater formulation of the spectral version of Theorem 1 (and to stress the analogy with the main theorem of [25]), we write

$$\Theta_f(\kappa, r; T, \Delta)$$
$$= \int_0^\infty \nu((\pi f/T)(2x+1)) x^{-1/2} (x+1)^{-1} \sin\left(T \log\left(1 + 1/x\right)\right)$$
$$\times \left(\log\left(1 + 1/x\right)\right)^{-1} \Lambda(x; \kappa, r) \exp\left(-\tfrac{1}{4}\Delta^2 \log^2\left(1 + 1/x\right)\right) dx, \quad (5.23)$$

$$\Lambda(x; \kappa, r) = \operatorname{Re}\left\{ (x + \tfrac{1}{2} + \delta(x))^{-ir} \right.$$
$$\left. \times \left(1 + i\frac{\cosh(2\pi\kappa)}{\sinh(\pi r)}\right) \frac{\Gamma(\tfrac{1}{2} + ir + 2i\kappa)\Gamma(\tfrac{1}{2} + ir - 2i\kappa)}{e(\kappa, r)\Gamma(1 + 2ir)} \right\} \quad (5.24)$$

with

$$\delta(x) = \frac{1}{2}(\sqrt{x(x+1)} - x) - \frac{1}{4} \quad (5.25)$$

and

$$e(\kappa, r) = \frac{\Gamma(\tfrac{1}{2}(\tfrac{1}{2} + ir) + i\kappa)\Gamma(\tfrac{1}{2}(\tfrac{1}{2} + ir) - i\kappa)}{\Gamma^2(\tfrac{1}{2}(\tfrac{1}{2} + ir))}.$$

Analogously, define $\Theta_f(r; T, \Delta)$ as in (5.23) but with $\Lambda(\cdots)$ replaced by

$$\Lambda(x; r) = \operatorname{Re}\left\{ (x + \tfrac{1}{2} + \delta(x))^{-ir} (1 + 1/x)^\alpha \right.$$
$$\left. \times \left(\frac{i}{\sinh(\pi r)}\right) \frac{\Gamma(\tfrac{1}{2} + ir)}{\Gamma(k - \tfrac{1}{2} - ir)\Gamma(1 + 2ir)} \right\}. \quad (5.26)$$

Note that $\delta(x) = O(1/x)$ as $x \to \infty$. Also, trivially $e(0, r) = 1$, and otherwise, for fixed κ, Stirling's formula (2.14) gives $e(\kappa, r) = 1 + O(r^{-1})$ as $r \to \infty$. Another useful observation is that $\Theta_f(\kappa, r; T, \Delta)$ and $\Theta_f(r; T, \Delta)$ decay rapidly as r exceeds T/Δ. The reason is that these quantities are much similar to $M_f(\overline{z}_j)$, so the assertion follows from the remark after the corollary to Lemma 3. Therefore the spectral series (and likewise the integrals) in the following theorem may be truncated. In that theorem, we are going to use the notation $S_{01}(T)$, defined in (4.2), to stand for the error term of $S_0(T)$ with the understanding that the corresponding function $C_0(s)$ is clear from the context. Recall that $f_0 = (T/\Delta) \log T$.

Theorem 2 *Let $E_{2,\varphi}(T, \Delta)$ be defined by (5.5). Then, for $T^{2/5+\varepsilon} \leq \Delta \leq T^{2/3}$, we have*

$$E_2(T, \Delta) = S_{01}(T) + 2 \sum_{j=1}^{\infty} \alpha_j H_j^2(\tfrac{1}{2}) \sum_{1 \leq f \leq f_0} t_j(f) f^{-1/2} \Theta_f(0, \kappa_j; T, \Delta)$$

$$+ \frac{2}{\pi} \int_{-\infty}^{\infty} \frac{|\zeta(\tfrac{1}{2} + iu)|^4}{|\zeta(1 + 2iu)|^2} \sum_{1 \leq f \leq f_0} \sigma_{2iu}(f) f^{-1/2-iu} \Theta_f(0, u; T, \Delta) \, du$$

$$+ O(\Delta \log^C T), \tag{5.27}$$

$$E_{2,F}(T, \Delta) = S_{01}(T)$$

$$+ 2(4\pi)^k \sum_{j=1}^{\infty} c_j \overline{\rho_j} \sum_{1 \leq f \leq f_0} t_j(f) f^{-1/2} \Theta_f(\kappa_j; T, \Delta)$$

$$+ 4^{k-1/2} \pi^{k-1/2} \int_{-\infty}^{\infty} \frac{c(u)}{\pi^{iu} \Gamma(\tfrac{1}{2} - iu) \zeta(1 - 2iu)}$$

$$\times \sum_{1 \leq f \leq f_0} \sigma_{2iu}(f) f^{-1/2-iu} \Theta_f(u; T, \Delta) \, du + O(\Delta \log^C T), \tag{5.28}$$

and

$$E_{2,H}(T, \Delta) = S_{01}(T)$$

$$+ \frac{16\pi}{|\rho(1)|^2} \sum_{j=1}^{\infty} \frac{\tilde{c}_j \overline{\rho_j}}{\cosh(\pi \kappa_j) |\Gamma(z_j/2)|^4} \sum_{1 \leq f \leq f_0} t_j(f) f^{-1/2} \Theta_f(\kappa, \kappa_j; T, \Delta)$$

$$+ \frac{8}{|\rho(1)|^2 \sqrt{\pi}} \int_{-\infty}^{\infty} \frac{\tilde{c}(u) \Gamma(\tfrac{1}{2} + iu)}{\pi^{iu} |\Gamma(\tfrac{1}{4} + \tfrac{iu}{4})|^4 \zeta(1 - 2iu)}$$

$$\times \sum_{1 \leq f \leq f_0} \sigma_{2iu}(f) f^{-1/2-iu} \Theta_f(\kappa, u; T, \Delta) \, du + O(\Delta \log^C T). \tag{5.29}$$

Proof The proofs of these formulae are closely analogous, so it suffices to consider (5.29) in more detail as an example. We decompose the formula (5.5) in the present case as follows:

$$E_{2,H}(T, \Delta) = S_{01}(T) + S^*(T, \Delta) + S^{**}(T, \Delta), \tag{5.30}$$

where

$$S^*(T, \Delta) = \sum_{1 \leq f \leq f_0} S_f^*(T, \Delta), \tag{5.31}$$

with $S_f^*(T, \Delta)$ as in (5.19), and

$$S^{**}(T, \Delta) = \sum_{1 \le f \le f_0} \frac{1}{2\pi i} \int_{(a)} (C_f(s) - C_f^*(s)) M_f(s) \, ds \qquad (a > 1). \qquad (5.32)$$

The explicit terms in (5.29) will arise from $S^*(T, \Delta)$, while $S^{**}(T, \Delta)$ will give only error terms.

The sum $S^*(T, \Delta)$: The function $C_f^*(s)$ was defined in (5.14), where our main concern will be the discrete part. Recall that the integral in (5.19) may be truncated to $|t| \ll T^{1+\varepsilon}/\Delta$. The contribution of the jth term in (5.14) to $S^*(T, \Delta)$ will be calculated in different ways depending on the size of κ_j relative to the number

$$K_0 = T^{1/5+\varepsilon}. \qquad (5.33)$$

Let first $\kappa_j \le K_0$, and move the integration in (5.19) to the line $\sigma = -1/2 + \varepsilon$ passing simple poles at z_j and $\overline{z}_j$. Then, for given s on the new line of integration, the integrand is

$$\ll e^{\pi|t|}(|t|^3 + 1) \sum_{\kappa_j \le K_0} |\Gamma(s - z_j)\Gamma(s - \overline{z}_j)||\tilde{c}_j| \exp(\pi\kappa_j/2)\alpha_j^{1/2}$$

$$\times \left| \sum_{1 \le f \le f_0} t_j(f) f^{1/2-s} M_f(s) \right|.$$

We apply this to estimate the integral over $|t| \le 2K_0$.

Let $K < K_0$, and consider first the local sum over $K \le \kappa_j \le K + 1$ using Cauchy's inequality, the large sieve inequality (5.15), and the estimate (5.7) for $M_f(s)$. It is easy to verify that

$$\sum_{K \le \kappa_j \le K+1} \alpha_j \left| \sum_{1 \le f \le f_0} t_j(f) f^{1/2-s} M_f(s) \right|^2 \ll K_0 T^{-1+\varepsilon};$$

split up the f-sum first into parts over $1 \le f < K$ and $2^\nu K \le f < 2^{\nu+1}K$ ($\nu = 0, 1, \ldots$). Next integrate over t, and finally sum over K applying Cauchy's inequality and (5.18). The final result is $\ll K_0^{5/2} T^{-1/2+\varepsilon} \ll T^\varepsilon$. The integral over $2K_0 \le |t| \ll T^{1+\varepsilon}/\Delta$ is treated on taking the integration sufficiently far to the left. The integrals over horizontal lines are estimated as above, and the new integral is small because the integrand decays exponentially as σ decreases.

Consider next the residual terms. Let R be the residue of the integrand in (5.20) at $s = \overline{z}_j$. Then

$$\cosh(\pi\kappa_j)e(\kappa, \kappa_j)R = \frac{\Gamma(z_j + 2i\kappa)\Gamma(z_j - 2i\kappa)}{4|\Gamma(z_j/2)|^4\Gamma(2z_j)} \left(1 + i\frac{\cosh(2\pi\kappa)}{\sinh(\pi\kappa_j)}\right) X^{1/2-i\kappa_j},$$

which may be verified by use of the formula $\Gamma(s)\Gamma(1-s) = \pi/\sin(\pi s)$ and the duplication formula for the gamma-function. We see that the contribution of the pair of the poles z_j, $\overline{z}_j$ to $S^*(T,\Delta)$ is almost identical with the corresponding term in (5.29), the only deviation being that instead of the factor $(x+1/2+\delta(x))^{-i\kappa_j}$ in (5.24) (for $r = \kappa_j$) we now have $(x+1)^{-i\kappa_j}$. Since $\kappa_j \leq K_0$ and we may suppose that $x \gg \Delta/\log T$, the relative error here is $1 + O(K_0\Delta^{-1}\log T)$. To clarify the effect of this replacement, let us first estimate a partial sum of the spectral sum in (5.29). Note that $\Theta_f(\kappa,\kappa_j;T,\Delta)$ is closely related to $M_f(z_j)$, so it may be estimated by Lemma 3, or at least by the same method. Then by the spectral large sieve and (5.18), we see that the sum over $\kappa_j \leq K$ in (5.29) is

$$\ll T^{1/2+\varepsilon}K^{1/2}. \tag{5.34}$$

A minor complication here is the dependence of $\Theta_f(\ldots)$ on κ_j, but a version of the large sieve allowing such a dependence is available [19], and it applies here to establish the estimate (5.34). Now the difference of the two sums under consideration is analytically of the same nature as the sum in (5.29), expect that there is now a diminishing factor $\ll K_0\Delta^{-1}\log T$. This scaling factor goes through the calculation outlined above (such a comparison device will be applied again below), so the approximation error to be estimated is at most the bound (5.34) multiplied by this factor, thus it is $\ll T^{1/2+\varepsilon}K_0^{3/2}\Delta^{-1}$. By (5.33) and (3.2), this is $\ll \Delta$.

Next we turn to the terms with $\kappa_j > K_0$. Again we start from the integral (5.20), but this time it is approximated by the integral (5.21) for $k = 0$, divided by 8π. The value of the latter integral may be read from (5.22). Leaving aside for a moment the estimation of the approximation error, let us proceed with the new integral. The hypergeometric function is transformed by the formula (see [22], eq. (9.6.12))

$$F(a,b;2b;z) = \left(\tfrac{1}{2}(1+\sqrt{1-z})\right)^{-2a}$$
$$\times F\left(a, a-b+\tfrac{1}{2}; b+\tfrac{1}{2}; \left(\frac{1-\sqrt{1-z}}{1+\sqrt{1-z}}\right)^2\right). \tag{5.35}$$

The new hypergeometric series is rapidly convergent, so we may approximate it by its leading term 1. Also, the exponent $-2a = -3 - 2i\kappa_j$ at the factor in front of it is simplified to $-2i\kappa_j$; all this can be done with a relative accuracy $1 + O(\Delta^{-1}\log T)$. Note also that

$$(x+1)^{-ir}\left(\tfrac{1}{2}(1+\sqrt{x/(x+1)})\right)^{-2ir} = (x+\tfrac{1}{2}+\delta(x))^{-ir}. \tag{5.36}$$

After this cosmetics, the jth term is as in (5.29) except that in place of $\Lambda(x; \kappa, r)$ we have the factor

$$(2\pi^2)^{-1} \cosh^2(\pi r) |\Gamma(\tfrac{1}{4} + \tfrac{1}{2}ir)|^4 \operatorname{Re}\left[(x + \tfrac{1}{2} + \delta(x))^{-ir} \frac{\Gamma(\tfrac{1}{2} + ir)\Gamma(-2ir)}{\Gamma(-\tfrac{1}{2} - ir)}\right].$$

It is easily seen that this coincides with $\Lambda(x; \kappa, r)$ with a relative accuracy $1 + O(r^{-1})$. The same comparison device as above shows that the consequent error due to the preceding approximations is

$$\ll T^{1/2+\varepsilon}(K_0^{-1/2} + (T/\Delta)^{1/2}\Delta^{-1}) \ll \Delta;$$

recall that the series in (5.29) may be truncated to $\kappa_j \leq T^{1+\varepsilon}\Delta^{-1}$.

The effect of the change of the gamma-factors made above is still to be discussed. Stirling's formula (in its more accurate form as an asymptotic expansion) shows that the ratio of the respective integrands is asymptotically $1 + a_1/s + a_2/s^2 + ..$, which may also be written as $1 + b_1/(s-1) + b_2/(s-1)s + ..$, where the coefficients depend on κ. Thus the difference of the integrals can be written in terms of integrals of the type (5.21) for $k = 0, 1, \ldots$; and the values of the latter integrals are given in (5.22). Thus, after all, the argument goes back to similar integrals as above, for bigger values for k however, which decreases their order at least by a factor κ_j^{-1}. As we saw already, such a diminishing factor turns main terms into error terms.

So far we have worked out the discrete part of the formula (5.29). The corresponding part of (5.27) emerges similarly; we have now $\kappa = 0$, and the estimate (5.16) plays the same role as (5.18) above. Further, the contribution of the continuous spectrum may be reduced to integrals of the type (5.20)-(5.21), with $1/2 + iu$ in place of z_j, so the integrals in (5.27) and (5.29) drop out by analogy. The proof of (5.28) is similar but simpler because we are now working with the integral (5.21) which is explicitly given by (5.22). This completes our analysis of the term $S^*(T, \Delta)$ in (5.30) and its analogues.

The sum $S^{**}(T, \Delta)$: We quote from [15], Lemma 6 and [21], Lemma 1, a formula for $C_f^*(s)$ indicating its similarity with $C_f(s)$: for $\sigma > 1$, we have

$$C_f^*(s) = \sum_{n=1}^{\infty} t(n)t(n+f)\,(1 + f/n)^{1/2}\,(n+f)^{-s}(1 + g_\kappa(s; n, f))$$

$$+ \frac{1}{2}f^{-s}\sum_{n=1}^{f-1} t(n)t(f-n)h_\kappa(s; n, f), \tag{5.37}$$

where

$$g_\kappa(s; n, f) = \frac{\pi 2^{3-2s} \Gamma(s)}{|\Gamma(\frac{1}{2} + i\kappa)|^2 \Gamma^2(s/2) \Gamma(s/2 + i\kappa) \Gamma(s/2 - i\kappa)}$$

$$\times \frac{1}{2\pi i} \int_{(b)} \Gamma(\xi + \tfrac{1}{2} + i\kappa) \Gamma(\xi + \tfrac{1}{2} - i\kappa) \Gamma(s - \xi - \tfrac{1}{2} + i\kappa)$$

$$\times \Gamma(s - \xi - \tfrac{1}{2} - i\kappa) \frac{\Gamma(-\xi)}{\Gamma(s - \xi)} \left((1 + f/n)^\xi - 1 \right) d\xi \qquad (5.38)$$

with $-1/2 < b < \min(\sigma - 1/2, 0)$, and

$$h_\kappa(s; n, f) = \frac{2^{3-s} \sqrt{\pi} \Gamma^3(s)}{\Gamma(s + \frac{1}{2}) \Gamma^2(s/2) \Gamma(s/2 + i\kappa) \Gamma(s/2 - i\kappa)}$$

$$\times \int_0^\infty (1 + \alpha(u))^{-s} F\left(s, \tfrac{1}{2}; s + \tfrac{1}{2}; \frac{1 - \alpha(u)}{1 + \alpha(u)} \right) \cos(\kappa u) \, du \qquad (5.39)$$

with

$$\alpha(u) = f^{-1} \sqrt{f^2 + 2n(f - n)(\cosh u - 1)}.$$

Since $g_\kappa(\cdots)$ in (5.37) appers to be usually a small "perturbation" term of order $\ll n^{-1}$ for given f and s, the first approximation of the difference $c_f(s) - c_f^*(s)$ involves series in $(n + f)^{-s-1}$. Hence this difference can be expressed in terms of series converging absolutely in the half-plane $\sigma > 0$. Note that the integral in (5.38) converges rapidly, so that it may be truncated to the segment $|\text{Im } \xi| \leq \log^2 T$. Then, for $|t| \gg T^\varepsilon$, the order of the whole gamma-expression in (5.38) is $\ll |t|^{-b}$ uniformly for the relevant values of ξ bounded away from the poles of the integrand.

We now show, for $a > 1$, that

$$\sum_{1 \leq f \leq f_0} \int_{(a)} M_f(s) \left(\sum_{n=1}^\infty t(n) t(n + f) \left(1 + f/n \right)^{1/2} g_\kappa(s; n, f)(n + f)^{-s} \right) ds$$

$$\ll (T/\Delta)^{1/2} T^\varepsilon, \qquad (5.40)$$

and that

$$\sum_{1 \leq f \leq f_0} \int_{(a)} M_f(s) \left(f^{-s} \sum_{n=1}^{f-1} t(n) t(f - n) h_\kappa(s; n, f) \right) ds \ll T^\varepsilon. \qquad (5.41)$$

For a proof of (5.40), we choose $b = -1/2 + \varepsilon$ in (5.38) and take the s-integration to the line $\sigma = 2\varepsilon$. The integral over $|t| \ll T^\varepsilon$ is easily estimated by (5.7) and Stirling's formula. Next we write the last factor of the integrand in (5.38) as

$$(1 + f/n)^{\xi} - 1 = \left(1 - \frac{f}{n+f}\right)^{-\xi} - 1 = \frac{\xi f}{n+f} + r(\xi; n, f), \qquad (5.42)$$

and split up $g_{\kappa}(s; n, f)$ into two parts, and likewise also for the integral (5.40) (over $|t| \geq T^{\varepsilon}$).

Consider first the contribution of $r(\xi; n, f)$. Note that

$$r(\xi; n, f) \ll (|\xi|(|\xi| + 1) \left(\frac{f}{n+f}\right)^2 \quad \text{for} \quad \frac{f}{n+f} \leq \frac{1}{2(|\xi| + 1)};$$

otherwise $r(\xi; n, f)$ does not make sense as a "remainder term", and it is then estimated trivially from (5.42).

We take the s-integration further left to the line $\sigma = -1 + 2\varepsilon$; the integrals over the horizontal segments $t = \pm T^{\varepsilon}$, $-1 + 2\varepsilon \leq \sigma \leq 2\varepsilon$ may be estimated by use of (5.7) and Stirling's formula as above. Next we move the ξ-integration to the line $b = 1/2$ noting that the integrand is regular at $\xi = 0$. Now, by (5.7) and the above remark on gamma-factors, we get a contribution $\ll T^{1+\varepsilon}\Delta^{-2}$ to (5.40); this is smaller than the bound in (5.40).

To estimate the contribution of the first term on the right of (5.42) to (5.40), we keep the s-integration on the line $\sigma = 2\varepsilon$, but take the ξ-integration to the line $b = 1 - \varepsilon$. Henceforth the argument is as above, and the contribution is $\ll (T/\Delta)^{1/2}T^{\varepsilon}$. This completes the proof of (5.40).

Turning to the proof of (5.41), we first move the integration to $a = \varepsilon$, and note that the integral over $|t| \leq T^{\varepsilon}$ gives $\ll T^{\varepsilon}$. Next we transform the hypergeometric series in (5.39) by the formula

$$F(a, b; c; z) = (1 - z)^{-a} F\left(a, c - b; c; \frac{z}{z-1}\right), \qquad (5.43)$$

valid for $|\arg(1 - z)| < \pi$ and $c \neq 0, -1, \ldots$ (see [22], eq. (9.5.1)). The new hypergeometric series $F(\frac{1}{2}, \frac{1}{2}; s + \frac{1}{2}; (\alpha(u) - 1)/(2\alpha(u)))$ is rapidly convergent, and its leading term gives rise to the integral

$$\frac{1}{\sqrt{2}} \int_0^{\infty} (1 + \alpha(u))^{1/2-s} \alpha(u)^{-1/2} \cos(\kappa u) \, du.$$

If $|t| \gg T^{\varepsilon}$, the present integrand, like those arising from the other terms of the hypergeometric series, is rapidly oscillating in a range where $u \sinh u \gg t^{-1} f^2 (n(f - n))^{-1} T^{\varepsilon}$. Therefore, by a familiar principle in the theory of exponential integrals, this part of the integral (equipped with a smooth weight function) is very small and may be omitted. The truncated version of $h_\kappa(s; n, f)$ is now substituted into (5.41), and the integration is moved sufficiently far to the left. In virtue of (5.7), the new integral will be small, and the integrals over the respective horizontal segments on the lines $t = \pm T^\varepsilon$ are easy to estimate to complete the proof (5.41).

In view of (5.40) and (5.41), we now have, by (5.32) and (5.37),

$$S^{**}(T, \Delta) = O(\Delta)$$
$$+ \sum_{1 \le f \le f_0} \frac{1}{2\pi i} \int_{(a)} M_f(s) \sum_{n=1}^{\infty} t(n)t(n+f) \left(1 - (1 + f/n)^{1/2}\right)(n+f)^{-s} \, ds.$$

Since $M_f(s)$, by definition, is the Mellin transform of $W_f((x-f)/f)$, this means that

$$S^{**}(T, \Delta) = \sum_{1 \le f \le f_0} \sum_{n=1}^{\infty} t(n)t(n+f)W_f(n/f)\left(1 - (1 + f/n)^{1/2}\right) + O(\Delta).$$

The significant terms here are those with $n \gg \Delta f / \log T$; otherwise $W_f(n/f)$ is very small. Therefore the first approximation to this sum is

$$-\frac{1}{2} \sum_{1 \le f \le f_0} f \sum_{n=1}^{\infty} t(n)t(n+f)n^{-1}W_f(n/f). \tag{5.44}$$

An analytic treatment of this amounts to using Dirichlet series with the argument $s + 1$ instead of s. Compared with the earlier situation, the terms in the new sum are now smaller by a factor about Δ^{-1} at least. The spectral contribution to the original sum was $\ll T^{1/2+\varepsilon}(T/\Delta)^{1/2}$ by (5.34), if estimated simply by absolute values, and it is therefore evident that the corresponding contribution to the sum (5.44) may be estimated by this bound divided by Δ; this gives $\ll T^{1+\varepsilon}\Delta^{-3/2} \ll \Delta$. Thus the sum (5.44), though explicit, is small enough to be treated as an error term.

Now, all in all, we have shown that $S^{**}(T, \Delta) \ll \Delta$, and together with our previous discussion of the sum $S^*(T, \Delta)$, this completes the proof of (5.29).

The proof of (5.27) is analogous to the above except that in the formula (5.37) for $C_f^*(s)$ there is the additional term (see [15], eq. (4.2) and [14], Lemma 1)

$$\frac{\sqrt{\pi} d(f) 2^{2-2s} \Gamma^3(s)}{f^s \Gamma^4(s/2) \Gamma(s+1/2)}.$$

Its contribution to $S^{**}(T, \Delta)$ is $\ll T^{1+\varepsilon}\Delta^{-2} \ll \Delta$, as one may verify on moving the integration in the corresponding integral to the line $\sigma = -1 + \varepsilon$ and estimating $M_f(s)$ again by (5.7).

6. Comparison with Motohashi's formulae It is interesting to compare the formula (5.27) with the fundamental formula of Motohashi [25] for the local weighted fourth moment

$$\frac{1}{\Delta\sqrt{\pi}} \int_{-\infty}^{\infty} |\zeta(\tfrac{1}{2} + i(T + t))|^4 e^{-(t/\Delta)^2}\, dt. \tag{6.1}$$

Likewise, (5.28) corresponds to the main theorem in [27].

In Motohashi's formula for the above mean value, the contribution of the discrete spectrum consists of the sum

$$\sum_{j=1}^{\infty} \alpha_j H_j^3(\tfrac{1}{2})\Theta(\kappa_j; T, \Delta), \tag{6.2}$$

where

$$\Theta(r; T, \Delta) = \int_0^\infty (x(x+1))^{-1/2} \cos(T\log(1 + 1/x))$$
$$\times \Lambda(x, r) \exp\left(-\tfrac{1}{4}\Delta^2 \log^2(1 + 1/x)\right) dx \tag{6.3}$$

with

$$\Lambda(x, r) = \operatorname{Re}\left[x^{-1/2-ir}\left(1 + \frac{i}{\sinh(\pi r)}\right) \frac{\Gamma^2(\tfrac{1}{2} + ir)}{\Gamma(1 + 2ir)}\right.$$
$$\left. \times F(\tfrac{1}{2} + ir, \tfrac{1}{2} + ir; 1 + 2ir; -1/x)\right]. \tag{6.4}$$

An apparent analogy is visible between (6.3)–(6.4) on one hand, and (5.23)–(5.24) on the other hand, the latter for $\kappa = 0$. (We follow here the notation of [25], and the present $\Lambda(\cdots)$ should not be confused with our earlier notation.)

To make the correspondence more precise, we transform here the hypergeometric function by the formula (5.35). Approximating the new hypergeometric series by its leading term 1, we may analyze the dependence of $\Lambda(x, r)$ on x as follows:

$$x^{-ir} F(\tfrac{1}{2} + ir, \tfrac{1}{2} + ir; 1 + 2ir; -1/x) \approx x^{-ir}\left(\tfrac{1}{2}(1 + \sqrt{(x+1)/x})\right)^{-1-2ir}$$
$$\approx x^{-ir}\left(\tfrac{1}{2}(1 + \sqrt{(x+1)/x})\right)^{-2ir} = (x+1)^{-ir}\left(\tfrac{1}{2}(1 + \sqrt{x/(x+1)})\right)^{-2ir}$$
$$= (x + \tfrac{1}{2} + \delta(x))^{-ir};$$

the last step by (5.36).

If the equation (6.1) is integrated with respect to T, then the result is essentially $I_{2,\varphi}(T, \Delta)$ for $\varphi(s) = \zeta^2(s)$. On the other hand, integration of (6.2) leads essentially to the corresponding spectral sum in (5.27). To see this, note that the relevant range for f in the formulae of Theorem 2 is $[1, \kappa_j/2\pi]$ because then the integral $\Theta(\cdots)$ has a saddle point (if the weight function ν is temporarily viewed as the characteristic function of the unit interval for simplicity), and that

$$2 \sum_{f \le \kappa_j/2\pi} t_j(f) f^{-1/2} \approx H_j(\tfrac{1}{2}) \tag{6.5}$$

by the approximate functional equation for $H_j(s)$.

In Motohashi's formula, there is also a contribution from holomorphic cusp forms, but nothing like that is seen in our theorem 2. The explanation is that this ingredient, which is small in practice, is hidden somewhere in the error term.

Turning to (5.28), we compare it with the formula of Motohashi [27] for the mean value of the type (6.1) for $|F(\tfrac{1}{2} + it)|^2$. In that formula, there is a main term and contributions from the discrete and continuous spectrum, but none from the holomorphic cusp forms. Let us restrict ourselves again to the contribution of the discrete spectrum, which reads as follows:

$$(4\pi)^k \sum_{j=1}^{\infty} c_j \overline{\rho_j} H_j(\tfrac{1}{2}) \Theta(\kappa_j; T, \Delta), \tag{6.6}$$

where

$$\Theta(r; T, \Delta) = \int_0^{\infty} x^{-k/2-1/2}(x+1)^{k/2-1} \cos\left(T \log\left(1 + 1/x\right)\right)$$
$$\times \exp\left(-\tfrac{1}{4}\Delta^2 \log^2\left(1 + 1/x\right)\right) \Lambda(x, r) dx,$$

$$\Lambda(x, r) = \operatorname{Re}\left[x^{-ir}\left(\frac{i}{\sinh(\pi r)}\right) \frac{\Gamma(\tfrac{1}{2} + ir)}{\Gamma(k - \tfrac{1}{2} - ir)\Gamma(1 + 2ir)}\right.$$
$$\left. \times F\left(k - \tfrac{1}{2} + ir, \tfrac{1}{2} + ir; 1 + 2ir; -1/x\right)\right].$$

On the other hand, the precise form of (5.28), given by the argument of its proof if no approximations are made, should involve the factor

$$(x+1)^{-ir} F\left(\tfrac{1}{2} + ir, \tfrac{3}{2} - k + ir; 1 + 2ir; 1/(x+1)\right)$$

in place of $(x + \tfrac{1}{2} + \delta(x))^{-ir}$. Now, by (5.43), this factor equals

$$(1 + 1/x)^{1/2} x^{-ir} F\left(k - \tfrac{1}{2} + ir, \tfrac{1}{2} + ir; 1 + 2ir; -1/x\right),$$

whence an integrated version of (6.6) amounts to the explicit part of (5.28), up to the approximation (6.5).

7. The mean square of the error term   As an application of Theorem 2, we extend the mean square estimate (1.7) for $E_2(T)$ to the more general error term $E_{2,\varphi}(T)$. Since the function $E_{2,\varphi}(T)$ can decrease only relatively slowly, the occurrence of a large value V of $|E_{2,\varphi}(T)|$ implies that this function must be large in an interval of length $\gg V/\log^C T$ after or before T (depending on the sign $E_{2,\varphi}(T)$); here C (1 or 4) is the degree of the polynomial P_φ in (1.5). This argument, in combination with the next theorem, yields a generalization of (1.6) to be formulated as a corollary below.

Theorem 3   *For $T \geq 2$, we have*

$$\int_0^T E_{2,\varphi}^2(t)\, dt \ll T^{2+\varepsilon}. \tag{7.1}$$

Moreover, for $E_2(T)$ and $E_{2,F}(T)$, the estimate holds in the stronger form $\ll T^2 \log^C T$.

Corollary   *We have*

$$E_{2,\varphi}(T) \ll T^{2/3+\varepsilon}.$$

Proof of Theorem 3:   Obviously it follows from the definition (3.3) of $I_{2,\varphi}(T)$ that

$$I_{2,\varphi}(T - \Delta \log T, \Delta) + O(1) \leq I_{2,\varphi}(T) \leq I_{2,\varphi}(T + \Delta \log T, \Delta) + O(1)$$

since $I_{2,\varphi}(T)$ is increasing. The respective main terms for $I_{2,\varphi}(\cdots)$ here differ at most by amounts $\ll \Delta \log^C T$ from each other, so for the error terms we have

$$E_{2,\varphi}(T - \Delta \log T, \Delta) + O(\Delta \log^C T)$$
$$\leq E_{2,\varphi}(T)$$
$$\leq E_{2,\varphi}(T + \Delta \log T, \Delta) + O(\Delta \log^C T).$$

Thus for a proof of (7.1) it suffices to show that

$$\int_0^\infty v(t) E_{2,\varphi}^2(t, \Delta)\, dt \ll T^{2+\varepsilon}, \tag{7.2}$$

where $\Delta = T^{1/2}$ and $v(t)$ is a smooth weight function of support in $[T/2, 3T]$ such that $v(t) \gg 1$ for $t \sim T$.

Let us consider the proof of (7.2) for $\varphi(s) = H(s)$ as an example. In the spectral formula (5.29) for $E_{2,H}(T)$, the error term is of admissible order by our choice of Δ. As to the leading term, we have

$$\int_0^T S_{01}^2(T)\, dt \ll T^{2+\varepsilon}.$$

This may be verified by using the formula (4.2) for $S_{01}(T)$ and a mean square estimate for the function $C_0(s)$ on the critical line (see [15], eqs. (3.5) and (3.7)).

We are now left with the spectral terms in (5.29). The contributions of the discrete and continuous spectra are analogous, the latter being however somewhat more straightforward to deal with, for instead of the spectral large sieve one may appeal to the ordinary one. Therefore we may henceforth focus our attention to the discrete part.

It is convenient to restrict the summations in (5.29) to "dyadic" intervals: $\kappa_j \sim K$, $f \sim F$. By our choice of Δ, we may suppose that K, F, and the variable x in (5.23) satisfy

$$K \ll T^{1/2+\varepsilon}, \quad F \ll T^{1/2}\log T, \quad T^{1/2}/\log T \ll x \ll T/F. \qquad (7.3)$$

The quantity $\Theta_f(\kappa, \kappa_j; T, \Delta)$ in (5.29) involves an integral, given in (5.23)–(5.24), which should be first simplified. For this purpose, the following general inequality is helpful. Let $\theta_0(r)$ and $\theta_f(r)$ be functions for $r \sim K$, $f \sim F$, the latter continuously differentiable in r, such that

$$\theta_0(r) \ll \theta_0, \quad |\theta_f(r)| + |\theta_f'(r)| \ll \theta. \qquad (7.4)$$

Then we have

$$\sum_{\kappa_j \sim K} \tilde{c}_j \exp(\pi\kappa_j/2)\alpha_j^{1/2}\theta_0(\kappa_j) \sum_{f \sim F} t_j(f)f^{-1/2}\theta_f(\kappa_j)$$
$$\ll K^{1/2}(K+F)^{1/2}\theta_0\theta T^\varepsilon. \qquad (7.5)$$

Indeed, if κ_j is first restricted to a short sum $[k, k+1]$ for an integer $k \asymp K$, then the dependence of $\theta_f(\kappa_j)$ on κ_j may be eliminated (as in [19]) by use of Sobolev's inequality familiar from the large sieve method. After that, Cauchy's inequality and (5.15) are applied to this short spectral sum, and finally the k-sum is estimated by Cauchy's inequality and (5.18).

Now the spectral sum over $\kappa_j \sim K$ in (5.29) can be written as the left hand side of (7.5), where $\theta_f(\kappa_j)$ represents the x-integral in (5.23), and $\theta_0(\kappa_j)$ with $\theta_0(\kappa_j) \ll \theta_0 = K^{1/2}$ consists of certain factors depending on κ_j.

The right hand side of (7.5) is $\ll T^{1/2+\varepsilon}$ if

$$\theta \ll T^{1/2+\varepsilon} K^{-1} (K+F)^{-1/2}. \tag{7.6}$$

This gives a bound for the accuracy of admissible approximations of the x-integral mentioned above.

That integral is now simplified as follows: we replace the weight function ν by the characteristic function of the unit interval $[0,1]$, omit the small correction function $\delta(x)$, and use the approximations

$$x^{-1/2}(x+1)^{-1} \left(\log\left(1+1/x\right)\right)^{-1} = \sqrt{2}(2x+1)^{-1/2} + O(x^{-3/2}),$$

$$\log\left(1+1/x\right) = \frac{2}{2x+1} + O(x^{-3}) \quad (x \to \infty).$$

These approximations are readily admissible by (7.5) and (7.6). Then, in the variable $y = (\pi f/T)(2x+1)$, the simplified integral becomes

$$\left(\frac{T}{2\pi f}\right)^{1/2 - i\kappa_j} \int_0^1 y^{-1/2 - i\kappa_j}$$
$$\times \sin(2\pi f/y) \exp(-\Delta^2 \pi^2 f^2 (Ty)^{-2})\, dy; \tag{7.7}$$

the lower limit of integration should actually be $\pi f/T$, but this may be replaced by 0 with a small error.

The exponential integral (7.7) could be evaluated approximately by the saddle point method. However, because its precise main term is irrelevant for our present purposes, we prefer to follow a more elementary line of argument: we are using the "second derivative test" (see [32], Lemma 4.5) in the form of the identity

$$\int_a^b G(x) e^{iF(x)}\, dx = \int_{c-\delta}^{c+\delta} G(x) e^{iF(x)}\, dx + \left(\Big|_a^{c-\delta} + \Big|_{c+\delta}^b\right) \frac{G(x)}{iF'(x)} e^{iF(x)}$$
$$- \left(\int_a^{c-\delta} + \int_{c+\delta}^b\right) \left(\frac{G(x)}{iF'(x)}\right)' e^{iF(x)}\, dx, \tag{7.8}$$

where $F'(c) = 0$, and $|F''(c)| \asymp r$, $\delta \asymp r^{-1/2}$. Here it is assumed that $a \le c - \delta < c + \delta \le b$; otherwise the equation should be modified in an obvious way. If no saddle point exists in the interval $[a-\delta, a+\delta]$, where δ is the typical order of $|F''(x)|^{-1/2}$, then (7.8) amounts simply to partial integration, or essentially to the "first derivative test".

The saddle point for the integral (7.7) is $y_0 = 2\pi f/\kappa_j$, which lies in the unit interval for $f \le \kappa_j/2\pi$. Let us suppose first the $F \ll K$. To begin with, we apply partial integration in (7.7) with respect to the first factor in the integrand, observing that the integrated terms vanish. Then we use the device (7.8), where the most critical term is the leading integral. We consider its contribution to (7.7), noting that the other terms lead to analogous calculations. Accordingly, we substitute y in a neighbourhood of the saddle point as $y = 2\pi f/\kappa_j + \xi$ with $\xi \ll FK^{-3/2}$. The integral over ξ gives a contribution to $\Theta_f(\kappa, \kappa_j; t, \Delta)$ in (5.29), which is squared out and integrated over t with the same weight $v(t)$ as in (7.2). The expression to be integrated involves sums over $f, g \sim F$, $\kappa_h, \kappa_j \sim K$, and integrals over $\xi, \eta \ll FK^{-3/2}$, say. Now the t-integral is very small by its oscillatory and smoothness properties unless

$$\kappa_h - \kappa_j \ll T^\varepsilon, \qquad (7.9)$$

in which case it is estimated trivially.

Each of the pairs κ_h, κ_j satisfying (7.9) is counted at least once if we first sum over

$$(\mu - 1)T^\varepsilon \le \kappa_h, \kappa_j \le (\mu + 1)T^\varepsilon$$

for an integer $\mu \asymp K$, and then sum over μ. The argument now amounts essentially to that used in the proof of (7.5) above. After having estimated these spectral sums, again by use of Cauchy's inequality, (5.15), and (5.18), we integrate trivially over ξ and η to end up with the desired estimate $\ll T^{2+\varepsilon}$.

In the case $F \gg K$ (with the implied constant sufficiently large) the argument is more straightforward because there is no saddle point for the integral (7.7), and the same estimate as above is obtained. This completes the proof of (7.2) for $\varphi(s) = H(s)$.

The preceding argument is slightly too wasteful to give (1.7) and its analogue for $E_{2,F}(T)$ with the logarithmic factor instead of T^ε. However, because now (5.15) or (5.17) is applied instead of (5.18), such a sharpening is possible in principle, and also in practice if the procedure is refined appropriately.

References

[1] T.M. Apostol. *Modular Functions and Dirichlet Series in Number Theory.* Springer-Verlag, Berlin, 1976.

[2] F.V. Atkinson. The mean value of the zeta-function on the critical line. *Proc. London Math. Soc.*, (2) **47** (1941), 174–200.

[3] —. The mean value of the Riemann zeta-function. *Acta Math.*, **81** (1949), 353–376.

[4] A. Good. Cusp Forms and eigenfunctions of the Laplacian. *Math. Ann.*, **255** (1981), 523–548.

[5] J.L. Hafner and A. Ivić. On sums of Fourier coefficients of cusp forms. *L'Enseignement Math.*, **35** (1989), 375–382.

[6] D.R. Heath-Brown. The fourth power moment of the Riemann zeta function. *Proc. London Math. Soc.*, **38** (1979), 385–422.

[7] A. Ivić. *Mean Values of the Riemann-Zeta Function.* Lect. Math. Phys., **82**, Tata Inst. Fund. Res.–Springer-Verlag, Bombay, 1991.

[8] —. On the Mellin transform and the Riemann zeta-function. To appear.

[9] A. Ivić and Y. Motohashi. On the fourth moment of the Riemann zeta-function. *J. Number Th.*, **51** (1995), 16–45.

[10] —. The mean square of the error term for the fourth power mean of the zeta-function. *Proc. London Math. Soc.*, (3) **69** (1994), 309–329.

[11] H. Iwaniec. *Introduction to the Spectral Theory of Automorphic Forms.* Bibl. Rev. Mat. Iberoamericana, Madrid, 1995.

[12] M. Jutila. *A Method in the Theory of Exponential Sums.* Lect. Math. Phys., **80**, Tata Inst. Fund. Res.–Springer-Verlag, Bombay, 1987.

[13] —. On exponential sums involving the divisor function. *J. reine angew. Math.*, **355** (1985), 173–190.

[14] —. The additive divisor problem and exponential sums. *Advances in Number Theory*, Clarendon Press, Oxford, 1993, pp. 113–135.

[15] —. The additive divisor problem and its analogs for Fourier coefficients of cusp forms. I. *Math. Z.*, **233** (1996), 435–461.

[16] —. The additive divisor problem and its analogs for Fourier coefficients of cusp forms. II. *Math. Z.*, to appear.

[17] —. The fourth moment of Riemann's zeta-function and the additive divisor problem. *Analytic Number Theory, Proceedings of a Conference in Honor of Heini Halberstam* (B.C. Berndt, H.G. Diamond and A.J. Hildebrand, Eds.) Birkhäuser, Boston, 1996, pp. 515–536.

[18] —. Atkinson's formula revisited. To appear.

[19] —. On spectral large sieve inequalities. To appear.

[20] —. The correspondence between a Dirichlet series and its coefficients. To appear.

[21] M. Jutila and Y. Motohashi. Mean value estimates for exponential sums and L-functions: a spectral theoretic approach. *J. reine angew. Math.*, **459** (1995), 61–87.

[22] N.N. Lebedev. *Special functions and their applications.* Dover Publ. Inc., New York, 1972.

[23] T. Meurman. On exponential sums involving the Fourier coefficients of Maass wave forms. *J. reine angew. Math.*, **384** (1988), 192–207.

[24] Y. Motohashi. Spectral mean values of Maass waveform L-functions. *J. Number Th.*, **42** (1992), 258–284.

[25] —. An explicit formula for the fourth power mean of the Riemann zeta-function. *Acta Math.*, **170** (1993), 181–220.

[26] —. The binary additive divisor problem. *Ann. Sci. École Norm. Sup.*, (4) **27** (1994), 529–572.

[27] —. The mean square of Hecke *L*-series attached to holomorphic cusp forms. *RIMS Kyoto Univ. Kokyuroku*, **886** (1994), 214–227.

[28] —. A relation between the Riemann zeta-function and the hyperbolic Laplacian. *Ann. Scuola Norm. Sup. Pisa, Cl. Sc.*, (4) **22** (1995), 299–313.

[29] —. *Spectral Theory of the Riemann Zeta-Function.* Cambridge Univ. Press, in press.

[30] L.A. Tahtadjan and A.I. Vinogradov. The zeta-function of the additive divisor problem and the spectral decomposition of the automorphic Laplacian. *Zap. Nauch. Sem. LOMI, AN SSSR*, **134** (1984), 84–116. (Russian)

[31] E.C. Titchmarsh. The mean-value of the zeta-function on the critical line. *Proc. London Math. Soc.*, (2) **27** (1928), 137–150.

[32] —. *The Theory of the Riemann Zeta-function.* Clarendon Press, Oxford, 1951.

[33] J.R. Wilton. An approximate functional equation with applications to a problem of Diophantine approximation. *J. reine angew. Math.*, **169** (1933), 219–237.

[34] N.I. Zavorotnyi. On the fourth moment of the Riemann zeta-function. *Automorphic Functions and Number Theory 2*, Computation Center of the Far East Branch of the Science Academy of USSR, Vladivostok, 1989, pp. 69–125. (Russian)

Matti Jutila
Department of Mathematics, University of Turku
20500 Turku, Finland

13

The Mean Square of the Error Term in a Genelarization of the Dirichlet Divisor Problem

KAI-YAM LAM and KAI-MAN TSANG*

1. Introduction and main results Let $\zeta(s)$ denote the Riemann zeta-function and let $\sigma_a(n) = \sum_{d|n} d^a$. For $y > 0$ we define

$$D_a(y) = \sideset{}{'}\sum_{n \le y} \sigma_a(n)$$

and

$$\Delta_a(y) = D_a(y) - \zeta(1-a)y - \frac{\zeta(1+a)}{1+a}y^{1+a} + \frac{1}{2}\zeta(-a),$$

where the symbol Σ' means that the last term in the sum is to be halved if y is an integer. In these definitions, a may be any complex number with the convention that $\Delta_a(y)$ is taken as $\lim_{u \to a} \Delta_u(y)$ when $a = 0$ or -1.

In this paper we shall be concerned with an asymptotic formula for the mean square of $\Delta_a(y)$. In 1996, Meurman [3] proved that, for $x \ge 1$,

$$\int_1^x \Delta_a(y)^2 dy = \begin{cases} c_1 x^{3/2+a} + O(x) & \text{if} \quad -1/2 < a < 0, \\ c_2 x \log x + O(x) & \text{if} \quad a = -1/2, \\ O(x) & \text{if} \quad -1 < a < -1/2, \end{cases} \tag{1.1}$$

where

$$c_1 = ((6+4a)\pi^2\zeta(3))^{-1}\zeta(3/2)^2\zeta(3/2-a)\zeta(3/2+a),$$
$$c_2 = (24\zeta(3))^{-1}\zeta(3/2)^2. \tag{1.2}$$

Here and in the sequel, the constants implied in the O-symbols depend only on a and ϵ, where ϵ is an arbitrarily small positive number whose choice eventually depends on a. Defining

$$F_a(x) = \int_1^x \Delta_a(y)^2 dy - c_1 x^{3/2+a} \tag{1.3}$$

---

*Research of the second author was supported in part by NSF Grant DMS 9304580 and HKU CRCG Grant 335/024/0006.

for $x \geq 1$ and $-1/2 < a < 0$, we shall prove below the following.

Theorem *For $X \geq 2$ and $-1/2 < a < 0$ we have*

$$\int_1^X F_a(x)dx = cX^2 + O(X^{2+a(\frac{1}{2}+a)/(2+a)} \log X),$$

where

$$c = -(2\pi)^{-1-2a}(12\zeta(2-2a))^{-1}\zeta(1-a)^2\zeta(1+2a)\Gamma(1+2a)\sin(\pi a). \quad (1.4)$$

Note that the constant c is negative for $-1/2 < a < 0$. As an immediate consequence of the theorem, we obtain the following

Corollary *For $-1/2 < a < 0$ we have*

$$F_a(x) = \Omega_-(x).$$

The corollary shows that the upper bound $F_a(x) = O(x)$ obtained by Meurman (cf. (1.1)) is sharp. Of course, we cannot exclude the possibility that, actually, $F_a(x) \sim 2cx$ for $-1/2 < a < 0$.

It is worthwhile to mention the interesting case $a = 0$, to which a great deal of work has been devoted. In this case,

$$\Delta(y) := \Delta_0(y) = D_0(y) - y(\log y + 2\gamma - 1) - 1/4,$$

where γ is Euler's constant. The study of $\Delta(y)$ dates back at least to Dirichlet, who first obtained in 1838 the upper bound $\Delta(y) = O(y^{1/2})$ by an elementary argument. Since then, the determination of the best bound for $\Delta(y)$ is called Dirichlet's divisor problem. In the context of the statistical aspect of the problem, Tong [9] obtained, in 1956, the following classical result:

$$\int_1^x \Delta(y)^2 dy = (6\pi^2\zeta(3))^{-1}\zeta(3/2)^4 x^{3/2} + F_0(x)$$

with $F_0(x) = O(x \log^5 x)$. Preissmann [6] then proved in 1988 the slightly better upper bound

$$F_0(x) = O(x \log^4 x). \quad (1.5)$$

This improvement, though small, is not insignificant. It remains the best known upper bound for $F_0(x)$ to date. On the other hand, Lau and Tsang [2] proved in 1995 that

$$\int_2^X F_0(x)dx = -(8\pi^2)^{-1}X^2 \log^2 X + c'X^2 \log X + O(X^2) \quad (1.6)$$

for some constant c' and, as an immediate consequence, that

$$F_0(x) = \Omega_-(x \log^2 x). \tag{1.7}$$

The asymptotic formula (1.6) shows that the average order of $F_0(x)$ is $x \log^2 x$. The remaining gap between the upper bound (1.5) and the Ω_--result (1.7) seems very difficult to close up. In strong contrast, our result in the present paper settled completely the problem of the order of $F_a(x)$ for $-1/2 < a < 0$.

For the proof of our theorem, the starting point is a formula for $\Delta_a(y)$, obtained in Lemma 1 of [3]. To deal with the main term of the integral of $F_a(x)$, we follow the idea of Lau–Tsang in [2] and make use of an asymptotic formula for

$$\sum_{m \leq y} \sigma_a(m)\sigma_a(m+h) \quad (-1/2 < a < 0, \ y > 0, \ h > 0).$$

(See (2.24)–(2.26) below.)

2. Proof of the Theorem From now on we assume always $-1/2 < a < 0$. Let $X \geq 2$, $1 \leq x \leq X$ and set $M = X^{1/(1+a/2)}$. From Lemma 1 of [3], we have, when $1 \leq y \leq x$ and y is not an integer,

$$\begin{aligned}
\Delta_a(y)^2 &= \Delta_a(y, 4M)^2 + 2\Delta_a(y, M)R_a(y, 4M, 2X) \\
&\quad + 2R_a(y, 4M, 2X)R_a(y, M, 2X) - R_a(y, 4M, 2X)^2 \\
&\quad + O(y^{-1/4+a/2}(|\Delta_a(y, 4M)| + |R_a(y, 4M, 2X)|) + y^{-1/2+a}),
\end{aligned}$$

where

$$\begin{aligned}
\Delta_a(y, Y) &= (\pi\sqrt{2})^{-1} y^{1/4+a/2} \\
&\quad \times \int_1^2 \sum_{n \leq uY} \sigma_a(n) n^{-3/4-a/2} \cos(4\pi\sqrt{ny} - \pi/4) du
\end{aligned} \tag{2.1}$$

and

$$\begin{aligned}
R_a(y, Y, Z) &= (2\pi)^{-1} \sum_{n \leq Z} \sigma_a(n) \\
&\quad \times \int_1^2 \int_{uY}^{\infty} t^{-1} \sin(4\pi(\sqrt{y} - \sqrt{n})\sqrt{t}) dt\, du.
\end{aligned} \tag{2.2}$$

Integrating with respect to y and then using Cauchy's inequality, we obtain

$$\begin{aligned}
\int_1^x \Delta_a(y)^2 dy &= I_1(x) + 2I_2(x) + O(\sqrt{I_3(x)I_4(x)} + I_3(x)) \\
&\quad + O(x^{1/4+a/2}(\sqrt{I_1(x)} + \sqrt{I_3(x)}) + x^{1/2+a}),
\end{aligned} \tag{2.3}$$

where

$$I_1(x) = \int_1^x \Delta_a(y, 4M)^2 dy, \quad I_2(x) = \int_1^x \Delta_a(y, M) R_a(y, 4M, 2X) dy,$$

$$I_3(x) = \int_1^x R_a(y, 4M, 2X)^2 dy, \quad I_4(x) = \int_1^x R_a(y, M, 2X)^2 dy. \quad (2.4)$$

By Lemma 2 of [3] and Cauchy's inequality, we get

$$I_3(x), \; I_4(x)$$

$$\ll \int_1^x \Big(\sum_{n \le 2X} \sigma_a(n) \min(1, XM^{-1}(y-n)^{-2}) \Big)^2 dy$$

$$\ll \int_1^x \sum_{n \le 2X} \sigma_a(n)^2 \min(1, XM^{-1}(y-n)^{-2}) dy \ll X^{3/2} M^{-1/2}. \quad (2.5)$$

For the right-hand side of (2.1), on interchanging the summation and integration, we get

$$\Delta_a(y, 4M) = (\pi\sqrt{2})^{-1} y^{1/4+a/2}$$

$$\times \sum_{n \le 8M} \sigma_a(n) n^{-3/4-a/2} w_1(n) \cos(4\pi\sqrt{ny} - \pi/4),$$

where

$$w_1(t) = \begin{cases} 1 & \text{if } 0 \le t \le 4M, \\ 2 - t/(4M) & \text{if } 4M < t \le 8M, \\ 0 & \text{if } t > 8M. \end{cases} \quad (2.6)$$

Squaring this and then integrating with respect to y, we have

$$I_1(x) = ((6 + 4a)\pi^2)^{-1}(x^{3/2+a} - 1)$$

$$\times \sum_{n \le 8M} \sigma_a(n)^2 n^{-3/2-a} w_1(n)^2 + S_1(x) + S_2(x),$$

where

$$S_1(x) = (4\pi^2)^{-1} \int_1^x y^{1/2+a} \sum_{\substack{m,n \le 8M \\ m \ne n}} \sigma_a(m)\sigma_a(n)(mn)^{-3/4-a/2}$$

$$\times w_1(m) w_1(n) \cos(4\pi(\sqrt{n} - \sqrt{m})\sqrt{y}) dy \quad (2.7)$$

and

$$S_2(x) = (4\pi^2)^{-1} \int_1^x y^{1/2+a} \sum_{m,n \le 8M} \sigma_a(m)\sigma_a(n)(mn)^{-3/4-a/2}$$

$$\times w_1(m) w_1(n) \sin(4\pi(\sqrt{n} + \sqrt{m})\sqrt{y}) dy. \quad (2.8)$$

In view of (2.6) and equation (1.3.3) in [8], we find that

$$\sum_{n \leq 8M} \sigma_a(n)^2 n^{-3/2-a} w_1(n)^2$$

$$= \zeta(3)^{-1} \zeta(3/2)^2 \zeta(3/2 - a) \zeta(3/2 + a) + O(M^{-1/2-a}).$$

Thus,

$$I_1(x) = c_1 x^{3/2+a} + S_1(x) + S_2(x) + O(x^{3/2+a} M^{-1/2-a} + 1), \qquad (2.9)$$

where c_1 is defined in (1.2).

Next we proceed to obtain an upper bound for $I_1(x)$. By the second mean value theorem, there exists $\xi \in [1, x]$ such that

$$S_1(x) \ll x^{1+a} \left| \int_{\xi}^{x} y^{-1/2} \sum_{\substack{m,n \leq 8M \\ m \neq n}} \cdots \cos(4\pi(\sqrt{n} - \sqrt{m})\sqrt{y}) dy \right|$$

$$\ll x^{1+a} \max_{\xi \leq y \leq x} \left| \sum_{\substack{m,n \leq 8M \\ m \neq n}} \cdots (\sqrt{n} - \sqrt{m})^{-1} e^{4\pi i (\sqrt{n} - \sqrt{m})\sqrt{y}} \right|.$$

By the Montgomery-Vaughan inequality (see [4]), we thus find that

$$S_1(x) \ll x^{1+a} \sum_{n \leq 8M} \sigma_a(n)^2 n^{-1-a} \ll x^{1+a} M^{-a}.$$

For $S_2(x)$, the second mean value theorem and trivial estimate will suffice, and we get

$$S_2(x) \ll x^{1+a} \sum_{m,n \leq 8M} \sigma_a(m) \sigma_a(n) (mn)^{-1-a/2} \ll x^{1+a} M^{-a}.$$

Substituting these estimates for $S_1(x)$ and $S_2(x)$ into (2.9), we get

$$I_1(x) \ll X^{3/2+a} + X^{1+a} M^{-a}. \qquad (2.10)$$

Finally, in view of (1.3), (2.5), (2.9), and (2.10), the formula (2.3) reduces to

$$F_a(x) = S_1(x) + S_2(x) + 2I_2(x) + O(X^{3/2+a} M^{-1/2-a}).$$

Integrating with respect to x then yields

$$\int_1^X F_a(x) dx = \int_1^X S_1(x) dx + \int_1^X S_2(x) dx$$

$$+ 2 \int_1^X I_2(x) dx + O(X^{5/2+a} M^{-1/2-a}).$$

To complete the proof of the theorem it is sufficient to prove

$$\int_1^X S_1(x)dx = cX^2 + O(X^{5/2+a}M^{-1/2-a}\log X), \qquad (2.11)$$

$$\int_1^X S_2(x)dx \ll XM^{-a} + X^{3/2+a} \qquad (2.12)$$

and

$$\int_1^X I_2(x)dx \ll X^{3/2}. \qquad (2.13)$$

Proof of (2.12). Integrating the right-hand side of (2.8) with respect to x and reversing the order of integrations, we have

$$\int_1^X S_2(x)dx = (2\pi^2)^{-1}X^{5/2+a} \sum_{m,n\le 8M} \sigma_a(m)\sigma_a(n)(mn)^{-3/4-a/2}w_1(m)w_1(n)$$

$$\times \int_{1/\sqrt{X}}^1 (1-\nu^2)\nu^{2+2a}\sin(4\pi(\sqrt{n}+\sqrt{m})\sqrt{X}\nu)d\nu.$$

Applying partial integration twice, we find that the integral on the right-hand side is $\ll X^{-3/2-a}(\sqrt{n}+\sqrt{m})^{-1} + X^{-1}(\sqrt{n}+\sqrt{m})^{-2}$. Thus,

$$\int_1^X S_2(x)dx \ll X \sum_{m,n\le 8M} \sigma_a(m)\sigma_a(n)(mn)^{-1-a/2}$$

$$+ X^{3/2+a} \sum_{m,n\le 8M} \sigma_a(m)\sigma_a(n)(mn)^{-5/4-a/2} \ll XM^{-a} + X^{3/2+a},$$

as desired.

Proof of (2.13). Let $m \le 2M$, $n \le 2X$, $V \ge 4M$ and $t \ge 4M$. Put

$$J(m,n,V) = \int_1^X (X-y)y^{\frac{1}{4}+\frac{a}{2}}\cos(4\pi\sqrt{my}-\pi/4)$$

$$\times \int_1^2 \int_{uV}^\infty t^{-1}\sin(4\pi(\sqrt{y}-\sqrt{n})\sqrt{t})dtdudy, \qquad (2.14)$$

$$K_\pm(m,n,t) = \int_1^X (X-y)y^{\frac{1}{4}+\frac{a}{2}}$$

$$\times \sin(4\pi(\sqrt{t}\pm\sqrt{m})\sqrt{y} - 4\pi\sqrt{nt} \mp \pi/4)dy. \qquad (2.15)$$

By a simple change of variable and then integrating by parts twice, we find that

$$K_\pm(m,n,t)$$

$$= 2X^{\frac{9}{4}+\frac{a}{2}} \int_{1/\sqrt{X}}^{1} (1-\nu^2)\nu^{\frac{3}{2}+a} \sin(4\pi(\sqrt{t}\pm\sqrt{m})\sqrt{X}\nu - 4\pi\sqrt{nt} \mp \pi/4)d\nu$$

$$= X(1-X^{-1})(2\pi(\sqrt{t}\pm\sqrt{m}))^{-1}\cos(4\pi(\sqrt{t}\pm\sqrt{m}) - 4\pi\sqrt{nt} \mp \pi/4)$$
$$+ O(X^{5/4+a/2}(\sqrt{t}\pm\sqrt{m})^{-2}).$$

Consequently,

$$\int_1^2 \int_{4uM}^{uV} t^{-1}K_\pm(m,n,t)dtdu = (2\pi)^{-1}X(1-X^{-1})$$

$$\times \int_1^2 \int_{4uM}^{uV} t^{-1}(\sqrt{t}\pm\sqrt{m})^{-1}\cos(4\pi(\sqrt{t}\pm\sqrt{m}) - 4\pi\sqrt{nt} \mp \pi/4)dtdu$$
$$+ O(X^{5/4+a/2}M^{-1})$$

$$\ll X\left|\int_1^2 \int_{4uM}^{uV} t^{-1}(\sqrt{t}\pm\sqrt{m})^{-1}e^{4\pi i(1-\sqrt{n})\sqrt{t}}dtdu\right| + X^{5/4+a/2}M^{-1}.$$

Note that $\sqrt{t}\pm\sqrt{m} \gg \sqrt{t}$ for $m \le 2M$ and $t \ge 4M$. When $n=1$, the double integral is $\ll M^{-1/2}$. Otherwise, by Lemma 4.3 of [8], the double integral is $\ll M^{-1}(\sqrt{n}-1)^{-1}$. Hence, by (2.14) and (2.15), we have

$$J(m,n,4M) - J(m,n,V) = \frac{1}{2}\int_1^2 \int_{4uM}^{uV} t^{-1}(K_+(m,n,t) + K_-(m,n,t))dtdu$$
$$\ll XM^{-1/2}\min(1,M^{-1/2}(\sqrt{n}-1)^{-1}) + X^{5/4+a/2}M^{-1},$$

uniformly in V. However, by Lemma 2 of [3], $\lim_{V\to\infty} J(m,n,V) = 0$. Therefore,

$$J(m,n,4M) \ll XM^{-1/2}\min(1,M^{-1/2}(\sqrt{n}-1)^{-1}) + X^{5/4+a/2}M^{-1}. \quad (2.16)$$

We now use this estimate to bound $\int_1^X I_2(x)dx$. We have, by (2.1),

$$\Delta_a(y,M) = (\pi\sqrt{2})^{-1}y^{1/4+a/2}$$
$$\times \sum_{m\le 2M} \sigma_a(m)m^{-3/4-a/2}w_2(m)\cos(4\pi\sqrt{my} - \pi/4), \quad (2.17)$$

where

$$w_2(t) = \begin{cases} 1 & \text{if } 0 \le t \le M, \\ 2-t/M & \text{if } M < t \le 2M, \\ 0 & \text{if } t > 2M. \end{cases} \quad (2.18)$$

Thus, in view of (2.2), (2.4), (2.14), and (2.17), we get

$$\int_1^X I_2(x)dx = \int_1^X (X - y)\triangle_a(y, M)R_a(y, 4M, 2X)dy$$

$$= (2\sqrt{2}\pi^2)^{-1} \sum_{m \le 2M} \sigma_a(m)m^{-3/4-a/2}w_2(m)$$

$$\times \sum_{n \le 2X} \sigma_a(n)J(m, n, 4M).$$

By (2.16) and (2.18), we then have

$$\int_1^X I_2(x)dx \ll \sum_{m \le 2M} \sigma_a(m)m^{-3/4-a/2} \sum_{n \le 2X} \sigma_a(n) |J(m, n, 4M)|$$

$$\ll XM^{-1/4-a/2} \sum_{n \le 2X} \sigma_a(n) \min(1, M^{-1/2}(\sqrt{n} - 1)^{-1})$$

$$+ X^{9/4+a/2}M^{-3/4-a/2}$$

$$\ll X^{3/2},$$

since $M \ge X$. This completes the proof of (2.13).

Proof of (2.11). By (2.7), reversing the order of integrations and making the change of variable $y = X\nu^2$, we have

$$\int_1^X S_1(x)dx = \pi^{-2}X^{5/2+a} \sum_{m < n \le 8M} \sigma_a(m)\sigma_a(n)(mn)^{-3/4-a/2}w_1(m)w_1(n)$$

$$\times \int_{1/\sqrt{X}}^1 (1 - \nu^2)\nu^{2+2a} \cos(4\pi(\sqrt{n} - \sqrt{m})\sqrt{X}\nu)d\nu. \qquad (2.19)$$

Firstly we replace the lower integration limit of the inner integral by zero. By an integration by parts, we have

$$\int_0^{1/\sqrt{X}} (1 - \nu^2)\nu^{2+2a} \cos(4\pi(\sqrt{n} - \sqrt{m})\sqrt{X}\nu)d\nu \ll X^{-3/2-a}(\sqrt{n} - \sqrt{m})^{-1}.$$

Hence, the contribution of this error term to the right-hand side of (2.19) is

$$\ll X \sum_{m < n \le 8M} \sigma_a(m)\sigma_a(n)(mn)^{-3/4-a/2}(\sqrt{n} - \sqrt{m})^{-1}$$

$$\ll XM^\epsilon \left\{ \sum_{n \le 8M} \sum_{m \le n/2} (mn)^{-\frac{3}{4}-\frac{a}{2}}n^{-1/2} \right.$$

$$\left. + \sum_{n \le 8M} \sum_{n/2 < m < n} (mn)^{-\frac{3}{4}-\frac{a}{2}}n^{1/2}(n - m)^{-1} \right\}$$

$$\ll XM^\epsilon \left\{ \sum_{n \le 8M} n^{-1-a} + \sum_{n \le 8M} n^{-1-a} \log n \right\} \ll XM^{-a+\epsilon}.$$

Thus,

$$\int_1^X S_1(x)dx = \pi^{-2}X^{5/2+a}\sum_{m<n\leq 8M}\phi_{m,n} + O(XM^{-a+\epsilon})$$
$$= \pi^{-2}X^{5/2+a}T + O(XM^{-a+\epsilon}), \qquad (2.20)$$

say, where

$$\phi_{m,n} = \sigma_a(m)\sigma_a(n)(mn)^{-3/4-a/2}w_1(m)w_1(n)g(4\pi(\sqrt{n}-\sqrt{m})\sqrt{X}) \quad (2.21)$$

and, for any $\theta \geq 0$,

$$g(\theta) = \int_0^1 (1-\nu^2)\nu^{2+2a}\cos(\theta\nu)d\nu. \qquad (2.22)$$

Applying partial integration twice and, otherwise, by trivial estimation, we have

$$g(\theta) \ll \min(1,\theta^{-2}). \qquad (2.23)$$

We then split T into the following two sub-sums:

$$T = \sum_{n\leq 8M}\sum_{m<n/2}\phi_{m,n} + \sum_{n\leq 8M}\sum_{n/2\leq m<n}\phi_{m,n}.$$

By (2.21) and (2.23), the first sub-sum is

$$\ll X^{-1}\sum_{n\leq 8M}\sum_{m<n/2}\sigma_a(m)\sigma_a(n)(mn)^{-3/4-a/2}(\sqrt{n}-\sqrt{m})^{-2}$$

$$\ll X^{-1}\sum_{n\leq 8M}\sigma_a(n)n^{-7/4-a/2}\sum_{m<n/2}\sigma_a(m)m^{-3/4-a/2} \ll X^{-1}.$$

In the second sub-sum, we write $n = m+h$ with $1 \leq h \leq \min(8M-m,m)$. Then

$$T = \sum_{h\leq\sqrt{M}}\sum_{h\leq m\leq 8M-h}\phi_{m,m+h} + \sum_{\sqrt{M}<h\leq 4M}\sum_{h\leq m\leq 8M-h}\phi_{m,m+h} + O(X^{-1}).$$

By (2.21) and (2.23), the second sum here is

$$\ll M^\epsilon X^{-1}\sum_{\sqrt{M}<h\leq 4M}\sum_{h\leq m\leq 8M-h}m^{-3/2-a}(\sqrt{m+h}-\sqrt{m})^{-2}$$

$$\ll M^\epsilon X^{-1}\sum_{\sqrt{M}<h\leq 4M}h^{-2}\sum_{h\leq m\leq 8M-h}m^{-1/2-a} \ll X^{-1}M^{-a+\epsilon}.$$

Let

$$Q_h = h^2 \sqrt{M}.$$

Then we can further write

$$T = \sum_{h \leq \sqrt{M}} \quad \sum_{h \leq m \leq \min(Q_h, 8M)} \phi_{m,m+h}$$

$$+ \sum_{h \leq \sqrt{8}M^{1/4}} \quad \sum_{Q_h < m \leq 8M} \phi_{m,m+h} + O(X^{-1}M^{-a+\epsilon}).$$

Similar to the above estimation, we find that the first sum on the right-hand side is

$$\ll X^{-1} \sum_{h \leq \sqrt{M}} h^{-2} \sum_{h \leq m \leq \min(Q_h, 8M)} \sigma_a(m)\sigma_a(m+h)m(m(m+h))^{-3/4-a/2}$$

$$\ll X^{-1} \sum_{h \leq \sqrt{M}} h^{-2}(\min(Q_h, 8M))^{1/2-a} \ll X^{-1}M^{1/4-a},$$

where an obvious application of Cauchy's inequality was made to the summands of the first line. Hence,

$$T = \sum_{h \leq \sqrt{8}M^{1/4}} \quad \sum_{Q_h < m \leq 8M} \phi_{m,m+h} + O(X^{-1}M^{1/4-a}).$$

To evaluate this last sum, we will need an asymptotic formula for

$$\psi_h(y) = \sum_{m \leq y} \sigma_a(m)\sigma_a(m+h), \quad \text{for } -1/2 < a < 0, \ y > 0, \ h > 0. \quad (2.24)$$

For $y > 0$ and $h > 0$, let

$$P_h(y) = \alpha_0(h)y + \alpha_1(h)y^{1+a} + \alpha_2(h)y^{1+2a}, \quad\quad\quad (2.25)$$

where

$$\alpha_0(h) = \zeta(1-a)^2\zeta(2-2a)^{-1}\sigma_{2a-1}(h) \ll 1,$$
$$\alpha_1(h) = 2(1+a)^{-1}\zeta(1+a)\zeta(1-a)\zeta(2)^{-1}\sigma_{-1}(h) \ll h^\epsilon,$$
$$\alpha_2(h) = (1+2a)^{-1}\zeta(1+a)^2\zeta(2+2a)^{-1}\sigma_{-1-2a}(h) \ll h^\epsilon.$$

Following closely the proof of Theorem 2 of [1], we can show that

$$\psi_h(y) = P_h(y) + E_h(y)$$

with

$$E_h(y) \ll y^{(1+a)/2+1/(3-2a)+\epsilon}, \tag{2.26}$$

uniformly for $h \ll y^{(1+a)/2+1/(3-2a)}$.

We remark that Motohashi [5, Theorem 1] has a sharper estimate for the sum $\sum_{m \le y} d(m)d(m + h)$ than Theorem 2 of [1]. Following his argument, one can get a correspondingly better bound for $E_h(y)$. Nonetheless, the estimate (2.26) is sufficient for the present purpose of obtaining our main results.

For convenience, let

$$\theta_{y,h} = 4\pi(\sqrt{y+h} - \sqrt{y})\sqrt{X} \qquad (y, \ h > 0).$$

Then by (2.21), (2.24), and partial summation, we have

$$\sum_{Q_h < m \le 8M} \phi_{m,m+h}$$

$$= \int_{Q_h}^{8M} w_1(y)w_1(y+h)(y(y+h))^{-3/4-a/2}g(\theta_{y,h})P_h'(y)dy$$

$$+ \left[w_1(y)w_1(y+h)(y(y+h))^{-3/4-a/2}g(\theta_{y,h})E_h(y) \right]_{Q_h}^{8M}$$

$$- \int_{Q_h}^{8M} E_h(y)\frac{d}{dy}\{w_1(y)w_1(y+h)(y(y+h))^{-3/4-a/2}g(\theta_{y,h})\}dy$$

$$= W_1(h) + W_2(h) + W_3(h),$$

say. Accordingly, we decompose T as:

$$T = \sum_{h \le \sqrt{8}M^{1/4}} (W_1(h) + W_2(h) + W_3(h)) + O(X^{-1}M^{1/4-a})$$

$$= T_1 + T_2 + T_3 + O(X^{-1}M^{1/4-a}), \tag{2.27}$$

say. As $y^{(1+a)/2+1/(3-2a)} \ge Q_h^{1/2} \ge h$ for $y \ge Q_h$, we may apply (2.26) to bound $E_h(y)$ in $W_2(h)$ and $W_3(h)$. Hence from (2.22) we have $W_2(h) \ll M^{1/3-a/2+\epsilon}h^{-2}X^{-1}$, so that

$$T_2 \ll X^{-1}M^{1/3-a/2+\epsilon}. \tag{2.28}$$

Next, by partial integration, we have

$$g'(\theta) = -\int_0^1 (1-\nu^2)\nu^{3+2a}\sin(\theta\nu)d\nu \ll \theta^{-2}$$

for $\theta \neq 0$. Hence

$$\frac{d}{dy} g(\theta_{y,h}) \ll h^{-1} y^{-1/2} X^{-1/2}.$$

Combining this with (2.23) yields

$$\frac{d}{dy} \{ w_1(y) w_1(y+h)(y(y+h))^{-3/4-a/2} g(\theta_{y,h}) \}$$

$$\ll y^{-3/2-a} h^{-2} X^{-1} + y^{-2-a} h^{-1} X^{-1/2}$$

for $Q_h \leq y \leq 8M$. Thus, in view of (2.26),

$$W_3(h)$$

$$\ll \int_{Q_h}^{8M} y^{(1+a)/2+1/(3-2a)+\epsilon} (y^{-3/2-a} h^{-2} X^{-1} + y^{-2-a} h^{-1} X^{-1/2}) dy$$

$$\ll h^{-2} X^{-1} M^{-a/2+1/(3-2a)+\epsilon} + h^{-1} X^{-1/2},$$

and so

$$T_3 \ll X^{-1} M^{1/3-a/2+\epsilon} + X^{-1/2} M^\epsilon. \tag{2.29}$$

Finally, as $P_h'(y) = \alpha_0(h) + O(h^\epsilon y^a)$, we see that

$$T_1 = \sum_{h \leq \sqrt{8} M^{1/4}} \alpha_0(h) \int_{Q_h}^{8M} w_1(y) w_1(y+h)(y(y+h))^{-3/4-a/2} g(\theta_{y,h}) dy$$

$$+ O(X^{-1} M^{1/2}).$$

Combining this with (2.27), (2.28), and (2.29), we find that

$$T = \sum_{h \leq \sqrt{8} M^{1/4}} \alpha_0(h) \int_{Q_h}^{8M} w_1(y) w_1(y+h)(y(y+h))^{-3/4-a/2} g(\theta_{y,h}) dy$$

$$+ O(X^{-1} M^{(1-a)/2}).$$

Using the integral representation (2.22) for $g(\theta)$, changing the order of integrations and making the substitution $y \to y - h/2$, we find that

$$T = \int_0^1 (1-\nu^2)\nu^{2+2a} \sum_{h \leq \sqrt{8} M^{1/4}} \alpha_0(h)$$

$$\times \int_{Q_h+h/2}^{8M+h/2} w_1(y-h/2) w_1(y+h/2)(y^2-h^2/4)^{-3/4-a/2}$$

$$\times \cos(4\pi(\sqrt{y+h/2} - \sqrt{y-h/2})\sqrt{X}\nu) dy d\nu$$

$$+ O(X^{-1} M^{(1-a)/2}). \tag{2.30}$$

Now, for $Q_h + h/2 \le y \le 8M + h/2$ and $0 \le \nu \le 1$, we have

$$(y^2 - h^2/4)^{-3/4-a/2} = y^{-3/2-a}(1 + O(h^2 y^{-2})),$$

$$\cos(4\pi(\sqrt{y+h/2} - \sqrt{y-h/2})\sqrt{X}\nu)$$
$$= \cos(2\pi h\sqrt{X}y^{-1/2}\nu) + O(h^3\sqrt{X}y^{-5/2})$$

and

$$w_1(y - h/2)w_1(y + h/2) = w_1(y)^2 + O(hM^{-1}).$$

Whence

$$w_1(y - h/2)w_1(y + h/2)(y^2 - h^2/4)^{-3/4-a/2}$$
$$\times \cos(4\pi(\sqrt{y+h/2} - \sqrt{y-h/2})\sqrt{X}\nu)$$
$$= w_1(y)^2 y^{-3/2-a}\cos(2\pi h\sqrt{X}y^{-1/2}\nu)$$
$$+ O(y^{-4-a}h^3\sqrt{X} + y^{-3/2-a}hM^{-1}).$$

If we replace the integrand of the inner integral in (2.30) by

$$w_1(y)^2 y^{-3/2-a}\cos(2\pi h\sqrt{X/y}\nu),$$

the error introduced is

$$\ll \sum_{h \le \sqrt{8}M^{1/4}} \int_{Q_h+h/2}^{8M+h/2} (y^{-4-a}h^3\sqrt{X} + y^{-3/2-a}hM^{-1})dy$$
$$\ll M^{-(3+a)/2}\sqrt{X} + M^{-1-a} \ll M^{-1-a}.$$

Consequently,

$$T = \int_0^1 (1 - \nu^2)\nu^{2+2a} \sum_{h \le \sqrt{8}M^{1/4}} \alpha_0(h)$$
$$\times \int_{Q_h+h/2}^{8M+h/2} w_1(y)^2 y^{-3/2-a}\cos(2\pi h\sqrt{X}y^{-1/2}\nu)dyd\nu + O(X^{-1}M^{(1-a)/2})$$
$$= \sum_{h \le \sqrt{8}M^{1/4}} \alpha_0(h) \int_{Q_h}^{8M} w_1(y)^2 y^{-3/2-a}g(2\pi h\sqrt{X}y^{-1/2})dy$$
$$+ O(X^{-1}M^{(1-a)/2}).$$

Then, making the change of variable $\theta = 2\pi h \sqrt{X} y^{-1/2}$ and interchanging the order of summation and integration, we get

$$T = 2(2\pi)^{-1-2a} X^{-1/2-a} \int_A^B \theta^{2a} g(\theta) \sum_{h \le \theta/A} \alpha_0(h) h^{-1-2a} w_1(4\pi^2 h^2 X \theta^{-2})^2 d\theta$$

$$+ O(X^{-1} M^{(1-a)/2}),$$

where

$$A = \pi \sqrt{X}/\sqrt{2M} \quad \text{and} \quad B = 2\pi \sqrt{X} M^{-1/4}. \tag{2.31}$$

To evaluate the sum in the integrand of the last integral, we begin by considering the summatory function

$$U(t) = \sum_{h \le t} \alpha_0(h) h^{-1-2a}, \quad t > 0.$$

By an elementary argument, we find that

$$U(t) = -(2a)^{-1} \zeta(1-a)^2 t^{-2a} + \zeta(1-a)^2 \zeta(2-2a)^{-1} \zeta(1+2a) \zeta(2) + R(t)$$

with

$$R(t) = O(t^{-1-2a}), \tag{2.32}$$

as $t \to \infty$. Moreover,

$$\lim_{t \to 1-} R(t) = (2a)^{-1} \zeta(1-a)^2 - \zeta(1-a)^2 \zeta(2-2a)^{-1} \zeta(1+2a) \zeta(2). \tag{2.33}$$

Then partial summation yields, for $\theta > 0$,

$$\sum_{h \le \theta/A} \alpha_0(h) h^{-1-2a} w_1(4\pi^2 h^2 X \theta^{-2})^2$$

$$= \zeta(1-a)^2 \int_1^{\theta/A} t^{-1-2a} w_1(4\pi^2 t^2 X \theta^{-2})^2 dt + [w_1(4\pi^2 t^2 X \theta^{-2})^2 R(t)]_{1-}^{\theta/A}$$

$$- \int_1^{\theta/A} R(t) \frac{d}{dt} \{w_1(4\pi^2 t^2 X \theta^{-2})^2\} dt$$

$$= J_1 + J_2 + J_3,$$

say. By (2.6), (2.31), and (2.32) we find that

$$J_3 = 4\pi^2 X M^{-1} \theta^{-2} \int_{\theta/(\sqrt{2}A)}^{\theta/A} t(2 - \pi^2 t^2 X M^{-1} \theta^{-2}) R(t) dt$$

$$\ll (M/X)^{-1/2-a} \theta^{-1-2a},$$

and, by (2.31) and (2.33),

$$J_2 = -R(1-)w_1(4\pi^2 X\theta^{-2})^2$$
$$= (\zeta(2-2a)^{-1}\zeta(1+2a)\zeta(2) - (2a)^{-1})\zeta(1-a)^2 w_1(4\pi^2 X\theta^{-2})^2.$$

Moreover, by direct computation,

$$J_1 = \zeta(1-a)^2 \int_1^{\theta/(\sqrt{2}A)} t^{-1-2a}dt$$
$$+ \zeta(1-a)^2 \int_{\theta/(\sqrt{2}A)}^{\theta/A} t^{-1-2a}(2 - \pi^2 t^2 X M^{-1}\theta^{-2})^2 dt$$
$$= \lambda\theta^{-2a}(M/X)^{-a} + (2a)^{-1}\zeta(1-a)^2,$$

where λ is a constant depending only on a. Thus,

$$T = 2(2\pi)^{-1-2a}\zeta(1-a)^2\zeta(2-2a)^{-1}\zeta(1+2a)\zeta(2)X^{-1/2-a}L_3$$
$$+ O(X^{-1/2}M^{-a}|L_1| + X^{-1/2-a}|L_2| + M^{-1/2-a}L_4)$$
$$+ O(X^{-1}M^{(1-a)/2}) \tag{2.34}$$

where

$$L_1 = \int_A^B g(\theta)d\theta, \quad L_2 = \int_A^B \theta^{2a}(1 - w_1(4\pi^2 X\theta^{-2})^2)g(\theta)d\theta,$$

$$L_3 = \int_A^B \theta^{2a}w_1(4\pi^2 X\theta^{-2})^2 g(\theta)d\theta, \quad L_4 = \int_A^B \theta^{-1}|g(\theta)|d\theta.$$

We now evaluate the above four integrals $L_j (1 \le j \le 4)$. Since

$$\int_0^\infty g(\theta)d\theta = \lim_{t\to 0+} \int_0^\infty \int_0^1 (1 - \nu^2)\nu^{2+2a}e^{-t\theta}\cos(\theta\nu)d\nu d\theta$$
$$= \lim_{t\to 0+} t\int_0^1 (1 - \nu^2)\nu^{2+2a}(\nu^2 + t^2)^{-1}d\nu = 0,$$

we find that

$$L_1 = -(\int_0^A + \int_B^\infty)g(\theta)d\theta \ll (X/M)^{1/2},$$

by (2.23) and (2.31). Also, by (2.6) and (2.23) we have

$$L_2 = \int_A^{\sqrt{2}A} \theta^{2a}(1 - w_1(4\pi^2 X\theta^{-2})^2)g(\theta)d\theta \ll (X/M)^{1/2+a},$$

$$L_3 = (\int_0^\infty - \int_0^{\sqrt{2}A} - \int_B^\infty)\theta^{2a}g(\theta)d\theta + \int_A^{\sqrt{2}A}\theta^{2a}w_1(4\pi^2X\theta^{-2})^2g(\theta)d\theta$$

$$= \int_0^\infty \theta^{2a}g(\theta)d\theta + O((X/M)^{1/2+a}),$$

and $L_4 \ll \log X$. Moreover, by the formula (see [7, §3.127])

$$\int_0^\infty \theta^{p-1}\cos(\nu\theta)d\theta = \Gamma(p)\nu^{-p}\cos(\pi p/2) \quad \text{for} \quad 0 < p < 1, \quad \nu > 0,$$

we find that

$$\int_0^\infty \theta^{2a}g(\theta)d\theta = \int_0^1 (1-\nu^2)\nu^{2+2a}\int_0^\infty \theta^{2a}\cos(\theta\nu)d\theta d\nu$$

$$= -\frac{1}{4}\Gamma(1+2a)\sin(\pi a).$$

Collecting all these into (2.34), we therefore deduce that

$$T = c\pi^2 X^{-1/2-a} + O(M^{-1/2-a}\log X),$$

where c is defined by (1.4). Then the formula (2.11) now follows from this and (2.20). The proof of the theorem is thus complete.

Acknowledgement The authors wish to thank the referee and the editor for many valuable comments and suggestions on this paper.

References

[1] D.R. Heath-Brown. The fourth power moment of the Riemann zeta-function. *Proc. London Math. Soc.*, (3)**38** (1979), 385–422.

[2] Y.K. Lau and K.M. Tsang. Mean square of the remainder term in the Dirichlet divisor problem. *J. Théorie des Nombres de Bordeaux*, **7** (1995), 75–92.

[3] T. Meurman. The mean square of the error term in a generalization of Dirichlet's divisor problem. *Acta Arith.*, **74** (1996), 351–364.

[4] H.L. Montgomery and R.C. Vaughan. Hilbert's inequality. *J. London Math. Soc.*, (2)**8** (1974), 73–82.

[5] Y. Motohashi. The binary additive divisor problem. *Ann. Sci. Ec. Norm. Sup.*, **27** (1994), 529–572.

[6] E. Preissmann. Sur la moyenne quadratique du terme de reste du problème du cercle. *C. R. Acad. Sci. Paris, Sér. I*, **306** (1988), 151–154.

[7] E.C. Titchmarsh. *The Theory of Functions* (2nd ed). Oxford Univ. Press, Oxford, 1939.

[8] —. *The Theory of the Riemann Zeta-Function.* Clarendon Press, Oxford, 1951.

[9] K.C. Tong. On divisor problems III. *Acta Math. Sinica,* **6** (1956), 515–541.

Kai-Yam Lam and Kai-Man Tsang
Department of Mathematics, The University of Hong Kong
Pokfulam Road, Hong Kong

14

The Goldbach Problem
with Primes in Arithmetic Progressions

MING-CHIT LIU and TAO ZHAN

1. Introduction and statement of results The Goldbach conjecture states that every even integer larger than 2 is a sum of two primes. One of the most significant contributions to this unsolved problem is due to J.-R. Chen [2], who proved in 1966 that every large even integer is a sum of one prime and an almost-prime with at most two prime factors. On the other hand, from different direction Montgomery and Vaughan [10] in 1975 considered the exceptional set in Goldbach problem and showed that,

$$E(X) := \#\{n \leq X : 2|n, \, n \neq p_1 + p_2$$
$$\text{for any primes } p_1 \text{ and } p_2\} < X^{1-\delta} \qquad (1.1)$$

for some computable absolute constant $\delta > 0$, where and throughout the paper the letter p with or without subscript always denotes a prime, n denotes a positive integer and $X > 0$ is sufficiently large. The present paper deals with a variant of the Goldbach conjecture, namely, the problem of representing even integers as a sum of two primes with one of which in an arithmetic progression. Let b and r be any positive integers with $(b, r) = 1$ and let

$$E(X; r, b) := \#\{n \leq X : 2|n, \, (n - b, r) = 1, n \neq p_1 + p_2$$
$$\text{for any } p_1 \equiv b \pmod{r} \text{ and } p_2\}.$$

Corresponding to (1.1), we establish the following

Theorem 1 *There exists an effectively computable absolute constant $\delta > 0$ such that for any $1 \leq r \leq X^{\delta}$ and $(b, r) = 1$ one has*

$$E(X; r, b) < X^{1-\delta}.$$

Remark 1 Let μ, ϕ and τ denote the Möbius function, Euler function and divisor function respectively. Since

$$\#\{n \leq X : 2|n \text{ and } (n - b, r) = 1\} = X\frac{\phi(2r)}{2r} + O(\tau(r))$$

for $(b,r) = 1$, Theorem 1 shows that almost all even integers n satisfying $(n - b, r) = 1$ can be written as

$$n = p_1 + p_2 \text{ with } p_1 \equiv b \pmod{r} \tag{1.2}$$

for any $1 \leq r \leq n^\delta$ and $(b, r) = 1$.

Remark 2 Denote by $N(n; r, b)$ the number of solutions of (1.2). The proof of Theorem 1 actually shows that for all the even integers $n \leq X$ satisfying $(n - b, r) = 1$, with at most $O(X^{1-\delta})$ exceptions, there holds for any $1 \leq r \leq X^\delta$ and $(b, r) = 1$,

$$N(n; r, b) \geq r^{-1} X^{1-10\delta}.$$

By using completely the same method we can obtain a corresponding result in Theorem 2 below for the prime twin problem.

Theorem 2 *Let*

$$T(X, k; r, b) := \#\{p \leq X : p \equiv b \pmod{r} \text{ and } p + k \text{ is also a prime}\}.$$

Then there exists an effectively computable absolute constant $\delta > 0$ such that for all the even positive integers $k \leq X$ satisfying $(k + b, r) = 1$, but at most $O(X^{1-\delta})$ exceptions, there holds

$$T(X, k; r, b) \geq r^{-1} X^{1-10\delta}$$

whenever $1 \leq r \leq X^\delta$ and $(b, r) = 1$.

It is evident that both Theorem 1 and Theorem 2 contain Linnik's theorem on the least primes in arithmetic progressions [6]. In fact, if we denote by $P(r, b)$ the least prime p satisfying $p \equiv b \pmod{r}$, then the solubility of (1.2) implies that $P(r, b) \leq X \ll r^c$ for any $(b, r) = 1$ where the constant c may be taken as $c = 1/\delta$.

From Theorem 1 we can derive a similar result for sum of two primes with both of them in arithmetic progressions and a corresponding result for the Goldbach-Vinogradov theorem.

Theorem 3 (a) *Let*

$$E(X; r, b_1, b_2) := \#\{n \leq X : 2|n, \ n \equiv b_1 + b_2 \pmod{r}, \ n \neq p_1 + p_2$$
$$\text{for any } p_i \equiv b_i \pmod{r}, \ i = 1, 2\}.$$

Then there exists an effectively computable absolute constant $\delta > 0$ such that

$$E(X; r, b_1, b_2) < X^{1-\delta}/\phi(r)$$

holds for any $1 \leq r \leq X^{\delta}$ *and* $(b_i, r) = 1$, $i = 1, 2$. (b) *For any positive integers* N *and* r, *let* $\mathbb{B}(r, N) = \{(b_1, b_2, b_3) : 1 \leq b_i \leq r$ *and* $(b_i, r) = 1$, $i = 1, 2, 3$, *and* $b_1 + b_2 + b_3 \equiv N \pmod{r}\}$. *Then there exists an effectively computable absolute constant* $\delta > 0$ *such that every large odd integer* N *can be written as*

$$N = p_1 + p_2 + p_3, \quad p_i \equiv b_i \pmod{r}, \quad 1 \leq i \leq 3$$

for any $1 \leq r \leq N^{\delta}$ *and* $(b_1, b_2, b_3) \in \mathbb{B}(r, N)$.

It should be noted that for any odd integer $N \geq 3$ the set $\mathbb{B}(r, N)$ is not empty, in fact one has, for $2 \nmid N$,

$$\#\mathbb{B}(r, N) = r^2 \prod_{p \mid N, p \mid r} \frac{(p-1)(p-2)}{p^2} \prod_{p \mid r, p \nmid N} \frac{p^2 - 3p + 3}{p^2} > 0$$

(see [7, (4.2) with all $a_i = 1$ and Lemma 4.3(2)]).

The problems of the kind in Theorems 1 and 2 have been considered by a number of authors. Their results were usually stated in the form on the prime twin problem, as in our Theorem 2. In the present article we shall concentrate mainly on the Goldbach problem. In fact, there are no essential differences between them. The first result in this direction was given by Lavrik [5] in 1961 who proved that, in the form on the Goldbach problem, for any positive constants C, A, and $1 \leq r \leq (\log X)^C$ and $(b, r) = 1$

$$D(X, r) := \sum_{\substack{n \leq X \\ 2 \mid n, (n-b, r) = 1}} \left| \sum_{\substack{n = m_1 + m_2 \\ m_1 \equiv b \pmod{r}}} \Lambda(m_1)\Lambda(m_2) - \frac{\sigma(n, r)}{\phi(r)} n \right|$$
$$\ll X^2 (\log X)^{-A-C},$$

where Λ denotes the von Mangoldt function, and

$$\sigma(n, r) := \frac{nr}{\phi(nr)} \prod_{p \nmid nr} \left(1 - \frac{1}{(p-1)^2}\right). \tag{1.3}$$

It follows that

$$E(X; r, b) < X(\log X)^{-A} \text{ for } 1 \leq r \leq (\log X)^C. \tag{1.4}$$

Further results were obtained quite recently by several authors. In a joint paper of Maier and Pomerance [8] on the large gaps between consecutive primes, they enlarged in 1990 the range of r to some power of X, but in average sense. Their result is

$$\sum_{r \leq R} D(X, r) \ll X^2 (\log X)^{-A} \tag{1.5}$$

for $R = X^\delta$. In other words, there exists $\delta > 0$ such that for $R = X^\delta$

$$E(X; r, b) < X(\log X)^{-A} \quad \text{for almost all } 1 \le r \le R. \tag{1.6}$$

By his weighted circle method and estimation of exponential sums over primes in arithmetic progressions, Balog [1] succeeded in showing (1.5), and in turn (1.6) with $R = X^{1/3}(\log X)^{-B}$ where $B > 0$ is a constant depending on A only. Afterwards Mikawa [9] further extended the range of r in (1.5) and (1.6) to $1 \le r \le R = X^{1/2}(\log X)^{-B}$ by applying Linnik's dispersion method and the Bombieri theorem. Our Theorem 1 is a direct improvement on (1.4) and (1.6), namely, we replace the bound $X(\log X)^{-A}$ by $X^{1-\delta}$, and "almost all $1 \le r \le R$" in (1.6) by "any $1 \le r \le R$".

The proof of Theorem 1, like that of Maier-Pomerance's result, depends on the method of Montgomery and Vaughan in [10]. The new ingredients in our proof, apart from the much more involvement, are mainly the following: The starting point of our method is Lemma 1 (the basic lemma) in Section 3 which interprets $S(\alpha; r, b)$, the exponential sum over primes in an arithmetic progression, as a new form involving the Dirichlet characters. Consequently, as the circle method is applied, a generalization of the Gaussian character sum, namely $H(m, \chi_q, \chi_{h_1})$ defined in Section 3, appears, which brings us more difficulties in establishing some useful inequalities, for example, Lemma 5 that plays an important role in estimating the error terms I_j $(5 \le j \le 9)$. Moreover, because of the appearance of r, more accurate estimates will be needed in the treatment of I_j $(2 \le j \le 4)$ resulting from the possible existence of the exceptional zero of Dirichlet L-functions.

2. Notations and treatment of the minor arcs In what follows all the implied constants in the Vinogradov symbol $\ll$ and the constants denoted by η, δ and c_j are effectively computable and positive.

Let $X > 0$ be sufficiently large, $\eta > 0$ be a constant sufficiently small, $c_1 > 0$ be a large constant compared with η, $Q := X^{4\eta}$, $R := X^\eta$, $T := (QR)^{c_1} < X^{1/10}$, $1 \le r \le R$ and $X/2 < n \le X$. For real α define

$$S(\alpha) := \sum_{QR < p \le X} \log p \, e(p\alpha) \,,$$

$$S(\alpha; r, b) := \sum_{\substack{QR < p \le X \\ p \equiv b \pmod{r}}} \log p \, e(p\alpha) \,,$$

$$S(\alpha, \chi_q) := \sum_{QR < p \le X} \log p \, \chi_q(p) e(p\alpha),$$

where and in the sequel we use $\chi_q = \chi \pmod q$ to denote a Dirichlet character modulo q, and χ_q^0 the principal character. It is known that there is a small absolute constant $c_2 > 0$ such that there exists at most one primitive character $\tilde{\chi}$ to a modulus $\tilde{q} \leq QR$ for which the corresponding L-function $L(\sigma + it, \tilde{\chi})$ has a zero in the region: $\sigma \geq 1 - c_2(\log T)^{-1}$, $|t| \leq T$; and if such an exceptional character exists, the corresponding modulus $\tilde{q}$ and zero $\tilde{\beta}$ are called the *exceptional character* and the *exceptional zero* (or *Siegel zero*) of order QR respectively. Moreover, $\tilde{\beta}$ must be simple, *real* and unique, $\tilde{\chi}$ must be *real* and $\tilde{q}$ has the form

$$\tilde{q} = 2^{\alpha} p_1 \cdots p_s \text{ with } \alpha = 0, 2 \text{ or } 3, \text{ and } 3 \leq p_1 < \cdots < p_s. \qquad (2.1)$$

Further, we define

$$T(\alpha) := \sum_{QR < m \leq X} e(m\alpha), \quad \tilde{T}(\alpha) := \sum_{QR < m \leq X} m^{\tilde{\beta}-1} e(m\alpha),$$

$$W(\alpha, \chi_q) := \begin{cases} S(\alpha, \chi_q) - T(\alpha) & \text{if } \chi_q = \chi_q^0, \\ S(\alpha, \chi_q) - \tilde{T}(\alpha) & \text{if } \chi_q = \tilde{\chi}\chi_q^0 \text{ (so } \tilde{q}|q), \\ S(\alpha, \chi_q) & \text{otherwise,} \end{cases}$$

and

$$R(n; r, b) := \sum_{\substack{n = p_1 + p_2 \\ QR < p_i \leq X \\ p_1 \equiv b \pmod{r}}} \log p_1 \log p_2.$$

Then by the circle method we have

$$R(n; r, b) = \int_0^1 S(\alpha) S(\alpha; r, b) e(-n\alpha) d\alpha$$

$$= \left(\int_{E_1} + \int_{E_2} \right) \cdots d\alpha = R_1(n) + R_2(n),$$

say, where E_1 is the *major arcs* defined as

$$E_1 := \bigcup_{q \leq Q} \bigcup_{(a,q)=1} \left[\frac{a}{q} - \frac{1}{\tau}, \frac{a}{q} + \frac{1}{\tau} \right], \quad \tau := XQ^{-1}$$

and $E_2 := [-Q^{-1}, 1 - Q^{-1}] \setminus E_1$ is the *minor arcs*. Clearly, all major arcs are disjoint. By Dirichlet's theorem on rational approximation (see [3, p. 150]), every $\alpha \in E_2$ can be written as

$$\alpha = (a/q) + \lambda \text{ with } (a, q) = 1, |\lambda| \leq 1/(q\tau), Q \leq q \leq \tau.$$

Then by Vinogradov's theorem on exponential sum over primes (see [3, p. 143]) we get

$$S(\alpha) \ll XQ^{-1/2}L^5, \quad \alpha \in E_2, \tag{2.2}$$

where and in the sequel L stands for $\log X$. It follows by Parseval's identity and (2.2) that

$$\sum_{X/2 < n \le X} |R_2(n)|^2 = \int_{E_2} |S(\alpha)|^2 |S(\alpha; r, b)|^2 d\alpha \ll X^3 L^{12}/(rQ).$$

We may now conclude that the number of even integers in $(X/2, X]$ for which

$$|R_2(n)| > X/(rQ^{1/3})$$

holds is $\ll XQ^{-1/12}L^{12}$. Therefore, if we can prove that apart from at most $O(X^{1-(\eta/7)})$ exceptions of the even integers $n \in (X/2, X]$ satisfying $(n-b, r) = 1$, there holds

$$R_1(n) > 2X/(rQ^{1/3}) \tag{2.3}$$

for $\eta > 0$ sufficiently small, then consequently, $E(X; r, b) \ll X^{1-(\eta/7)}$ for any $1 \le r \le R$ and $(r, b) = 1$. Taking $\delta = \eta/7$, we get immediately Theorem 1 and its Remark 2.

In the following sections we shall use some standard notations. For example,

$$G(m, \chi_q) := \sum_{a=1}^{q} \chi_q(a)e(am/q), \quad \tau(\chi) := G(1, \chi), \quad C_q(m) := G(m, \chi_q^0),$$

$\sum_{\chi_q}^{*}$ denotes the summation over all primitive characters modulo q, and $p^\alpha \| q$ means $p^\alpha | q$ and $p^{\alpha+1} \nmid q$. For clarity, sometimes, we use χ_q^* to denote a primitive character.

3. Lemmas and outline of treatment of the major arcs In order to interpret $S(\alpha; r, b)$ in terms of the Dirichlet characters we introduce the "factorization of an integer q with respect to r": for fixed $r \ge 1$, an integer $q \ge 1$ can be written uniquely as

$$q = uhs \text{ with } h = (q, r), \ (s, r) = 1, \ u|h^\infty, \tag{3.1}$$

where $u|h^\infty$ means that $p|u$ always implies $p|h$. Once we have the factorization (3.1), then the integer qr/h can be factorized as

$$qr/h = h_1 h_2 s \text{ with } h_1 = \prod_{p^\alpha \| q, p|u} p^\alpha, \ h_2 = \prod_{p^\alpha \| r, p \nmid u} p^\alpha. \tag{3.2}$$

The representations in (3.2) are also unique and $(h_1, h_2) = 1$.

For any fixed $r \geq 1$, the h, s, u, h_1, and h_2 defined by (3.1) and (3.2) are functions of q. Sometimes we write, for example, $q = u(q)h(q)s(q)$ to avoid any confusion. We further define $\bar{s} \equiv \bar{s}(q) \pmod{r^3}$ by

$$\bar{s}s \equiv 1 \pmod{r^3}. \tag{3.3}$$

It seems more natural to define $\bar{s}$ by $\bar{s}s \equiv 1 \pmod{r}$ instead of (3.3). However, (3.3) may simplify our treatment of I_4 in Section 4. By (3.1)–(3.3) we have the assertions:

(i) h, u, s, h_1, h_2 and $\bar{s} \pmod{r^3}$ are multiplicative, for example, if $(v_1, v_2) = 1$ then $h(v_1 v_2) = h(v_1)h(v_2)$, $\bar{s}(v_1 v_2) \equiv \bar{s}(v_1)\bar{s}(v_2) \pmod{r^3}$, etc.

(ii) $h_1 = 1$ if and only if $u = 1$; $h_1 = 1$ if q is square-free or $(q, r) = 1$; and if $h_1 > 1$, it must be square-full.

We now define a generalization of the Gaussian sum $G(m, \chi)$ based on the factorization of q with respect to r. For any characters χ_q, $\chi_{h_1(q)}$ and integer m let

$$F(u, \chi_{h_1}, a) := \sum_{\ell=1}^{u} \chi_{h_1}(bs\bar{s} + hs\ell)e\left(\frac{a\ell}{u}\right);$$

$$H(m, \chi_q, \chi_{h_1}) := \sum_{a=1}^{q} \chi_q(a)F(u, \chi_{h_1}, a)e\left(\frac{ab\bar{s}}{uh} + \frac{a}{q}m\right).$$

Clearly, if $u = 1$ then $h_1 = 1$ and one has

$$H(m, \chi_q, \chi_{h_1}) = G(m + bs\bar{s}, \chi_q), \quad H(m, \chi_q^0, \chi_{h_1}) = C_q(m + bs\bar{s}).$$

If further assume that $(q, r) = 1$ then $H(m, \chi_q, \chi_{h_1}) = G(m, \chi_q)$.

We are now ready to state

Lemma 1 (Basic lemma)   *Let*

$$E(q, r) := \begin{cases} 1 & \text{if } (q/h, r) = 1, \text{ i.e., if } h_1 = 1, \\ 0 & \text{otherwise,} \end{cases}$$

and

$$\tilde{E}(q, r) := \begin{cases} 1 & \text{if } \tilde{q}|qr/h \text{ and } h_1|\tilde{q}, \\ 0 & \text{otherwise.} \end{cases}$$

We have that for $\alpha = (a/q) + \lambda$ with $1 \leq q \leq Q$ and $(a,q) = 1$

$$
S(\alpha; r, b) = \frac{e(ab\overline{s}/uh)}{\phi(qr/h)} \sum_{\chi_{h_1}}^* F(u, \overline{\chi}_{h_1}, a) \sum_{\chi_s} \tau(\overline{\chi}_s) \overline{\chi}_s(uh) \chi_s(a)
$$

$$
\times \sum_{\chi_{h_2}} \overline{\chi}_{h_2}(b) S(\lambda, \chi_{h_1} \chi_{h_2} \chi_s)
$$

$$
= E(q, r) \frac{\mu(q/h) e(ab\overline{s}/h)}{\phi(qr/h)} T(\lambda)
$$

$$
+ \tilde{E}(q, r) \frac{e(ab\overline{s}/uh)}{\phi(qr/h)} \tau(\chi_s) \chi_{h_2}(b) \chi_s(uha) F(u, \chi_{h_1}^*, a) \tilde{T}(\lambda)
$$

$$
+ \frac{e(ab\overline{s}/uh)}{\phi(qr/h)} \sum_{\chi_{h_1}}^* F(u, \overline{\chi}_{h_1}, a) \sum_{\chi_s} \tau(\overline{\chi}_s) \overline{\chi}_s(uh) \chi_s(a)
$$

$$
\times \sum_{\chi_{h_2}} \overline{\chi}_{h_2}(b) W(\lambda, \chi_{h_1} \chi_{h_2} \chi_s). \tag{3.4}
$$

In particular, if $r = 1$, one gets the well-known formula in the circle method

$$
S(\alpha) = \frac{\mu(q)}{\phi(q)} T(\lambda) + \tilde{E}(q, 1) \frac{\tau(\tilde{\chi} \chi_q^0)}{\phi(q)} \tilde{\chi}(a) \tilde{T}(\lambda)
$$

$$
+ \frac{1}{\phi(q)} \sum_{\chi_q} \tau(\overline{\chi}_q) \chi_q(a) W(\lambda, \chi_q). \tag{3.5}
$$

Remark In the second term on the right-hand side of (3.4), one should note (a) the character $\chi_{h_1}^* \chi_{h_2} \chi_s = \chi_{qr/h}$ is induced by $\tilde{\chi}(= \tilde{\chi}_{\tilde{q}})$ and (b) if there is $p \geq 3$, $p | h_1$, then by the property of h_1 given in (ii) above one has $p^2 | h_1$ and $p^2 | \tilde{q}$. This is a contradiction to (2.1). Hence the only possible values of h_1 are: 1, 4 and 8. Furthermore, if $2 \nmid r$, one has $h_1 = 1$.

Proof The second equality in (3.4) follows from the definition of $W(\lambda, \chi_{qr/h})$ and $\tau(\chi_s^0) = \mu(s)$ where $s = q/h$ (see Lemma 2 below). So it suffices to prove the first equality in (3.4). By (3.1) and (3.2) we know that for $\ell \equiv b \pmod{h}$, the simultaneous congruent relations

$$
n \equiv \ell \pmod{uh} \quad \text{and} \quad n \equiv b \pmod{r} \tag{3.6}
$$

is equivalent to

$$
n \equiv \ell \pmod{h_1} \quad \text{and} \quad n \equiv b \pmod{h_2}. \tag{3.7}
$$

Then

$$S(\alpha; r, b) = \sum_{\substack{QR < p \leq X \\ p \equiv b \ (\mathrm{mod}\ r) \\ p \nmid q}} \log p \ e(p\alpha)$$

$$= \sum_{\substack{\ell = 1 \\ \ell \equiv b \ (\mathrm{mod}\ h) \\ (\ell, q) = 1}}^{q} e\left(\frac{a}{q}\ell\right) \sum_{\substack{QR < p \leq X \\ p \equiv b \ (\mathrm{mod}\ r) \\ p \equiv \ell \ (\mathrm{mod}\ q)}} \log p \ e(p\lambda).$$

By (3.1) we can rewrite the condition, $p \equiv \ell$ (mod q) in the last sum as $p \equiv \ell$ (mod uh) and $p \equiv \ell$ (mod s). Then by the equivalence of (3.6) and (3.7), and the orthogonality relation of characters we have

$$S(\alpha; r, b) = \sum_{\substack{\ell = 1, (\ell, q) = 1 \\ \ell \equiv b \ (\mathrm{mod}\ h)}}^{q} e(a\ell/q) \sum_{\substack{QR < p \leq X, p \equiv \ell \ (\mathrm{mod}\ h_1) \\ p \equiv b \ (\mathrm{mod}\ h_2), p \equiv \ell \ (\mathrm{mod}\ s)}} \log p \ e(p\lambda)$$

$$= \frac{1}{\phi(qr/h)} \sum_{\chi_{h_1}} \sum_{\chi_s} \sum_{\chi_{h_2}} \overline{\chi}_{h_2}(b)$$

$$\times \sum_{\substack{\ell = 1, (\ell, q) = 1 \\ \ell \equiv b \ (\mathrm{mod}\ h)}}^{q} \overline{\chi}_{h_1}(\ell) \overline{\chi}_s(\ell) e(a\ell/q) S(\lambda, \chi_{h_1} \chi_{h_2} \chi_s). \tag{3.8}$$

Let $\ell = s\ell_2 + uh\ell_1$ in the last summation over ℓ. It follows that

$$\sum_{\substack{\ell = 1, (\ell, q) = 1 \\ \ell \equiv b \ (\mathrm{mod}\ h)}}^{q} \overline{\chi}_{h_1}(\ell) \overline{\chi}_s(\ell) e(a\ell/q)$$

$$= \sum_{\ell_1 = 1}^{s} \overline{\chi}_s(uh\ell_1) e(a\ell_1/s) \sum_{\substack{\ell_2 = 1 \\ \ell_2 \equiv \overline{s}b \ (\mathrm{mod}\ h)}}^{uh} \overline{\chi}_{h_1}(s\ell_2) e(a\ell_2/uh)$$

$$= e(ab\overline{s}/uh) \overline{\chi}_s(uh) \sum_{\ell_1 = 1}^{s} \overline{\chi}_s(\ell_1) e(a\ell_1/s) \sum_{t=1}^{u} \overline{\chi}_{h_1}(bs\overline{s} + sht) e(at/u)$$

$$= e(ab\overline{s}/uh) \overline{\chi}_s(uh) \chi_s(a) \tau(\overline{\chi}_s) F(u, \overline{\chi}_{h_1}, a). \tag{3.9}$$

By (3.8) and (3.9) we obtain that

$$S(\alpha; r, b) = \frac{e(ab\overline{s}/uh)}{\phi(qr/h)} \sum_{\chi_{h_1}} F(u, \overline{\chi}_{h_1}, a)$$

$$\times \sum_{\chi_s} \overline{\chi}_s(uh) \chi_s(a) \tau(\overline{\chi}_s) \sum_{\chi_{h_2}} \overline{\chi}_{h_2}(b) S(\lambda, \chi_{h_1} \chi_{h_2} \chi_s).$$

To prove the lemma we now only have to show that if χ_{h_1} is not primitive then

$$F(u, \chi_{h_1}, a) = 0. \tag{3.10}$$

Suppose that χ_{h_1} is induced by $\chi_{h'}^*$ (so $h'|h_1$) and $1 \le h' < h_1$. If $h' = 1$, i.e., $\chi_{h_1} = \chi_{h_1}^0$, then we have trivially that $F(u, \chi_{h_1}^0, a) = 0$, where we used the facts that $(u(q), a) = 1$, and the inequality $h_1 > 1$ indicates $u > 1$. Now assume $1 < h' < h_1$. Since we only consider those ℓ with $(bs\bar{s} + hs\ell, h_1) = 1$ we see that, by [3, p. 65, (2)]

$$
\begin{aligned}
F(u, \chi_{h_1}, a) &= \sum_{\ell=1}^{u} \chi_{h'}^*(bs\bar{s} + hs\ell)e(a\ell/u) \\
&= \frac{1}{\tau(\overline{\chi}_{h'}^*)} \sum_{\ell=1}^{u} \sum_{t=1}^{h'} \overline{\chi}_{h'}^*(t)e\left(\frac{bs\bar{s} + hs\ell}{h'}t\right)e(a\ell/u) \\
&= \frac{1}{\tau(\overline{\chi}_{h'}^*)} \sum_{t=1}^{h'} \overline{\chi}_{h'}^*(t)e(bs\bar{s}t/h') \sum_{\ell=1}^{u} e\left(\frac{a + uh(h')^{-1}st}{u}\ell\right).
\end{aligned}
$$

Note that there exists $p|h_1/h'$, which implies $p|u$. However, $p \nmid a$ since $(a, q) = 1$. Then we have that $u \nmid a + (uhst/h')$ for any $1 \le t \le h'$, from which it then follows $F(u, \chi_{h_1}, a) = 0$. This finishes the proof of (3.10) and the basic lemma.

Applying (3.4) and (3.5) to the integral over major arcs, we obtain that

$$
\begin{aligned}
R_1(n) &= \sum_{q \le Q} \sum_{a=1,(a,q)=1}^{q} \int_{-1/\tau}^{1/\tau} S\left(\frac{a}{q} + \lambda\right) S\left(\frac{a}{q} + \lambda; r, b\right) e\left(-\left(\frac{a}{q} + \lambda\right)n\right) d\lambda \\
&:= \sum_{j=1}^{9} I_j, \tag{3.11}
\end{aligned}
$$

where

$$I_1 := \sum_{q \le Q, (q/h,r)=1} \frac{\mu(q)\mu(q/h)}{\phi(q)\phi(qr/h)} C_q(bs\bar{s} - n) \int_{-1/\tau}^{1/\tau} T^2(\lambda)e(-n\lambda)d\lambda,$$

$$
\begin{aligned}
I_2 := \sum_{q \le Q} \frac{\tilde{E}(q,r)\mu(q)}{\phi(q)\phi(qr/h)} \tau(\chi_s)\chi_s(uh)\chi_{h_2}(b) H(-n, \chi_s\chi_q^0, \chi_{h_1}^*) \\
\times \int_{-1/\tau}^{1/\tau} \tilde{T}(\lambda)T(\lambda)e(-n\lambda)d\lambda,
\end{aligned}
$$

$$I_3 := \sum_{\substack{q \leq Q, \, \tilde{q}|q \\ (q/h,r)=1}} \frac{\mu(q/h)\tau(\tilde{\chi}\chi_q^0)}{\phi(q)\phi(qr/h)} G(bs\overline{s} - n, \tilde{\chi}\chi_q^0) \int_{-1/\tau}^{1/\tau} T(\lambda)\tilde{T}(\lambda)e(-n\lambda)d\lambda,$$

$$I_4 := \sum_{q \leq Q, \, \tilde{q}|q} \frac{\tilde{E}(q,r)\tau(\tilde{\chi}\chi_q^0)\tau(\chi_s)}{\phi(q)\phi(qr/h)}\chi_{h_2}(b)\chi_s(uh)H(-n, \tilde{\chi}\chi_s\chi_q^0, \chi_{h_1}^*)$$

$$\times \int_{-1/\tau}^{1/\tau} (\tilde{T}(\lambda))^2 e(-n\lambda)d\lambda,$$

$$I_5 := \sum_{q \leq Q} \frac{\mu(q)}{\phi(q)\phi(qr/h)} \sum_{\chi_{h_1}}^* \sum_{\chi_s} \tau(\overline{\chi}_s)\overline{\chi}_s(uh) \sum_{\chi_{h_2}} \overline{\chi}_{h_2}(b)H(-n, \chi_s\chi_q^0, \overline{\chi}_{h_1})$$

$$\times \int_{-1/\tau}^{1/\tau} T(\lambda)W(\lambda, \chi_{h_1}\chi_{h_2}\chi_s)e(-n\lambda)d\lambda,$$

$$I_6 := \sum_{q \leq Q, \, \tilde{q}|q} \frac{\tau(\tilde{\chi}\chi_q^0)}{\phi(q)\phi(qr/h)} \sum_{\chi_{h_1}}^* \sum_{\chi_s} \tau(\overline{\chi}_s)\overline{\chi}_s(uh) \sum_{\chi_{h_2}} \overline{\chi}_{h_2}(b)H(-n, \tilde{\chi}\chi_s\chi_q^0, \overline{\chi}_{h_1})$$

$$\times \int_{-1/\tau}^{1/\tau} \tilde{T}(\lambda)W(\lambda, \chi_{h_1}\chi_{h_2}\chi_s)e(-n\lambda)d\lambda,$$

$$I_7 := \sum_{q \leq Q, \, (q/h,r)=1} \frac{\mu(q/h)}{\phi(q)\phi(qr/h)} \sum_{\chi_q} \tau(\overline{\chi}_q)G(bs\overline{s} - n, \chi_q)$$

$$\times \int_{-1/\tau}^{1/\tau} T(\lambda)W(\lambda, \chi_q)e(-n\lambda)d\lambda,$$

$$I_8 := \sum_{q \leq Q} \frac{\tilde{E}(q,r)}{\phi(q)\phi(qr/h)} \sum_{\chi_q} \tau(\overline{\chi}_q^{(1)})\tau(\chi_s^{(2)})\chi_{h_2}^{(2)}(b)\chi_s^{(2)}(uh)H(-n, \chi_s^{(2)}, \chi_{h_1}^{(2)})$$

$$\times \int_{-1/\tau}^{1/\tau} \tilde{T}(\lambda)W(\lambda, \chi_q^{(1)})e(-n\lambda)d\lambda,$$

$$I_9 := \sum_{q \leq Q} \frac{1}{\phi(q)\phi(qr/h)} \sum_{\chi_{h_1}}^* \sum_{\chi_{h_2}} \overline{\chi}_{h_2}^{(2)}(b) \sum_{\chi_s} \tau(\overline{\chi}_s^{(2)})\overline{\chi}_s^{(2)}(uh)$$

$$\times \sum_{\chi_q} \tau(\overline{\chi}_q^{(1)})H(-n, \chi_q^{(1)}\chi_s^{(2)}, \overline{\chi}_{h_1}^{(2)})$$

$$\times \int_{-1/\tau}^{1/\tau} W(\lambda, \chi_q^{(1)})W(\lambda, \chi_{h_1}^{(2)}\chi_{h_2}^{(2)}\chi_s^{(2)})e(-n\lambda)d\lambda.$$

Here it should be noted that χ_{h_1} is always primitive; that in I_2, I_4, and I_8 the character $\chi_{h_1}\chi_{h_2}\chi_s = \chi_{qr/h}$ is induced by $\tilde{\chi}$ and is therefore unique for every q satisfying $\tilde{q}|qr/h$ and $h_1|\tilde{q}$; and that in I_8 and I_9 we use $\chi_q^{(1)}$ and

$\chi_{qr/h}^{(2)} = \chi_{h_1}^{(2)}\chi_{h_2}^{(2)}\chi_s^{(2)}$ to denote the characters appearing in (3.4) and (3.5) respectively, in order to avoid any ambiguity.

In order to estimate I_j $(1 \le j \le 9)$ we need some more preliminary lemmas.

Lemma 2 *If $\chi_{\bmod q}$ is primitive then $|\tau(\chi)| = q^{1/2}$. If χ (mod q) is a primitive quadratic character then $\tau(\chi)^2 = \chi(-1)q$.*

Lemma 3 *Let χ_q be a character induced by the primitive character $\chi_{q^*}^*$; we denote it by $\chi_q \leftrightarrow \chi_{q^*}^*$. For an arbitrary integer m put $q' = q/(m,q)$. If $q^* \nmid q'$ then $G(m,\chi_q) = 0$. If $q^*|q'$ then*

$$G(m,\chi_q) = \overline{\chi}_{q^*}^*(m/(m,q))\phi(q)\phi^{-1}(q')\mu(q'/q^*)\chi_{q^*}^*(q'/q^*)\tau(\chi_q^*).$$

In particular, $G(m,\chi_q^0) = C_q(m) = \mu(q/(q,m))\phi(q)\phi^{-1}(q/(q,m))$, $\tau(\chi) = \mu(q/q^*)\chi_{q^*}(q/q^*)\tau(\chi_{q^*}^*)$, and if χ_q is primitive then $G(m,\chi_q) = \overline{\chi}_q(m)\tau(\chi_q)$.

The above two lemmas may be found in [10, Section 5] or [7, (4.9) and (4.4)]. For any integers m, t and any characters χ_q and χ_{h_1} we define

$$F^*(u,\chi_{h_1},a,t) := \sum_{\ell=1}^{u} \chi_{h_1}(bs\overline{s}t + sh\ell)e(a\ell/u)$$

and

$$H^* := H^*(m,\chi_q,\chi_{h_1},t) := \sum_{a=1}^{q} \chi_q(a)F^*(u,\chi_{h_1},a,t)e\left(\frac{ab\overline{s}t}{uh} + \frac{a}{q}m\right).$$

Lemma 4 *Let $s\overline{s}(q)$ stand for $s(q)\overline{s}(q)$. Then we have the following three assertions:*
(1) *If $u(q) = 1$, $q = q_1q_2$ and $(q_1,q_2) = 1$ then*

$$C_q(m + bs\overline{s}(q)) = C_{q_1}(m + bs\overline{s}(q_1))C_{q_2}(m + bs\overline{s}(q_2));$$
$$G(m + bs\overline{s}(q),\chi_q) = \chi_{q_1}(q_2)\chi_{q_2}(q_1)G(m+bs\overline{s}(q_1),\chi_{q_1})G(m+bs\overline{s}(q_2),\chi_{q_2}).$$

(2) $H(m,\chi_q,\chi_{h_1}) = H^*(m,\chi_q,\chi_{h_1},1)$.
If $(q,r) = 1$ then $H^* = G(m,\chi_q)$; if $u(q) = 1$ then $H^* = G(m+bs\overline{s}(q)t,\chi_q)$.
(3) *If $q = q_1q_2$ and $(q_1,q_2) = 1$ then*

$$H^*(m,\chi_q,\chi_{h_1(q)},t) = \chi_{q_1}(q_2)\chi_{q_2}(q_1)H^*(m,\chi_{q_1},\chi_{h_1(q_1)},ts\overline{s}(q_2))$$
$$\times H^*(m,\chi_{q_2},\chi_{h_1(q_2)},ts\overline{s}(q_1)).$$

In particular, if $u(q_1) = 1$ then

$$H^*(m,\chi_q,\chi_{h_1(q)},t) = \chi_{q_1}(q_2)\chi_{q_2}(q_1)G(m+bs\overline{s}(q_1)t,\chi_{q_1})$$
$$\times H^*(m,\chi_{q_2},\chi_{h_1(q_2)},ts\overline{s}(q_1)).$$

The lemma can be proved in a standard way (see, for example, the proof of [11, Lemma 1.1]), by recalling (3.1), (3.2), and (3.3).

Lemma 5 *Let*

$$Y := Y(n, r, v, v_1, \chi_v^{(1)}, \chi_{v_1}^{(2)}, \chi_{h_1})$$

$$:= \sum_{q=1, v|q, v_1|s}^{\infty} \frac{1}{\phi(q)\phi(qr/h)} |\tau(\chi_v^{(1)}\chi_q^0)\tau(\chi_{v_1}^{(2)}\chi_s^0)| \, |H(-n, \chi_v^{(1)}\chi_{v_1}^{(2)}\chi_q^0, \chi_{h_1})|.$$

For any primitive characters $\chi_v^{(1)}, \chi_{v_1}^{(2)}$, and χ_{h_1} one has

$$Y \le c_3 nr/(\phi(r)\phi(nr)) \, ,$$

where $c_3 > 0$ is an absolute constant.

Proof By Lemma 2 we know that $\tau(\chi_v \chi_q^0)\tau(\chi_{v_1}\chi_s^0) = 0$ unless

$$(qv^{-1}, v) = (sv_1^{-1}, v_1) = 1 \text{ and } \mu^2(qv^{-1}) = \mu^2(sv_1^{-1}) = 1, \quad (3.12)$$

which we henceforth assume. Let $d = (v, v_1)$, $v = dv'$, $v_1 = dv_1'$. From $(v_1, r) = 1$ and (3.12) it follows that

$$(d, v') = (d, v_1') = (v', v_1') = (d, r) = 1.$$

Let $q = dv'v_1'k$. By (3.12) we have that d, v', v_1', and k are pairwise co-prime and $\mu^2(v_1') = \mu^2(k) = 1$. Then $s = s(d)s(v')s(v_1')s(k) = dv_1's(v')k/(k, r)$. By (3.12) we see that $u(q)s(v') = v'/(v', r)$ and then $qr/h = dv_1'(rv'/(v', r)) \times k/(k, r)$. Applying Lemma 3 with $m = 1$, and Lemmas 4(2) and 4(3) with $u(k) = 1$ and $t = 1$ we get that

$$Y \le \frac{|\tau(\chi_v^{(1)})\tau(\chi_{v_1}^{(2)})|}{\phi^2(d)\phi^2(v_1')\phi(v')\phi(v'r/(v', r))} \sum_{k=1, (k, dv'v_1')=1}^{\infty} \frac{\mu^2(k)|C_k(bs\overline{s}(k) - n)|}{\phi(k)\phi(k/(k, r))}$$

$$\times \left| H^*\left(-n, \chi_d^{(1)}\chi_d^{(2)}\chi_{v'}^{(1)}\chi_{v_1'}^{(2)}, \chi_{h_1(dv'v_1')}, s\overline{s}(k) \right) \right|.$$

Applying Lemmas 4(3) and 4(2) with $(r, d) = 1 = (r, v_1')$ the above $|H^*|$ can be written as

$$|G(-n, \chi_d^{(1)}\chi_d^{(2)})G(-n, \chi_{v_1'}^{(2)})H^*(-n, \chi_{v'}^{(1)}, \chi_{h_1(v')}, s\overline{s}(dv_1'k))|.$$

On the other hand, by Lemma 2 and Lemma 3 we have

$$|G(-n, \chi_{v_1'}^{(2)})| = |\chi_{v_1'}^{(2)}(-n)\tau(\chi_{v_1'}^{(2)})| \le \sqrt{v_1'},$$

$$|G(-n, \chi_d^{(1)}\chi_d^{(2)})| \le \phi(d)\phi(d/(n, d))^{-1}\sqrt{d/(n, d)}.$$

Then

$$
Y \le \frac{d}{\phi(d)} \cdot \frac{(d/(n,d))^{1/2}}{\phi(d/(n,d))} \cdot \frac{v_1'}{\phi^2(v_1')} \cdot \frac{\sqrt{v'}}{\phi(v')\phi(v'r/(v',r))}
$$

$$
\times \sum_{k=1,(k,dv'v_1')=1}^{\infty} \frac{\mu^2(k)|C_k(b s\bar{s}(k)-n)|}{\phi(k)\phi(k/(k,r))}
$$

$$
\times \left| H^*\left(-n, \chi_{v'}^{(1)}, \chi_{h_1(v')}, s\bar{s}(dv_1'k)\right) \right|. \tag{3.13}
$$

We now show that

$$
\left| H^*\left(-n, \chi_{v'}^{(1)}, \chi_{h_1(v')}, s\bar{s}(dv_1'k)\right) \right| \le \sqrt{v'} u(v') \tag{3.14}
$$

under the factorization of v' with respect to r (see (3.1)). Denote by K the left-hand side of (3.14), and let $q' = (v',r)u(v')$. Appealing again to Lemmas 4(3) and 4(2) with $(r, s(v')) = 1$, we get that

$$
K \le |G(-n, \chi_{s(v')}^{(1)})| \, |H^*(-n, \chi_{q'}^{(1)}, \chi_{h_1(q')}, s\bar{s}(dv'v_1'k))|
$$

$$
\le (s(v'))^{1/2} \prod_{p^\alpha \| q', p \nmid u(v')} |G(bv_1' - n, \chi_{p^\alpha}^{(1)})|
$$

$$
\times \prod_{p^\alpha \| q', p | u(v')} |H^*(-n, \chi_{p^\alpha}^{(1)}, \chi_{h_1(p^\alpha)}, m)|
$$

$$
\le s(v')^{1/2} \prod_{p^\alpha \| q', p \nmid u(v')} p^{\alpha/2} \prod_{p^\alpha \| q', p | u(v')} |H^*(-n, \chi_{p^\alpha}^{(1)}, \chi_{h_1(p^\alpha)}, m)| \tag{3.15}
$$

where m satisfies $(m, u(v')) = 1$. For $p^\alpha \| q'$ and $p | u(v')$ we suppose that $p^\beta \| u(v')$. Then $0 < \beta < \alpha$, $p^{\alpha-\beta} \| (v',r)$ and $p^\alpha \| h_1(v')$. In this case $uh(p^\alpha) = h_1(p^\alpha) = p^\alpha$ and

$$
H^*\left(-n, \chi_{p^\alpha}^{(1)}, \chi_{h_1(p^\alpha)}, m\right) = \sum_{a=1}^{p^\alpha} \chi_{p^\alpha}^{(1)}(a) F^*(p^\beta, \chi_{p^\alpha}, a, m) e\left(\frac{bm-n}{p^\alpha}a\right)
$$

$$
= \sum_{\ell=1}^{p^\beta} \chi_{p^\alpha}(bm + p^{\alpha-\beta}\ell) \sum_{a=1}^{p^\alpha} \chi_{p^\alpha}^{(1)}(a) e\left(\frac{bm-n+\ell p^{\alpha-\beta}}{p^\alpha}a\right)
$$

$$
= \tau(\chi_{p^\alpha}^{(1)}) \sum_{\ell=1}^{p^\beta} \chi_{p^\alpha}(bm + p^{\alpha-\beta}\ell)\overline{\chi}_{p^\alpha}^{(1)}(bm - n + \ell p^{\alpha-\beta}).
$$

Obviously

$$
|H^*(-n, \chi_{p^{\alpha_1}}^{(1)}, \chi_{h_1(p^\alpha)}, m)| \le \sqrt{p^\alpha} \cdot p^\beta. \tag{3.16}
$$

The bound (3.14) now follows from (3.15) and (3.16). With (3.13) and (3.14) we have that

$$Y \leq \frac{d}{\phi(d)} \cdot \frac{(d/(n,d))^{1/2}}{\phi(d/(n,d))} \cdot \frac{v_1'}{\phi^2(v_1')} \cdot \frac{v'u(v')}{\phi(v')\phi(v'r/(v,r))}$$

$$\times \sum_{k=1,(k,dv'v_1')=1}^{\infty} \frac{\mu^2(k)|C_k(bs\bar{s}(k)-n)|}{\phi(k)\phi(k/(k,r))}.$$

The last sum is

$$\prod_{p \nmid dv'v_1'} \left(1 + \frac{|C_p(bs\bar{s}(p)-n)|}{\phi(p)\phi(p/(p,r))}\right).$$

By Lemma 3,

$$C_p(bs\bar{s}(p)-n) = \begin{cases} C_p(-n) = p-1 & \text{if } p|n \text{ and } p \nmid r , \\ C_p(-n) = -1 & \text{if } p \nmid nr , \\ C_p(b-n) = -1 & \text{if } p|r \ (\Rightarrow p \nmid (b-n)). \end{cases} \quad (3.17)$$

On writing $f = p/(p-1)$, a straightforward calculation yields

$$Y \ll \frac{1}{\phi(r)} \prod_{p|nr} f \cdot \left(\frac{(d,n)}{d}\right)^{1/2} \cdot \prod_{p|d/(n,d)} f^2 \cdot \frac{1}{v_1'}$$

$$\times \prod_{p|v_1'} f^2 \cdot \frac{u(v')(v',r)}{v'} \prod_{p|v',p\nmid r} f^2 \leq c_3 \cdot \frac{1}{\phi(r)} \cdot \frac{nr}{\phi(nr)}.$$

In the last step we use $v' = u(v')(v',r)s(v')$ and the inequality $q^{3/2}\phi^{-2}(q) \ll 1$ for any $q \geq 1$. This finishes the proof for Lemma 5.

Lemma 6 (Gallagher) *For any $y \geq \sqrt{X}$ we have*

$$\sum_{d \leq QR} \sum_{\chi_d}^* \left(\int_{-1/\tau}^{1/\tau} |W(\lambda, \chi_d)|^2 d\lambda\right)^{1/2} \ll \Omega^3 X^{1/2} \exp(-c_5/\eta)$$

where

$$\Omega := \left\{ \begin{array}{ll} (1-\tilde{\beta})\log T \ (<1) & \text{if } \tilde{\beta} \text{ exists}, \\ 1 & \text{otherwise}, \end{array} \right\} \quad (3.18)$$

and $c_5 > 0$ is an absolute constant.

This lemma is essentially contained in [10]. The detailed proof of it may also be found in [11, Lemmas 11.12 and 11.13]. Here the additional factor Ω^3 in our Lemma 9 is obtained as in the proof of [7, Lemma 2.1].

4. Estimation of I_j $(1 \leq j \leq 9)$ In what follows we shall use another small quantity $\varepsilon > 0$ which can be taken arbitrarily small, for example, we may write $0 < \varepsilon < \delta/100$.

Estimation of I_1: By [11, Lemma 11.11] one has that

$$
I_1 = n \sum_{\substack{q \leq Q \\ (q/h,r)=1}} \frac{\mu(q)\mu(q/h)}{\phi(q)\phi(qr/h)} C_q(bs\bar{s}(q) - n)
$$

$$
+ O\left(\tau \sum_{\substack{q \leq Q \\ (q/h,r)=1}} \frac{\mu^2(q)}{\phi(q)\phi(qr/h)} |C_q(bs\bar{s}(q) - n)|\right)
$$

$$
= n \sum_{\substack{q=1 \\ (q/h,r)=1}}^{\infty} \frac{\mu(q)\mu(q/h)}{\phi(q)\phi(qr/h)} C_q(bs\bar{s}(q) - n)
$$

$$
+ O\left(\left[X \sum_{\substack{q > Q \\ (q/h,r)=1}} + \tau \sum_{\substack{q \leq Q \\ (q/h,r)=1}}\right] \frac{\mu^2(q)}{\phi(q)\phi(qr/h)} |C_q(bs\bar{s}(q) - n)|\right)
$$

$$
:= I_1^{(m)} + O(I_1^{(\ell_1)} + I_1^{(\ell_2)}),
$$

say, in an obvious order. Since we may suppose that $\mu^2(q) = 1$, it then follows that $u = h_1 = 1$ and $s = q/h$. By (3.17) and (1.3) it is easy to check that

$$
I_1^{(m)} = \frac{n}{\phi(r)} \prod_p \left(1 - \frac{\mu(p/(p,r))}{(p-1)\phi(p/(p,r))} C_p(bs\bar{s}(p) - n)\right)
$$

$$
= \frac{n}{\phi(r)} \prod_{p\nmid r} \left(1 + \frac{C_p(-n)}{(p-1)^2}\right) \prod_{p|r} \left(1 - \frac{C_p(b-n)}{p-1}\right) = \frac{\sigma(n,r)}{\phi(r)} n.
$$

Let $q = hq'$ in the first error term. We get by Lemma 4(1) and [11, Lemma 11.5]

$$
I_1^{(\ell_1)} \ll \frac{X}{\phi(r)} \sum_{h|r} \frac{\mu^2(h)}{\phi(h)} \sum_{\substack{q' > Q/h \\ (q',r)=1}} \frac{\mu^2(q')}{\phi^2(q')} |C_h(b-n)| \, |C_{q'}(-n)|
$$

$$
\ll \frac{X}{\phi(r)} \sum_{h|r} \mu^2(h) \sum_{q' > Q/h} \frac{\mu^2(q')}{\phi^2(q')} |C_{q'}(-n)| \ll X^{1+\varepsilon} Q^{-1}.
$$

Similarly, the same upper bound holds for $I_1^{(\ell_2)}$. Then

$$
I_1 = \frac{\sigma(n,r)}{\phi(r)} n + O(X^{1+\varepsilon} Q^{-1}). \tag{4.1}
$$

Estimation of I_2 and I_3: We may suppose that $\mu^2(q) = 1$. In this case by property (ii) of $u(q)$ in Section 3 (below (3.3)), we have $u = 1 = h_1$ and hence $q = hs$, $h_2 = r$. Define

$$\tilde{q}' = \prod_{p^\alpha \| \tilde{q}, \, p \nmid r} p^\alpha \quad \text{and} \quad \tilde{q}'' = \tilde{q}/\tilde{q}'. \tag{4.2}$$

So $(\tilde{q}', r) = 1$ and $(\tilde{q}', \tilde{q}'') = 1$. From $\bar{E}(q, r)$, i.e., $\tilde{q} | qr/h \ (= sr)$, we know that $\tilde{q}'' | r$ which yields that $(\tilde{q}/(\tilde{q}, r), r) = 1$ and then by (4.2), we have

$$\tilde{q}'' = (\tilde{q}, r) \quad \text{and} \quad \tilde{q}' = \tilde{q}/(\tilde{q}, r).$$

Since $\tilde{q} | sr$ if and only if $\tilde{q}' | s$ and $\tilde{q}'' | r$, by Remark (a) on (3.4) we obtain that $\chi_{s(q)} \leftrightarrow \chi_{\tilde{q}'}$ and $\chi_{h_2(q)} \leftrightarrow \chi_{\tilde{q}''}$ where $\chi_{\tilde{q}'}$ and $\chi_{\tilde{q}''}$ are the two unique characters satisfying $\chi_{\tilde{q}'} \cdot \chi_{\tilde{q}''} = \tilde{\chi}$. Define $\tilde{J}(n) := \sum_{QR < m \leq n - QR} m^{\tilde{\beta}-1}$. By [11, Lemma 11.11]

$$I_2 = \sum_{q \leq Q, \, \tilde{q}' | s} \frac{\mu(q) \chi_{\tilde{q}'}(h) \chi_{\tilde{q}''}(b) \tau(\chi_{\tilde{q}'} \chi_s^0)}{\phi(q) \phi(qr/h)} G(bs\bar{s}(q) - n, \chi_{\tilde{q}'} \chi_q^0)(\tilde{J}(n) + O(\tau)).$$

Let $q = \tilde{q}'k$. Then $(\tilde{q}', k) = 1$ by $\mu(q) \neq 0$. By Lemma 3 (where $m = 1$) with $s/\tilde{q}' = k/(k, r)$ and Lemma 4(1) with $\tilde{q}' = s(\tilde{q}')$ we have

$$\tau(\chi_{\tilde{q}'} \chi_s^0) = \tau(\chi_{\tilde{q}'}) \mu(k/(k, r)) \chi_{\tilde{q}'}(k/(k, r)),$$
$$G(bs\bar{s}(q) - n, \chi_{\tilde{q}'} \chi_q^0) = \chi_{\tilde{q}'}(k) G(-n, \chi_{\tilde{q}'}) C_k(bs\bar{s}(k) - n).$$

Then

$$I_2 = (\tilde{J}(n) + O(\tau)) \frac{\mu(\tilde{q}') \tau(\chi_{\tilde{q}'}) \chi_{\tilde{q}''}(b)}{\phi(\tilde{q}') \phi(\tilde{q}'r)} G(-n, \chi_{\tilde{q}'})$$
$$\times \sum_{k \leq Q/\tilde{q}', \, (k, \tilde{q}')=1} \frac{\mu(k) \mu(k/(k, r)) \chi_{\tilde{q}'}(k/(k, r)) \chi_{\tilde{q}'}((k, r)) \chi_{\tilde{q}'}(k) C_k(bs\bar{s}(k) - n)}{\phi(k) \phi(k/(k, r))}.$$

By Lemma 3 and Lemma 2 we have $\tau(\chi_{\tilde{q}'}) G(-n, \chi_{\tilde{q}'}) = \tau(\chi_{\tilde{q}'}) \chi_{\tilde{q}'}(-n) \tau(\chi_{\tilde{q}'}) = \chi_{\tilde{q}'}(n) \tilde{q}'$. Then

$$I_2 = (\tilde{J}(n) + O(\tau)) \frac{\mu(\tilde{q}') \chi_{\tilde{q}'}(n) \chi_{\tilde{q}''}(b) \tilde{q}'}{\phi(r) \phi^2(\tilde{q}')}$$
$$\times \sum_{k \leq Q/\tilde{q}', \, (k, \tilde{q}')=1} \frac{\mu(k) \mu(k/(k, r)) C_k(bs\bar{s}(k) - n)}{\phi(k) \phi(k/(k, r))}.$$

Arguing similarly as in the estimation for I_1, we get

$$I_2 = \frac{\sigma(n,r)}{\phi(r)}\tilde{J}(n)\triangle_2 + O(X^{1+\varepsilon}Q^{-1}),$$

$$I_3 = \frac{\sigma(n,r)}{\phi(r)}\tilde{J}(n)\triangle_3 + O(X^{1+\varepsilon}Q^{-1}),$$

where

$$\left.\begin{array}{c} \triangle_2 := \chi_{\tilde{q}'}(n)\chi_{\tilde{q}''}(b)\triangle \;, \quad \triangle_3 := \chi_{\tilde{q}'}(n)\chi_{\tilde{q}''}(n-b)\triangle, \\[2mm] \triangle := \mu(\tilde{q}')\phi^{-1}(\tilde{q}') \displaystyle\prod_{p|\tilde{q}',p\nmid nr} \frac{p-1}{p-2}. \end{array}\right\} \tag{4.3}$$

Note that for $i = 2$, 3 we have $\triangle_i = I_i = 0$ if $(n,\tilde{q}') > 1$ or $u(\tilde{q}) > 1$.
Estimation of I_4: Define

$$\tilde{I}(n) := \sum_{QR < m \le n - QR} (m(n-m))^{\tilde{\beta}-1}.$$

By [11, Lemma 11.11]

$$I_4 = (\tilde{I}(n) + O(\tau)) \sum_{q \le Q, \, \tilde{q}|q, \, h_1|\tilde{q}} \frac{\tau(\tilde{\chi}\chi_q^0)\tau(\chi_s)}{\phi(q)\phi(qr/h)} \chi_{h_2}(b)\chi_s(uh)H(-n, \tilde{\chi}\chi_s\chi_q^0, \chi_{h_1(q)}^*).$$

Let $q = \tilde{q}k$. We may assume that $(\tilde{q},k) = 1$ and $\mu^2(k) = 1$, otherwise by Lemma 3 we have $\tau(\tilde{\chi}\chi_q^0) = 0$. Similarly, as in the case of I_2, by Remark (a) on Lemma 1 we have

$$\chi_{s(q)} \leftrightarrow \chi_{\tilde{q}'}, \quad \chi_{h_1(q)}\chi_{h_2(q)} \leftrightarrow \chi_{\tilde{q}''}. \tag{4.4}$$

However, $h_1(q)$ is no longer always equal to 1 in the present case, which makes the estimation of I_4 more involved. First note that $h_1(\tilde{q}k) = h_1(\tilde{q}) = h_1(\tilde{q}'')$ depending only on $\tilde{q}$ and r, and that the only possible values of h_1 are 1, 4 and 8. Moreover, $h_1(\tilde{q}) = 1$ if $2 \nmid \tilde{q}$ or $2 \nmid r$. Consider two cases separately:
Firstly we assume that $u(\tilde{q}) = 1$, i.e., $h_1(\tilde{q}) = 1$ and $u(\tilde{q}') = 1$. In this case, $\tilde{q}'' = (\tilde{q},r)$ and $\tilde{q}' = \tilde{q}/(\tilde{q},r)$. By (4.4), i.e., $\tilde{\chi}\chi_s\chi_q^0 = \chi_{\tilde{q}''}\chi_{\tilde{q}'}^2\chi_k^0$, Lemma 3, and Lemmas 4(2) and 4(3),

$$\tau(\tilde{\chi}\chi_q^0) = \tau(\tilde{\chi})\mu(k)\tilde{\chi}(k), \quad \tau(\chi_s) = \tau(\chi_{\tilde{q}'})\mu(k/(k,r))\chi_{\tilde{q}'}(k/(k,r)),$$

$$H(-n, \tilde{\chi}\chi_s\chi_q^0, \chi_{h_1(q)}^*) = \chi_{\tilde{q}''}(k)\chi_{\tilde{q}''}(\tilde{q}')G(b-n, \chi_{\tilde{q}''})C_{\tilde{q}'}(-n)C_k(bs\overline{s}(k) - n).$$

Note that $\chi_{h_2}(b) = \chi_{\tilde{q}''}(b)$ as $(b,r) = 1$ and $h_1(q) = 1$. Then

$$I_4 = (\tilde{I}(n) + O(\tau))\frac{\chi_{\tilde{q}''}(n-b)\chi_{\tilde{q}''}(b)C_{\tilde{q}'}(-n)\chi_{\tilde{q}'}(-1)\tilde{q}}{\phi(r)\phi(\tilde{q}')\phi(\tilde{q})}$$

$$\times \sum_{k \le Q/\tilde{q}, (k,\tilde{q})=1} \frac{\mu(k)\mu(k/(k,r))C_k(bs\overline{s}(k) - n)}{\phi(k)\phi(k/(k,r))}.$$

Arguing similarly as for I_1, we get

$$I_4 = \tilde{I}(n)\frac{\sigma(n,r)}{\phi(r)}\Delta_4 + O(X^{1+\varepsilon}Q^{-1}),$$

where

$$\Delta_4 := \chi_{\tilde{q}''}(b)\chi_{\tilde{q}''}(n-b)\chi_{\tilde{q}'}(-1)\phi^{-1}(\tilde{q}')C_{\tilde{q}'}(-n)\prod_{p|\tilde{q},p\nmid nr}\frac{p-1}{p-2}.$$

Note that if $(n,\tilde{q}') = 1$ then we have $C_{\tilde{q}'}(-n) = \mu(\tilde{q}')$.
Secondly we assume that $u(\tilde{q}) > 1$. By (2.1), (3.1), and (3.2), we may suppose that $h_1(\tilde{q}) = h_1(\tilde{q}'') = 2^\alpha$ and $u(\tilde{q}) = u(\tilde{q}'') = 2^\beta$, $0 < \beta < \alpha \le 3$. Then we have

$$\left.\begin{array}{l}\tilde{q}'' = 2^\alpha\tilde{q}_1'', \ 2\nmid\tilde{q}_1'' \text{ with } \mu(\tilde{q}_1'')\neq 0; \text{ and } r = 2^{\alpha-\beta}r_1, \ 2\nmid r_1; \\ \text{thus } \tilde{q}'' = 2^\beta(\tilde{q},r)\end{array}\right\} \quad (4.5)$$

and

$$\chi_{h_1(q)} \leftrightarrow \chi_{2^\alpha}, \ \chi_{h_2(q)} \leftrightarrow \chi_{\tilde{q}_1''}, \ \chi_{s(q)} \leftrightarrow \chi_{\tilde{q}'}$$

where χ_{2^α}, $\chi_{\tilde{q}_1''}$, and $\chi_{\tilde{q}'}$ are determined uniquely by $\chi_{2^\alpha}\chi_{\tilde{q}_1''}\chi_{\tilde{q}'} = \tilde{\chi}$. By Lemma 3 and Lemmas 4(2) and 4(3) it follows that

$$I_4 = (\tilde{I}(n) + O(\tau))\frac{\tau(\tilde{\chi})\tau(\chi_{\tilde{q}'})}{\phi(\tilde{q})\phi(\tilde{q}r/(\tilde{q},r))}\chi_{\tilde{q}''}(b)\chi_{\tilde{q}'}(\tilde{q}'')\chi_{\tilde{q}''}(\tilde{q}')\chi_{\tilde{q}_1''}(2^\alpha)\chi_{2^\alpha}(\tilde{q}_1'')$$

$$\times C_{\tilde{q}'}(-n)G(b-n,\chi_{\tilde{q}_1''})\sum_{\substack{k\le Q/\tilde{q} \\ (k,\tilde{q})=1}}\frac{\mu(k)\mu(k/(k,r))C_k(bs\bar{s}(k)-n)}{\phi(k)\phi(k/(k,r))}$$

$$\times H^*(-n,\chi_{2^\alpha},\chi_{h_1(2^\alpha)},s\bar{s}(k\tilde{q})).$$

By (3.3) one has $s\bar{s}(k\tilde{q}) \equiv 1 \pmod{r^3}$, $\equiv 1 \pmod{2^\alpha}$. Then

$$H^*(-n,\chi_{2^\alpha},\chi_{h_1(2^\alpha)}s\bar{s}(k\tilde{q})) = H(-n,\chi_{2^\alpha},\chi_{h_1(2^\alpha)})$$

is independent of k, and moreover, by (3.14) it follows $|H(-n,\chi_{2^\alpha},\chi_{h_1(2^\alpha)})| \le 2^{\alpha/2}\cdot 2^\beta$. On noting that

$$\tilde{q}r/(\tilde{q},r) = 2^\alpha\tilde{q}'r_1, \quad G(b-n,\chi_{\tilde{q}_1''}) = \chi_{\tilde{q}_1''}(b-n)\tau(\chi_{\tilde{q}_1''}),$$
$$C_{\tilde{q}'}(-n) = \phi(\tilde{q}')\phi^{-1}(\tilde{q}'/(n,\tilde{q}'))\mu(\tilde{q}'/(n,\tilde{q}')),$$

we obtain by similar argument as for I_2 that

$$|I_4| \leq (n + O(\tau)) \frac{\sqrt{\tilde{q}}\sqrt{\tilde{q}'}\phi(\tilde{q}')\sqrt{\tilde{q}_1''}\sqrt{2^{\alpha}}2^{\beta}}{\phi(\tilde{q})\phi(r_1)\phi(2^{\alpha})\phi(\tilde{q}')\phi(\tilde{q}'/(n,\tilde{q}'))}$$

$$\times \left| \sum_{\substack{k \leq Q/\tilde{q} \\ (k,\tilde{q})=1}} \frac{\mu(k)\mu(k/(k,r))C_k(bs\bar{s}(k)-n)}{\phi(k)\phi(k/(k,r))} \right|$$

$$\leq \frac{\sigma(n,r)}{\phi(r)}n\triangle_4' + O(X^{1+\varepsilon}Q^{-1}) \tag{4.6}$$

where

$$\triangle_4' := \frac{1}{\phi(\tilde{q}'/(n,\tilde{q}'))} \prod_{p|\tilde{q}, p\nmid nr} \frac{p-1}{p-2} = \frac{1}{\phi(\tilde{q}'/(n,\tilde{q}'))} \prod_{p|\tilde{q}'/(n,\tilde{q}')} \frac{p-1}{p-2} \leq 1. \tag{4.7}$$

Estimation of I_j $(5 \leq j \leq 9)$: The estimation of I_j $(5 \leq j \leq 9)$ is quite similar to each other. We take I_9 as an example. We have

$$I_9 = \sum_{v \leq Q} \sum_{\chi_v}^* \sum_{d \leq Qr} \sum_{\chi_d}^* \int_{-1/\tau}^{1/\tau} W(\lambda, \chi_v)W(\lambda, \chi_d)e(-n\lambda)d\lambda \times$$

$$\times \sum_{q \leq Q, v|q, d|qr/h, h_1|d} \frac{1}{\phi(q)\phi(qr/h)} \tau(\overline{\chi}_q^{(1)})\tau(\overline{\chi}_s^{(2)})\chi_{h_2}^{(2)}(b)\overline{\chi}_s^{(2)}(uh)$$

$$\times H(-n, \chi_q^{(1)}\chi_s^{(2)}, \chi_{h_1}^{(2)}).$$

Let $v_1 = \prod_{p^{\alpha}\|d, p\nmid r} p^{\alpha}$. Then by $d|s(q)h_1(q)h_2(q)$ $(= qr/h)$ one gets $v_1|s$ and $(s, d/v_1) = 1$. Hence $\chi_s \leftrightarrow \chi_{v_1}^*$. On noting that v_1 depends on d and r only, we have by Lemma 5,

$$I_9 \ll \sum_{v \leq Q} \sum_{\chi_v}^* \sum_{d \leq Qr} \sum_{\chi_d}^* \int_{-1/\tau}^{1/\tau} |W(\lambda, \chi_v)| \, |W(\lambda, \chi_d)|d\lambda$$

$$\times \sum_{q=1, v|q, v_1|s}^{\infty} \frac{1}{\phi(q)\phi(qr/h)} |\tau(\chi_v\chi_q^0)\tau(\chi_{v_1}^*\chi_s^0)| |H(-n, \chi_v\chi_{v_1}^*\chi_q^0, \chi_{h_1(q)}^{(2)})|$$

$$\ll \frac{1}{\phi(r)} \cdot \frac{nr}{\phi(nr)} W^2 \, ,$$

where

$$W = \sum_{d \leq Qr} \sum_{\chi_d}^* \left(\int_{-1/\tau}^{1/\tau} |W(\lambda, \chi_d)|^2 d\lambda \right)^{1/2}.$$

By Lemma 6 and (1.3) one obtains that

$$I_9 \ll \frac{\sigma(n,r)}{\phi(r)} X\Omega^6 \exp(-c_5/\eta).$$

Similarly, the same estimate also holds for I_j $(5 \le j \le 8)$. Then

$$\sum_{j=5}^{9} |I_j| \ll \frac{\sigma(n,r)}{\phi(r)} X\Omega^3 \exp(-c_5/\eta). \tag{4.8}$$

5. Proof of Theorem 1 In this section the constant η is taken sufficiently small. Consider the three possible cases separately.

Case 1: There is no exceptional zero $\tilde{\beta}$. Then $I_2 = I_3 = I_4 = 0$ and $\Omega = 1$ (see (3.18)). By (3.11), (4.1), and (4.8) we obtain

$$R_1 = \frac{\sigma(n,r)}{\phi(r)} n + O\Big(X^{1+\varepsilon}Q^{-1} + \frac{\sigma(n,r)X}{\phi(r)} \exp(-c_5/\eta)\Big) \ge \frac{\sigma(n,r)}{2\phi(r)} n.$$

In what follows we always assume that there is an exceptional zero $\tilde{\beta}$.

Case 2: $(n, \tilde{q}') > 1$ or $u(\tilde{q}) > 1$. In this case $\triangle_2 = \triangle_3 = 0$ or $I_2 = I_3 = 0$. Then similar to Case 1, by (4.6)

$$R_1(n) \ge \frac{\sigma(n,r)}{\phi(r)} n - \frac{\sigma(n,r)}{\phi(r)} \tilde{I}(n)|\triangle_4'|$$
$$+ O\Big(\frac{\Omega^3\sigma(n,r)}{\phi(r)} X \exp(-c_5/\eta)\Big) + O(X^{1+\varepsilon}Q^{-1}).$$

If there is a prime $p > 3$ satisfying $p|\tilde{q}'$ and $p \nmid n$, then we have, by (4.7), $|\triangle_4'| \le 1/3$. From this and the inequalities $\tilde{I}(n) \le n$ and $X \le 2n$, it follows that

$$R_1(n) \ge \frac{2\sigma(n,r)}{3\phi(r)} n$$
$$+ O\Big(\frac{\sigma(n,r)}{\phi(r)} X \exp(-c_5/\eta)\Big) + O(X^{1+\varepsilon}Q^{-1}) \ge \frac{\sigma(n,r)}{4\phi(r)} n. \tag{5.1}$$

If there is no $p > 3$ satisfying $p|\tilde{q}'$ and $p \nmid n$, then by (2.1) we have

$$\tilde{q}' \le 2^3 \cdot 3(n, \tilde{q}'). \tag{5.2}$$

Since $\tilde{I}(n) \le n^{\tilde{\beta}}$, we have

$$R_1(n) \ge \frac{\sigma(n,r)}{\phi(r)} (n - n^{\tilde{\beta}}) + O\Big(\frac{\Omega^3\sigma(n,r)}{\phi(r)} X \exp(-c_5/\eta)\Big) + O(X^{1+\varepsilon}Q^{-1}).$$

If $\tilde{\beta} < 1 - (\log 2 / \log n)$, i.e., $n^{\tilde{\beta}-1} < 1/2$, we obtain again (5.1). Otherwise one has, by $\Omega = (1 - \tilde{\beta}) \log T$ in (3.18), that

$$
\begin{aligned}
R_1(n) &\geq \sigma(n,r)(2\phi(r))^{-1} n(1 - \tilde{\beta}) \log n \\
&\quad + O(\sigma(n,r)\phi(r)^{-1}\Omega^3 X \exp(-c_5/\eta)) + O(X^{1+\varepsilon}Q^{-1}) \\
&\geq \sigma(n,r)(4\phi(r))^{-1} n(1 - \tilde{\beta}) \log n + O(X^{1+\varepsilon}Q^{-1}).
\end{aligned}
$$

We note that the number of $n \leq X$ satisfying $(n, \tilde{q}) > Q^{1/24}$ is at most

$$
\ll \sum_{d | \tilde{q}, d > Q^{1/24}} \sum_{n \leq X, d | n} 1 \ll XQ^{-1/24}\tau(\tilde{q}) \ll X^{1-(\eta/7)}.
$$

We then only consider those even integers satisfying $(n, \tilde{q}) \leq Q^{1/24}$ which, together with (5.2) and (4.5), gives $\tilde{q}' \leq 24Q^{1/24}$ and $\tilde{q} = \tilde{q}'\tilde{q}'' \leq 96rQ^{1/24}$. Then by [3, p. 96, (12)] and (1.3) one gets, except for at most $\ll X^{1-(\eta/7)}$ even $n \in (X/2, X]$,

$$
\begin{aligned}
R_1(n) &\geq \frac{c_4}{4} \cdot \frac{\sigma(n,r)}{\phi(r)} \frac{n \log n}{\tilde{q}^{1/2} \log^2 \tilde{q}} + O(X^{1+\varepsilon}Q^{-1}) \\
&\geq \frac{c_4}{40} \cdot \frac{\sigma(n,r)}{\phi(r)} (n \log n) Q^{-1/48} r^{-1/2} (\log X)^{-2} + O(X^{1+\varepsilon}Q^{-1}) \\
&> 2X/(rQ^{1/3}).
\end{aligned}
$$

Case 3:　$(n, \tilde{q}') = 1$ and $u(\tilde{q}) = 1$. If there exists $p \geq 5$, $p | \tilde{q}'$, by (4.3) we have $|\Delta| \leq 1/3$ and then $|\Delta_j| \leq 1/3$, $j = 2, 3, 4$. From $\tilde{J}(n) \leq n$ it follows that

$$
\begin{aligned}
R_1(n) &\geq \frac{\sigma(n,r)}{\phi(r)} \left(\frac{1}{3}(n - \tilde{I}(n)) + \frac{2}{3}(n - \tilde{J}(n)) \right) \\
&\quad + O\left(\frac{\Omega^3 \sigma(n,r)}{\phi(r)} X \exp(-c_5/\eta) \right) + O(X^{1+\varepsilon}Q^{-1}) \\
&\geq \frac{\sigma(n,r)}{3\phi(r)}(n - n^{\tilde{\beta}}) + O\left(\frac{\Omega^3 \sigma(n,r)}{\phi(r)} X \exp(-c_5/\eta) \right) + O(X^{1+\varepsilon}Q^{-1}).
\end{aligned}
$$

Similar to the argument in Case 2 we can obtain that

$$
R_1(n) > 2X/(rQ^{1/3}). \tag{5.3}
$$

Finally, we consider the case where $p | \tilde{q}'$ implies $p \leq 3$. Since $(n, \tilde{q}') = 1$ and $2 | n$, it follows that the only prime factor of $\tilde{q}'$ is 3. So $\tilde{q}' = 3$. In this case $|\Delta| = 1$ and then $|\Delta_i| = 1$ for $i = 2, 3, 4$. If $\Delta_2 + \Delta_3 \geq 0$ then

$$
R_1(n) \geq \frac{\sigma(n,r)}{\phi(r)}(n - n^{\tilde{\beta}}) + O\left(\frac{\Omega^3 \sigma(n,r)}{\phi(r)} X \exp(-c_5/\eta) \right) + O(X^{1+\varepsilon}Q^{-1}).
$$

We may again get (5.3) by similar method as in Case 2.

Now assume that $\triangle_2 + \triangle_3 < 0$. Then by the fact that $\triangle_2$ and $\triangle_3$ are real, we must have $\triangle_2 = \triangle_3 = -1$ and then by (4.3) we have $\chi_{\tilde{q}''}(b)\chi_{\tilde{q}''}(n-b) > 0$. Hence $\triangle_4 = \mu(3)\chi_3(-1)\chi_{\tilde{q}''}(b)\chi_{\tilde{q}''}(n-b) = 1$ since $\chi_3(-1) = -1$. Then

$$R_1(n) = \frac{\sigma(n,r)}{\phi(r)}(n - 2\tilde{J}(n) + \tilde{I}(n))$$

$$+ O\left(\frac{\Omega^3 \sigma(n,r)}{\phi(r)} X \exp(-c_5/\eta)\right) + O(Q^{-1}X^{1+\varepsilon}). \qquad (5.4)$$

By $\tilde{I}(n) = (n^{2\tilde{\beta}-1}\Gamma^2(\tilde{\beta})/\Gamma(2\tilde{\beta})) + O(QR)$ and $\tilde{J}(n) = (n^{\tilde{\beta}}/\tilde{\beta}) + O(QR)$ we have

$$n - 2\tilde{J}(n) + \tilde{I}(n) = n - \frac{2n^{\tilde{\beta}}}{\tilde{\beta}} + n^{2\tilde{\beta}-1}\frac{\Gamma^2(\tilde{\beta})}{\Gamma(2\tilde{\beta})} + O(QR). \qquad (5.5)$$

If $1 - \tilde{\beta} \geq (\log(3/2))/\log n$, then by $1 - \tilde{\beta} \ll 1/\log n$ in (3.18) we have trivially

$$n - 2\tilde{J}(n) + \tilde{I}(n) = n(1 - 2n^{\tilde{\beta}-1} + n^{2(\tilde{\beta}-1)} + O(1 - \tilde{\beta})) + O(QR)$$

$$= n(1 - n^{\tilde{\beta}-1})^2 + O(n(1 - \tilde{\beta})) + O(QR) \geq n/10.$$

By this and (5.4), (5.5) one gets

$$R_1(n) \geq \frac{\sigma(n,r)}{10\phi(r)}n + O\left(\frac{\sigma(n,r)}{\phi(r)} X \exp(-c_5/\eta)\right) + O(Q^{-1}X^{1+\varepsilon}) \geq \frac{\sigma(n,r)}{20\phi(r)}n.$$

Now assume that $1 - \tilde{\beta} < (\log(3/2))/\log n$. Then $1 - m^{\tilde{\beta}-1} \geq \frac{2}{3}(1 - \tilde{\beta})\log m$ for any $1 < m \leq n$, and

$$n - 2\tilde{J}(n) + \tilde{I}(n) = \sum_{QR < m \leq n - QR} (1 - m^{\tilde{\beta}-1})(1 - (n - m)^{\tilde{\beta}-1}) + O(QR)$$

$$\geq \frac{4}{9}(1 - \tilde{\beta})^2 \sum_{\sqrt{n} < m < n - \sqrt{n}} \log m \log(n - m) + O(QR)$$

$$\geq \frac{1}{10}n(1 - \tilde{\beta})^2 \log^2 n + O(QR). \qquad (5.6)$$

By (5.4), (5.6), and (3.18) we get that

$$R_1(n) \geq \frac{\sigma(n,r)}{10\phi(r)}n(1 - \tilde{\beta})^2 \log^2 n$$

$$+ O\left(\frac{\Omega^3 \sigma(n,r)}{\phi(r)} X \exp(-c_5/\eta)\right) + O(X^{1+\varepsilon}Q^{-1})$$

$$\geq \frac{\sigma(n,r)}{11\phi(r)}n(1 - \tilde{\beta})^2 \log^2 n + O(Q^{-1}X^{1+\varepsilon}).$$

Finally, invoking [3, p. 96, (12)], $\tilde{q} = 3(\tilde{q}, r) \leq 3r$, and $r \leq Q^{1/4}$, we have

$$R_1(n) \geq c_6 \cdot \frac{\sigma(n, r)}{\phi(r)} n Q^{-1/4} (\log X)^{-2} + O(Q^{-1} X^{1+\varepsilon}) > 2X/(rQ^{1/3}).$$

After the above discussion for all the three cases we conclude that apart from at most $O(X^{1-(\eta/7)})$ exceptions of the even integers $n \in (X/2, X]$ we have (2.3) and therefore $R(n; r, b) \geq X/(rQ^{1/3})$. This finishes the proof of Theorem 1.

6. Proof of Theorem 3   *The assertion* (a): Suppose that $(b, r) = (b_i, r) = 1$, $i = 1, 2$. Let $\mathcal{A}(r, b) = \{n : 2|n, (n - b, r) = 1$ and $n \neq p_1 + p_2$ for any $p_1 \equiv b$ (mod r) and $p_2\}$, and $\mathcal{B}(r, b_1, b_2) = \{n : 2|n, n \equiv b_1 + b_2$ (mod r) and $n \neq p_1 + p_2$ for any $p_i \equiv b_i$ (mod r), $i = 1, 2\}$. We have

$$\mathcal{B}(r, b_1, b_2) \subset \mathcal{A}(r, b_i), \quad i = 1, 2. \tag{6.1}$$

In fact, if (6.1) is not true for $i = 1$, then there exists an even integer $n \in \mathcal{B}(r, b_1, b_2)$ but $n \notin \mathcal{A}(r, b_1)$. Since $n \equiv b_1 + b_2$ (mod r) implies $(n - b_1, r) = (b_2, r) = 1$, it follows that

$$n = p_1 + p_2 , \quad p_1 \equiv b_1 \text{ (mod } r) \tag{6.2}$$

is soluble. However, from $n \equiv b_1 + b_2$ (mod r) one actually has that $p_2 = n - p_1 \equiv b_2$ (mod r) in (6.2). This is a contradiction to $n \in \mathcal{B}(r, b_1, b_2)$, and (6.1) is then proved. Take $\delta' = \delta/2$ where $\delta > 0$ is the constant in Theorem 1. We obtain by (6.1) and Theorem 1 that $E(X; r, b_1, b_2) \leq E(X; r, b_1) \ll X^{1-\delta} \leq r X^{1-\delta}/\phi(r) \leq X^{1-(\delta/2)}/\phi(r) = X^{1-\delta'}/\phi(r)$ for any $1 \leq r \leq X^{\delta'}$ and $(b_i, r) = 1$, $i = 1, 2$.

The assertion (b): This may be derived directly from (a) and the following known result on $\pi(N; r, b)$, i.e., the lower bound for the number of primes $p \leq N$ with $p \equiv b$ (mod r).

$$\pi(N; r, b) > N/\phi^2(r) \quad \text{if} \quad r^c \leq N \quad \text{and} \quad (b, r) = 1 \tag{6.3}$$

where N is sufficiently large and $c > 0$ is an effectively computable absolute constant. In fact, the recent work of Wang [12] or Heath-Brown [4] for example, clearly shows that (6.3) holds for $c = 8$ or even 5.5. Since (6.3) and part (a) indicate that $\#\{N - p_1 : 3 \leq p_1 \leq N$ and $p_1 \equiv b_1$ (mod $r)\} > N\phi^{-2}(r)E(N; r, b_2, b_3)$ for $r \leq \min(N^\delta, N^{1/c})$, it then follows that there must exist $p_1 \equiv b_1$ (mod r) satisfying

$$N - p_1 = p_2 + p_3 \text{ with } p_i \equiv b_i \text{ (mod } r), \, i = 2, 3.$$

Acknowledgement   The paper was written while the second named author was holding a Croucher Foundation Visitorship at the Department of Mathematics, The University of Hong Kong. He wants to thank the host department for providing him with excellent conditions for doing researches. He is also very grateful to the conference organizers for supporting him to attend the very interesting meeting.

References

[1] A. Balog. The prime k-tuples conjecture on average. *Analytic Number Theory (Allerton Park, IL, 1989)*, Birkhäuser, Boston, 1990, pp. 47–75.

[2] J.R. Chen. On the representation of a large even integer as the sum of a prime and the product of at most two primes. *Chin. Sci. Bull.*, **17** (1966), 385–386; *Sci. Sin.*, **16** (1973), 157–176.

[3] H. Davenport. *Multiplicative Number Theory* (2nd ed). Springer-Verlag, Berlin, 1980.

[4] D.R. Heath-Brown. Zero-free region for Dirichlet L-function, and the least prime in an arithmetic progression. *Proc. London Math. Soc.*, **64** (1992), 265–338.

[5] A.F. Lavrik. The number of k-twin primes lying on an interval of a given length. *Dokl. Akad. Nauk SSSR*, **136** (1961), 281–283; English transl., *Soviet Math. Dokl.*, **2** (1961), 52–55.

[6] Yu.V. Linnik. On the least prime in an arithmetic progression. II. The Deuring–Heilbronn phenomenon. *Rec. Math.*, **15**(57) (1944), 347–368.

[7] M.C. Liu and K.M. Tsang. Small prime solutions of linear equations. *Théorie des Nombres* (J.-M. de Koninck and C. Levesque, Eds.), Walter de Gruyter, Berlin, 1989, pp. 595–624.

[8] H. Maier and C. Pomerance. Unusually large gaps between consecutive primes. *Trans. Amer. Math. Soc.*, **322** (1990), 201–237.

[9] H. Mikawa. On prime twins in arithmetic progressions. *Tsukuba J. Math.*, **16** (1992), 377–387.

[10] H.L. Montgomery and R.C. Vaughan. The exceptional set in Goldbach's problem. *Acta Arith.*, **27** (1975), 353–370.

[11] C.D. Pan and C.B. Pan. *Goldbach Conjecture*. Sci. Press, Beijing, 1981; English version, 1992.

[12] W. Wang. On the least prime in an arithmetic progression. *Acta Math. Sin.*, **7** (1991), 279–289.

Ming-Chit Liu
Department of Mathematics, The University of Hong Kong
Pokfulam Road, Hong Kong

Tao Zhan
Deapartment of Mathematics, Shandong University
250100 Jinan, China

15

On the Sum of Three Squares of Primes

HIROSHI MIKAWA

1. Introduction Shortly after I.M.Vinogradov's proof in 1937 of the ternary Goldbach conjecture, L.K. Hua [5] investigated the solubility of the equation

$$p_1^2 + p_2^2 + p_3^2 = n \qquad (1.1)$$

with primes p_j, j=1,2,3. A necessary condition on n is that the congruence

$$x_1^2 + x_2^2 + x_3^2 \equiv n \pmod{q}, \qquad (1.2)$$

subject to $(x_j, q) = 1$, j=1,2,3, is soluble for any q. An elementary argument implies that this local solubility is equivalent to the condition that n belongs to the set

$$\mathbb{H} = \{n : n \equiv 3 \pmod{24}, \quad n \not\equiv 0 \pmod{5}\},$$

the density of which is 1/30. Hua proved that almost all $n \in \mathbb{H}$ can be put in the form (1.1). Here "almost all" means that the number of exceptional n's not exceeding x, say $E(x)$, is $o(x)$ as $x \to \infty$. In 1961 W. Schwarz [12] replaced the bound $o(x)$ by $O(x(\log x)^{-A})$ for any fixed $A > 0$. Then in 1993 M.-Ch. Leung and M.-Ch. Liu [8] proved that there exists an absolute constant $\delta > 0$ such that

$$E(x) \ll x^{1-\delta}.$$

The explicit value of δ that is implicit in their argument is very small.

Recently, T. Zhan [13] considered the short interval version of this problem. The aim of the present paper is to give a genuine improvement upon his result. To state his and our results, we need first to introduce some notation: Let $R(n)$ denote the weighted number of representations of n in the form (1.1); the weight is equal to $\prod_{j=1,2,3} \log p_j$. It is conjectured that as $n \in \mathbb{H}$ tends to infinity we should have

$$R(n) \sim \frac{\pi}{4}\mathfrak{S}(n)\sqrt{n}, \qquad (1.3)$$

where

$$\mathfrak{S}(n) = \frac{8r(8,n)}{\varphi(8)^3} \prod_{\substack{p>2 \\ p\,:\,\text{prime}}} \frac{pr(p,n)}{(p-1)^3} \qquad (1.4)$$

with $r(q, n)$ being the number of solutions of (1.2). Zhan proved that the asymptotic formula (1.3) holds true for almost all $n \in \mathbb{H} \cap [x, x + x^\Theta]$, provided $\Theta > 3/4$. In particular it is implied that if $\theta > 3/4$ then

$$E(x + x^\theta) - E(x) \ll x^\theta (\log x)^{-A} \qquad (1.5)$$

for any fixed $A > 0$. Our improvement is embodied in

Theorem *Let $\Theta > 1/2$ and $A > 0$ be given. Then we have*

$$\sum_{\substack{x < n \le x + x^\Theta \\ n \in \mathbb{H}}} \left| R(n) - \frac{\pi}{4} \mathfrak{S}(n) \sqrt{n} \right|^2 \ll x^{\Theta+1} (\log x)^{-A}$$

where the implied constant may depend on Θ and A. Hence (1.5) holds for $\theta > 1/2$.

In his work Zhan proved and used an inequality [13; Theorem 3] to estimate the mean values for exponential sums over squares of primes in short intervals. We shall not require such a result but our argument is, instead, somewhat tricky, for it depends on a certain peculiarity of the sequence of squares. That is, our main tool is Lemma 1 below, which is not as strong as Zhan's inequality. So his argument would probably be able to yield a further reduction of the value of θ in (1.5). Our assertion $\Theta > 1/2$ appears to be the best possible that our argument can attain. In another respect it should be worth remarking that in dealing with the singular series $\mathfrak{S}(n)$ we shall appeal to the argument of M.B. Barban [1; §5]. Also we record that we were inspired by C. Hooley ([4; Chap.II]).

We are indebted to Professors Zhan and Liu for their preprints and kind comments. We would like to thank the organizer for his kind invitation to the symposium.

2. Lemmas We begin with the fundamental

Lemma 1 *Suppose that $|\alpha - a/q| \le q^{-2}$ with $(a, q) = 1$. Then we have*

$$\sum_{p^2 \le x} e(\alpha p^2) \log p \ll x^{1/2} \left(q^{-1} + x^{-1/4} + q x^{-1} \right)^C (\log qx)^D$$

where $e(x) = \exp(2\pi i x)$ as usual; and $0 < C < 1/4$, $D \ge 8$ as well as the implied constants are absolute.

This assertion goes back to I.M. Vinogradov, and has been refined by several people; see for example [3; Lemma 2].

We then collect some facts about the singular series; for the proof see, e.g., [6; Chap. VIII]. Thus let us put

$$g(q,a) = \sum_{\substack{1 \le m \le q \\ (m,q)=1}} e\left(\frac{m^2}{q}a\right) \quad ((a,q)=1) \tag{2.1}$$

and

$$G(q,n) = \sum_{\substack{1 \le a \le q \\ (a,q)=1}} \left(\frac{g(q,a)}{\varphi(q)}\right)^3 e\left(-\frac{a}{q}n\right), \tag{2.2}$$

which is a multiplicative function of q. Either if $p > 2$ and $m > 1$ or if $p = 2$ and $m > 3$ then $g(p^m, a) = 0$ and thus $G(p^m, n) = 0$. For $p > 2$, $g(p,a) = \left(\frac{-a}{p}\right)G - 1$, where $\left(\frac{*}{p}\right)$ is the Legendre symbol and G is the Gaussian sum attached to it. When $p > 2$, we have

$$(p-1)^3 G(p,n) = \begin{cases} -(p-1)\left(3\left(\frac{-1}{p}\right)p+1\right) & \text{if } p|n, \\ \left(\frac{-n}{p}\right)p^2 + 3\left(\frac{-1}{p}\right)p + 3\left(\frac{n}{p}\right)p + 1 & \text{if } p \nmid n. \end{cases} \tag{2.3}$$

Thus we have

$$g(q,a) \ll \tau(q)q^{1/2}, \tag{2.4}$$

where τ is the divisor function, and

$$G(q,n) \ll q^{-1}\log q. \tag{2.5}$$

On the other hand we have

$$\begin{cases} 1 + G(p,n) = pr(p,n)(p-1)^{-3} & \text{if } p > 2, \\ 1 + G(2,n) + G(4,n) + G(8,n) = 8r(8,n)\varphi(8)^{-3}. \end{cases} \tag{2.6}$$

and

$$\begin{cases} r(p,n) > 0 & \text{for all } p > 5 \text{ and all } n, \\ r(5,n) > 0 & \text{if and only if } n \not\equiv 0 \pmod 5, \\ r(3,n) > 0 & \text{if and only if } n \equiv 0 \pmod 3, \\ r(8,n) > 0 & \text{if and only if } n \equiv 3 \pmod 8. \end{cases} \tag{2.7}$$

Also we shall use the arithmetic function defined by

$$\Psi(m,z) = \begin{cases} 1 & \text{either if } p|m \text{ implies } p \le z \text{ or if } m = 1, \\ 0 & \text{otherwise.} \end{cases} \tag{2.8}$$

The following assertion is well-known; see, e.g., [11; Kap.V, Lemma 5.2]:

Lemma 2 *Suppose that $z = z(x) \to \infty$ as $x \to \infty$, and $(\log x)/(\log z) \gg \log\log x$. Then*

$$\sum_{m \leq x} \Psi(m, z) \ll x \exp\left(-(\log x)/(\log z)\right).$$

3. Proof of Theorem   Let $\Theta > 1/2$ and $A > 0$ be given. Let x be sufficiently large and $H = x^{\Theta}$. We may assume $H \leq x$. Put $P = L^B$ where $L = \log x$, and $B = B(A) > 0$ is to be specified later. Using the Farey fractions of order $Q = xP^{-2}$, we dissect the unit interval $\mathbf{U} = [1/Q, 1 + 1/Q]$. Let the major arc $\mathbf{M}$ be the union of all intervals $\mathbf{I}(q, a) = [a/q - 1/qQ, a/q + 1/qQ]$ with $1 \leq a \leq q \leq P$ and $(a, q) = 1$. Then $\mathbf{I}(q, a)$'s are mutually disjoint and contained in $\mathbf{U}$. Write $\mathbf{m} = \mathbf{U} \setminus \mathbf{M}$ which we designate as the minor arc.

We have, for $x < n \leq 2x$,

$$R(n) = \int_{\mathbf{U}} S(\alpha)^3 e(-n\alpha) d\alpha$$

where $S(\alpha) = \sum_{p^2 \leq 2x} e(\alpha p^2) \log p$. A familiar argument transforms this into

$$R(n) = \frac{\pi}{4} \sum_{q \leq P} G(q, n)\sqrt{n} + \int_{\mathbf{m}} S(\alpha)^3 e(-n\alpha) d\alpha + O(x^{1/2} L P^{-1}). \qquad (3.1)$$

In fact we are at the situation to be able to appeal to the Siegel-Walfisz theorem [11; Kap.IV, Satz 8.3].

We then state our main lemmas:

Lemma 3 *The infinite product defining $\mathfrak{S}(n)$ converges for any $n \in \mathbb{H}$, and we have*

$$V = \sum_{\substack{x < n \leq x+H \\ n \in \mathbb{H}}} \left| \mathfrak{S}(n) - \sum_{q \leq P} G(q, n) \right|^2 \ll x^{1/2} L^2 P + H L^{17} P^{-1}.$$

Lemma 4 *Suppose that $\sup_{\alpha \in \mathbf{m}} |S(\alpha)| \ll x^{1/2} L^{-E}$. Then we have*

$$W = \int_{\mathbf{m}} \int_{\mathbf{m}} |S(\alpha_1)|^3 |S(\alpha_2)|^3 \min\left(H, \frac{1}{\|\alpha_1 - \alpha_2\|}\right) d\alpha_1 d\alpha_2$$
$$\ll x^{3/2} L^{10} + H x L^{10-E},$$

where $\|x\|$ is the distance from x to the nearest integer.

Postponing the proofs of these assertions to Sections 4 and 6, respectively, we shall proceed to our proof of the theorem.

For this purpose we consider the integral in (3.1) with the aid of Lemma 4. We note first that for any $n \in \mathbf{m}$ there exists a rational number a/q such that $|\alpha - a/q| \leq q^{-2}$, $(a, q) = 1$ and $P < q \leq Q$. In view of Lemma 1, the assumption of Lemma 4 is then fulfilled with $-E = -BC + D$, provided that B is sufficiently large. So we take $B = B(A) > 0$ to satisfy $10 - BC + D = -A$. Note that $18 - B \leq -A$. Then we have, by Lemma 4, that

$$\sum_{x < n \leq x+H} \left| \int_{\mathbf{m}} S(\alpha)^3 e(-n\alpha) d\alpha \right|^2$$

$$\ll \int_{\mathbf{m}} \int_{\mathbf{m}} |S(\alpha_1)|^3 |S(\alpha_2)|^3 \left| \sum_{x < n \leq x+H} e((\alpha_1 - \alpha_2)n) \right| d\alpha_1 d\alpha_2$$

$$\ll W \ll HxL^{-A}. \tag{3.2}$$

Collecting (3.1), (3.2), and Lemma 3, we infer that

$$\sum_{\substack{x < n \leq x+H \\ n \in \mathbb{H}}} \left| R(n) - \mathfrak{S}(n)\frac{\pi}{4}\sqrt{n} \right|^2 \ll Vx + W + HxL^2 P^{-2}$$

$$\ll (x^{1/2}L^2P + HL^{17}P^{-1})x + HxL^{-A}$$

$$\ll x^{3/2}L^{2+B} + HxL^{-A} \ll HxL^{-A},$$

since $H = x^{\Theta}$, $\Theta > 1/2$. This proves the theorem.

4. Singular series In this section we shall prove Lemma 3. To this end we consider the truncation of the infinite product for $\mathfrak{S}(n)$, which we may put, in view of (2.6),

$$s_n \prod_{2 < p \leq P} (1 + G(p, n)),$$

where $s_n = 1 + G(2, n) + G(4, n) + G(8, n)$. We modify this as

$$s_n \prod_{\substack{p | n \\ 2 < p \leq P}} (1 + G(p, n)) \prod_{\substack{p \nmid n \\ 2 < p \leq P}} (1 + G(p, n)) \left(1 - \left(\frac{-n}{p}\right) p^{-1}\right)$$

$$\times \prod_{\substack{p \nmid n \\ 2 < p \leq P}} \left(1 - \left(\frac{-n}{p}\right) p^{-1}\right)^{-1} = s_n Y_1(n) Y_2(n) Y_3(n),$$

say. It follows from (2.6) and (2.7) that if $n \in \mathbb{H}$ then this finite product is positive. And, by (2.3) , we find that, uniformly for $x < n \leq 2x$, $n \in \mathbb{H}$,

$$s_n Y_1(n) Y_2(N) Y_3(n) \gg \prod_{3 < p \leq P} \left(1 - \frac{3}{p}\right) \gg (\log \log x)^{-3}. \qquad (4.1)$$

In the definition of $Y_j(n)$, $j=1,2,3$, we drop the restriction $p \leq P$ and denote the resulting infinite products by $Z_j(n)$, respectively. We note that $Z_1(n)$ is actually a finite product. The product $Z_2(n)$ is absolutely convergent, for we have, by (2.3),

$$(1 + G(p,n)) \left(1 - \left(\frac{-n}{p}\right) p^{-1}\right) = 1 + O(p^{-2}). \qquad (4.2)$$

Also the product $Z_3(n)$ converges; it is the value at $s = 1$ of the Dirichlet series $L(s, \chi)$ corresponding to the quadratic field $\mathbb{Q}(\sqrt{-n})$. Hence, we have $\mathfrak{S}(n) = s_n Z_1(n) Z_2(n) Z_3(n)$ for any $n \in \mathbb{H}$.

On the other hand we obviously have

$$s_n Y_1(n) Y_2(n) Y_3(n) = \sum_{q=1}^{\infty} G(q,n) \Psi(q,P).$$

On noting that $\Psi(q,P) = 1$ for all $q \leq P$ and that $Z_i(n), Y_j(n) \ll \log L$ for $i = 1, 2$, $j = 1, 2, 3$, we have

$$\left| \mathfrak{S}(n) - \sum_{q \leq P} G(q,n) \right|^2 \ll \sum_{j=1,2,3} |Z_j(n) - Y_j(n)|^2 L + \left| \sum_{q > P} G(q,n) \Psi(q,P) \right|^2.$$

Thus we have, for the V defined in Lemma 3,

$$V \ll V_1 + V_2 + V_3 + V_4 \qquad (4.3)$$

in an obvious arrangement of terms.

Let us consider V_1 first. By (2.3) we have

$$\frac{Z_1(n)}{Y_1(n)} = \prod_{\substack{p|n \\ p>P}} (1 + G(p,n)) = \prod_{\substack{p|n \\ p>P}} \left(1 - 3\left(\frac{-1}{p}\right) p^{-1}\right) (1 + O(p^{-2}))$$

$$= 1 + O\left(\sum_{\substack{d|n \\ d>P}} \frac{\tau_3(d)}{d} + \sum_{\substack{d|n \\ d>P}} \frac{\tau_K(d)}{d^2}\right)$$

where $\tau_\nu(n)$ is the number of representations of n as a product of ν factors, and K is an absolute constant. This gives

$$
V_1 \ll L \sum_{x < n \leq x+H} \left| \frac{Z_1(n)}{Y_1(n)} - 1 \right|
$$

$$
\ll L \sum_{P < d \leq 2x} \frac{\tau_3(d)}{d} \left(\frac{H}{d} + 1 \right) + HL \sum_{P < d \leq 2x} \frac{\tau_K(d)}{d^2}
$$

$$
\ll HL^2 P^{-1}. \tag{4.4}
$$

Similarly we have, by (4.2),

$$
\frac{Z_2(n)}{Y_2(n)} = \prod_{\substack{p \mid n \\ p > P}} (1 + O(p^{-2})) = 1 + O(LP^{-1}),
$$

and hence

$$
V_2 \ll L \sum_{x < n \leq x+H} \left| \frac{Z_2(n)}{Y_2(n)} - 1 \right| \ll HL^2 P^{-1}. \tag{4.5}
$$

We proceed to V_3. If $n \in \mathbb{H}$ then $n \equiv 3 \pmod 4$. Thus the Jacobi symbol $\left(\dfrac{-n}{*} \right)$ is a non-principal character to the modulo $\leq n$ (see [7; I.Teil, Kap.6]). Then the Pólya–Vinogradov inequality [2; §23] yields that for any $M > P$

$$
\sum_{\substack{m=1 \\ 2 \nmid m \\ (m,n)=1}}^{\infty} \left(\frac{-n}{m} \right) \frac{1}{m} = \sum_{\substack{m \leq M \\ 2 \nmid m \\ (m,n)=1}} \left(\frac{-n}{m} \right) \frac{1}{m} + O \left(\frac{x^{1/2} L}{M} \right).
$$

That is, we have

$$
Z_3(n) - Y_3(n)
$$

$$
= \sum_{\substack{m \leq M \\ 2 \nmid m \\ (m,n)=1}} \left(\frac{-n}{m} \right) \frac{1 - \Psi(m, P)}{m} + O \left(\frac{x^{1/2} L}{M} \right) + O \left(\sum_{m > M} \frac{\Psi(m, P)}{m} \right).
$$

We put $M = Px^{1/2}$. Then the first O-term is $O(LP^{-1})$. By partial summation and Lemma 2, the second one is at most

$$
\exp\left(-(\log M)/(\log P) \right) \log P \ll LP^{-1}.
$$

Since $\Psi(m, P) = 1$ for all $m \le P$, it follows that

$$
V_3 \ll L \sum_{\substack{x < n \le x+H \\ n \equiv 3 \ (\text{mod } 4)}} \left| \sum_{\substack{P < m \le M \\ 2 \nmid m, \ (m,n)=1}} \left(\frac{-n}{m} \right) \frac{a_m}{m} \right|^2 + HL(LP^{-1})^2
$$

$$
= LU + HL^3 P^{-2}, \tag{4.6}
$$

say, where $a_m = 1$ or 0. Expanding the square and exchanging the order of summation we get

$$
U \ll \sum_{\substack{P < m_1 \le M \\ 2 \nmid m_1}} \sum_{\substack{P < m_2 \le M \\ 2 \nmid m_2}} (m_1 m_2)^{-1} \left| \sum_{\substack{x < n \le x+H, \ n \equiv 3 \ (\text{mod } 4) \\ (n, m_1 m_2)=1}} \left(\frac{-n}{m_1} \right) \left(\frac{-n}{m_2} \right) \right|.
$$

By the law of quadratic reciprocity we have $\left(\dfrac{-n}{m_1} \right) \left(\dfrac{-n}{m_2} \right) = \left(\dfrac{m_1 m_2}{n} \right)$. Unless $m_1 m_2$ is a square, the Jacobi symbol $\left(\dfrac{m_1 m_2}{*} \right)$ is a non-principal character to the modulo $\le 4 m_1 m_2$. Thus we have, again by the Pólya–Vinogradov inequality, that

$$
U \ll H \sum_{P < l \le M} \frac{\tau(l^2)}{l^2} + \sum_{P < m_1 \le M} \sum_{P < m_2 \le M} (m_1 m_2)^{-1/2} L
$$

$$
\ll HP^{-1} \sum_{l \le M} \frac{\tau(l)^2}{l} + ML \ll HL^4 P^{-1} + x^{1/2} LP.
$$

Substituting this into (4.6) , we have

$$
V_3 \ll HL^5 P^{-1} + x^{1/2} L^2 P. \tag{4.7}
$$

We then turn to V_4. Put $N = HP^{-1}$. As before, we have

$$
\sum_{q > N} G(q, n) \Psi(q, P) \ll \sum_{q > N} \frac{\log q}{q} \Psi(q, P)
$$

$$
\ll (\log N)(\log P) \exp \left(-(\log N)/(\log P) \right) \ll P^{-1}
$$

by (2.3), partial summation and Lemma 2. The definition (2.2) of G gives

$$
\sum_{P < q \le N} G(q, n) \Psi(q, P) = \sum_{P < q \le N} \sum_{\substack{1 \le a \le q \\ (a,q)=1}} \Psi(q, P) \left(\frac{g(q, a)}{\varphi(q)} \right)^3 e \left(-\frac{a}{q} n \right).
$$

Hence the dual form of the additive large sieve inequality [9], [10] gives that

$$\sum_{x<n\leq x+H}\left|\sum_{q>P}G(q,n)\Psi(q,P)\right|^2 \ll \sum_{P<q\leq N}\sum_{\substack{1\leq a\leq q\\(a,q)=1}}(H+qN)\left|\frac{g(q,a)}{\varphi(q)}\right|^6 + HP^{-2}.$$

Then by (2.2) we find that

$$V_4 \ll \sum_{P<q\leq N}(H+qN)\frac{\tau(q)^4 q^2}{\varphi(q)^6}\sum_{a=1}^{q}|g(q,a)|^2 + HP^{-2}$$

$$\ll \left(\frac{H}{P}+N\right)L\sum_{q\leq N}\frac{\tau(q)^4}{q} + HP^{-2} \ll HL^{17}P^{-1}. \qquad (4.8)$$

Combining (4.3), (4.4), (4.5), (4.7), and (4.8), we obtain the assertion of Lemma 3.

5. Minor arc In this section we shall prove Lemma 4. Our task is to estimate

$$W_l = \int_{\mathbf{m}}\int_{\mathbf{m}}|S(\alpha_1)|^l|S(\alpha_2)|^l\min\left(H,\frac{1}{\|\alpha_1-\alpha_2\|}\right)d\alpha_1 d\alpha_2$$

for $l = 3$. To this end we first consider the case of $l = 4$: We note that

$$\int_{\mathbf{U}}|S(\xi)|^4 d\xi \ll L^4|\{m_1^2+m_2^2=m_3^2+m_4^2\leq 2x\}|$$

$$\ll L^4\sum_{l\leq 2x}\tau(l)^2 \ll xL^7. \qquad (5.1)$$

Since

$$|S(\alpha_1)|^4|S(\alpha_2)|^4 \ll |S(\alpha_1)|^6|S(\alpha_2)|^2 + |S(\alpha_1)|^2|S(\alpha_2)|^6.$$

we have that

$$W_4 \ll \int_{\mathbf{m}}\int_{\mathbf{m}}|S(\alpha_1)|^6|S(\alpha_2)|^2\min\left(H,\frac{1}{\|\alpha_1-\alpha_2\|}\right)d\alpha_1 d\alpha_2$$

$$\ll \int_{-\frac{1}{2}}^{\frac{1}{2}}\int_{\mathbf{m}}|S(\alpha)|^6|S(\alpha+\beta)|^2\min\left(H,\frac{1}{|\beta|}\right)d\alpha d\beta.$$

Expanding the square out and performing the integration with respect to β, we get

$$W_4 \ll \sum_{|h|\leq 2x}J(h)I(h)\left|\int_{\mathbf{m}}|S(\alpha)|^6 e(h\alpha)d\alpha\right|, \qquad (5.2)$$

Here we have

$$J(h) = \sum_{\substack{p_1^2 - p_2^2 = h \\ p_1^2, p_2^2 \le 2x}} \log p_1 \log p_2 \ll L^2 \begin{cases} x^{1/2} & \text{if } h = 0, \\ \tau(h) & \text{if } h > 0, \end{cases} \tag{5.3}$$

and

$$I(h) = \left| \int_{-1/2}^{1/2} \min\left(H, \frac{1}{|\beta|}\right) e(h\beta) d\beta \right| \ll \min\left(\log H, \frac{H}{|h|}\right). \tag{5.4}$$

Taking the absolute value inside in (5.2), we have that

$$W_4 \ll \int_{\mathbf{m}} |S(\alpha)|^6 d\alpha \left(J(0)I(0) + \sum_{0 < h \le 2x} J(h)I(h) \right)$$

$$\ll \sup_{\alpha \in \mathbf{m}} |S(\alpha)|^2 \int_{\mathbf{U}} |S(\xi)|^4 d\xi \left(x^{1/2}L + H \sum_{0 < h \le 2x} \frac{\tau(h)}{h} \right) L^2$$

$$\ll (x^{1/2} + H) x^2 L^{11 - 2E} \tag{5.5}$$

by the assumption given in Lemma 4, (5.1), (5.3), and (5.4).

We may now turn to W_3. The initial step is the same as above; in place of (6.2) we have

$$W_3 \ll \sum_{|h| \le 2x} J(h)I(h) \left| \int_{\mathbf{m}} |S(\alpha)|^4 e(h\alpha) d\alpha \right|.$$

Separating the term with $h = 0$, we split up the remaining range of h according as $(k-1)H < |h| \le kH$ $(1 \le k \ll x/H)$. Then by (5.3), (5.4), and Cauchy's inequality we have that

$$W_3 \ll x^{1/2} L^3 \int_{\mathbf{m}} |S(\alpha)|^4 d\alpha$$

$$+ L^2 \left(L + \sum_{2 \le k \ll x/H} \frac{1}{k-1} \right) \max_{0 \le y \le 2x} \sum_{y < h \le y+H} \tau(h) \left| \int_{\mathbf{m}} |S(\alpha)|^4 e(h\alpha) d\alpha \right|$$

$$\ll x^{1/2} L^3 \int_{\mathbf{U}} |S(\xi)|^4 d\xi$$

$$+ L^3 \max_{0 \le y \le 2x} \left(\sum_{y < h \le y+H} \tau(h)^2 \right)^{1/2} \left(\sum_{y < h \le y+H} \left| \int_{\mathbf{m}} |S(\alpha)|^4 e(h\alpha) d\alpha \right|^2 \right)^{1/2}.$$

Hence via (5.1) and (5.5) we obtain

$$W_3 \ll x^{3/2}L^{10} + L^3(HL^3)^{1/2}(W_4)^{1/2}$$
$$\ll x^{3/2}L^{10} + (H(x^{1/2} + H))^{1/2}xL^{10-E}$$
$$\ll x^{3/2}L^{10} + HxL^{10-E}.$$

This obviously completes the proof of the bound stated in our theorem.

6. Exceptional set We now prove (1.5) in the range $\theta > 1/2$. It should be stressed that for this particular purpose we do not need Lemma 3.

We put $\mathfrak{S}'(n) = s_n \prod_{2<p\leq P}(1 + G(p,n))$. Then we have, by (4.1),

$$\mathfrak{S}'(n) \gg (\log\log x)^{-3}$$

uniformly for $x < n \leq 2x$, $n \in \mathbb{H}$. Also it follows from (3.1), (3.2), and (4.8) that

$$\sum_{\substack{x<n\leq x+H \\ n\in\mathbb{H}}} \left| R(n) - \frac{\pi}{4}\mathfrak{S}'(n)\sqrt{n} \right|^2$$

$$\ll \sum_{x<n\leq x+H} \left| \sum_{q>P} G(q,n)\Psi(n,P) \right|^2 + W + HxL^2P^{-1}$$

$$\ll HxL^{17-B} + HxL^{-A} + x^{3/2}L^{10} \ll HxL^{-A}$$

for any $A > 0$, provided $H = x^{1/2+\varepsilon}$ with an arbitrary small but fixed $\varepsilon > 0$. We thus have

$$\sum_{\substack{x<n\leq x+H \\ n\in\mathbb{H},\, R(n)=0}} |\mathfrak{S}'(n)|^2 x \ll HxL^{-A}.$$

This proves our claim.

References

[1] M.B. Barban. The large sieve method and its applications to number theory. *Uspekhi Mat. Nauk*, **21** (1966), 51–102.

[2] H. Davenport. *Multiplicative Number Theory* (2nd ed., revised by H.L. Montgomery). Springer-Verlag, Berlin, 1980.

[3] A. Fujii. Some additive problems of numbers. *Banach Center Publ.*, **17** (1985), 121–141.

[4] C. Hooley. On a new approach to various problems of Waring's type. *Recent Progress in Analytic Number Theory* (H. Halberstam and C. Hooley, Eds.), vol. 1, Academic Press, London, 1981, pp. 127–191.

[5] L.K. Hua. Some results in the additive prime number theory. *Quart. J. Math. Oxford*, 9 (1938), 68–80.

[6] —. *Additive Theory of Prime Numbers*. A.M.S., Providence, 1965.

[7] E. Landau. *Vorlesungen über Zahlentheorie*. Hirzel, Berlin, 1927.

[8] M.-Ch. Leung and M.-Ch. Liu. On generalized quadratic equations in three prime variables. *Monatsh. Math.*, 115 (1993), 113–169.

[9] H.L. Montgomery. The analytic principle of the large sieve. *Bull. Amer. Math. Soc.*, 84 (1978), 547–567.

[10] H.L. Montgomery and R.C. Vaughan. Hilbert's inequality. *J. London Math. Soc.*, (2)8 (1974), 73–82.

[11] K. Prachar. *Primzahlverteilung*. Springer-Verlag, Berlin, 1957.

[12] W. Schwarz. Zur Darstellung von Zahlen durch Summen von Primzahl-potenzen, II. *J. reine angew. Math.*, 206 (1961), 78–112.

[13] T. Zhan. On a theorem of Hua. To appear.

Hiroshi Mikawa
Institute of Mathematics, University of Tsukuba
Tsukuba-305, Japan

16

Trace Formula over the Hyperbolic Upper Half Space

Yoichi Motohashi

1. Introduction The aim of the present paper is to establish an analogue of the Kuznetsov trace formula for $\mathcal{H}^3$ the hyperbolic upper half space, i.e., the Beltrami model of the three dimensional Lobachevsky geometry. Although we shall restrict ourselves to the case of the Picard group, our argument readily extends to general Bianchi groups defined over arbitrary imaginary quadratic number fields.

Let us first introduce the basics: Geometrically $\mathcal{H}^3$ is embedded in the three dimensional Euclidean space $\mathbb{C} \times \mathbb{R}$ so that its points are denoted by $z = (x, y) = (x_1 + x_2 i, y)$ with $x_1, x_2 \in \mathbb{R}$ and $y > 0$; and algebraically it is embedded in the Hamiltonian algebra so that we have $z = x + y\jmath$ where $\jmath^2 = -1$ and $i\jmath = -\jmath i$. These two notations for the points of $\mathcal{H}^3$ will be used interchangeably. As a Riemannian manifold, $\mathcal{H}^3$ carries the metric $((dx_1)^2 + (dx_2)^2 + (dy)^2)^{\frac{1}{2}}/y$ and the corresponding volume element $d\mu(z) = y^{-3} dx_1 dx_2 dy$. The Laplace–Beltrami operator of the manifold or rather its negative is denoted by Δ. We have

$$\Delta = -y^2 \left((\partial/\partial x_1)^2 + (\partial/\partial x_2)^2 + (\partial/\partial y)^2 \right) + y(\partial/\partial y).$$

As to the motions of the points in $\mathcal{H}^3$, let us put

$$\mathbb{T}(\mathcal{H}^3) = \left\{ z \mapsto (az + b) \cdot (lz + h)^{-1} : a, b, l, h \in \mathbb{C} \text{ with } ah - bl = 1 \right\}, \quad (1.1)$$

where the algebraic operation is that of the Hamiltonian. In the coordinate notation this generic map is

$$z = (x, y) \mapsto \left(\frac{(ax + b)\overline{(lx + h)} + a\bar{l}y^2}{|lx + h|^2 + |ly|^2}, \frac{y}{|lx + h|^2 + |ly|^2} \right). \quad (1.2)$$

We see readily that

$$\mathbb{T}(\mathcal{H}^3) \cong SL(2, \mathbb{C})/\{\pm 1\}.$$

The metric and thus the volume element as well as the Laplacian are all invariant with respect to the maps in $\mathbb{T}(\mathcal{H}^3)$; that is, these maps are rigid motions of $\mathcal{H}^3$.

The concept corresponding to the full modular group acting over $\mathcal{H}^2$ the two dimensional hyperbolic space is here the Picard group Γ composed of those elements in (1.1) with $a, b, l, h \in \mathbb{Z}[i]$, i.e., Gaussian integers. It is known that Γ acts discontinuously over $\mathcal{H}^3$, having the fundamental domain

$$\mathcal{F} = \left\{ z : x_1 \leq \tfrac{1}{2}, \, x_2 \leq \tfrac{1}{2}, \, x_1 + x_2 \geq 0, \, x_1^2 + x_2^2 + y^2 \geq 1 \right\}$$

(see Picard [8]). We may take $\mathcal{F}$ for a three-dimensional manifold carrying the above metric. It should be noted that the volume of $\mathcal{F}$ is finite:

$$|\mathcal{F}| = 2\pi^{-2}\zeta_K(2), \tag{1.3}$$

where ζ_K is the Dedekind zeta-function of the field $K = \mathbb{Q}(i)$.

Moving to the analytical structure, let $L^2(\mathcal{F}, d\mu)$ stand for the set of all Γ-invariant functions over $\mathcal{H}^3$ which are square integrable with respect to $d\mu$ over $\mathcal{F}$. This is a Hilbert space equipped with the inner-product

$$\langle f_1, f_2 \rangle = \int_{\mathcal{F}} f_1(z)\overline{f_2(z)}d\mu(z),$$

which is obviously well-defined for any $f_1, f_2 \in L^2(\mathcal{F}, d\mu)$. The spectral resolution of Δ over $L^2(\mathcal{F}, d\mu)$ is analogous to the case of the full modular group over $\mathcal{H}^2$. What we need in our discussion is the Parseval formula thus arising, and to state it we have to introduce further concepts: The non-trivial discrete spectrum of Δ is denoted by $\left\{ \lambda_j = 1 + \kappa_j^2 : j = 1, 2, \ldots \right\}$ in non-decreasing order, and the corresponding orthonormal system of eigenfunctions by $\{\psi_j\}$. It is known that the number of the λ_j's is infinite and that $\kappa_j > 0$ for all $j \geq 1$. Here the relevant multiplicities, which are always finite, are counted. We have the Fourier expansion

$$\psi_j(z) = y \sum_{\substack{n \in \mathbb{Z}[i] \\ n \neq 0}} \rho_j(n) K_{i\kappa_j}(2\pi|n|y)e([n, x]).$$

Here K_ν is the K-Bessel function of order ν, $e(a) = \exp(2\pi i a)$ as usual, and $[n, x] = \operatorname{Re}(n\bar{x})$. Also we need to define the Eisenstein series

$$E(z, s) = \sum_{\gamma \in \Gamma_\infty \backslash \Gamma} y^s(\gamma(z)),$$

where $y(z)$ denotes the third coordinate of z, and Γ_∞ is the stabilizer subgroup in Γ of the cusp at infinity. We note that

$$\Gamma_\infty = \Gamma_t \cup \imath\Gamma_t$$

where Γ_t is the translation group $\{z \mapsto z + b : b \in \mathbb{Z}[i]\}$, and $\imath$ is the involution $\imath(z) = -x + yj$. In another expression we have

$$\Gamma_\infty \cong \left\{ \pm \begin{pmatrix} 1 & b \\ & 1 \end{pmatrix}, \pm \begin{pmatrix} -i & b' \\ & i \end{pmatrix} : b, b' \in \mathbb{Z}[i] \right\} \Big/ \{\pm 1\}.$$

That is, we have

$$E(z, s) = \frac{1}{4} \sum_{\substack{l, h \in \mathbb{Z}[i] \\ (l, h) = 1}} \frac{y^s}{(\|lx + h\|^2 + \|ly\|^2)^s}.$$

It is easy to check that the series converges absolutely for $\mathrm{Re}\, s > 2$. As we shall briefly show in the third section, we have the Fourier expansion

$$E(z, s) = y^s + \frac{\pi}{s - 1} \frac{\zeta_K(s - 1)}{\zeta_K(s)} y^{2-s}$$

$$+ \frac{2\pi^s y}{\Gamma(s)\zeta_K(s)} \sum_{\substack{n \in \mathbb{Z}[i] \\ n \neq 0}} |n|^{s-1} \sigma_{1-s}(n) K_{s-1}(2\pi|n|y) e([n, x]), \qquad (1.4)$$

where ζ_K is as above, and $\sigma_\nu(n) = \frac{1}{4} \sum_{d|n} |d|^{2\nu}$ $(d \in \mathbb{Z}[i])$. Thus $E(z, s)$ is meromorphic over $\mathbb{C}$ as a function of s, being regular for $\mathrm{Re}\, s \geq 1$ except for the simple pole at $s = 2$, the residue of which has a relation with the assertion (1.3).

With the above notation we may state the Parseval formula: For any $f_1, f_2 \in L^2(\mathcal{F}, d\mu)$ we have

$$\langle f_1, f_2 \rangle = \sum_{j=0}^\infty \langle f_1, \psi_j \rangle \overline{\langle f_2, \psi_j \rangle} + \frac{1}{2\pi} \int_{-\infty}^\infty \mathcal{E}(t, f_1) \overline{\mathcal{E}(t, f_2)} dt. \qquad (1.5)$$

Here $\psi_0 \equiv |\mathcal{F}|^{-\frac{1}{2}}$ and

$$\mathcal{E}(t, f) = \int_{\mathcal{F}} f(z) E(z, 1 - it) \, d\mu(z),$$

where the integral is to be taken in the sense of the limit in mean (cf. Theorem 1.1 of our book [7]).

The spectral expansion (1.5) can be proved by following closely the discussion in [7, Chapter 1]. Although the latter is developed for the full modular group over $\mathcal{H}^2$, the argument carries over into our present situation with minor modifications. To some extent we have a simpler task here, for the free-space resolvent kernel of $\Delta + \lambda$ has an explicit algebraic expression in terms of the non-Euclidean distance (see also Selberg [11, p. 76]).

Having said these, we are now able to state our trace formula:

Theorem *Let*

$$S(m,n;l) = \sum_{\substack{v \bmod l \\ (v,l)=1}} e([m,v/l])e([n,v^*/l]) \quad (vv^* \equiv 1 \bmod l)$$

be the Kloosterman sum for Gaussian integers l, m, n. Let us assume that the function $h(r)$, $r \in \mathbb{C}$, is regular in the horizontal strip $|\operatorname{Im} r| < \frac{1}{2} + \varepsilon$ and satisfies

$$h(r) = h(-r), \quad h(r) \ll (1 + |r|)^{-3-\varepsilon}$$

for an arbitrary fixed $\varepsilon > 0$. Then we have, for any non-zero m, $n \in \mathbb{Z}[i]$,

$$\sum_{j=1}^{\infty} \frac{\overline{\rho_j(m)}\rho_j(n)}{\sinh(\pi\kappa_j)}\kappa_j h(\kappa_j) + 2\pi \int_{-\infty}^{\infty} \frac{\sigma_{ir}(m)\sigma_{ir}(n)}{|mn|^{ir}|\zeta_K(1+ir)|^2}h(r)dr$$

$$= \pi^{-2}(\delta_{m,n} + \delta_{m,-n})\int_{-\infty}^{\infty} r^2 h(r)dr + \sum_{\substack{l \in \mathbb{Z}[i] \\ l \neq 0}} |l|^{-2}S(m,n;l)\check{h}(2\pi\varpi) \quad (1.6)$$

with $\varpi^2 = \overline{mn}/l^2$. Here $\delta_{m,n}$ is the Kronecker delta, and

$$\check{h}(t) = i \int_{-\infty}^{\infty} \frac{r^2}{\sinh(\pi r)} \mathcal{J}_{ir}(t)h(r)dr, \quad (1.7)$$

$$\mathcal{J}_\nu(t) = 2^{-2\nu}|t|^{2\nu}J_\nu^*(t)J_\nu^*(\bar{t}), \quad (1.8)$$

where $J_\nu^*(t)$ is the entire function equal to $J_\nu(t)(\frac{1}{2}t)^{-\nu}$ when $t > 0$ with J_ν being the J-Bessel function of order ν.

It should be stressed that the choice of the sign of ϖ is immaterial, for $J_\nu^*(t)$ is a function of t^2. We could put $J_\nu(t)J_\nu(\bar{t})$ in place of $\mathcal{J}_\nu(t)$; the above formulation is to avoid the possible ambiguity pertaining to the branching of the values of J_ν.

This trace formula was announced in [6] and [7, Section 2.7]; what is developed below is a full proof of our claim. Because of our past experience with the full modular group over $\mathcal{H}^2$, it appears reasonable to suppose that (1.6) should have applications to various analytical problems involving Gaussian integers. Especially interesting is the possible application to the mean value problem of the zeta-function ζ_K. For, if everything develops as we expect, then the result should have a relevance to the eighth power moment of the Riemann zeta-function. In order to fully discuss such an application, we shall, however, need to devise a theory of the inversion of the integral transform (1.7), and to establish an analogue of Kuznetsov's sum formula [7, Theorem 2.3]. It is also conceivable that trace formulas twisted with Grössencharakters would have more relevances to the problem related to ζ_K. To these topics we shall return elsewhere.

We note in passing that the trace formula of the Selberg type for the Picard group is given in Tanigawa [14], Venkov [15], and Szmidt [13]. The theory of various arithmetic Eisenstein series over $\mathcal{H}^3$ is developed in Kubota [4] and in Elstrodt et al [1]. Further references can be found in these articles.

Acknowledgement The author is much indebted to Prof. M. Jutila for his kind comments and corrections.

2. Bessel gear As a preparation of our proof of the theorem we shall collect here integral formulas involving Bessel functions. Some of them are redundant for our purpose, but we list them because we want to emphasize a close relationship between our argument and classical works in the theory of the Bessel functions.

Lemma 1 *Let us suppose that f is sufficiently smooth on the positive real axis and decays rapidly at both ends of the line. Then we have*

$$f(y) = \pi^{-2} \int_{-\infty}^{\infty} r \sinh(\pi r) K_{ir}(y) \int_{0}^{\infty} f(v) K_{ir}(v) v^{-1} dv dr, \qquad (2.1)$$

where K_ν is the K-Bessel function of order ν.

Proof An equivalent assertion is stated in Lebedev [5, (5.14.14)] with a more precise condition but without proof. In [7, Section 2.6] is given a proof, which depends on the observation that this identity is nothing else but the result of applying the differential operator

$$D_s = -y^2 (d/dy)^2 + s(s-1) + y^2$$

to the integral transform of $y^{-\frac{3}{2}} f(y)$ with the resolvent kernel of $y^{-2} D_s$, where the s is a complex parameter to be chosen appropriately.

Lemma 2 *Let $h(r)$ be an even function which is regular in the horizontal strip $|\operatorname{Im} r| \leq \alpha$ for a certain $\alpha > 0$, and let us assume that $h(r) = O(e^{-2\pi|r|})$ there. Then we have, for $|\operatorname{Im} t| < \alpha$,*

$$h(t) = \pi^{-2} \int_{0}^{\infty} K_{it}(v) v^{-1} \int_{-\infty}^{\infty} r \sinh(\pi r) K_{ir}(v) h(r) dr dv. \qquad (2.2)$$

Proof This assertion is obviously the dual of the previous lemma but can be proved independently. The decay condition on h could of course be replaced by a less stringent one. A proof is given in [7, Section 2.6]. It rests on the idea to replace the factor v^{-1} in the integrand by v^{s-1} with $s \in \mathbb{C}$.

Lemma 3 *We have*

$$\frac{i}{2\pi^2} \int_{-i\infty}^{i\infty} \lambda \sin(2\pi\lambda)\Gamma(\omega_1 + \lambda)\Gamma(\omega_2 + \lambda)\Gamma(\omega_3 + \lambda)$$
$$\times \Gamma(\omega_1 - \lambda)\Gamma(\omega_2 - \lambda)\Gamma(\omega_3 - \lambda)d\lambda$$
$$= \Gamma(\omega_1 + \omega_2)\Gamma(\omega_1 + \omega_3)\Gamma(\omega_2 + \omega_3). \tag{2.3}$$

Here the path is curved to ensure that the poles of $\Gamma(\omega_1 + \lambda)\Gamma(\omega_2 + \lambda)\Gamma(\omega_3 + \lambda)$ lie to the left of it, and those of $\Gamma(\omega_1 - \lambda)\Gamma(\omega_2 - \lambda)\Gamma(\omega_3 - \lambda)$ to the right. It is assumed that parameters $\omega_1, \omega_2, \omega_3$ are such that the path can be drawn.

Proof This is the formula (2.6.6) of [7], where it is proved by the combination of (2.1) and the well-known Mellin transform of the product of two K-Bessel functions.

Lemma 4 *We have, for any positive b and complex η with $\operatorname{Re}\eta > 0$,*

$$\int_0^\infty e^{-a}J_0(ab)(a/2)^{2\eta-1}da$$
$$= \frac{1}{2\pi^{\frac{3}{2}}i} \int_{(\alpha)} \frac{\Gamma(\xi + \eta)\Gamma(\xi + \eta + \frac{1}{2})\Gamma(-\xi)}{\Gamma(\xi + 1)} b^{2\xi}d\xi. \tag{2.4}$$

On the right side the path is the vertical line $\operatorname{Re}\xi = \alpha$ with $-\operatorname{Re}\eta < \alpha < 0$.

Proof The left side is equal to $2^{1-2\eta}\Gamma(2\eta)F(\eta, \eta + \frac{1}{2}; 1; -b^2)$, where F is the hypergeometric function (see [16, p. 385]). Then the Mellin–Barnes formula [17, pp. 286–288] for F yields (2.4). It should be observed that both integrals in (2.4) are rapidly convergent.

Lemma 5 *If $a > 0$ and $\frac{1}{2}\operatorname{Re}(\rho - \frac{1}{2}) > \operatorname{Re}\mu > -1$, then we have*

$$\int_0^\infty \frac{J_\rho(\sqrt{u^2 + a^2})}{(u^2 + a^2)^{\frac{1}{2}\rho}}u^{2\mu+1}du = 2^\mu a^{\mu+1-\rho}\Gamma(\mu + 1)J_{\rho-\mu-1}(a). \tag{2.5}$$

Proof This is due to Sonine. For the proof see either [16, pp. 415–417] or Sonine's original account [12].

Lemma 6 *If $a, b > 0$ and $\operatorname{Re}\lambda > -\frac{1}{2}$, then we have*

$$\int_0^\pi \frac{J_\lambda(\sqrt{a^2 + b^2 - 2ab\cos\tau})}{(a^2 + b^2 - 2ab\cos\tau)^{\lambda/2}}(\sin\tau)^{2\lambda}d\tau$$
$$= \sqrt{\pi}\Gamma(\lambda + 1)(ab/2)^{-\lambda}J_\lambda(a)J_\lambda(b). \tag{2.6}$$

Proof   This is due to Gegenbauer (see [16, p. 367]). An accessible proof can be found in Sonine [12, p. 37].

Lemma 7   *If $a, b > 0$ and $\operatorname{Re}\lambda > -\frac{1}{2}$, then we have*

$$\int_0^{\frac{1}{2}\pi} \cos(a\cos\tau)J_{2\lambda}(b\sin\tau)d\tau$$

$$= \frac{1}{2}\pi J_\lambda\big(\tfrac{1}{2}(\sqrt{a^2+b^2}+a)\big)J_\lambda\big(\tfrac{1}{2}(\sqrt{a^2+b^2}-a)\big). \tag{2.7}$$

Proof   This can probably be attributed to Gegenbauer. It is tabulated as the formula 6.688(1) in [2] and mentioned in other tables as well. But its proof is hard to find in literature. Because of its importance in our argument, we shall give two proofs: In the first proof we begin with the expansion of the integrand by means of the defining series for the J-Bessel and the cosine functions. The order of the integral and the resulting double sum can obviously be exchanged. Computing the integral termwise we have

$$\int_0^{\frac{1}{2}\pi} \cos(a\cos\tau)J_{2\lambda}(b\sin\tau)d\tau$$

$$= \frac{1}{2}\sqrt{\pi}(b/2)^{2\lambda}\sum_{m,n=0}^{\infty} \frac{(-1)^{m+n}(a/2)^{2m}(b/2)^{2n}\Gamma(n+\lambda+\frac{1}{2})}{\Gamma(m+1)\Gamma(n+1)\Gamma(n+2\lambda+1)\Gamma(m+n+\lambda+1)}$$

$$= \frac{\sqrt{\pi}(b/2)^{2\lambda}}{2\Gamma(\lambda+\frac{1}{2})}\sum_{N=0}^{\infty} \frac{(-1/4)^N}{\Gamma(N+\lambda+1)}\sum_{m+n=N}\frac{a^{2m}b^{2n}}{m!n!}\int_0^1 u^{n+\lambda-\frac{1}{2}}(1-u)^{\lambda-\frac{1}{2}}du$$

$$= \frac{\sqrt{\pi}(b/2)^{2\lambda}}{2\Gamma(\lambda+\frac{1}{2})}\sum_{N=0}^{\infty} \frac{(-1/4)^N}{\Gamma(N+1)\Gamma(N+\nu+1)}\int_0^1 (a^2+b^2u)^N(u(1-u))^{\lambda-\frac{1}{2}}du$$

$$= \frac{\sqrt{\pi}(b/2)^{2\lambda}}{2\Gamma(\lambda+\frac{1}{2})}\int_0^1 \frac{J_\lambda(\sqrt{a^2+b^2u})}{(\frac{1}{2}\sqrt{a^2+b^2u})^\lambda}(u(1-u))^{\lambda-\frac{1}{2}}du. \tag{2.8}$$

It is easy to see that the last integral can be computed by means of (2.6); and we get the identity (2.7). The second proof is more direct: We consider instead the proof of the equivalent identity

$$\int_0^{\frac{1}{2}\pi} \cos((a-b)\cos\tau)J_{2\lambda}(2\sqrt{ab}\sin\tau)d\tau = \frac{1}{2}\pi J_\lambda(a)J_\lambda(b), \tag{2.9}$$

where a, b, λ are as above (see [2, 6.688(3)]). We differentiate twice the left side with respect to a, and rearrange the result using the relation

$$\big[u^2(d/du)^2 + u(d/du) + u^2 - 4\lambda^2\big]J_{2\lambda}(u) = 0.$$

Then we find that the left side of (2.9) is a solution of the differential equation for $J_\lambda(a)$. On noting that the integral is obviously of order $O(a^{\text{Re}\,\lambda})$ as a tends to 0, we may conclude, by the basic theory of cylinder functions, that it should be a constant multiple of $J_\lambda(a)$. The computation of this constant is immediate. This ends our discussion of (2.7). It should be worth remarking that the second argument yields, via (2.7)–(2.8), a proof of (2.6) that is different from both Sonine's and Gegenbauer's.

Lemma 8   *Let $p > 0$ and $|\text{Re}\,\lambda| < \frac{1}{2}$. Then we have, for any real ϑ,*

$$\int_0^{\frac{1}{2}\pi} \cos(2p\cos\vartheta\cos\tau)K_{2\lambda}(2p\sin\tau)d\tau$$

$$= \frac{\pi^2}{4\sin(2\pi\lambda)}\left[\mathfrak{J}_{-\lambda}(pe^{i\vartheta}) - \mathfrak{J}_\lambda(pe^{i\vartheta})\right], \qquad (2.10)$$

where $\mathfrak{J}_\nu$ is as above.

Proof   This is a consequence of the previous lemma. We shall, however, use rather (2.9) than (2.7): Since

$$K_\nu(v) = \frac{\pi}{2\sin\pi\nu}\{I_{-\nu}(v) - I_\nu(v)\}, \qquad (2.11)$$

where I_ν is the I-Bessel function of order ν, it is sufficient to show that

$$\int_0^{\frac{1}{2}\pi} \cos(2p\cos\vartheta\cos\tau)I_{2\lambda}(2p\sin\tau)d\tau = \frac{1}{2}\pi\mathfrak{J}_\lambda(pe^{i\vartheta}). \qquad (2.12)$$

To this end we put (2.9) in the form

$$\int_0^{\frac{1}{2}\pi} \cos((a-b)\cos\tau)(\sin\tau)^{2\lambda}J_{2\lambda}^*(2\sqrt{ab}\sin\tau)d\tau = 2^{-2\lambda-1}\pi J_\lambda^*(a)J_\lambda^*(b),$$

where J_ν^* is as in (1.8). By analytic continuation this holds for all $a, b \in \mathbb{C}$. So we may put, for instance, $a = pe^{i\vartheta}$, $b = -pe^{-i\vartheta}$, getting (2.12) immediately.

The following basic integral representations are listed here for the sake of convenience:

$$J_\nu(u) = \frac{(u/2)^\nu}{\sqrt{\pi}\Gamma(\nu + \frac{1}{2})}\int_0^\pi \cos(u\cos\tau)(\sin\tau)^{2\nu}d\tau$$

$$(|\arg u| < \pi, \text{ Re}\,\nu > -\tfrac{1}{2}), \qquad (2.13)$$

$$J_\nu(u) = \frac{1}{2\pi i}\int_{(\alpha)} \frac{\Gamma(\eta + \frac{1}{2}\nu)}{\Gamma(1 - \eta + \frac{1}{2}\nu)}(u/2)^{-2\eta}d\eta$$

$$(u > 0, \text{ Re}\,\nu > -2\alpha > 0), \qquad (2.14)$$

$$K_\nu(u) = \frac{1}{4\pi i}\int_{(\alpha)} \Gamma(\eta + \tfrac{1}{2}\nu)\Gamma(\eta - \tfrac{1}{2}\nu)(u/2)^{-2\eta}d\eta$$

$$(u > 0, 2\alpha > |\text{Re}\,\nu|). \qquad (2.15)$$

3. Poincaré series   Now we introduce the Poincaré series: We put

$$P_m(z,s) = \sum_{\gamma \in \Gamma_t \backslash \Gamma} y^s(\gamma(z)) \exp\left\{ -2\pi|m|y(\gamma(z)) + 2\pi i[m, x(\gamma(z))] \right\} \quad (3.1)$$

with Γ_t being as above and with an obvious convention (cf. Sarnak [10] modulo many blemishes). Since $|P_m(z,s)| \leq 2E(z, \mathrm{Re}\,s)$, the series is absolutely convergent for $\mathrm{Re}\,s > 2$; and $P_m(z,s)$ is Γ-invariant for such an s.

We need to expand $P_m(z,s)$ into a double Fourier series with respect to the variables x_1, x_2. To this end we note that in view of (1.2) we have, for $\mathrm{Re}\,s > 2$,

$$P_m(z,s) = 2y^s \exp(-2\pi|m|y) \cos(2\pi[m, x])$$

$$+ \frac{1}{2} y^s \sum_{\substack{l \in \mathbb{Z}[i] \\ l \neq 0}} \sum_{\substack{h \in \mathbb{Z}[i] \\ (h,l)=1}} (|lx + h|^2 + |ly|^2)^{-s}$$

$$\times \exp\left(-\frac{2\pi|m|y}{|lx+h|^2 + |ly|^2} + 2\pi i \mathrm{Re}\left(\frac{\overline{m}h^*}{l} - \frac{\overline{m(lx+h)}}{l(|lx+h|^2 + |ly|^2)} \right) \right),$$

where $h^*h \equiv 1 \bmod l$. The inner sum is divided into parts according to $h \bmod l$; and to each of the sub-sums we apply Poisson's sum formula. After a simple rearrangement we get, for $\mathrm{Re}\,s > 2$,

$$P_m(z,s) = 2y^s \exp(-2\pi|m|y) \cos(2\pi[m, x])$$

$$+ \frac{1}{2} y^{2-s} \sum_{n \in \mathbb{Z}[i]} e([n, x]) \sum_{\substack{l \in \mathbb{Z}[i] \\ l \neq 0}} |l|^{-2s} S(m, n; l) A_s(m, n; l; y), \quad (3.2)$$

where $S(m, n; l)$ is as above, and

$$A_s(m, n; l; y) = \int_{-\infty}^{\infty} \int_{-\infty}^{\infty} (|\mu|^2 + 1)^{-s}$$

$$\times \exp\left(-2\pi i y[n, \mu] - \frac{2\pi|m|}{|l|^2(|\mu|^2+1)y} - \frac{2\pi i[l^{-2}, m\mu]}{(|\mu|^2+1)y} \right) d\mu_1 d\mu_2 \quad (3.3)$$

with $\mu = \mu_1 + i\mu_2$. In deriving (3.2) we exchanged the order of sums. This is legitimate. For, as can be seen by shifting appropriately the two paths in (3.3), we have, uniformly for $\mathrm{Re}\,s > 1$,

$$A_s(m, n; l; y) \ll \exp(-|n|y) \quad (3.4)$$

provided y is not too small; the implied constant may depend on m but does not on l, n.

We note here that the assertion (1.4) follows from (3.2). In fact we have $E(z,s) = \frac{1}{2}P_0(z,s)$, and

$$A_s(0,n;l;y) = \begin{cases} \pi(s-1)^{-1} & \text{if } n = 0, \\ 2\pi^s|ny|^{s-1}K_{s-1}(2\pi|n|y)/\Gamma(s) & \text{if } n \neq 0 \end{cases}$$

as well as

$$\sum_{\substack{l\in\mathbb{Z}[i] \\ l\neq 0}} |l|^{-2s}S(0,n;l) = \begin{cases} 4\zeta_K(s-1)/\zeta_K(s) & \text{if } n = 0, \\ 4\sigma_{1-s}(n)/\zeta_K(s) & \text{if } n \neq 0. \end{cases}$$

Proofs of these four identities are standard, so we skip them.

Now we have the bound:

$$S(m,n;l) \ll |l||(m,n,l)|\sigma_0(l), \tag{3.5}$$

which can be proved by following Gundlach [3] (cf. Sarnak [10]). This and (3.2), (3.4) imply that $P_m(z,s)$ exists at least for $\operatorname{Re} s > \frac{3}{2}$ and there it is an element of $L^2(\mathcal{F}, d\mu)$ whenever $m \neq 0$. Hence we can apply the Parseval formula (1.5) to the inner product of two Poincaré series. The result is stated in

Lemma 9 Let $m, n \in \mathbb{Z}[i]$, $mn \neq 0$, and let $\operatorname{Re} s_1, \operatorname{Re} s_2 > \frac{3}{2}$. Then we have

$$\langle P_m(\cdot, s_1), P_n(\cdot, \overline{s_2}) \rangle$$
$$= \pi \frac{(4\pi)^{2-s_1-s_2}|m|^{1-s_1}|n|^{1-s_2}}{\Gamma(s_1 - \frac{1}{2})\Gamma(s_2 - \frac{1}{2})} \cdot \left\{ \sum_{j=1}^{\infty} \overline{\rho_j(m)}\rho_j(n)\Lambda(s_1, s_2; i\kappa_j) \right.$$
$$\left. + 2\pi^2 \int_{-\infty}^{\infty} \frac{\sigma_{ir}(m)\sigma_{ir}(n)}{|mn|^{ir}|\Gamma(1+ir)\zeta_K(1+ir)|^2}\Lambda(s_1, s_2; ir)dr \right\}, \tag{3.6}$$

where

$$\Lambda(s_1, s_2; \lambda) = \Gamma(s_1 - 1 + \lambda)\Gamma(s_1 - 1 - \lambda)\Gamma(s_2 - 1 + \lambda)\Gamma(s_2 - 1 - \lambda) \tag{3.7}$$

We do not give the proof, for the reasoning is essentially the same as in the case of the full modular group over $\mathcal{H}^2$ (see [7, Section 2.1]). It would, however, be expedient to note this: The argument involves an application of the unfolding method. It requires a bound for the sum of the absolute values of the terms in (3.1), and the necessary estimation can be accomplished with the expansion (1.4).

4. Transforming the inner-product Next we shall develop a geometric–arithmetic treatment of the inner-product in (3.6). This is obviously the core of our discussion.

First we modify (3.3) a little. Moving to the polar coordinate and invoking Poisson's formula (2.13), we readily find that

$$A_s(m,n;l;y) = 2\pi \int_0^\infty \frac{u}{(u^2+1)^s} J_0\Big(2\pi u\big|ny + \frac{\overline{m}}{l^2(u^2+1)y}\big|\Big)$$

$$\times \exp\Big(-\frac{2\pi|m|}{|l|^2(u^2+1)y}\Big)du.$$

On noting the remark at the end of the previous section, we have, by the unfolding method,

$$\langle P_m(\cdot,s_1), P_n(\cdot,\overline{s_2})\rangle = (\delta_{m,n} + \delta_{m,-n})\Gamma(s_1+s_2-2)(4\pi|m|)^{2-s_1-s_2}$$

$$+ \frac{1}{2}\int_0^\infty e^{-2\pi|n|y} y^{s_2-s_1-1} \sum_{\substack{l\in\mathbb{Z}[i]\\l\neq0}} |l|^{-2s_1} S(m,n;l) A_{s_1}(m,n;l;y)dy,$$

provided $\operatorname{Re} s_1 > \frac{3}{2}$ and $\operatorname{Re} s_2 > 2$. Then we assume temporarily that $\operatorname{Re} s_2 > \operatorname{Re} s_1 > 2$. Given this, we may take the integral inside the sum. For $A_{s_1}(m,n; l;y)$ is now bounded uniformly for all involved parameters; and thus the last expression is absolutely convergent. As a consequence we have that if

$$\operatorname{Re} s_1, \operatorname{Re} s_2 > \tfrac{3}{2} \tag{4.1}$$

then

$$\langle P_m(\cdot,s_1), P_n(\cdot,\overline{s_2})\rangle = (\delta_{m,n} + \delta_{m,-n})\Gamma(s_1+s_2-2)(4\pi|m|)^{2-s_1-s_2}$$

$$+ \pi(|m|/|n|)^{\frac{1}{2}(s_2-s_1)}$$

$$\times \sum_{\substack{l\in\mathbb{Z}[i]\\l\neq0}} |l|^{-s_1-s_2} S(m,n;l) B(2\pi\sqrt{|mn|}/|l|, \vartheta_0; s_1, s_2). \tag{4.2}$$

Here $\vartheta_0 = \arg\varpi$ with ϖ as in (1.6), and

$$B(p,\vartheta;s_1,s_2) = \int_0^\infty y^{s_2-s_1-1} C(p,\vartheta; \tfrac{1}{2}(s_1+s_2);y)dy\,;$$

$$C(p,\vartheta;\tau;y) = \int_0^\infty \frac{u}{(u^2+1)^\tau} J_0\Big(\frac{pu}{\sqrt{u^2+1}}\big|ye^{i\vartheta} + (ye^{i\vartheta})^{-1}\big|\Big)$$

$$\times \exp\Big(-\frac{p(y+y^{-1})}{\sqrt{u^2+1}}\Big)du.$$

To confirm this assertion we only need to show that the sum in (4.2) converges in the region (4.1). To this end we observe that

$$
\begin{aligned}
B(p,\vartheta; s_1, s_2) \\
\ll \int_0^\infty y^{\sigma_2 - \sigma_1 - 1} \int_0^\infty \frac{u}{(u^2+1)^{\frac{1}{2}(\sigma_1+\sigma_2)}} \exp\left(-\frac{p(y+y^{-1})}{\sqrt{u^2+1}}\right) du\, dy \\
= \int_0^\infty \frac{u}{(u^2+1)^{\frac{1}{2}(\sigma_1+\sigma_2)}} K_{\sigma_1-\sigma_2}\left(\frac{2p}{\sqrt{u^2+1}}\right) du,
\end{aligned} \tag{4.3}
$$

where $\operatorname{Re} s_j = \sigma_j$ $(j = 1, 2)$, and the implied constant is absolute. So we have, via (2.11),

$$
B(2\pi\sqrt{|mn|}/|l|, \vartheta_0; s_1, s_2) \ll |l|^{|\sigma_1-\sigma_2|} \log(|l|+1) \quad (\sigma_1, \sigma_2 > 1)
$$

uniformly in l. This and (3.5) end the verification.

Our task is now to transform $B(p, \vartheta; s_1, s_2)$ into an integral involving the Λ defined by (3.7). It should be stressed that we are now discussing the transformation of $B(p, \vartheta; s_1, s_2)$, and thus the conditions on s_1, s_2 to be introduced in the sequel as far as (4.15) are independent of (4.1).

We begin with a separation of the parameters in the integrand of $C(p, \vartheta; \tau; y)$. By means of the Mellin inversion of the relation (2.4) we have

$$
\begin{aligned}
\exp(\cdots)J_0(\cdots) = -\frac{1}{4\pi^{\frac{5}{2}}} \int_{(\alpha)} \left(\frac{p(y+y^{-1})}{2\sqrt{u^2+1}}\right)^{-2\eta} \\
\times \int_{(\beta)} \frac{\Gamma(\xi+\eta)\Gamma(\xi+\eta+\frac{1}{2})\Gamma(-\xi)}{\Gamma(\xi+1)} \left(1 - \left(\frac{2\sin\vartheta}{y+y^{-1}}\right)^2\right)^\xi u^{2\xi} d\xi\, d\eta,
\end{aligned}
$$

where $\alpha > 0$, $\beta < 0$ are small while satisfying $\alpha + \beta > 0$; note that the naming of the path in the integral on the right side of (2.4) has been changed. This double integral is, however, not absolutely convergent. To gain the absolute convergence we shift the contour $\operatorname{Re}\xi = \beta$ to $\operatorname{Re}\xi = \beta'$ with a small $\beta' > 0$. We have

$$
\begin{aligned}
\exp(\cdots)J_0(\cdots) = \exp\left(-\frac{p(y+y^{-1})}{\sqrt{u^2+1}}\right) \\
-\frac{1}{4\pi^{\frac{5}{2}}} \int_{(\alpha)} (\cdots)^{-2\eta} \int_{(\beta')} \frac{\Gamma(\xi+\eta)\Gamma(\xi+\eta+\frac{1}{2})\Gamma(-\xi)}{\Gamma(\xi+1)} (\cdots)^\xi u^{2\xi} d\xi\, d\eta.
\end{aligned}
$$

We replace the integrand of this new double integral by its absolute value, and see, by Stirling's formula, that the integral is $O\big(u^{2\beta'}(u^2+1)^\alpha\big)$ with the implied constant depending only on α, β'. Thus, inserting the last identity into the integral for $C(p, \vartheta; \tau; y)$, we have a triple integral which is absolutely convergent, provided $\operatorname{Re} \tau > 1 + \alpha + \beta'$. Given this condition, we may perform the u-integral first, getting

$$C(p, \vartheta; \tau; y) = R - \frac{1}{8\pi^{\frac{5}{2}}} \int_{(\alpha)} \frac{1}{\Gamma(\tau - \eta)}$$

$$\times \int_{(\beta')} \Gamma(\xi + \eta) \Gamma(\xi + \eta + \tfrac{1}{2}) \Gamma(\tau - \xi - \eta - 1) \Gamma(-\xi)$$

$$\times (\tfrac{1}{2} p(y + y^{-1}))^{-2\eta} \left(1 - \left(\frac{2 \sin \vartheta}{y + y^{-1}}\right)^2\right)^\xi d\xi d\eta, \qquad (4.4)$$

where

$$R = \int_0^\infty \frac{u}{(u^2 + 1)^\tau} \exp\left(-\frac{p(y + y^{-1})}{\sqrt{u^2 + 1}}\right) du.$$

We insert (4.4) with $\tau = \tfrac{1}{2}(s_1 + s_2)$ into the integral for $B(p, \vartheta; s_1, s_2)$. We readily see that the triple integral thus obtained is absolutely convergent provided

$$|\operatorname{Re}(s_1 - s_2)| < 2\alpha, \quad \operatorname{Re}(s_1 + s_2) > 2 + 2\alpha + 2\beta'. \qquad (4.5)$$

We arrange the order of integration by putting the ξ-integral inner, the y-integral middle, and the η-integral outer so that we have, given (4.5),

$$B(p, \vartheta; s_1, s_2) = R^* - \frac{1}{8\pi^{\frac{5}{2}}} \int_{(\alpha)} \frac{(p/2)^{-2\eta}}{\Gamma(\tfrac{1}{2}(s_1 + s_2) - \eta)}$$

$$\times \int_0^\infty \frac{y^{s_2 - s_1 - 1}}{(y + y^{-1})^{2\eta}} \int_{(\beta')} \Gamma(\xi + \eta) \Gamma(\xi + \eta + \tfrac{1}{2})$$

$$\times \Gamma(\tfrac{1}{2}(s_1 + s_2) - \xi - \eta - 1) \Gamma(-\xi) \left(1 - \left(\frac{2 \sin \vartheta}{y + y^{-1}}\right)^2\right)^\xi d\xi dy d\eta, \quad (4.6)$$

where R^* is the contribution of R above. We have

$$R^* = 2 \int_0^\infty \frac{u}{(u^2 + 1)^{\frac{1}{2}(s_1 + s_2)}} K_{s_1 - s_2}\left(\frac{2p}{\sqrt{u^2 + 1}}\right) du \qquad (4.7)$$

(cf. (4.3)). We shall use a somewhat roundabout argument to transform the ξ-integral in (4.6); this is for the sake of completeness. We expand the factor $(1 - (2 \sin \vartheta/(y + y^{-1}))^2)^\xi$ into a binomial series, and take the integration inner; that is

$$\int_{(\beta')} \cdots d\xi = \sum_{\nu=0}^\infty \frac{1}{\Gamma(\nu + 1)} \left(\frac{2 \sin \vartheta}{y + y^{-1}}\right)^{2\nu}$$

$$\times \int_{(\beta')} \Gamma(\xi + \eta) \Gamma(\xi + \eta + \tfrac{1}{2}) \Gamma(\tfrac{1}{2}(s_1 + s_2) - \xi - \eta - 1) \Gamma(-\xi + \nu) d\xi. \quad (4.8)$$

We have to verify this. So we shall estimate the generic νth term in the sum. The part of the integral corresponding to $|\mathrm{Im}\,\xi| < \nu$ contributes, by Stirling's formula,

$$\ll \nu^{-\nu-\frac{1}{2}}A^\nu \int_{-\nu}^{\nu} (|u|+1)^c (|u|+\nu+1)^{\nu-\beta'-\frac{1}{2}} e^{-\frac{3}{2}\pi|u|}du \ll \nu^{-\beta'-1}A^\nu,$$

where $A = (2\sin\vartheta/(y+y^{-1}))^2 \le 1$, and the implied constant as well as c are independent of ν. The remaining part contributes

$$\ll \frac{A^\nu}{\Gamma(\nu+1)} \int_\nu^\infty u^c (u+\nu)^\nu e^{-2\pi u}du,$$

which is obviously negligible in comparison with the first part. Thus (4.8) has been confirmed. Then by Barnes' integral formula [17, 14·52] we have, in place of (4.8),

$$\int_{(\beta')} \cdots d\xi = -2\pi i \Gamma(\eta)\Gamma(\eta+\tfrac{1}{2})\Gamma(\tfrac{1}{2}(s_1+s_2)-\eta-1)$$

$$+ 2\pi i \Gamma(\tfrac{1}{2}(s_1+s_2)-1)\Gamma(\tfrac{1}{2}(s_1+s_2)-\tfrac{1}{2})$$

$$\times \sum_{\nu=0}^\infty \frac{\Gamma(\eta+\nu)\Gamma(\eta+\nu+\tfrac{1}{2})}{\Gamma(\nu+1)\Gamma(\tfrac{1}{2}(s_1+s_2)+\eta+\nu-\tfrac{1}{2})} \left(\frac{2\sin\vartheta}{y+y^{-1}}\right)^{2\nu}.$$

The νth summand is $O(\nu^{\alpha-\frac{1}{2}\mathrm{Re}\,(s_1+s_2)}A^\nu)$ by Stirling's formula.

Thus, given (4.5), we may insert the last result into (4.6), and exchange the order of the sum and the y-integral. We get, after a little rearrangement,

$$\int_0^\infty \frac{y^{s_2-s_1-1}}{(y+y^{-1})^{2\eta}} \int_{(\beta')} \cdots d\xi \, dy$$

$$= -i2^{1-2\eta}\pi^{\frac{3}{2}}\Gamma(\tfrac{1}{2}(s_1+s_2)-\eta-1)\Gamma(\tfrac{1}{2}(s_1-s_2)+\eta)\Gamma(\tfrac{1}{2}(s_2-s_1)+\eta)$$

$$+ 2^{4-2\eta-s_1-s_2}\pi^2 i \Gamma(s_1+s_2-2)$$

$$\times \sum_{\nu=0}^\infty \frac{\Gamma(\tfrac{1}{2}(s_1-s_2)+\eta+\nu)\Gamma(\tfrac{1}{2}(s_2-s_1)+\eta+\nu)}{\Gamma(\nu+1)\Gamma(\tfrac{1}{2}(s_1+s_2)+\eta+\nu-\tfrac{1}{2})} (\sin\vartheta)^{2\nu}. \qquad (4.9)$$

If we insert this into (4.6), we would not be able to exchange the order of the η-integral and the sum over ν, for the resulting expression does not converge absolutely. To overcome this difficulty we use Gauss' integral representation of the hypergeometric function. So we have, in (4.9),

$$\sum_{\nu=0}^\infty \cdots = \frac{\Gamma(\tfrac{1}{2}(s_1-s_2)+\eta)}{\Gamma(s_1-\tfrac{1}{2})}$$

$$\times \int_0^1 u^{\frac{1}{2}(s_2-s_1)+\eta-1}(1-u)^{s_1-\frac{3}{2}}(1-(\sin\vartheta)^2 u)^{\frac{1}{2}(s_2-s_1)-\eta}du. \qquad (4.10)$$

We then insert (4.9) together with (4.10) into (4.6). The contribution of the first term on the right side of (4.9) cancels R^* out. In fact it is equal to

$$-\frac{1}{4\pi i}\int_{(\alpha)}\frac{\Gamma(\frac{1}{2}(s_1-s_2)+\eta)\Gamma(\frac{1}{2}(s_2-s_1)+\eta)}{\frac{1}{2}(s_1+s_2)-\eta-1}p^{-2\eta}d\eta$$

$$=-\frac{1}{2\pi i}\int_{(\alpha)}\Gamma(\tfrac{1}{2}(s_1-s_2)+\eta)\Gamma(\tfrac{1}{2}(s_2-s_1)+\eta)p^{-2\eta}$$

$$\times\int_0^\infty\frac{u}{(u^2+1)^{\frac{1}{2}(s_1+s_2)-\eta}}du d\eta.$$

We exchange the order of integration, which is obviously legitimate, and use the Mellin–Barnes formula (2.15). Then by (4.7) we see that our claim is correct.

We now have

$$B(p,\vartheta;s_1,s_2)=-i2^{1-s_1-s_2}\pi^{-\frac{1}{2}}\frac{\Gamma(s_1+s_2-2)}{\Gamma(s_1-\frac{1}{2})}\int_{(\alpha)}\frac{\Gamma(\frac{1}{2}(s_1-s_2)+\eta)}{\Gamma(\frac{1}{2}(s_1+s_2)-\eta)}p^{-2\eta}$$

$$\times\int_0^1 u^{\frac{1}{2}(s_2-s_1)+\eta-1}(1-u)^{s_1-\frac{3}{2}}(1-(\sin\vartheta)^2 u)^{\frac{1}{2}(s_2-s_1)-\eta}du d\eta. \qquad (4.11)$$

Given the condition

$$|\mathrm{Re}\,(s_1-s_2)|<2\alpha, \quad \mathrm{Re}\,(s_1+s_2)>2+2\alpha$$

with a small α, the double integral in (4.11) is absolutely convergent. The η-integral is taken inner, and we have, by the Mellin–Barnes formula (2.14),

$$B(p,\vartheta;s_1,s_2)=4\sqrt{\pi}\frac{p^{1-s_2}\Gamma(s_1+s_2-2)}{2^{s_1+s_2}\Gamma(s_1-\frac{1}{2})}$$

$$\times\int_0^1 u^{s_2-\frac{1}{2}s_1-\frac{3}{2}}(1-u)^{s_1-\frac{3}{2}}(1-(\sin\vartheta)^2 u)^{\frac{1}{2}(1-s_1)}$$

$$\times J_{s_1-1}(2p\sqrt{u^{-1}-(\sin\vartheta)^2})du.$$

Further, replacing u by $(u^2+1)^{-1}$, we get

$$B(p,\vartheta;s_1,s_2)=8\sqrt{\pi}\frac{p^{1-s_2}\Gamma(s_1+s_2-2)}{2^{s_1+s_2}\Gamma(s_1-\frac{1}{2})}\int_0^\infty u^{2s_1-2}(u^2+1)^{\frac{1}{2}-s_2}$$

$$\times(u^2+(\cos\vartheta)^2)^{\frac{1}{2}(1-s_1)}J_{s_1-1}(2p\sqrt{u^2+(\cos\vartheta)^2})du.$$

To relate this with Sonine's formula (2.5) we use

$$(u^2+1)^{\frac{1}{2}-s_2}=\frac{1}{2\pi i\Gamma(s_2-\frac{1}{2})}\int_{(\delta)}\Gamma(\xi)\Gamma(s_2-\tfrac{1}{2}-\xi)u^{-2\xi}d\xi,$$

where $0 < \delta < \operatorname{Re} s_2 - \frac{1}{2}$. So we are led to

$$
B(p, \vartheta; s_1, s_2) = -i \frac{p^{1-s_2}\Gamma(s_1 + s_2 - 2)}{2^{s_1 + s_2 - 2}\sqrt{\pi}\,\Gamma(s_1 - \frac{1}{2})\Gamma(s_2 - \frac{1}{2})}
$$
$$
\times \int_{(\delta)} \Gamma(\xi)\Gamma(s_2 - \tfrac{1}{2} - \xi) \int_0^\infty u^{2s_1 - 2\xi - 2} \frac{J_{s_1 - 1}(2p\sqrt{u^2 + (\cos\vartheta)^2}\,)}{(u^2 + (\cos\vartheta)^2)^{\frac{1}{2}(s_1 - 1)}}\, du\, d\xi.
$$

For the double integral converges absolutely, provided

$$
\operatorname{Re} s_1, \operatorname{Re} s_2 > \tfrac{1}{2} + \delta > \tfrac{1}{2}(\operatorname{Re} s_1 + \tfrac{1}{2}),
$$

which we may suppose temporarily. Then we have, by means of (2.5),

$$
B(p, \vartheta; s_1, s_2) = \frac{(2p)^{2 - s_1 - s_2}\Gamma(s_1 + s_2 - 2)}{2i\sqrt{\pi}\,\Gamma(s_1 - \frac{1}{2})\Gamma(s_2 - \frac{1}{2})}
$$
$$
\times \int_{(\delta)} \Gamma(\xi)\Gamma(s_1 - \tfrac{1}{2} - \xi)\Gamma(s_2 - \tfrac{1}{2} - \xi) J_{\xi - \frac{1}{2}}(2p\cos\vartheta)\left(\frac{p}{\cos\vartheta}\right)^{\xi - \frac{1}{2}} d\xi. \quad (4.12)
$$

It is easy to see that the integral is absolutely convergent, provided

$$
0 < \delta < \min\{\operatorname{Re} s_1, \operatorname{Re} s_2\} - \tfrac{1}{2}.
$$

This means that the interim restriction on s_1, s_2 have been dropped.

We replace the J-factor in (4.12) by the representation (2.13), getting

$$
\int_{(\delta)} \cdots d\xi = \frac{2}{\sqrt{\pi}} \int_0^{\frac{1}{2}\pi} \cos(2p\cos\vartheta\cos\tau)
$$
$$
\times \int_{(\delta)} \Gamma(s_1 - \tfrac{1}{2} - \xi)\Gamma(s_2 - \tfrac{1}{2} - \xi)(p\sin\tau)^{2\xi - 1}\, d\xi\, d\tau. \quad (4.13)
$$

The inner integral is essentially a value of the K-Bessel function of order $s_1 - s_2$. But we rather appeal to (2.3). In it we put $\omega_1 = s_1 - 1$, $\omega_2 = s_2 - 1$, $\omega_3 = \frac{1}{2} - \xi$ assuming that $\operatorname{Re}\xi = \delta$ is small and that

$$
\operatorname{Re} s_1, \operatorname{Re} s_2 > 1. \quad (4.14)
$$

We have that

$$
\Gamma(s_1 - \tfrac{1}{2} - \xi)\Gamma(s_2 - \tfrac{1}{2} - \xi) = \frac{i}{2\pi^2 \Gamma(s_1 + s_2 - 2)}
$$
$$
\times \int_{(0)} \lambda \sin(2\pi\lambda)\Lambda(s_1, s_2; \lambda)\Gamma(\tfrac{1}{2} - \xi + \lambda)\Gamma(\tfrac{1}{2} - \xi - \lambda)\, d\lambda
$$

with Λ defined by (3.7). Thus we see, by this and (2.15), that the inner integral in (4.13) is equal to

$$-\frac{2}{\pi\Gamma(s_1 + s_2 - 1)} \int_{(0)} \lambda \sin(2\pi\lambda)\Lambda(s_1, s_2; \lambda)K_{2\lambda}(2p\sin\tau)d\lambda$$

We then find that

$$B(p, \vartheta; s_1, s_2) = i\frac{2\pi^{-2}(2p)^{2-s_1-s_2}}{\Gamma(s_1 - \frac{1}{2})\Gamma(s_2 - \frac{1}{2})} \int_{(0)} \lambda \sin(2\pi\lambda)\Lambda(s_1, s_2; \lambda)$$

$$\times \int_0^{\frac{1}{2}\pi} \cos(2p\cos\vartheta\cos\tau)K_{2\lambda}(2p\sin\tau)d\tau d\lambda.$$

By virtue of (2.10) we now have, given (4.14),

$$B(p, \vartheta; s_1, s_2) = -i\frac{(2p)^{2-s_1-s_2}}{\Gamma(s_1 - \frac{1}{2})\Gamma(s_2 - \frac{1}{2})} \int_{(0)} \lambda\Lambda(s_1, s_2; \lambda)\mathcal{J}_\lambda(pe^{i\vartheta})d\lambda. \quad (4.15)$$

Since (4.1) is contained in (4.14), the expression (4.15) can be used in (4.2). Hence we have proved

Lemma 10 *Let $m, n \in \mathbb{Z}[i]$, $mn \neq 0$, and let $\operatorname{Re} s_1, \operatorname{Re} s_2 > \frac{3}{2}$. Further, let ϖ be as in (1.6). Then we have*

$$\langle P_m(\cdot, s_1), P_n(\cdot, \overline{s_2}) \rangle = (\delta_{m,n} + \delta_{m,-n})\Gamma(s_1 + s_2 - 2)(4\pi|m|)^{2-s_1-s_2}$$

$$- i\frac{(4\pi)^{3-s_1-s_2}|m|^{1-s_1}|n|^{1-s_2}}{4\Gamma(s_1 - \frac{1}{2})\Gamma(s_2 - \frac{1}{2})}$$

$$\times \sum_{\substack{l \in \mathbb{Z}[i] \\ l \neq 0}} |l|^{-2}S(m, n; l) \int_{(0)} \lambda\Lambda(s_1, s_2; \lambda)\mathcal{J}_\lambda(2\pi\varpi)d\lambda. \quad (4.16)$$

5. A spectral estimate We are now going to prove a spectro-statistical estimate of the Fourier coefficients $\rho_j(n)$ introduced in the first section. The result will play a crucial rôle in our proof of the theorem in much the same way as Kuznetsov's corresponding estimate [7, (2.3.2)] does in all known proofs of his trace formula over the space $\mathcal{H}^2$. We shall try to extend the argument developed in [7, Section 2.3] to our present situation. The somewhat elaborated discussion given below is to be taken for a preparation for the future extension of our spectral bound to the large sieve of the Iwaniec type. It will yield a result stronger than what our proof of the theorem actually requires.

First we transform the infinite sum in (4.9) in a way different from (4.10). So, on noting that

$$\Gamma(\tfrac{1}{2}(s_1 - s_2) + \eta + \nu)\Gamma(\tfrac{1}{2}(s_2 - s_1) + \eta + \nu)$$

$$= \Gamma(2\eta + 2\nu) \int_0^1 (u(1 - u))^{\eta+\nu-1}(u/(1 - u))^{\frac{1}{2}(s_1-s_2)}du,$$

we have, in (4.9),

$$\sum_{\nu=0}^{\infty}\cdots = \frac{2^{2\eta-1}}{\sqrt{\pi}}\int_0^1 (u(1 - u))^{\eta-1}(u/(1 - u))^{\frac{1}{2}(s_1-s_2)}$$

$$\times \sum_{\nu=0}^{\infty}\frac{\Gamma(\eta + \nu)\Gamma(\eta + \frac{1}{2} + \nu)}{\Gamma(\nu + 1)\Gamma(\frac{1}{2}(s_1 + s_2) + \eta + \nu - \frac{1}{2})}(2\sin\vartheta\sqrt{u(1 - u)})^{2\nu}du.$$

We have used the duplication formula for the Γ-function. Also the exchange of the sum and the integral that we have performed is easy to verify; note that the relevant condition on s_1, s_2 is maintained. The sum in the integrand can be expressed in terms of Gauss' integral representation of the hypergeometric function as before. Thus the above is equal to

$$\frac{2^{2\eta-1}\Gamma(\eta)}{\sqrt{\pi}\Gamma(\frac{1}{2}(s_1 + s_2) - 1)}\int_0^1\int_0^1 u^{\frac{1}{2}(s_1-s_2)-1}(1 - u)^{\frac{1}{2}(s_2-s_1)-1}$$

$$\times v^{-\frac{1}{2}}(1 - v)^{\frac{1}{2}(s_1+s_2)-2}\left(\frac{u(1 - u)v}{1 - (2\sin\vartheta)^2u(1 - u)v}\right)^{\eta}dudv.$$

This gives, in place of (4.11),

$$B(p,\vartheta;s_1,s_2) = \frac{\Gamma(\frac{1}{2}(s_1 + s_2 - 1))}{8\pi^{\frac{3}{2}}i}$$

$$\times \int_0^1\int_0^1 u^{\frac{1}{2}(s_1-s_2)-1}(1 - u)^{\frac{1}{2}(s_2-s_1)-1}v^{-\frac{1}{2}}(1 - v)^{\frac{1}{2}(s_1+s_2)-2}$$

$$\times \int_{(\alpha)}\frac{\Gamma(\eta)}{\Gamma(\frac{1}{2}(s_1 + s_2) - \eta)}\left(\frac{4p^{-2}u(1 - u)v}{1 - (2\sin\vartheta)^2u(1 - u)v}\right)^{\eta}d\eta dudv, \qquad (5.1)$$

provided

$$|\mathrm{Re}\,(s_1 - s_2)| < 2\alpha, \quad \mathrm{Re}\,(s_1 + s_2) > 2 + 4\alpha$$

with a small α. The verification of absolute convergence is immediate. The innermost integral could be expressed in terms of the $J^*_{\frac{1}{2}(s_1+s_2)-1}$ (see (2.14)).

We now specialize the above discussion by putting

$$s_1 = 2 + it, s_2 = 2 - it \quad (t \in \mathbb{R}).$$

We insert (5.1) into (4.2) and equate the result with (3.6). Further, we put $m = n$. Then we get, for any real t and non-zero $m \in \mathbb{Z}[i]$,

$$\sum_{j=1}^{\infty} |\rho_j(m)|^2 \Psi(t, \kappa_j) + 2\pi^2 \int_{-\infty}^{\infty} \frac{|\sigma_{ir}(m)|^2}{|\Gamma(1 + ir)\zeta_K(1 + ir)|^2} \Psi(t, r) dr$$

$$= 1 + (\pi|m|)^2 \sum_{\substack{l \in \mathbb{Z}[i] \\ l \neq 0}} |l|^{-4} S(m, m; l) \Phi(t; m, l). \tag{5.2}$$

Here we have

$$\Psi(t, r) = \pi \Lambda(2 + it, 2 - it; ir) |\Gamma(\tfrac{3}{2} + it)|^{-2},$$

$$\Phi(t; m, l) = -i \int_0^1 \int_0^1 u^{it-1}(1 - u)^{-it-1} v^{-\frac{1}{2}}$$

$$\times \int_{(\alpha)} \frac{\Gamma(\eta)}{\Gamma(2 - \eta)} \Big(\frac{|l|^2 u(1 - u)v}{(\pi|m|)^2(1 - (2\sin\vartheta_0)^2 u(1 - u)v)} \Big)^{\eta} d\eta du dv,$$

where ϑ_0 is as before, and $0 < \alpha < \frac{1}{2}$.

We multiply both sides of (5.2) by the non-negative factor

$$\exp(-(t/K)^2) - \exp(-(2t/K)^2)$$

with a large parameter $K > 0$, and integrate with respect to t over the real axis. We readily find that this procedure yields the inequality

$$\sum_{\frac{1}{2}K \leq \kappa_j \leq K} |\rho_j(m)|^2 e^{-\pi\kappa_j} \ll K^2 + K|m|^2 \sum_{\substack{l \in \mathbb{Z}[i] \\ l \neq 0}} |l|^{-4} |S(m, m; l)| |P(m, l; K)|,$$

where the implied constant is absolute, and

$$P(m, l; K) = \int_0^1 \int_0^1 \frac{v^{\frac{1}{2}} Q(u, K)}{u(1 - u)} J_1^* \Big(\frac{2\pi|m|}{|l|} \Big(\frac{1 - (2\sin\vartheta_0)^2 u(1 - u)v}{u(1 - u)v} \Big)^{\frac{1}{2}} \Big) du dv$$

with

$$Q(u, K) = K \exp\Big(-\Big(\frac{K}{2} \log \frac{u}{1 - u}\Big)^2 \Big) - \frac{1}{2} K \exp\Big(-\Big(\frac{K}{4} \log \frac{u}{1 - u}\Big)^2 \Big).$$

We then observe that the representation (2.13) implies that the J_1^* factor and thus $P(m, l; K)$ are absolutely bounded. Hence we have proved

Lemma 11 *We have, uniformly in non-zero $m \in \mathbb{Z}[i]$ and $K > 0$,*

$$\sum_{\frac{1}{2}K \leq \kappa_j \leq K} |\rho_j(m)|^2 e^{-\pi\kappa_j} \ll K^2 + K|m|^2. \tag{5.3}$$

It should be remarked that our argument can be refined at least in two immediate respects outside the aforementioned possibility of its large sieve extension: The factor $|m|^2$ could be replaced by $|m|^{1+\varepsilon}$ for any $\varepsilon > 0$. Also an asymptotic equality could be proved instead of the inequality. We note that Raghavan and Sengupta [9] considered a sum that is essential the same as ours, and proved an asymptotic expression for it. However, it appears to us that their argument is hard to extend to the large sieve situation.

6. Proof of Theorem We may now finish the proof of our theorem. But we shall show only salient points, for the reasoning is very much similar to our third proof of Kuznetsov's trace formulas over $\mathcal{H}^2$ that is given in [7, Section 2.6].

First we observe that by virtue of the bound (5.3) we may consider the case where we have $e^{-\delta r^2} h(r)$, $\delta > 0$, instead of $h(r)$ in the statement of the theorem. That is, we may assume that $h(r)$ is very rapidly decaying as $|r| \to \infty$ in the horizontal strip where it is regular. We then put $s_2 = 2$, $s_1 = s$ in (3.6) and (4.16). Equating the results we get, for $m, n \in \mathbb{Z}[i]$, $mn \neq 0$, $\operatorname{Re} s > \frac{3}{2} + \varepsilon$ with a small $\varepsilon > 0$,

$$\sum_{j=1}^{\infty} \frac{\kappa_j \overline{\rho_j(m)} \rho_j(n)}{\sinh \pi \kappa_j} \Gamma(s - 1 + i\kappa_j) \Gamma(s - 1 - i\kappa_j)$$

$$+ 2\pi \int_{-\infty}^{\infty} \frac{\sigma_{ir}(m) \sigma_{ir}(n)}{|mn|^{ir} |\zeta_K(1 + ir)|^2} \Gamma(s - 1 + ir) \Gamma(s - 1 - ir) dr$$

$$= \pi^{-1} 2^{1-2s} \Gamma(2s - 1)(\delta_{m,n} + \delta_{m,-n})$$

$$+ i \sum_{\substack{l \in \mathbb{Z}[i] \\ l \neq 0}} |l|^{-2} S(m, n; l) \int_{[\varepsilon]} \frac{r^2}{\sinh \pi r} \mathcal{J}_{ir}(2\pi \varpi) \Gamma(s - 1 + ir) \Gamma(s - 1 - ir) dr.$$

Here $[\varepsilon]$ indicates that the path is $\operatorname{Im} r = -\frac{1}{2} - \varepsilon$. The shift of path involved here is to gain absolute convergence throughout in this and subsequent expressions. We multiply both sides by the factor $(v/2)^{2-2s}$, $v > 0$, and integrate with respect to s over the line $\operatorname{Re} s = 2$. It is easy to check the absolute convergence, and by (2.15) we have, for any $v > 0$,

$$\sum_{j=1}^{\infty} \frac{\kappa_j \overline{\rho_j(m)} \rho_j(n)}{\sinh \pi \kappa_j} K_{2i\kappa_j}(v) + 2\pi \int_{-\infty}^{\infty} \frac{\sigma_{ir}(m) \sigma_{ir}(n)}{|mn|^{ir} |\zeta_K(1 + ir)|^2} K_{2ir}(v) dr$$

$$= \frac{1}{8\pi} (\delta_{m,n} + \delta_{m,-n}) v e^{-v}$$

$$+ i \sum_{\substack{l \in \mathbb{Z}[i] \\ l \neq 0}} |l|^{-2} S(m, n; l) \int_{[\varepsilon]} \frac{r^2}{\sinh \pi r} \mathcal{J}_{ir}(2\pi \varpi) K_{2ir}(v) dr.$$

Further, we multiply both sides of this identity by the factor

$$g(v) = \frac{1}{\pi^2 v} \int_{-\infty}^{\infty} \xi \sinh(\pi\xi) K_{i\xi}(v) h(\xi/2) d\xi,$$

and integrate with respect to x over the positive real axis. The absolute convergence is again not hard to check. It should, however, be expedient to note that the width of the horizontal strip where $h(r)$ is regular becomes relevant when we check the convergence of the Kloosterman-sum part. By means of (2.2) we get (1.6), save for the verification of

$$\int_0^{\infty} v e^{-v} g(v) dv = 8\pi^{-1} \int_{-\infty}^{\infty} r^2 h(r) dr.$$

But this is a simple consequence of the well-known Mellin transform of $K_\nu(v)$ $\times e^{-v}$. This ends the proof of the theorem.

References

[1] J. Elstrodt, F. Grunewald and J. Mennicke. Eisenstein series on the three-dimensional hyperbolic space and imaginary quadratic number fields. *J. reine angew. Math.*, **360** (1986), 160–213.

[2] I.S. Gradshteyn and I.M. Ryzhik. *Table of Integrals, Series and Products.* Academic Press, London, 1980.

[3] K.-B. Gundlach. Über die Darstellung der ganzen Spitzenformen zu den Idealstufen der Hilbertschen Modulgruppe und die Abschätzung ihrer Fourierkoeffizienten. *Acta Math.*, **92** (1954), 309–345.

[4] T. Kubota. Über diskontinuierliche Gruppen Picardschen Typus und zugehörige Eisensteinsche Reihen. *Nagoya J. Math.*, **32** (1968), 259–271.

[5] N.N. Lebedev. *Special Functions and their Applications.* Dover Publications, New York, 1972.

[6] Y. Motohashi. A trace formula for the Picard group. I. *Proc. Japan Acad. Ser. A.*, **72** (1996), 183–186.

[7] —. *Spectral Theory of the Riemann Zeta-Function.* Cambridge Univ. Press, in press.

[8] M.E. Picard. Sur un group de transformations des points de l'espace situés du même côté d'un plan. *Bull. Soc. Math. de France*, **12** (1884), 43–47.

[9] S. Raghavan and J. Sengupta. On Fourier coefficients of Maass cusp forms in 3-dimensional hyperbolic space. *Proc. Steklov Inst. Math.*, **207** (1995), 251–257.

[10] P. Sarnak. The arithmetic and geometry of some hyperbolic three manifolds. *Acta Math.*, **151** (1983), 253–295.

[11] A. Selberg. Harmonic analysis and discontinuous groups in weakly symmetric Riemannian spaces with applications to Dirichlet series. *J. Indian Math. Soc.*, **20** (1956), 47–87.

[12] N. Sonine. Recherches sur les fonctions cylindriques et le développement des fonctions continues en sèries. *Math. Ann.*, **16** (1880), 1–80.

[13] J. Szmidt. The Selberg trace formula for the Picard group SL(2, $Z[i]$). *Acta Arith.*, **42** (1983), 391–424.

[14] Y. Tanigawa. Selberg trace formula for Picard groups. *Proc. Int. Symp. Algebraic Number Theory, Tokyo, 1977*, pp. 229–242.

[15] A.B. Venkov. Expansion in automorphic eigenfunctions of the Laplace–Beltrami operator in classical symmetric spaces of rank one, and the Selberg trace formula. *Proc. Steklov Inst. Math.*, **125** (1973), 1–48.

[16] G.N. Watson. *A Treatise on the Theory of Bessel Functions*. Cambridge Univ. Press, London, 1944.

[17] E.T. Whittaker and G.N. Watson. *A Course of Modern Analysis*. Cambridge Univ. Press, London, 1927.

Yoichi Motohashi
Department of Mathematics, College of Science and Technology, Nihon University
Surugadai, Tokyo-101, Japan

Modular Forms and
the Chebotarev Density Theorem, II

V. KUMAR MURTY[*]

1. Introduction Let $\pi(x, q, a)$ denote the number of primes less than x which are congruent to $a(\bmod q)$. The classical Siegel–Walfisz theorem states that given $N > 0$, there exists $c = c(N) > 0$ such that

$$\pi(x, q, a) = \frac{1}{\phi(q)} \text{Li } x + O(x \exp(-c \ \overline{\log x}))$$

holds uniformly for $q \leq (\log x)^N$.

In trying to generalize this to the non-Abelian context, one encounters difficulties from possible poles of the Artin L-functions. One of our aims here is to formulate and prove a non-Abelian version assuming Artin's conjecture on the holomorphy of Artin L-functions, and show how it can be used to deduce results independent of this conjecture. In particular, we present unconditional analogues of some results of [8].

A second aim of this paper is to apply the above results to modular forms. As in [8], we employ group theory to put ourself in a situation where Artin's conjecture is known. Thus we are able to apply the results mentioned in the previous paragraph without introducing any hypotheses. In particular, we get sharper results than those of Serre [11] and Daqing Wan [13] on the frequency with which the Fourier coefficients of a non-CM modular form take a given value. We also apply our estimates to deduce lower bounds for these Fourier coefficients which are valid on a set of density one. These make precise the results stated in [8], §5.

To state the general result, let K be a number field and L a finite Galois extension of K with group G. Let C be a subset of G stable under conjugation. For a prime $\mathfrak{p}$ of K which does not ramify in L, denote by $\sigma_\mathfrak{p}$ the conjugacy class of Frobenius elements at primes of L dividing $\mathfrak{p}$. Denote by $\pi_C(x, L/K)$ the number of primes $\mathfrak{p}$ of K, unramified in L, for which $\sigma_\mathfrak{p} \subseteq C$ and $\mathsf{N}_{K/\mathbb{Q}}\mathfrak{p} \leq x$. (When there is no possibility of confusion, we shall simply write $\pi_C(x)$ for $\pi_C(x, L/K)$.) The Chebotarev density theorem tells us that as $x \to \infty$,

$$\pi_C(x) \sim \frac{|C|}{|G|} \text{Li } x$$

---

[*] E.W.R. Steacie Fellow.

(where vertical bars denote cardinality). Lagarias and Odlyzko [4] have given an effective version of this. Their result states that for $x \geq \exp(10n_L(\log d_L)^2)$,

$$\pi_C(x) = \frac{|C|}{|G|}\text{Li } x + O\left(\frac{|C|}{|G|}\text{Li } x^{\beta_1}\right) + O\left(||C||x \exp\left(-c\left(\frac{\log x}{n_L}\right)^{\frac{1}{2}}\right)\right) \quad (1.1)$$

where β_1 is a possible real zero near $s = 1$ of the Dedekind zeta function $\zeta_L(s)$ of L, $n_L = [L : \mathbb{Q}]$, d_L is the absolute value of the discriminant of $L/\mathbb{Q}$ and $||C||$ is the number of conjugacy classes contained in C, and c and the implied constants are positive, effective and absolute.

It is necessary to define what one means by the exceptional zero β_1 in this context. In [12], Lemma 3, it is shown that the Dedekind zeta function $\zeta_L(s)$ has at most one zero in the region

$$1 - \frac{1}{4\log d_L} \leq \sigma \leq 1, \quad |t| \leq \frac{1}{4\log d_L}.$$

This zero, if it exists, must be real and simple. It is this possible simple real zero which is denoted β_1 in (1.1).

The estimate (1.1) has been used by Serre [11] and by Murty, Murty and Saradha [8] to study the distribution of values of Fourier coefficients of modular forms. More precisely, let f be a cusp form which is not of CM-type, of weight $k \geq 2$ for the congruence subgroup $\Gamma_0(N)$. Suppose that f is a normalized eigenform for the Hecke operators T_p ($p \nmid N$), and the U_p (for $p|N$) and that f has a Fourier expansion

$$f(z) = \sum_{n \geq 1} a_n e^{2\pi i n z}.$$

Suppose for simplicity that the a_n are rational integers and that the weight k is equal to 2. For $a \in \mathbb{Z}$, set $\pi_f(x, a)$ to be the number of primes p such that $p \leq x$ and $a_p = a$. Using (1.1), Serre showed that for any $\epsilon > 0$, we have

$$\pi_f(x, a) \ll_\epsilon x/(\log x)^{3/2 - \epsilon}.$$

Daqing Wan [13] improved the exponent in the denominator to $2 - \epsilon$. We give a further refinement, namely we show that

$$\pi_f(x, a) \ll x(\log \log x)^2/(\log x)^2. \quad (1.2)$$

We do this by relating $\pi_f(x, a)$ to $\pi_C(x)$ in an extension where Artin's conjecture is known, and then use our improvement of (1.1). Our method is the unconditional analogue of the method of [8] which, contrary to the opinion expressed in [13, p.253], presents no new serious difficulties. Of course, (1.2) is still quite far from the expected truth. Lang and Trotter conjecture that

$$\pi_f(x, a) \ll \sqrt{x}/\log x.$$

We apply the above estimate (1.2) to deduce lower bounds for the a_p and the a_n which are valid on a set of density one. More precisely, we show that for any monotone increasing function F which tends to infinity, the bound

$$|a_p| \geq \frac{\log p}{(\log \log p)^2 F(p)}$$

holds for a set of primes of density 1. Also, for any $\epsilon > 0$, there is a set of n of density 1 such that

$$a_n = 0 \text{ or } |a_n| \geq \frac{\log n}{(\log \log n)^{3+\epsilon}}.$$

2. Preliminaries Let L/K, G and C be as above. For any character χ of G, set

$$\psi(x, \chi) = \sum_{\substack{\mathbb{N}\, \mathfrak{p}^m \leq x \\ \mathfrak{p} \nmid d_L}} \chi(\sigma_{\mathfrak{p}}^m) \log(\mathbb{N}\mathfrak{p}).$$

Set also

$$\psi_C(x) = \sum_{\substack{\mathbb{N}\, \mathfrak{p}^m \leq x \\ \sigma^m \in C \\ \mathfrak{p} \nmid d_L}} \log(\mathbb{N}\mathfrak{p}).$$

By the usual technique of partial summation, an estimate for $\psi_C(x)$ will also yield an estimate for $\pi_C(x)$. If C is a conjugacy class, we have by the orthogonality relations

$$\psi_C(x) = \frac{|C|}{|G|} \sum_{\chi} \bar{\chi}(C) \psi(x, \chi).$$

Here, the sum ranges over the irreducible characters χ of G. Thus, it suffices to estimate each $\psi(x, \chi)$.

Let χ be, as above, an irreducible character of $G = \mathrm{Gal}(L/K)$. Denote by $L(s, \chi)$ the associated Artin L-function, by $\mathfrak{f}_\chi$ the (Artin) conductor of χ, and by V_χ the underlying space of χ. For each infinite place ν of K, denote by $\chi_\nu^\pm(1)$ the dimension of the subspace on which σ_ν acts by ± 1. (Note that $\chi_\nu^-(1) = 0$ if ν is complex.) Let us set

$$a_\chi = \sum \chi_\nu^+(1), \quad b_\chi = \sum \chi_\nu^-(1)$$

where the sum ranges over the real infinite places ν of K. Let us also set

$$c_\chi = \chi(1)r_2(K)$$

where $r_2(K)$ is (as usual) the number of complex places of K. Finally, let

$$A_\chi = d_K^{\chi(1)} N_{K/\mathbb{Q}} \mathfrak{f}_\chi.$$

Set

$$\mathfrak{G}(s,\chi) = \left(\pi^{-s/2}\Gamma(s/2)\right)^{a_\chi} \left(\pi^{-(s+1)/2}\Gamma((s+1)/2)\right)^{b_\chi} \left((2\pi)^{-s}\Gamma(s)\right)^{c_\chi}$$

and

$$\Lambda(s,\chi) = A_\chi^{s/2}\mathfrak{G}(s,\chi)L(s,\chi).$$

Then, we have the functional equation (see for example, [7], Chapter 2)

$$\Lambda(s,\chi) = W(\chi)\Lambda(1-s,\bar\chi)$$

for some complex number $W(\chi)$ of magnitude 1. Let

$$\delta(\chi) = (\chi, 1)$$

be the multiplicity of the trivial representation in χ. Then $(s(s-1))^{\delta(\chi)}\Lambda(s,\chi)$ is a meromorphic function.

In the remainder of this section, we shall *assume* that $(s(s-1))^{\delta(\chi)}\Lambda(s,\chi)$ is in fact *holomorphic*. This is Artin's holomorphy conjecture.

Under this assumption, we have the Hadamard factorization

$$\Lambda(s,\chi) = (s(s-1))^{-\delta(\chi)} \exp\left(\alpha_\chi + \beta_\chi s\right) \prod \left(1 - \frac{s}{\rho}\right) e^{s/\rho}.$$

We know (for example, from [10], (2.8)) that

$$\mathrm{Re}(\beta_\chi) = -\mathrm{Re}\sum \frac{1}{\rho}$$

and

$$\frac{L'}{L}(s,\chi) + \frac{L'}{L}(\bar{s},\bar{\chi}) = \sum_{\rho}\left(\frac{1}{s-\rho} + \frac{1}{\bar{s}-\bar{\rho}}\right)$$

$$- \log A_\chi - 2\mathrm{Re}\,\frac{\mathfrak{G}'}{\mathfrak{G}}(s,\chi) - 2\mathrm{Re}\,\delta(\chi)\left(\frac{1}{s} + \frac{1}{s-1}\right).$$

Since

$$\mathrm{Re}\left(\frac{1}{s-\rho}\right) = \frac{\sigma - \beta}{|s-\rho|^2} > 0$$

for $\sigma > 1$ and any $\rho = \beta + i\gamma$ with $0 < \beta < 1$ we deduce that

$$-\mathrm{Re}\,\frac{L'}{L}(s,\chi) \le -\sum_{\rho}\mathrm{Re}\left(\frac{1}{s-\rho}\right) + \frac{1}{2}\log A_\chi$$

$$+ \mathrm{Re}\,\frac{\mathfrak{G}'}{\mathfrak{G}}(s,\chi) + \mathrm{Re}\,\delta(\chi)\left(\frac{1}{s} + \frac{1}{s-1}\right)$$

where the sum over ρ is over any subset of zeros of $\Lambda(s,\chi)$ with $0 < \mathrm{Re}\,\rho < 1$. ¿From Stirling's formula,

$$\frac{\mathfrak{G}'}{\mathfrak{G}}(s,\chi) \ll \chi(1)n_K \log(|s|+2).$$

Hence, for $\sigma > 1$,

$$-\mathrm{Re}\,\frac{L'}{L}(s,\chi) \le -\sum \mathrm{Re}\left(\frac{1}{s-\rho}\right) + \delta(\chi)\mathrm{Re}\left(\frac{1}{s} + \frac{1}{s-1}\right)$$

$$+ \frac{1}{2}\log A_\chi + O(\chi(1)n_K \log(|s|+2)). \tag{2.1}$$

Also, if we take $1 < \sigma < 3/2$ then $\frac{\Gamma'}{\Gamma}(\sigma/2)$, $\frac{\Gamma'}{\Gamma}((\sigma+1)/2)$, and $\frac{\Gamma'}{\Gamma}(\sigma) - \log 2$ are negative. Hence,

$$\mathrm{Re}\,\frac{\mathfrak{G}'}{\mathfrak{G}}(\sigma,\chi) = \frac{a_\chi}{2}\frac{\Gamma'}{\Gamma}(\sigma/2) + \frac{b_\chi}{2}\frac{\Gamma'}{\Gamma}((\sigma+1)/2)$$

$$+ c_\chi\left(\frac{\Gamma'}{\Gamma}(\sigma) - \log 2\right) - \frac{\chi(1)n_K}{2}\log\pi \le -\frac{\chi(1)n_K}{2}\log\pi < 0.$$

Hence, for $1 < \sigma < 3/2$,

$$-\mathrm{Re}\,\frac{L'}{L}(\sigma,\chi) \le -\sum_{\rho}\mathrm{Re}\left(\frac{1}{\sigma-\rho}\right) + \frac{1}{2}\log A_\chi + \delta(\chi)\left(\frac{1}{\sigma} + \frac{1}{\sigma-1}\right). \tag{2.2}$$

3. Zero-free regions   We continue to assume Artin's holomorphy conjecture. We shall prove the following.

Proposition 3.1   *Assume Artin's conjecture for L/K and let χ be a character which does not contain the trivial character. Set $r = (\chi, \chi)$. Then, there is an absolute constant $c > 0$ such that $L(s, \chi)$ has at most $\frac{1}{2}(r+1)$ zeros (counting multiplicities) in the region*

$$1 - \frac{c}{\chi(1)\log A_\chi} \le \sigma \le 1, \quad |t| \le \frac{c}{\chi(1)\log A_\chi}.$$

Proof   We consider the function

$$H(s, \chi) = L(s, (1+\chi) \otimes (1+\bar\chi)).$$

Expanding the product, we see that

$$H(s, \chi) = \zeta_K(s)L(s, \chi)L(s, \bar\chi)L(s, \chi \otimes \bar\chi).$$

Then, for $\sigma > 1$, we have

$$0 \le -\frac{H'}{H}(\sigma, \chi) = -\left(\frac{L'}{L}(\sigma, \chi) + \frac{L'}{L}(\sigma, \bar\chi)\right) - \frac{\zeta_K'}{\zeta_K}(\sigma) - \frac{L'}{L}(\sigma, \chi \otimes \bar\chi).$$

Now using (2.1) and (2.2), and the fact that if ρ is a zero of $L(s, \chi)$ then $\bar\rho$ is a zero of $L(s, \bar\chi)$, we deduce that for $1 < \sigma < 3/2$,

$$2\sum_\rho \operatorname{Re}\left(\frac{1}{\sigma - \rho}\right) \le \frac{r+1}{\sigma - 1} + c_1 r + \frac{1}{2}\log d_K + \frac{1}{2}\log A_\chi + \frac{1}{2}\log A_{\bar\chi} + \frac{1}{2}\log A_{\chi \otimes \bar\chi}$$

where the sum on the left is over zeros of $L(s, \chi)$ in the stated region. It is clear from the definition that

$$c_1 r + \frac{1}{2}\log d_K + \frac{1}{2}\log A_\chi + \frac{1}{2}\log A_{\bar\chi} + \frac{1}{2}\log A_{\chi \otimes \bar\chi} \le c_2 \chi(1)\log A_\chi$$

for some constant $c_2 > 0$. Now choose

$$\sigma = 1 + \frac{c_3}{\chi(1)\log A_\chi}$$

with $c_3 > c$, and assume that $L(s, \chi)$ has more than $\frac{1}{2}(r+1)$ zeros in the given rectangle. Let r' denote the greatest integer $\le \frac{1}{2}(r+1)$. Then

$$2(r'+1)\frac{c + c_3}{(c + c_3)^2 + c^2} \le \frac{r+1}{c_3} + c_2.$$

Now choosing c_3 and c sufficiently small we get a contradiction.

Corollary 3.2 *If χ is irreducible, then $L(s,\chi)$ has at most one zero in the region*

$$1 - \frac{c}{\chi(1)\log A_\chi} \leq \sigma \leq 1, \quad |t| \leq \frac{c}{\chi(1)\log A_\chi}.$$

Proof We apply Proposition 3.1 and observe that $r = 1$.

Corollary 3.3 *If χ_1 and χ_2 are two distinct irreducible characters, then there is an absolute constant $c > 0$ such that $L(s,\chi_1)L(s,\chi_2)$ has at most one zero in the region*

$$1 - \frac{c}{\chi(1)\log A} \leq \sigma \leq 1,$$

$$|t| \leq \frac{c}{\chi(1)\log A}$$

where $\chi(1) = \max(\chi_1(1),\chi_2(1))$ and $A = \max(A_{\chi_1}, A_{\chi_2})$.

Proof We apply Proposition 3.1 to $L(s,\chi_1 + \chi_2)$ and observe that $r = 2$.

Proposition 3.4 *Assume Artin's conjecture for L/K and set*

$$\mathcal{A} = \max A_\chi, \quad d = \max \chi(1)$$

and

$$L(s) = \prod L(s,\chi)$$

where the maximum and product range over all the irreducible characters of $\mathrm{Gal}(L/K)$. Then, there is a $c > 0$ such that $L(s)$ has at most one zero in the region

$$1 - \frac{c}{d\log\mathcal{A}} \leq \sigma < 1, \quad |t| \leq \frac{c}{d\log\mathcal{A}}.$$

Proof This follows immediately from Corollary 3.3.

Remark 3.5 In [4], pp. 452-453, the remark is made that if Artin's conjecture is assumed, one should be able to establish a zero-free region with A_χ in place of d_L. However, the exact form is left unclear. In particular, note that in the above, it is $A_\chi^{\chi(1)}$ which enters.

Remark 3.6 It is possible that the full Artin conjecture is not needed to prove the Propositions above and that a region free of poles near $s = 1$ would suffice.

Notice that Proposition 3.4 does not show that the exceptional zero, if it exists, is real, or that it comes from an Abelian character. It is possible to prove this if we narrow the zero-free region somewhat.

Proposition 3.7   *Assume Artin's conjecture for L/K and let χ be an irreducible character of degree larger than 1. Then, there is a constant $c > 0$ (effective and absolute) such that $L(s,\chi)$ does not vanish in the region*

$$1 - \frac{c}{\chi(1)^3(\log A_\chi + n_K \log(|t| + 2))} \leq \sigma \leq 1.$$

If $\chi(1) = 1$, then it has at most one zero in this region. Such a zero is necessarily real and simple.

Proof   Let ℓ be a prime so that K does not contain any ℓ-th roots of unity. (There exists such a prime which is $O(\log d_K)$.) Denote by χ_t the character $\chi \otimes \chi_{cycl}^{it}$ where χ_{cycl} is the complex character giving the action of $\mathrm{Gal}(\bar{K}/K)$ on ℓ-th roots of unity. Let ϕ be an irreducible constituent of $\chi \otimes \bar{\chi}$ which is not the identity character or χ. (This is possible as $\chi(1) > 1$.) Following an idea of Hoffstein and Ramakrishnan [3, p. 297], we consider the function

$$H(s,\chi) = L(s,(1 + \phi + \chi_\gamma) \otimes (1 + \bar{\phi} + \bar{\chi}_{-\gamma}))$$

with γ yet to be specified. Since $(\phi \otimes \bar{\chi}, \bar{\chi}) = (\phi, \chi \otimes \bar{\chi}) \geq 1$ and $(\bar{\phi} \otimes \chi, \chi) \geq 1$, we can write

$$L(s,\phi \otimes \bar{\chi}_{-\gamma})L(s,\bar{\phi} \otimes \chi_\gamma) = L(s + i\gamma, \chi)L(s - i\gamma, \bar{\chi})I_1(s + i\gamma)I_2(s - i\gamma)$$

with $I(s) = I_1(s + i\gamma)I_2(s - i\gamma)$ entire. Moreover, $I(s)$ is real for real s. Expanding the product for $H(s,\chi)$, we see that

$$H(s,\chi) = L(s+i\gamma,\chi)^2 L(s-i\gamma,\bar{\chi})^2 \zeta_K(s)L(s,\chi \otimes \bar{\chi})L(s,\phi \otimes \bar{\phi})L(s,\phi)L(s,\bar{\phi})I(s).$$

Now let $\rho = \beta + i\gamma$ be a zero of $L(s,\chi)$ with $0 < \beta < 1$, $\gamma > 0$. Then we get for $\sigma > 1$

$$0 \leq -\frac{H'}{H}(\sigma,\chi) = -2\left(\frac{L'}{L}(\sigma + i\gamma,\chi) + \frac{L'}{L}(\sigma - i\gamma,\bar{\chi})\right)$$

$$-\frac{\zeta_K'}{\zeta_K}(\sigma) - \frac{L'}{L}(\sigma,\chi \otimes \bar{\chi}) - \frac{L'}{L}(\sigma,\phi \otimes \bar{\phi})$$

$$-\left(\frac{L'}{L}(\sigma,\phi) + \frac{L'}{L}(\sigma,\bar{\phi})\right) - \frac{I'}{I}(\sigma + i\gamma).$$

Now applying (2.1) and (2.2), we deduce that

$$\frac{4}{\sigma - \beta} \leq \frac{3}{\sigma - 1} + c_1(\chi(1)^3 \log A_\chi + \chi(1)^3 n_K \log(|\gamma| + 2)).$$

Now choosing

$$\sigma = 1 + \frac{c_2}{\chi(1)^3(\log A_\chi + n_K \log(|\gamma| + 2))}$$

for a sufficiently small $c > 0$ and $c_2 > 0$ gives the desired zero-free region. Finally, the case of $\chi(1) = 1$ is proved by the classical method. (See, for example, [6], Proposition 3.4 or [13], Proposition 3.4.) This completes the proof of the Proposition.

Proposition 3.8 *Assume Artin's conjecture for L/K and set*

$$d = \max \chi(1)$$

and as before, set

$$\mathcal{A} = \max A_\chi.$$

Then, there is a $c > 0$ such that $\zeta_L(s)$ has at most one zero in the region

$$1 - \frac{c}{d^3(\log \mathcal{A} + n_K \log(|t| + 2))} \leq \sigma \leq 1.$$

This zero, β_1 say, is simple, real and belongs to a character χ_1 which is Abelian and real.

Remark 3.9 As the proof will show, we can broaden the region slightly to

$$1 - \frac{c}{\log \mathcal{B}(t)} \leq \sigma < 1$$

where

$$\mathcal{B}(t) = \max_\chi (A_\chi(|t| + 2)^{n_K})^{\chi(1)^3}.$$

Proof Let M be the maximal Abelian extension of K contained in L. If $\zeta_L(s)$ vanishes in the above region, then by Proposition 3.7, so does $\zeta_M(s)$. Hence, it suffices to prove the result for L/K Abelian. But this is well-known. (See for example, [6], Proposition 3.4 or [13], Proposition 3.4.)

4. The Chebotarev Density Theorem   Let us set

$$\log \mathcal{M} = \log \mathcal{M}(L/K) = \frac{1}{n_K} \log d_K + 2\{\sum_{p|d_{L/K}} \log p + \log(n_L/n_K)\}.$$

The main result of this section is the following.

Theorem 4.1 *Assume Artin's conjecture for L/K and suppose that*

$$d^4 n_K \log \mathcal{M} \ll \log x.$$

Let χ_1 be as in Proposition 3.8 and let C be a conjugacy class. Then

$$\pi_C(x) = \frac{|C|}{|G|}\mathrm{Li}\, x - \frac{|C|}{|G|}\chi_1(C)\mathrm{Li}\,(x^{\beta_1})$$

$$+ \, O\left(|C|^{\frac{1}{2}} n_K x(\log x \mathcal{M})^2 \exp\left(-c'(\log x)/\left(d^{3/2}\,\sqrt{d^3(\log \mathcal{A})^2 + n_K \log x}\right)\right)\right)$$

Remark 4.2 It is instructive to compare this with the estimate (1.1) of [4]. The condition on x in (1.1) is approximately that

$$\log x \gg n_L^3 (\log \mathcal{M})^2 = n_K^3 |G|^3 (\log \mathcal{M})^2.$$

On the other hand, the condition in Theorem 4.1 is satisfied if

$$\log x \gg n_K |G|^2 \log \mathcal{M}.$$

The error term of (1.1) depends only on the number of conjugacy classes in C while the result above depends on the size of C. Since

$$d^3(\log \mathcal{A})^2 + n_K \log x \ll n_K(n_K d^5 (\log \mathcal{M})^2 + \log x),$$

the above result is nontrivial provided

$$d^4 n_K (\log \mathcal{M}) = o(\log x).$$

On the other hand, for (1.1) to be nontrivial, one needs

$$n_L = n_K |G| = o(\log x).$$

If G is Abelian, then (1.1) is valid and nontrivial provided

$$\log x \gg n_K^3 |G|^3 (\log \mathcal{M})^2 \text{ and } n_K |G| = o(\log x)$$

while Theorem 4.1 is valid and nontrivial provided

$$n_K \log \mathcal{M} = o(\log x).$$

If G has an Abelian normal subgroup A, then Theorem 4.1 is valid and non-trivial provided

$$n_K[G:A]^4(\log\mathcal{M}) = o(\log x).$$

Example 4.3 Suppose $G = \mathrm{Gal}(L/K)$ is the semidirect product of an Abelian normal subgroup A and a subgroup H. Assume Artin's conjecture is true for the characters of H. Every irreducible character of G is of the form $Ind_J^G(\psi \otimes \rho)$ where $J \supseteq A$ and ψ is a one dimensional character of J and ρ is a character of J/A (viewed as a character of J by the natural projection $J \to J/A$.) We may write J as the semidirect product AH_1, where H_1 is a subgroup of H. Now the assumption of Artin's conjecture for H implies that the L-functions associated to characters of H_1 are also analytic. (Indeed, if χ' is a character of H_1, then

$$L(s,\chi') = L(s, \mathrm{Ind}_{H_1}^H \chi')$$

and the right hand side is analytic by assumption.) If we assume in addition that Artin's conjecture is true for the one-dimensional twists of characters of H, then Artin's conjecture holds for G. Hence, if C is a conjugacy class of G, then

$$\pi_C(x) = \frac{|C|}{|G|}\mathrm{Li}\, x - \frac{|C|}{|G|}\chi_1(C)\mathrm{Li}\,(x^{\beta_1})$$
$$+ O\left(|C|^{\frac{1}{2}}n_K x \exp\left(-c'(\log x)/\left(d^{3/2}\,\overline{d^3(\log\mathcal{A})^2 + n_K\log x}\right)\right)(\log x\mathcal{M})^2\right)$$

provided

$$\log x \gg d^4 n_K \log\mathcal{M}.$$

Note that $d \le |H|$. On the other hand, if every conjugacy class in C contains an element of A, then we can improve this estimate. This will be seen in Theorem 4.4 and Theorem 4.6. We shall return to the situation considered in this example when we briefly look at the Fourier coefficients of CM modular forms.

Now we begin the proof of Theorem 4.1. Under the assumption of Artin's holomorphy conjecture, one can derive the following explicit formula using standard methods. For any $2 \le T \le x$,

$$\psi(x,\chi) - \delta(\chi)x + \sum_{|\gamma|<T}\frac{x^\rho}{\rho} \ll \frac{x\log x}{T}(\log A_\chi) + \frac{x\log x}{T}n_K\chi(1)\log T$$
$$+ \chi(1)\log x(\frac{1}{|G|}\log d_L + n_K x^{\frac{1}{2}})$$

where $\rho = \beta + i\gamma$ runs over the zeros of $L(s, \chi)$ with $0 < \beta < 1$, $|\gamma| < T$. (For the details, see [5, Proposition 3.3] or [4], §7.)

Set $N(t, \chi)$ to be the number of ρ with $|\gamma - t| \leq 1$. In [5, Lemma 4] and [4, Lemma 5.4], it is shown that under the assumption of Artin's conjecture,

$$N(t, \chi) \ll \log A_\chi + n_K \chi(1) \log(|t| + 2).$$

It follows that

$$\sum_{|\gamma| < T} \frac{1}{\rho} \ll \sum_{j < T} \frac{N(j, \chi)}{j} \ll (\log A_\chi)(\log T) + \chi(1) n_K (\log T)^2.$$

By the zero-free region of the previous section (Proposition 3.8), we have

$$\psi(x, \chi) - \delta(\chi)x + \epsilon(\chi)\frac{x^{\beta_1}}{\beta_1} \ll x^{\sigma_1}(\log A_\chi + n_K \chi(1) \log T) \log T$$
$$+ \frac{x \log x}{T}(\log A_\chi + n_K \chi(1) \log T) + \chi(1)(\log x)\left(\frac{1}{|G|} \log d_L + n_K x^{\frac{1}{2}}\right)$$

where $\sigma_1 = 1 - c/d^3(\log \mathcal{A} + n_K \log T)$, d is the maximum of the character degrees, β_1 is the possible exceptional zero of Proposition 3.8, and

$$\epsilon(\chi) = \begin{array}{ll} 1 & \text{if } \chi = \chi_1 \\ 0 & \text{otherwise.} \end{array}$$

Recall that we have set

$$\log \mathcal{M} = \log \mathcal{M}(L/K) = \frac{1}{n_K} \log d_K + 2\left\{ \sum_{p | d_{L/K}} \log p + \log(n_L/n_K) \right\}$$

and that from [8], Proposition 2.5, we have the estimate

$$\log A_\chi \leq \chi(1) n_K \log \mathcal{M}.$$

Using this, we have

$$\sum_\chi \left| \psi(x, \chi) - \delta(\chi)x + \epsilon(\chi)\frac{x^{\beta_1}}{\beta_1} \right|^2 \ll x^{2\sigma_1} n_K^2 (\log T\mathcal{M})^4 |G|$$
$$+ \frac{x^2(\log x)^2}{T^2} n_K^2 |G|(\log T\mathcal{M})^2 + n_K^2 |G|(\log x)^2 (x^{\frac{1}{2}} + \log \mathcal{M})^2.$$

Then we choose

$$\log T = \left\{ \overline{d^2(\log \mathcal{A})^2 + 4c_2 \frac{n_K}{d} \log x} - (d \log \mathcal{A}) \right\} / 2d n_K.$$

(Recall that d is the maximum of the character degrees and $\log \mathcal{A}$ is the maximum of $\log A_\chi$.) The condition $T \geq 2$ is satisfied provided

$$d^4 n_K \log \mathcal{M} \ll \log x.$$

With this choice of T, we have

$$\sum_\chi \left| \psi(x, \chi) - \delta(\chi)x + \epsilon(\chi)\frac{x^{\beta_1}}{\beta_1} \right|^2 \ll x^{2\sigma_1} n_K^2 (\log x \mathcal{M})^4 |G|.$$

Then we have

$$\psi_C(x) - \frac{|C|}{|G|}x + \frac{|C|}{|G|}\chi_1(C)\frac{x^{\beta_1}}{\beta_1}$$

$$= \frac{|C|}{|G|} \sum_\chi \bar{\chi}(C) \left(\psi(x, \chi) - \delta(\chi)x + \epsilon(\chi)\frac{x^{\beta_1}}{\beta_1} \right).$$

Now applying the Cauchy–Schwarz inequality, we deduce that the right hand side is

$$\leq \frac{|C|}{|G|} \left(\sum |\bar{\chi}(C)|^2 \right)^{\frac{1}{2}} \left(\sum_\chi \left| \psi(x, \chi) - \delta(\chi)x + \epsilon(\chi)\frac{x^{\beta_1}}{\beta_1} \right|^2 \right)^{\frac{1}{2}}.$$

Using the above estimates, we see that this is

$$\ll \frac{|C|}{|G|} \left(\frac{|G|}{|C|} \right)^{\frac{1}{2}} x^{\sigma_1} n_K (\log x \mathcal{M})^2 |G|^{\frac{1}{2}}$$

$$\ll |C|^{\frac{1}{2}} n_K x \exp\left(-c \log x / \left(d^{3/2} \ \overline{d^3(\log \mathcal{A})^2 + n_K \log x} \right) \right) (\log x \mathcal{M})^2.$$

Now using partial summation, we deduce the statement of the Theorem.

We shall now discuss some variants of Theorem 4.1. In the following, if H is a subgroup of $G = \mathrm{Gal}(L/K)$, we shall write β_H and χ_H for the corresponding exceptional zero and exceptional character (respectively). Moreover, we define T_H by

$$2 d_H n_K |G| \log T_H = |H| \left\{ \overline{d_H^2 (\log \mathcal{A}_H)^2 + 4c_2 \frac{n_K |G|}{d_H |H|} \log x} - d_H (\log \mathcal{A}_H) \right\}$$

where d_H is the maximum of the character degrees of H and

$$\log \mathcal{A}_H = \max_{\phi \in \mathrm{Irr}(H)} \log A_\phi.$$

Now set

$$\sigma_H = 1 - c \left\{ d_H^3 (\log \mathcal{A}_H + \frac{|G|}{|H|} n_K \log T_H) \right\}^{-1}.$$

If C is a conjugacy class of G we denote by $C_H = C_H(h)$ the conjugacy class in H of any element $h \in H \cap C$. Also, if D is a union of conjugacy classes of G, we set

$$\chi_H(D) = \sum_{C \subseteq D} |C| \chi_H(C_H)$$

where the sum is over the conjugacy classes C contained in D. As a group-theoretic quantity, $\chi_H(D)$ depends on the choice of the elements h. However, the existence of β_H implies that $\chi_H(D)$ is well-defined.

Theorem 4.4 *Let D be a union of conjugacy classes in G, and H a subgroup of G satisfying*

(i) Artin's conjecture is true for the irreducible characters of H

(ii) H intersects nontrivially every class in D.

Suppose

$$\log x \gg d_H^4 n_K \frac{|G|}{|H|} \log \mathcal{M}.$$

Then

$$\pi_D(x) = \frac{|D|}{|G|} \text{Li } x - \frac{\chi_H(D)}{|G|} \text{Li } (x^{\beta_H})$$

$$+ O\left(x^{\sigma_H} \left(\sum_{C \subseteq D} \frac{|C|^2}{|C_H|} \right)^{\frac{1}{2}} n_K (\log x \mathcal{M}_H)^2 \right).$$

Proof As in Proposition 3.9 of [8], we find that

$$\pi_D(x) = \frac{|D|}{|G|} \text{Li } x - \frac{\chi_H(D)}{|G|} \text{Li } (x^{\beta_H}) + E_1 + E_2$$

where

$$E_1 \ll \left(\max \frac{|C|}{|C_H|} \right) (n_K x^{\frac{1}{2}} + n_K \log \mathcal{M} x)$$

and

$$E_2 \ll \frac{|H|}{|G|} \sum_{C \subseteq D} \frac{|C|}{|C_H|^{\frac{1}{2}}} \frac{1}{|C_H|^{\frac{1}{2}}}$$

$$\times \left| \pi_{C_H}(x) - \frac{|C_H|}{|H|} \text{Li } x + \frac{|C_H|}{|H|} \chi_H(C_H) \text{Li } x^{\beta_H} \right|.$$

Now applying the Cauchy–Schwarz inequality, we see that

$$E_2 \ll \frac{|H|}{|G|} \left(\sum_{C \subseteq D} \frac{|C|^2}{|C_H|} \right)^{\frac{1}{2}}$$

$$\times \left(\sum_{C \subseteq D} \frac{1}{|C_H|} |\pi_{C_H}(x) - \frac{|C_H|}{|H|} \mathrm{Li}\, x + \frac{|C_H|}{|H|} \chi_H(C_H) \mathrm{Li}\, x^{\beta_H} |^2 \right)^{\frac{1}{2}} .$$

Using (a small variant of) the previous result, we deduce that

$$E_2 \ll \frac{|H|}{|G|} \left(\sum_{C \subseteq D} \frac{|C|^2}{|C_H|} \right)^{\frac{1}{2}} \left(x^{\sigma_H} \frac{|G|}{|H|} n_K (\log x \mathcal{M}_H)^2 \right)$$

$$= \left(\sum_{C \subseteq D} \frac{|C|^2}{|C_H|} \right)^{\frac{1}{2}} x^{\sigma_H} n_K (\log x \mathcal{M}_H)^2 .$$

Corollary 4.5 *Under the same hypotheses as above*

$$\pi_D(x) = \frac{|D|}{|G|} \mathrm{Li}\, x - \frac{\chi_H(D)}{|G|} \mathrm{Li}\, (x^{\beta_H})$$

$$+ O \left(|D|^{\frac{1}{2}} \left(\frac{|G|}{|H|} \right)^{\frac{1}{2}} n_K x^{\sigma_H} (\log x \mathcal{M}_H)^2 \right) .$$

This follows from the observation that

$$\sum_{C \subseteq D} \frac{|C|^2}{|C_H|} \le \frac{|G|}{|H|} \sum_{C \subseteq D} |C| = \frac{|G|}{|H|} |D|.$$

Theorem 4.6 *Let D be a nonempty union of conjugacy classes in G and let H be a normal subgroup of G such that Artin's conjecture holds for the characters of G/H and $HD \subseteq D$. Let $\bar{D}$ be the image of D in G/H. If*

$$d_{G/H}^5 n_K (\log \mathcal{M})^2 \ll \log x$$

then

$$\pi_D(x) = \frac{|D|}{|G|} \mathrm{Li}\, x - \frac{|H|}{|G|} \chi_{G/H}(\bar{D}) \mathrm{Li}\, x^{\beta_{G/H}}$$

$$+ O \left(\frac{|D|}{|H|}^{\frac{1}{2}} n_K x \exp \left(-c'(\log x)^{1/2} / (d_{G/H}^{3/2} n_K^{1/2}) \right) (\log x \mathcal{M})^2 \right) .$$

Here,

$$d_{G/H} = \max_{\psi \in \mathrm{Irr}(G/H)} \psi(1).$$

Proof   We note that $\bar{D}$ is a union of conjugacy classes in G/H and

$$\pi_{\bar{D}}(x) = \frac{|\bar{D}||H|}{|G|}\mathrm{Li}\,x - \frac{|H|}{|G|}\chi_{G/H}(\bar{D})\mathrm{Li}\,x^{\beta_{G/H}}$$
$$+ O(|\bar{D}|^{\frac{1}{2}}n_K x(\log x \mathcal{M}_{G/H})^2$$
$$\times \exp\left(-c'(\log x)/\left(d_{G/H}^{3/2}(d_{G/H}^3(\log\mathcal{A}_{G/H})^2 + n_K\log x)^{1/2}\right)\right)$$

provided

$$\log x \gg d_{G/H}^4 n_K \log \mathcal{M}.$$

Now we observe that

$$\pi_D(x) = \pi_{\bar{D}}(x) + O\left(\frac{1}{n}\log d_L\right)$$

and that

$$\log\mathcal{A}_{G/H} = \max_{\phi \in \mathrm{Irr}(G/H)}\log A_\phi \leq d_{G/H}n_K\log\mathcal{M}_{G/H}$$

and

$$d_{G/H} \leq \left(\frac{|G|}{|H|}\right)^{\frac{1}{2}}.$$

From these estimates, it follows that

$$d_{G/H}^3(\log\mathcal{A}_{G/H})^2 + n_K\log x \ll n_K\{d_{G/H}^5 n_K(\log\mathcal{M}_G)^2 + \log x\}.$$

The Theorem follows when we use the lower bound on $\log x$ given in the statement.

5. Applications

We apply this estimate to the Fourier coefficients of modular forms. Fix an f as in the introduction. Our aim is to prove the following estimate.

Theorem 5.1   *We have*

$$\pi_f(x,a) \ll \frac{x(\log\log x)^2}{(\log x)^2}.$$

Remark 5.2 An even sharper form of the above is valid in the case that f has complex multiplication (CM). Recall that f is said to have CM if there is a Dirichlet character χ such that $f = f \otimes \chi$. Equivalently, f arises from a Hecke character of an imaginary quadratic field. In this case, one can in fact prove (using the estimate given in Example 4.3) that for *any* $A > 0$,

$$\pi_f(x, a) \ll \frac{x}{(\log x)^A}.$$

Thus in the CM case, the situation is very analogous to the classical Siegel-Walfisz theorem.

For the relation between $\pi_f(x, a)$ and the Chebotarev density theorem, we refer the reader to [8], Introduction and §4 or to [11]. We find that for each prime ℓ (sufficiently large), there is an extension $L_\ell/\mathbb{Q}$ unramified outside ℓN with group

$$G = \{g \in \mathrm{GL}_2(\mathbb{F}_\ell) : \det g \in (\mathbb{F}_\ell^\times)^{k-1}\},$$

and a conjugacy set

$$C = C_a = \{g \in G : tr\, g = a\}$$

so that

$$\pi_f(x, a) \leq \pi_C(x).$$

As in [8], §4, let $y > 0$ be arbitrary and $c_1 > 0$ sufficiently small. Let u satisfy

$$y \geq u > y \exp(-c_1 (\log y)^{\frac{1}{2}})$$

and set $I = [y, y + u]$. For a prime p, set

$$\omega_p = \overline{a_p^2 - 4p}.$$

Set

$$\pi_f(x, a; \ell) = \#\{p \leq x : a_p = a \text{ and } \ell \text{ splits in } \mathbb{Q}(\omega_p)\}.$$

Then

$$\sum_{\ell \in I} \pi_f(x, a; \ell) = \sum_{\substack{p \leq x \\ a_p = a}} \pi_p(I)$$

where

$$\pi_p(I) = \#\{\ell \in I : \ell \text{ splits in } \mathbb{Q}(\omega_p)\}.$$

Let us set

$$\pi_p(x) = \#\{\ell \leq x : \ell \text{ splits in } \mathbb{Q}(\omega_p)\}.$$

We have

$$\sum_{\substack{p \le x \\ a_p = a}} \pi_p(I) \; = \; \frac{1}{2}\pi(I)\pi_f(x,a) \; + \; \sum_{\substack{p \le x \\ a_p = a}} \left(\pi_p(I) - \frac{1}{2}\pi(I)\right)$$

where $\pi(I)$ is the number of primes in the interval I. Now

$$\sum_{\substack{p \le x \\ a_p = a}} \left(\pi_p(I) - \frac{1}{2}\pi(I)\right) \; \ll \; \pi_f(x,a)^{\frac{1}{2}} \big(\sum_{\substack{p \le x \\ a_p = a}} |\pi_p(I) - \frac{1}{2}\pi(I)|^2\big)^{\frac{1}{2}}.$$

We have

$$\sum_{\substack{p \le x \\ a_p = a}} |\pi_p(I) - \frac{1}{2}\pi(I)|^2 \; \le \; \sum_{n \le x} \left| \frac{1}{2}\sum_{\ell \in I} \left(\frac{a^2 - 4n}{\ell}\right) \right|^2.$$

Expanding the inner sum, we see that the right side is equal to

$$\frac{1}{4}\pi(I)x \; + \; \frac{1}{4}\sum_{\substack{\ell_1 \ne \ell_2 \\ \ell_1, \ell_2 \in I}} \sum_{n \le x} \left(\frac{a^2 - 4n}{\ell_1 \ell_2}\right).$$

Using the Polya-Vinogradov estimate on the second sum, we see that this is

$$\frac{1}{4}\pi(I)x \; + \; O(y^3 \log y).$$

Hence,

$$\sum_{\substack{p \le x \\ a_p = a}} \left(\pi_p(I) - \frac{1}{2}\pi(I)\right) \; \ll \; \pi_f(x,a)^{\frac{1}{2}}\pi(I)^{\frac{1}{2}}x^{\frac{1}{2}}.$$

Putting all this together, it follows that

$$\pi_f(x,a) \; \ll \; \max_{\ell \in I} \pi_f(x,a;\ell) + O\left(\frac{\pi_f(x,a)^{\frac{1}{2}}x^{\frac{1}{2}}}{\pi(I)^{\frac{1}{2}}}\right).$$

¿From this it follows easily that if

$$\pi_f(x,a;\ell) \; \ll \; \frac{x(\log\log x)^2}{(\log x)^2}$$

then the same estimate holds for $\pi_f(x,a)$.

Taking an $\ell \in I$, we estimate $\pi_f(x,a;\ell)$ as follows. Let B denote the intersection of the upper triangular matrices in $\mathrm{GL}_2(\mathsf{F}_\ell)$ with G, and let K_ℓ denote the subfield of L_ℓ fixed by B. Then

$$\ell \ll [K_\ell : \mathbb{Q}] \ll \ell.$$

By choice of ℓ, every conjugacy class in C intersects B. Choose a maximal set Γ of elements in B of trace a which are not conjugate in G. Then

$$C = \cup C_G(\gamma)$$

where the union runs over elements $\gamma \in \Gamma$ and $C_G(\gamma)$ denotes the conjugacy class of γ in G. Set

$$C_B = \cup C_B(\gamma).$$

Then

$$\pi_C(x, L_\ell/\mathbb{Q}) = \pi_{C_B}(x, L_\ell/K_\ell) + O(\log \ell N + x^{\frac{1}{2}}).$$

Let A be the unipotent elements of B. Then $AC_B \subseteq C_B$. Let M_ℓ denote the subfield of L_ℓ fixed by A. Then M_ℓ/K_ℓ is Abelian, and in particular, Artin's conjecture holds. Applying Theorem 4.6, we deduce that if

$$\log x \geq c_4 \ell (\log \ell N)^2$$

then

$$\pi_{C_B}(x, L_\ell/K_\ell) \ll \frac{1}{\ell}\mathrm{Li}\, x \;+\; O\left(\ell^{3/2} x \exp\left(-c' \; \overline{\frac{\log x}{\ell}} \right) (\log x \ell N)^2 \right).$$

In the above, we used the trivial bound $\beta < 1$. Now choose

$$y = c_5 (\log x)/(\log \log x)^2$$

for c_4 sufficiently large and a sufficiently small absolute constant $c_5 > 0$. Then

$$\pi_f(x, a) \;\ll\; \frac{x(\log \log x)^2}{(\log x)^2} + O\left(\frac{x(\log x)^{3/2}}{(\log \log x)^3}(\log x N)^2 \exp(-c_6 \log \log x) \right)$$

and this is seen to be

$$\pi_f(x, a) \;\ll\; \frac{x(\log \log x)^2}{(\log x)^2}$$

by taking c_5 sufficiently small. This proves the main result.

By summing over a, we can deduce a lower bound for the a_p valid on a set of primes of density 1.

Proposition 5.3 *Let g be a monotone increasing function satisfying $g(x) \to \infty$ as $x \to \infty$. Then*

$$|a_p| \geq (\log p)/(\log\log p)^2 g(p)$$

for a set of primes p of density 1.

Proof We see that

$$\# \left\{ p \leq x : |a_p| < \frac{\log p}{(\log\log p)^2 g(p)} \right\}$$

is majorized by

$$\sum_{|a| < \log x/((\log\log x)^2 g(x))} \pi_f(x, a)$$

and by Theorem 5.1, this is

$$\ll \frac{\log x}{(\log\log x)^2 g(x)} \frac{x(\log\log x)^2}{(\log x)^2} = \frac{x}{g(x)\log x} = o(\pi(x))$$

and this proves the result.

We now combine this with an elementary sieve argument to deduce a bound on the a_n. A result of this sort was stated in [8], Theorem 5.4, with an unspecified constant in the exponent. The result here makes this constant and the bound explicit.

Theorem 5.4 *The set*

$$\left\{ n : a_n = 0 \text{ or } |a_n| > \frac{\log n}{(\log\log n)^{3+\epsilon}} \right\}$$

has density 1.

Proof Let $\mathcal{D}$ denote the set of all prime powers p^m satisfying

$$|a_{p^m}| \geq \frac{\log p^m}{(\log\log p^m)^{3+\epsilon}}.$$

By Proposition 5.3, the set of primes p for which this bound holds has density 1. For $m = 2, 4$, the same result can be proved by considering the symmetric power representations. For other values of m an even stronger bound holds from [9]. Given n let us factor it as $n = n_1 n_2$ where n_2 is divisible only by prime powers p^m in $\mathcal{D}$ and n_1 is divisible only by prime powers which are not in $\mathcal{D}$. Then if $a_n \neq 0$,

$$|a_n| \geq |a_{n_2}| \geq \prod_{\substack{p^m \| n_2 \\ p^m \in \mathcal{D}}} \frac{\log p^m}{(\log \log p^m)^{3+\epsilon}}$$

and this is seen to imply

$$|a_n| \geq \frac{\log n_2}{(\log \log n_2)^{3+\epsilon}}.$$

Now the number of primes $\leq x$ which are in the complement of $\mathcal{D}$ satisfy the estimate

$$\ll \frac{x}{(\log x)(\log \log x)^{1+\epsilon}}.$$

Hence, in the notation introduced above, it follows that

$$\sum \frac{1}{n_1}$$

is convergent. (It is to ensure the convergence of this series that the lower bound for a_n is weaker than that for a_p given in Proposition 5.3 by a factor of $\log \log n$.) We deduce from this that

$$\{n : n_2 > n/\log n\}$$

is a set of density 1. Indeed, for any $y \leq x$, we have

$$\#\{n \leq x : n_2 \leq n/\log n\} \leq y + \sum_{\substack{y < n \leq x \\ n_2 \leq n/(\log n)}} 1 \leq y + \sum_{\substack{y < n \leq x \\ n_1 > \log y}} x/n_1.$$

Now, if we choose, say $y \to \infty$ and $y = o(x)$ we see that the above quantity is $o(x)$. Hence, for n which are not in this sequence of density zero, we have

$$|a_n| \geq \frac{\log n}{(\log \log n)^{3+\epsilon}}.$$

This completes the proof.

References

[1] H. Davenport. *Multiplicative Number Theory.* Springer-Verlag, New York, 2nd ed., 1980.

[2] H. Heilbronn. On real zeros of Dedekind zeta functions. *Canad. J. Math.*, 4 (1973), 870–873. (See also paper 45 in:*The Collected Papers of H.A. Heilbronn* (E. Kani and R.A. Smith, editors), Wiley–Interscience, Toronto, 1988.)

[3] J. Hoffstein and D. Ramakrishnan. Siegel zeros and cusp forms. *Intl. Math. Res. Not.*, 6 (1995), 279–308.

[4] J. Lagarias and A. Odlyzko. Effective versions of the Chebotarev density theorem. *Algebraic Number Fields*(A. Fröhlich, editor), Academic Press, New York, 1977, pp. 409–464,

[5] V. Kumar Murty. Explicit formulae and the Lang–Trotter conjecture. *Rocky Mountain J. Math.*, 15 (1985), 535–551.

[6] M. Ram Murty and V. Kumar Murty. A variant of the Bombieri–Vinogradov theorem. *Number Theory* (H. Kisilevsky and J. Labute, editors), Amer. Math. Soc., Providence, 1987, pp. 243–272.

[7] —. *Non-vanishing of L-functions and applications*, Birkhauser, 1997.

[8] M. Ram Murty, V. Kumar Murty and N. Saradha. Modular forms and the Chebotarev density theorem. *Amer. J. Math.*, 110 (1988), 253–281.

[9] M. Ram Murty, V. Kumar Murty and T. N. Shorey. Odd values of the Ramanujan τ-function. *Bull. Soc. Math. France*, 115 (1987), 391–395.

[10] A. Odlyzko. On conductors and discriminants. *Algebraic Number Fields* (A. Fröhlich, editor), Academic Press, New York, 1977, pp. 377–407.

[11] J.-P. Serre. Quelques applications du théorème de densité de Chebotarev. *Publ. Math. IHES*, 54 (1981), 123–201.

[12] H. Stark. Some effective cases of the Brauer–Siegel theorem. *Invent. Math.*, 23 (1974), 135–152.

[13] Daqing Wan. On the Lang–Trotter conjecture. *J. Number Th.*, 35 (1990), 247–268.

V. Kumar Murty
Department of Mathematics, University of Toronto
Toronto, Ontario M5S 1A1, Canada

18

Congruences between Modular Forms

M. RAM MURTY*

1. Introduction Recently, Goldfeld and Hoffstein [4] have shown, using the theory of L-functions that if f and g are two holomorphic Hecke newforms of weight k and *squarefree* levels N_1, N_2 respectively, then there is an $n = O(N \log N)$, with $N = \mathrm{lcm}(N_1, N_2)$ so that

$$a_f(n) \neq a_g(n).$$

(Here, $a_f(n)$ denotes the n-th Fourier coefficient in the Fourier expansion of f at $i\infty$.) In other words, there is a constant c so that the first $cN \log N$ Fourier coefficients determine the newform. They obtain an analogous result if the weights are distinct. Assuming the generalized Riemann hypothesis for the Rankin-Selberg L-functions attached to these eigenforms, they deduce that the bound above can be improved to $O((\log N)^2 (\log \log N)^4)$.

We will show that these results can be established without the use of L-functions. Our approach leads to sharper results and is applicable in the wider context of two arbitrary cusp forms of any weight and level. In fact, we will prove a more general and sharper:

Theorem 1 *Let f and g be two distinct holomorphic modular forms of weight k and levels N_1 and N_2 respectively. Let $N = \mathrm{lcm}(N_1, N_2)$. Then, for some*

$$n \leq \frac{k}{12} N \prod_{p \mid N} \left(1 + \frac{1}{p} \right)$$

we must have $a_f(n) \neq a_g(n)$. (Here, the product is over primes p dividing N.)

Remark Note that we do not assume that f and g are Hecke eigenforms nor that they are of squarefree levels as in the Goldfeld–Hoffstein [4] paper. Moreover, let us observe that if $\nu(N)$ denotes the number of prime factors of N, and p_i denotes the i-th prime, then

$$\prod_{p \mid N} \left(1 + \frac{1}{p} \right) \leq \prod_{1 \leq i \leq \nu(N)} \left(1 + \frac{1}{p_i} \right) \leq \prod_{1 \leq i \leq \nu(N)} \left(1 - \frac{1}{p_i} \right)^{-1} \ll \log \log N$$

* Research partially supported by NSERC, FCAR and CICMA.

which is a sharper bound than the one given in [4, p. 387].

Proof Let $\phi = f - g$, and suppose $a_f(n) = a_g(n)$ for $n \leq M$. Then, ϕ has a zero of order $\geq M$ at $i\infty$. If Δ denotes Ramanujan's cusp form, then ϕ^{12}/Δ^k is of weight zero and hence a meromorphic function on the compact Riemann surface $X_0(N)$. Since Δ does not vanish on the upper half-plane, and has a simple zero at $i\infty$, the number of zeroes of ϕ, which is at least $12M - k$, is equal to the number of poles, which cannot exceed k times the index of $\Gamma_0(N)$ in $\Gamma(1)$ minus one (to account for $i\infty$ which already contributed to the zero count). The index of $\Gamma_0(N)$ in $\Gamma(1)$ is equal to

$$N \prod_{p \mid N} \left(1 + \frac{1}{p} \right).$$

The result is now immediate.

We can make a few remarks. The first is that the method can be applied to any discrete subgroup contained in $\Gamma(1)$ to get an analogous result. One can also adapt it to deal with f and g of different weights.

The proof of Theorem 1 can be modified to handle different weights. Indeed, if

$$a_n(f) = a_n(g)$$

for all $n \leq M$, then

$$\phi = (f^{k_2} - g^{k_1})^{12}/\Delta^{k_1 k_2}$$

is a function on $X_0(N)$. The order of the zero at $i\infty$ is $\geq 12M - k_1 k_2$. The number of poles, on the other hand is $\leq k_1 k_2([\Gamma(1):\Gamma_0(N)]-1)$ by an analogous argument as before.

Let us introduce the following notation. Suppose

$$f(z) = \sum_{n=1}^{\infty} a_n(f) e^{2\pi i n z}$$

is the Fourier expansion of f at $i\infty$. Then, define

$$\mathrm{ord}_\infty(f) = \min\{n : a_n(f) \neq 0\}.$$

We have therefore proved:

Theorem 2 *If f and g are holomorphic modular forms such that*

$$\mathrm{ord}_\infty(f - g) > \frac{k_1 k_2}{12}([\Gamma(1):\Gamma_0(N)])$$

then $f = g$.

The result which we state below will be proved by a different method in Section 5 as an application of the Riemann-Roch theorem and is a small improvement of Theorem 2.

Theorem 3 *Suppose that f and g are two holomorphic cusp forms of weights k_1 and k_2 and levels N_1 and N_2 respectively. If*

$$\text{ord}_\infty(f - g) > k_1 k_2(\mu - 1),$$

where μ is the genus of $X_0(N)$ and $N = \text{lcm}(N_1, N_2)$, then $f = g$.

Let us note that the genus μ satisfies the inequality

$$\mu \le 1 + \frac{N}{12} \prod_{p|N} \left(1 + \frac{1}{p}\right)$$

and thus, the bound of Theorem 3 is comparable to the one of Theorem 1.

In case that f and g are normalized newforms of distinct weights k_1 and k_2 and levels N_1 and N_2 respectively, we can in fact do better and derive an estimate superior to the *conditional* estimate of Goldfeld and Hoffstein [4, p. 386].

Theorem 4 *Let f and g be two holomorphic Hecke newforms of distinct weights k_1 and k_2 on $\Gamma_0(N_1)$ and $\Gamma_0(N_2)$ respectively. Then, there is an $n < 4(\log N)^2$ with $N = \text{lcm}(N_1, N_2)$ so that*

$$a_n(f) \ne a_n(g).$$

Proof We can view f and g as cusp forms on $\Gamma_0(N)$. Let us first note that there is a $p < 2 \log N$ which is coprime to N for otherwise N would be divisible by all the primes $< 2 \log N$ and hence by their product. By a classical estimate of Chebycheff, this product is

$$\prod_{p < 2\log N} p = \exp\left(\sum_{p < 2\log N} \log p\right) > N,$$

which is a contradiction. Thus, fixing such a prime p, and observing that

$$a_{p^2}(f) = a_p^2(f) - p^{k_1 - 1}$$

and

$$a_{p^2}(g) = a_p^2(g) - p^{k_2 - 1}$$

we deduce that $k_1 = k_2$. Noting that $p^2 < 4 \log^2 N$, we have a contradiction.

If we view Theorems 1, 2 and 3 as statements at the "infinite prime" then it is natural to ask for analogous results at the "finite primes". That is, if we have a congruence between coefficients of modular forms (mod p) up to a certain natural number, can we conclude that the coefficients are always congruent? In the case the two forms have equal weight, such a result was first established by Sturm [12]. Sturm's proof is sketchy in some places and therefore, for the sake of clarity of exposition and emphasis with the analogy above, we give the complete proof in Sections 2, 3 and 4. Sturm's argument however cannot be easily modified to handle different weights. In Section 5, we therefore take a different approach through the Riemann-Roch theorem. This has the merit of being conceptually simple and at the same time working (mod p) (for p not dividing $N = \mathrm{lcm}(N_1, N_2)$, however) thanks to the algebro-geometric generalization of the Riemann-Roch theorem. Recently, K. Ono [7] applied the theorem of Sturm in investigating the parity of the partition function.

2. Preliminaries In our paper [8], we indicated how the celebrated ABC conjecture leads naturally to the problem of congruences between modular forms. We will not discuss this connection here, but refer the reader to the forthcoming paper [8] for a detailed derivation. The purpose of this paper is to determine the (finite) amount of calculation necessary in order to establish a congruence between two modular forms.

More precisely, let us fix an algebraic number field F with ring of integers $\mathcal{O}_F$. Fix a prime ideal $\mathfrak{p}$ of $\mathcal{O}_F$ and for a formal power series

$$s = \sum_{n=0}^{\infty} c_s(n) q^k \ , \ c_s(n) \in \mathcal{O}_F,$$

define

$$\mathrm{ord}_{\mathfrak{p}}(s) = \min\{n \ : \ \mathfrak{p} \nmid c_s(n)\}$$

with the convention that $\mathrm{ord}_{\mathfrak{p}}(s) = \infty$ if $\mathfrak{p}|c_s(n)$ for all n. Recall that $k[[q]]$ with $k = \mathcal{O}_F/\mathfrak{p}$ is a discrete valuation ring. In particular, this implies the following. Notice that if

$$f = \sum_{n \geq 0} c_f(n) q^n$$

$$g = \sum_{n \geq 0} c_g(n) q^n$$

then

$$c_{fg}(n) = \sum_{i+j=n} c_f(i) c_g(j),$$

so that, if $\text{ord}_\mathfrak{p}(f) = n_f < \infty$ and $\text{ord}_\mathfrak{p}(g) = n_g < \infty$, then from the above formula, we see

$$c_{fg}(n_f + n_g) \equiv c_f(n_f)c_g(n_g) \bmod \mathfrak{p}$$
$$\not\equiv 0 \bmod \mathfrak{p}.$$

We also note that $c_{fg}(n) \equiv 0 \bmod \mathfrak{p}$ if $n < n_f + n_g$. Hence,

$$\text{ord}_\mathfrak{p}(fg) = \text{ord}_\mathfrak{p}(f) + \text{ord}_\mathfrak{p}(g)$$

when each of the terms on the right hand side is finite. By the convention made above, the equality also holds if either one of $\text{ord}_\mathfrak{p}(f)$ or $\text{ord}_\mathfrak{p}(g)$ is infinity.

For each positive integer N, let

$$\Gamma(N) = \left\{ \begin{pmatrix} a & b \\ c & d \end{pmatrix} \in SL_2(\mathbb{Z}) \; : \; \begin{pmatrix} a & b \\ c & d \end{pmatrix} \equiv \begin{pmatrix} 1 & 0 \\ 0 & 1 \end{pmatrix} \bmod N \right\}$$

and fix Γ, a subgroup of $\Gamma(1)$ containing $\Gamma(N)$. As usual, $\mathfrak{h}$ will denote the upper half-plane, k will be a positive integer and we will consider functions

$$f : \mathfrak{h} \to \mathbb{C}$$

satisfying certain conditions. $M_k(\Gamma)$ will denote the $\mathbb{C}$-vector space of modular forms of weight k for Γ. To be precise, let us define for each $\gamma \in GL_2^+(\mathbb{R})$, the function

$$(f|\gamma)(z) = f\left(\frac{az+b}{cz+d}\right)(cz+d)^{-k}(ad-bc)^{k/2}$$

where

$$\gamma = \begin{pmatrix} a & b \\ c & d \end{pmatrix}$$

is a matrix of $GL_2(\mathbb{R})$ of positive determinant. Then $M_k(\Gamma)$ consists of holomorphic functions of the extended upper half-plane:

$$f : \mathfrak{h}^* = \mathfrak{h} \cup \mathbb{Q} \cup \{i\infty\} \to \mathbb{C}$$

satisfying $f|\gamma = f$ for all $\gamma \in \Gamma$.

Since

$$\begin{pmatrix} 1 & N \\ 0 & 1 \end{pmatrix} \in \Gamma,$$

such an $f \in M_k(\Gamma)$ has a Fourier expansion of the following type:

$$f(z) = \sum_{\substack{n \geq 0 \\ n \in N^{-1}\mathbb{Z}}} a_f(n)e(nz)$$

where $e(x) = e^{2\pi i x}$. If R is a subring of $\mathbb{C}$, we denote by $M_k(\Gamma, R)$ those forms of $M_k(\Gamma)$ whose Fourier coefficients lie in R. That is, $a_f(n) \in R$ for all n.

Our purpose now is to explain in some detail a fundamental result of Sturm [12] regarding congruences between modular forms. To this end, let us fix as before an algebraic number field F and let $\mathcal{O}_F$ be the ring of integers of F. Let $\mathfrak{p}$ be a prime ideal of $\mathcal{O}_F$. We will explain the following:

Theorem 5 (Sturm) *Let $f, g \in M_k(\Gamma, \mathcal{O}_F)$. Suppose $\mathrm{ord}_\mathfrak{p}(f - g) > k[\Gamma(1) : \Gamma]/12$. Then $f \equiv g$ mod $\mathfrak{p}$ with Γ a congruence subgroup.*

Notice that the bound does not depend on $\mathfrak{p}$.

3. Sturm's theorem: the level one case   To prove the theorem, we first consider the level one case. That is, $N = 1$. Recall that Ramanujan's cusp form

$$\Delta(z) = \sum_{n=1}^{\infty} \tau(n)e(nz) \in M_{12}(\Gamma(1), \mathbb{Z})$$

and can be written in terms of the standard Eisenstein series:

$$\Delta(z) = \frac{1}{1728}(E_4^3 - E_6^2)$$

with

$$E_4(z) = 1 + 240 \sum_{n=1}^{\infty} \sigma_3(n)e(nz),$$

$$E_6(z) = 1 - 504 \sum_{n=1}^{\infty} \sigma_5(n)e(nz)$$

and

$$\sigma_k(n) = \sum_{\substack{d \mid n \\ d > 0}} d^k.$$

Also, the modular function $j(z)$ is

$$j(z) = E_4^3/\Delta.$$

We reproduce below a result that is well-known for $M_k(\Gamma(1))$. We adapt it to the case $M_k(\Gamma(1), \mathcal{O}_F)$.

Proposition 6   *Let $\Phi \in M_{12k}(\Gamma(1), \mathcal{O}_F)$, satisfying $\mathrm{ord}_\mathfrak{p}(\Phi) > k$. Then*

$$\Phi/\Delta^k \in \mathfrak{p}[j]$$

is a polynomial in j of degree at most k, all of whose coefficients are divisible by $\mathfrak{p}$.

Remark This fact is stated without proof in [12]. It is more or less evident from the fact that any modular function with only a pole at infinity must be a polynomial in j. However, the divisibility of the coefficients is not so clear. One can see this by comparing q-expansions. For the sake of completeness, we give a proof that assumes minimal background.

Proof We induct on k. For $k = 1$, Φ has weight 12 and so we can write it as an $\mathcal{O}_F$-linear combination of E_4^3 and Δ, as is easily checked. Dividing by Δ gives the result. Since $\mathrm{ord}_\mathfrak{p}\Phi/\Delta > 0$, we observe that writing

$$\Phi/\Delta = \sum_{n \geq -1} c(n)e(nz),$$

we have $c(n) \in \mathcal{O}_F$ and $\mathfrak{p}|c(n)$ if $n \leq 0$.

For general k, let us find i and j so that

$$12k = 4i + 6j.$$

Then for some $c \in \mathcal{O}_F$,

$$\Phi - cE_4^i E_6^j$$

is a cusp form of weight $12k$.

Thus, we can write

$$\Phi = cE_4^i E_6^j + \Delta f_1$$

with $f_1 \in M_{12(k-1)}(\Gamma(1), \mathcal{O}_F)$. Dividing by Δ^k yields

$$\Phi/\Delta^k = cE_4^i E_6^j/\Delta^k + f_1/\Delta^{k-1}.$$

By induction hypothesis,

$$f_1/\Delta^{k-1} \in \mathfrak{p}[j].$$

Noting that $4i + 6j = 12k$ implies that $i \equiv 0 \bmod 3$ and $j \equiv 0 \bmod 2$ we can write $i = 3i_0$, $j = 2j_0$ so that

$$E_4^i E_6^j/\Delta^k = (E_4^3/\Delta)^{i_0}(E_6^2/\Delta)^{j_0}$$

and $E_4^3/\Delta = j$, $E_6^2/\Delta = j - 1728$. This completes the proof of the proposition.

We can now prove the theorem in the level one case. Let $\phi = f - g$. Then, $\mathrm{ord}_\mathfrak{p}(\phi^{12}) > k$ implies

$$\phi^{12}/\Delta^k = \sum_{n \geq -k} c(n)e(nz)$$

with $c(n) \in \mathcal{O}_F$ and $\mathfrak{p}|c(n)$ if $n \leq 0$. By the proposition, $\phi^{12}/\Delta^k \in \mathfrak{p}[j]$ is a polynomial in j of degree at most k. Thus, $\phi^{12} \in \Delta^k\mathfrak{p}[j]$ implies $\mathrm{ord}_\mathfrak{p}\phi^{12} = \infty$ so that $\mathrm{ord}_\mathfrak{p}\phi = \infty$, as desired.

4. Sturm's theorem: the general level case   We begin by discussing some preliminaries.

Let Γ contain $\Gamma(N)$. We want to reduce the proof of the Theorem to the level one case by constructing a map:

$$T : M_{12k}(\Gamma(N), \mathcal{O}_F) \rightarrow M_{12k}(\Gamma(1), \mathcal{O}_F)$$

such that
$$\mathrm{ord}_\mathfrak{p}(T(\phi)) \geq \mathrm{ord}_\mathfrak{p}(\phi).$$

Theorem 3.52 of Shimura [10, p. 85] assures us that the space $S_k(\Gamma(N))$, the space of cups forms of weight k for $\Gamma(N)$ has a basis of cusp forms whose Fourier coefficients at $i\infty$ are rational integers, provided $k \geq 2$. This means that any element of $S_k(\Gamma(N), F)$ has the bounded denominator property. That is, given an element $f \in S_k(\Gamma(N), F)$, there is an element $A \in F$ so that $Af \in S_k(\Gamma(N), \mathcal{O}_F)$.

By the theory of Eisenstein series, we conclude that $M_k(\Gamma(N))$ has a basis whose Fourier coefficients are rational over $\mathbb{Q}(\zeta_N)$ where ζ_N denotes a primitive N-th root of unity. (See also Theorems 6.6 and 6.9 of Shimura [10, pp. 136-140]).

Now we can prove the theorem in the general case. As in [12, p. 276], let $\phi = f - g$. Our aim is to show that under the hypotheses of the Theorem, $\mathrm{ord}_\mathfrak{p}(\phi) = \infty$. Thus, replacing ϕ by ϕ^{12} if necessary, we may suppose $12|k$. Then, $\phi\Delta^{-k/12}$ is a modular function of level N. Since $\Gamma(N)$ is a normal subgroup of $\Gamma(1)$, we note that for any $\gamma \in \Gamma(1)$,

$$\phi|\gamma \in M_k(\Gamma(N), F(\zeta_N)).$$

By what we have said in the previous paragraph, $\phi|\gamma$ has bounded denominators. Now let K be the Hilbert class field of $F(\zeta_N)$. Then $\mathfrak{p}\mathcal{O}_K$ is a principal ideal in $\mathcal{O}_K$. Let $\wp$ be a prime ideal of $\mathcal{O}_K$ dividing $\mathfrak{p}\mathcal{O}_K$.

For every $\gamma \in \Gamma(1)$, we can clearly find $A(\gamma) \in K^*$ such that $\mathrm{ord}_\wp A(\gamma)(\phi|\gamma)$ is finite, simply by dividing by a suitable power of $\wp$. Moreover, by the Chinese remainder theorem, we can arrange

$$A(\gamma)(\phi|\gamma) \in M_k(\Gamma(N), \mathcal{O}_K)$$

Now consider the "norm function from $\Gamma(N)$ to $\Gamma(1)$", namely,

$$\Phi = \phi \prod_{i=2}^{m} A(\gamma_i)(\phi|\gamma_i)$$

where $1 = \gamma_1$, γ_2, ..., γ_m is a set of coset representatives of $\Gamma(N)$ in $\Gamma(1)$
Note that

$$\Phi \in M_{km}(\Gamma(1))$$

because if $\sigma \in \Gamma(1)$,

$$\Phi|\sigma = \left(\prod_{i=2}^{m} A(\gamma_i)\right) \prod_{i=1}^{m} (\phi|\gamma_i \sigma)$$

and $\gamma_1 \sigma$, $\gamma_2 \sigma$, ... $\gamma_m \sigma$ is again a set of coset representatives of $\Gamma(N)$ in $\Gamma(1)$
so that $\Phi|\sigma = \Phi$. Note also that

$$\mathrm{ord}_{\wp}(\Phi) \geq \mathrm{ord}_{\wp}\phi = \mathrm{ord}_{\mathfrak{p}}\phi > km/12.$$

By the level one cases we deduce that $\mathrm{ord}_{\wp}\Phi = \infty$. Thus,

$$\mathrm{ord}_{\wp}\phi + \sum_{i=2}^{m} \mathrm{ord}_{\wp}(A(\gamma_i)\phi|\gamma_i) = \infty$$

Since $\mathrm{ord}_{\wp}(A(\gamma_i)\phi|\gamma_i) < \infty$ for $i = 2, ..., m$, we conclude that $\mathrm{ord}_{\wp}\phi = \infty$.
Hence $\mathrm{ord}_{\mathfrak{p}}\phi = \infty$. This completes the proof.

5. An application of the Riemann-Roch theorem

Theorem 7 *Suppose that f and g are holomorphic cusp forms on $\Gamma_0(N)$,
of weight k and levels N_1, N_2 respectively. Let N be the lcm of N_1 and N_2.
Suppose that*

$$\mathrm{ord}_{\infty}(f - g) > \frac{k}{2}(2\mu - 1)$$

where μ is the genus of $X_0(N)$. Then $f = g$.

Proof If $f \neq g$, then k is even since there are no non-zero odd weight forms.
$\omega = (f - g)(dz)^{k/2}$ is a holomorphic differential $k/2$-form on $X_0(N)$. Its degree is $(k/2)(2\mu - 2)$. On the other hand, the hypothesis means that at $i\infty$,
$\mathrm{ord}_{i\infty}(\omega) \geq (k/2)(2\mu - 1) - (k/2)$ where the extra $k/2$ comes from $(dz)^{k/2}$.
Thus, (see for example, Shimura [10, Prop. 2.16, p. 39])

$$\frac{k}{2}(2\mu - 2) = \deg(\omega) \geq \mathrm{ord}_{i\infty}(\omega) \geq \frac{k}{2}(2\mu - 1) - \frac{k}{2},$$

which is a contradiction.

Proof of Theorem 3 It is now clear that Theorem 3 can be proved in an exactly similar manner. Indeed, as before, let us consider the differential $(f^{k_2} - g^{k_1})(dz)^{k_1 k_2/2}$ and proceed as in the previous proof.

Theorem 8 *Suppose f and g are cusp forms of even weight k and level N with coefficients lying in the ring of integers $\mathcal{O}_F$ of some algebraic number field F. Suppose that for some prime ideal $\mathfrak{p}$ coprime to the level N, we have*

$$\mathrm{ord}_\mathfrak{p}(f - g) > \frac{k}{2}(2\mu - 1),$$

then, $f \equiv g \bmod \mathfrak{p}$.

Proof We apply the Riemann-Roch theorem valid in any field of characteristic p which is coprime to N. This uses the non-trivial fact that $X_0(N)$ has good reduction for all $p \nmid N$ and is due to Igusa [5] (see also Deligne–Rapoport [3]). Again, cusp forms of weight 2 can be interpreted as differentials on $X_0(N)$ over $\mathbb{F}_p$. The same argument as before is valid for arbitrary even weight k. (See Silverman [11, p. 39] and [2, p. 96], for example.)

This approach has the advantage that it can generalize to two different weights k_1 and k_2.

Theorem 9 *Suppose that f and g are cusp forms of weights k_1 and k_2 and levels N_1 and N_2 respectively. Suppose further that at least one of k_1 or k_2 is even. As before, let us suppose the coefficients lie in the ring of integers $\mathcal{O}_F$ of some algebraic number field. If $\mathfrak{p}$ is a prime ideal of $\mathcal{O}_F$, and*

$$\mathrm{ord}_\mathfrak{p}(f - g) > k_1 k_2(\mu - 1)$$

then $f \equiv g \bmod \mathfrak{p}$.

We can derive better variations of Theorems 8 and 9 which are better for small primes, if we are willing to assume the generalized Riemann hypothesis for certain Dedekind zeta functions. In fact, if f and g are normalized Hecke eigenforms of level N, we know by Deligne [1] that there exists a Galois extension $K_f/\mathbb{Q}$ and a representation

$$\rho_{\mathfrak{p},f} : \mathrm{Gal}(K_f/\mathbb{Q}) \to GL_2(O_{K_f}/\mathfrak{p})$$

such that for each prime $v \nmid \mathfrak{p}N$, we have

$$tr(\rho_{\mathfrak{p},f}(\sigma_v(K_f/\mathbb{Q}))) \equiv a_v(f) \bmod \mathfrak{p}$$

where σ_v denotes the Artin symbol of v. An identical result holds for g. Thus, we may consider the compositum $K_f K_g$ and deduce by the Chebotarev density theorem (see [6] or [9]) the following:

Theorem 10 *Assume that the Dedekind zeta function of $K_f K_g/\mathbb{Q}$ satisfies the analogue of the Riemann hypothesis and f and g are normalized Hecke eigenforms as above with Fourier coefficients lying in a field F. If $\mathfrak{p}$ is a prime ideal of O_F coprime to N and*

$$\mathrm{ord}_\mathfrak{p}(f - g) > (\log(\mathrm{Norm}_{F/\mathbb{Q}}(\mathfrak{p})) + \log N)^4,$$

then $f \equiv g \bmod \mathfrak{p}$.

Proof By the Chebotarev density theorem and the Riemann hypothesis for the Dedekind zeta function of $K_f K_g/\mathbb{Q}$, we deduce that for any given conjugacy class C of $\mathrm{Gal}(K_f K_g/\mathbb{Q})$, there is a prime v with

$$\mathrm{Norm}(v) \leq (\log(\mathrm{Norm}_{F/\mathbb{Q}}(\mathfrak{p})) + \log N)^4$$

so that $\sigma_v(K_f K_g/\mathbb{Q}) \in C$. Thus, if

$$a_v(f) \equiv a_v(g) \bmod \mathfrak{p}$$

for each v whose norm satisfies the last inequality, then by Deligne's theoremand the effective Chebotarev density theorem as cited above, we can conclude thedesired result.

We can also remark that even if f and g are not Hecke eigenforms, a similar result can still be established. Note however, these bounds depend on $\mathfrak{p}$. Also worthy of contrast is that Theorem 5 is valid for all primes $\mathfrak{p}$ whereas Theorems 8 and 9 are applicable only when $\mathfrak{p}$ is coprime to N.

References

[1] P. Deligne. Formes modulaires et répresentations ℓ-adiques. *Séminaire Bourbaki, 1968–69, exposé 355*, Lect. Notes in Math., **179**, Springer-Verlag, Berlin, 1971, pp. 139–172.

[2] K. Doi and M. Ohta. On some congruences between cusp forms on $\Gamma_0(N)$. *Modular Functions of One Variable. V, Bonn, 1976* (J.-P. Serre and D. Zagier, Eds.), Lect. Notes in Math., **601**, Springer-Verlag, Berlin, 1977.

[3] P. Deligne and M. Rapoport. Les schémas de modules de courbes elliptiques. *Modular Functions of One Variable. II.* Lect. Notes in Math., **349**, Springer-Verlag, Berlin, 1973, pp. 143–316.

[4] D. Goldfeld and J. Hoffstein. On the number of terms that determine a modular form. *Contemp. Math.*, AMS, **143** (1993), 385–393.

[5] J. Igusa. Kroneckerian model of fields of elliptic modular functions. *Amer. J. Math.*, **81** (1959), 561–577.

[6] K. Murty, R. Murty and N. Saradha. Modular forms and the Chebotarev density theorem. *Amer. Journal of Math.*, **110** (1988), 253-281.

[7] K. Ono. Parity of the Partition Function in Arithmetic Progressions. *J. reine angew. Math.*, **472** (1996), 1–15.

[8] M.R. Murty. Bounds for congruence primes. Preprint.

[9] J.-P. Serre. Quelques applications du théorème de densité de Chebotarev. *Inst. Hautes Etudes Sci. Publ. Math.*, **54** (1982), 123–201.

[10] G. Shimura. *Introduction to the Arithmetic Theory of Automorphic Functions*. Publ. Math. Soc. Japan, **11**, Tokyo, 1971.

[11] J. Silverman. *The Arithmetic of Elliptic Curves*. Graduate Texts in Mathematics, **106**, Springer-Verlag, Berlin, 1986.

[12] J. Sturm. On the congruence of modular forms. *Number Theory, New York, 1984–85*, Lect. Notes in Math., **1240**, Springer-Verlag, 1987, Berlin, pp. 275–280.

M. Ram Murty
Department of Mathematics and Statistics, Queen's University, Jeffery Hall
Kingston, Ontario K7L 3N6, Canada

19

Regular Singularities in G-Function Theory

MAKOTO NAGATA

The aim of the present paper is to generalize the theory of G-functions in one variable to the two-variable situation. To indicate our motivation, let us consider a G-function which is a solution of the Fuchsian differential equation

$$\left(\frac{d}{ds} - A \right) X = 0 \tag{1}$$

where $A \in M_n(K(s))$ with K being an algebraic number field of finite degree over the rationals. A principal obstacle in investigating the behavior of G-functions in one variable lies in the fact that such a solution is not expressible as a power series at any regular singular points of the equation (1). In the previous investigations [1], [8], [9] the derived differential equation

$$\frac{d}{ds} Y = AY - Y \frac{1}{s} \operatorname{Res} A,$$

where $\operatorname{Res} A$ is the residue matrix of $A \in M_n(K((s)))$ and Y the uniform part of (1), was utilized in this context. This leads us, however, to a more complicated situation than was experienced in [5, §1].

In order to overcome such a methodical difficulty we introduce, instead, a partial differential equation in two variables which is equivalent to the equation (1):

$$(\partial - B)Z = 0, \tag{2}$$

where $\partial = a\partial_x + b\partial_y$ with a, $b \in K(x,y)$ and $B \in M_n(K((x,y)))$. This is because solutions of (2) can be expressed as power series at any points, as we shall show in Proposition 4. Moreover, we have a natural generalization of a basic assertion in the one-variable theory (cf. Corollary 4.2 in [8, Chap.VIII]):

Theorem 1 *Consider the partial differential equation* (2) *which is equivalent to the Fuchsian differential equation* (1). *If the size of the matrix solution of the partial differential equation* (2) *is finite, then the size of the Fuchsian differential equation* (1) *is also finite.*

A quantitative version of Theorem 1 is to be given in Corollary 1.

The introduction of the differential operator ∂ implies the possibility of a separation of variables; and thus a certain change of variables is to be performed.

This is not an easy task in general. However, our Theorem 3 provides us with an explicit method to attain our purpose via a Frobenius transformation and a substitution of logarithmic factors.

Acknowledgment The author would like to express his gratitude to Professor Yoichi Motohashi for his warm encouragement.

Conventions and definitions Let K be as above. For a place v of K we normalize the absolute value $|\cdot|_v$ as in [1]. We put $|M|_v := \max_{i,j=1,\dots,n} |m_{i,j}|_v$ for $M = (m_{i,j})_{i,j=1,\dots,n} \in M_n(K)$, which is a pseudo valuation. We write $\log^+ a := \log\max(1,a)$ $(a \in \mathbb{R}_{\geq 0})$. For $Y = \sum_{i+j\geq 0}^{i,j\geq 0} Y_{i,j}x^i y^j \in M_n(K[[x,y]])$, we define the function $h_{v,\cdot}(\cdot)$ by

$$h_{v,0}(Y) := \log^+ |Y_{0,0}|_v , \quad h_{v,m}(Y) := \frac{1}{m} \max_{i+j\leq m} \log^+ |Y_{i,j}|_v \quad (m = 1,2,\dots).$$

The *size* of $Y \in M_n(K[[x,y]])$ means the quantity

$$\sigma(Y) := \varlimsup_{m\to\infty} \sum_v h_{v,m}(Y),$$

where v runs over all places of K. With this we shall call $Y \in M_n(K[[x,y]])$ a *G-function in the two variables* x,y whenever $\sigma(Y) < \infty$ holds. Next, suppose that x,y are formal variables. For $f = \sum_{i+j\leq N} f_{i,j}x^i y^j \in K[x,y]$ and for every place v of K, the *Gauss absolute value* is defined by $|f|_v := \max_{i+j\leq N} |f_{i,j}|_v$. With an arbitrary non-Archimedean valuation v the Gauss absolute value is defined in the quotient field of $K(x,y)$ of $K[x,y]$ as well; that is, we put $|f/g|_v := |f|_v/|g|_v$ for $g \neq 0$. This allows us to extend the above pseudo valuation on $M_n(K)$ to $M_n(K(x,y))$ in an obvious way. Then, for a sequence $\{F_i\}_{i=0,1,\dots} \subset M_n(K(x,y))$ and for each place $v \nmid \infty$, we put

$$h_{v,0}(\{F_i\}) := \log^+ |F_0|_v , \quad h_{v,m}(\{F_i\}) := \frac{1}{m} \max_{i\leq m} \log^+ |F_i|_v \quad (m = 1,2,\dots).$$

In this way the *size* of $\{F_i\}_{i=0,1,\dots} \subset M_n(K(x,y))$ is defined to be $\sigma(\{F_i\}) := \varlimsup_{m\to\infty} \sum_{v\nmid\infty} h_{v,m}(\{F_i\})$. In passing we remark that our definitions of two *sizes* are formal extensions of the corresponding concepts in [1] and [9]. In the one-variable situation, the inequality

$$\left| \frac{1}{n!}\left(\frac{d}{dx}\right)^n f \right|_v \leq |f|_v$$

holds for any $f \in K(x)$ and $n \in \mathbb{N}$. But such a simple inequality does not hold in the two-variable case. To avoid this disadvantage we introduce a factor γ

in the following way: For any $f, g \in K[x, y]$ with $\max(\deg_x f, \deg_x g) \leq l$ and $\max(\deg_y f, \deg_y g) \leq m$, we put

$$\left| \frac{\partial^n}{n!}(f/g) \right|_v \leq \gamma(\partial, l, m, n, v) \, |f/g|_v \quad (n = 0, 1, 2, \dots).$$

Next, let $\mathfrak{J}, \mathfrak{A} \in M_n(K((x,y)))$. We define the sequence $\{(\mathfrak{J}, \mathfrak{A})_\partial^{\langle i \rangle}\}_{i=0,1,\dots} \subset M_n(K((x,y)))$ by

$$(\mathfrak{J}, \mathfrak{A})_\partial^{\langle 0 \rangle} = \mathfrak{J}, \quad (\mathfrak{J}, \mathfrak{A})_\partial^{\langle i+1 \rangle} = \frac{1}{i+1} \left(\partial(\mathfrak{J}, \mathfrak{A})_\partial^{\langle i \rangle} - \mathfrak{A}(\mathfrak{J}, \mathfrak{A})_\partial^{\langle i \rangle} \right) \quad (i = 0, 1, \dots).$$

To treat $\{(\mathfrak{J}, \mathfrak{A})_\partial^{\langle i \rangle}\}_{i=0,1,\dots}$ is equivalent to considering the differential equation (2). This is an analogue of the relevant sequence introduced in [9]; we have replaced d/dx by ∂ and put $\mathfrak{B} = 0$ there. We write $\sigma_\partial(\mathfrak{J}, \mathfrak{A})$ as an abbreviation for $\sigma(\{(\mathfrak{J}, \mathfrak{A})_\partial^{\langle i \rangle}\})$; also $\sigma_\partial(A)$ is used in place of $\sigma_\partial(I, A)$. We remark that $\sigma_{d/dx}(A) = \sigma_{d/dx}(-{}^t A)$ (see [10]). The *size* of the equation $(\partial - A)X = 0$ is defined to be $\sigma_\partial(A)$. Further, we let a *change of basis on* ∂ mean the operation $\mathfrak{J}[A]_\partial := \mathfrak{J}A\mathfrak{J}^{-1} + (\partial\mathfrak{J})\mathfrak{J}^{-1}$, where $\mathfrak{J} \in GL_n(K(x,y))$ and $A \in M_n(K(x,y))$.

Having said these we have the following basic identities; proofs are similar to those of the corresponding assertions in [9]:

Proposition 1 *For* $\mathfrak{J}_1, \mathfrak{J}_2 \in GL_n(K(x,y))$, *and* $A \in M_n(K(x,y))$, *we have*

$$\mathfrak{J}_1[\mathfrak{J}_2[A]_\partial]_\partial = (\mathfrak{J}_1\mathfrak{J}_2)[A]_\partial.$$

Proposition 2 *For* $\mathfrak{J} \in GL_n(K(x,y))$, $A \in M_n(K(x,y))$, *and* $m = 0, 1, \dots$, *we have*

$$\mathfrak{J}(I, A)_\partial^{\langle m \rangle} = (\mathfrak{J}, \mathfrak{J}[A]_\partial)_\partial^{\langle m \rangle}.$$

Proposition 3 *For* $\mathfrak{J}_1, \mathfrak{J}_2, A \in M_n(K((x,y)))$, *and* $m = 0, 1, \dots$, *we have*

$$(\mathfrak{J}_1\mathfrak{J}_2, A)_\partial^{\langle m \rangle} = \sum_{i+j=m} (\mathfrak{J}_1, A)_\partial^{\langle i \rangle} (\mathfrak{J}_2, 0)_\partial^{\langle j \rangle}.$$

1. Estimation of the sizes

We shall show some properties of simultaneous approximations of a solution of a partial differential equation and estimate sizes.

Let us assume that $X \in GL_n(K[[x,y]])$ with $X(0,0) = I$ is a solution of the partial differential equation

$$\partial X - AX = 0 \tag{3}$$

for an $A \in M_n(K(x,y))$. Then we have

Lemma 1   *For any $q \in K[x,y]$, $P \in M_n(K[x,y])$, and integers $m \geq 0$ it holds that*

$$(P,A)_\partial^{\langle m \rangle} = \left(\frac{\partial^m}{m!}q\right) X - (R,A)_\partial^{\langle m \rangle}.$$

where $R := qX - P \in M_n(K[[x,y]])$.

Proof   Similar to that of Lemma 5.1 in [9].

Notation   Throughout this section the integers N, L are arbitrary but supposed to satisfy the inequalities $\max(\deg q, \deg P) < N$ and $\operatorname{ord} R \geq L$, where R is as above, $\deg q$ the minimum integer M such that $q = \sum_{i+j \leq M} q_{i,j} x^i y^j$, and $\operatorname{ord} R$ the maximum integer M such that $R = \sum_{i+j \geq M} R_{i,j} x^i y^j$. For the A in (3) we choose a $u \in \mathcal{O}_K[x,y]$ satisfying

$$uA \in M_n(K[x,y]), \quad \partial u \in K[x,y]$$

so that there exist $f, g \in K[x,y]$ such that

$$u\partial = f\partial_x + g\partial_y.$$

Here $\mathcal{O}_K$ is the integer ring of K as usual. We put

$$s := \max(\deg(uA), \deg \partial u, \deg f, \deg g).$$

Remark   We may set $u = \operatorname{den}(A) \cdot (\operatorname{den}(a,b))^2$.

Lemma 2   *We have, for $m = 0, 1, \ldots, L$,*

$$u^m \frac{\partial^m}{m!} q \in K[x,y], \quad \deg u^m \frac{\partial^m}{m!} q < N + ms,$$

$$u^m (P,A)_\partial^{\langle m \rangle} \in M_n(K[x,y]), \quad \deg u^m (P,A)_\partial^{\langle m \rangle} < N + ms,$$

$$u^m (R,A)_\partial^{\langle m \rangle} \in M_n(K[[x,y]]), \quad \operatorname{ord} u^m (R,A)_\partial^{\langle m \rangle} \geq L - m.$$

Proof   These assertions are proved by the inductive argument with respect to m (cf. Lemma 5.3 in [9]).

Notation For the power series $Z = \sum_{i+j\geq 0}^{\infty} Z_{i,j} x^i y^j \in M_n(K[[x,y]])$, we introduce the truncation $Z_{<m} := \sum_{i+j<m} Z_{i,j} x^i y^j$ $(m = 1, 2, \ldots)$. We then put

$$H_{f,m}(Z) := \sum_{\substack{v \nmid \infty}} \max_{i+j<m} \log^+ |Z_{i,j}|_v, \quad H_{\infty,m}(Z) := \sum_{\substack{v | \infty}} \max_{i+j<m} \log^+ |Z_{i,j}|_v$$

as well as $H_m(Z) := H_{f,m}(Z) + H_{\infty,m}(Z)$. In particular, if $Z \in M_n(K[x,y])$, then we write $H(Z)$ in place of $H_{1+\deg Z}(Z)$.

Now write $X = \sum_{i+j\geq 0} X_{i,j} x^i y^j \in M_n(K[[x,y]])$. We assume that

$$\left. \begin{array}{l} \text{there exist } W \in M_n(K[[x]]) \text{ and } V \in GL_n(K[y]) \\ \text{with } \det V = 1, \deg V < d \text{ and } V_{|y=0} = I \\ \text{such that } X = WV, \end{array} \right\} \tag{4}$$

and that

$$(1 + n^{-2})N - (d - 1) > L > N.$$

We then put

$$\delta := \frac{L}{N} - 1, \quad \alpha := \frac{(\delta + (d-1)/N)n^2}{1 - (\delta + (d-1)/N)n^2}, \quad \beta := \alpha \log(2N\kappa) + \log \kappa,$$

where $\kappa = 4[K : \mathbb{Q}]^{2[K:\mathbb{Q}]} \sqrt{|D_K|}$ with D_K being the discriminant of K and $|\cdot|$ means the usual Archimedean absolute value.

Lemma 3 *There exists a non-trivial $q \in K[x]$ of degree $\leq N - 1$ such that $(qX)_{<L} = (qX)_{<N}$. Moreover we may let this q satisfy the conditions $H_N(q) \leq \alpha H_L(X) + \beta$, $H_{f,N}(qX) \leq H_{f,N}(q) + H_{f,N}(X)$ and $H_{\infty,N}(qX) \leq H_{\infty,N}(q) + H_{\infty,N}(X) + 3\log(N + 2)$. In particular, we have $\det((qX)_{<N}) \in K[x]$.*

Proof We write $q = \sum_{i<N} q_i x^i$, $W = \sum_{i\geq 0} W_i x^i$ and $V = \sum_{i=0}^{d-1} V_i y^i$. Then $qW = \sum_{l\geq 0} \sum_{i+j=l}^{i<N} q_i W_j x^l$. We consider the system of linear equations:

$$\sum_{\substack{i<N \\ i+j=l}} q_i W_j = 0 \quad (l = N - (d-1), \ldots, L - 1).$$

The number of equations is $(L - N + (d - 1))n^2$ and the number of unknowns q_i is N. Thus, if $N > (L - N + (d - 1))n^2$, then the system has a non-trivial solution. Since $qX = qWV = \sum_{k=0}^{d-1}\sum_{l\geq0}\sum_{i+j=l}^{i<N} q_iW_jV_kx^ly^k$, we have, for $m = N,\ldots,L$,

$$(qWV)_{<m} = \sum_{k=0}^{d-1}\sum_{\substack{k\leq i+j+k<m\\i<N}} q_iW_jV_kx^{i+j}y^k$$

$$= \sum_{k=0}^{d-1}\sum_{\substack{0\leq i+j\leq N-d\\i<N}} q_iW_jV_kx^{i+j}y^k.$$

On the other hand, since $(qW)_{<m} = (qW)_{<N-(d-1)}$, we have

$$(qW)_{<m}V = (qW)_{<N-(d-1)}V = \sum_{\substack{0\leq i+j\leq N-d\\i<N}} q_iW_jx^{i+j}\sum_{k=0}^{d-1}V_ky^k$$

$$= \sum_{k=0}^{d-1}\sum_{\substack{0\leq i+j\leq N-d\\i<N}} q_iW_jV_kx^{i+j}y^k.$$

These yield that $(qX)_{<m} = (qWV)_{<m} = (qW)_{<m}V = (qW)_{<N}V = (qX)_{<N}$. Moreover $\det(qX)_{<m} = \det(qW)_{<m}\det V = (qW)_{<m} \in K[x]$ because of the assumption $\det V = 1$. Since

$$H_L(W) = \sum_v \max_{i<L}\log^+|W_i|_v \leq \sum_v \max_{i+j<L}\log^+|W_iV_j|_v = H_L(WV) = H_L(X),$$

we obtain $H_N(q) \leq \alpha H_L(X) + \beta$ by virtue of Siegel's lemma in [2]. As to $H_{f,N}(P)$ and $H_{\infty,N}(P)$ we can estimate them as in [9, Lemma 5.6].

Lemma 4 *Let $X = \sum_{i+j\geq0} X_{i,j}x^iy^j \in M_n(K[[x,y]])$ with $X(0,0) = I$, and let us assume that $\max(\deg q, \deg P) < \operatorname{ord} R$ and $qP \neq 0$. Then we have that $\det P \neq 0$.*

Proof If there exists a $t \in \mathbb{C}$ such that $\det P_{|y=tx} \neq 0$ then $\det P \neq 0$. Write $qP = \sum_{i+j<2N} U_{i,j}x^iy^j$. If $qP_{|y=tx} = \sum_{i+j<2N} U_{i,j}t^jx^{i+j} = 0$, then $\sum_{i+j=m} U_{i,j}t^j = 0$ for $m = 0,1,\ldots,2N-1$. The number of roots in $\mathbb{C}$ of these equations in t is at most $2N-1$, since there exists at least a $U_{i,j} \neq 0$. We fix a $t \in \mathbb{C}$ outside the set of these roots. Then $q_{|y=tx} \neq 0$, $P_{|y=tx} \neq 0$ and $\max(\deg q_{|y=tx}, \deg P_{|y=tx}) < \operatorname{ord} R_{|y=tx}$. Thus we conclude via Lemma 5.6 in [9] that $\det P \neq 0$.

Lemma 5 *For $f \in K[x,y]$, $f \neq 0$, we have*

$$\sum_{v\nmid\infty}\log|1/f|_v \leq \sum_{v|\infty}\log|f|_v + 2\log(2 + \deg f).$$

Proof We take a root of unity ξ such that $f_{|y=\xi x} \neq 0$. On noting that the number of terms of f is not greater than $(2 + \deg f)^2$, we proceed as in the proof of [9, Lemma 6.1].

We now turn to the estimation of the sizes:

Theorem 2 *We have that*

$$\sigma_\partial(A) \leq \left(2n^2(1+s)\left(2n + \frac{1}{2n} + 1\right) + s \right) \sigma(X) + H(u) + 2\log(s+2)$$

$$+ 2 \varlimsup_{m\to\infty} \frac{1}{m} \sum_{v\nmid\infty} \log^+ \max_{i\leq m} \gamma(\partial, (2n^3(1+s)-1)m, d_y, i, v), \tag{5}$$

where $d_y = n(d-1)$.

Proof Let the integer N be sufficiently large. We put $L := [(1 + \frac{1}{2}n^{-2})N] - d$. Let q be as in Lemma 3 and put $P = (qX)_{<N}$. If

$$N + ms \leq L - m,$$

then we see, by Proposition 3, Lemma 1, and Lemma 2, that

$$(I, A)_\partial^{\langle m\rangle} = \sum_{i+j=m} (P, A)_\partial^{\langle i\rangle} (P^{-1}, 0)_\partial^{\langle j\rangle} = \sum_{i+j=m} (P, A)_\partial^{\langle i\rangle} \frac{\partial^j}{j!} P^{-1}$$

$$= \sum_{i+j=m} u^{-m} \left(\left(u^i \frac{\partial^i}{i!} q \right) X \right)_{<N+is} \frac{\partial^j}{j!} P^{-1}.$$

This implies that we have, for $v \nmid \infty$,

$$\log^+ \max_{i\leq m} \left| (I, A)_\partial^{\langle i\rangle} \right|_v \leq \log \max_{i\leq m} \left| u^{-i} \right|_v$$

$$+ \log^+ \max_{i\leq m} \left| \left(\left(u^i \frac{\partial^i}{i!} q \right) X \right)_{<N+is} \right|_v + \log \max_{i\leq m} \left| \frac{\partial^i}{i!} P^{-1} \right|_v.$$

Let $\tilde{P}$ be the adjoint matrix of P. We write

$$d_x = \max(\deg_x \tilde{P}, \deg_x \det P)$$

and

$$d_y = \max(\deg_y \tilde{P}, \deg_y \det P).$$

We have that $d_x \leq n(N-1)$ and $d_y \leq n(d-1)$, and by Lemma 3 that $\det P \in K[x]$. By Lemma 5, we have

$$\sum_{v \nmid \infty} \max_{i \leq m} \log \left| u^{-i} \right|_v \leq m \sum_{v \nmid \infty} \log \left| u^{-1} \right|_v \leq m \left(\sum_{v \mid \infty} \log \left| u \right|_v + 2\log(s+2) \right),$$

$$\sum_{v \nmid \infty} \log \left| (\det P)^{-1} \right|_v \leq \sum_{v \mid \infty} \log \left| \det P \right|_v + 2\log(nN+1)$$

$$\leq n \sum_{v \mid \infty} \log \left| P \right|_v + 5n \log n!(N+1),$$

and

$$\sum_{v \nmid \infty} \log^+ \left| \tilde{P} \right|_v \leq (n-1) \sum_{v \nmid \infty} \log^+ \left| P \right|_v .$$

We note that $q \in K[x]$. Write $\tilde{d}_x = \deg_x q$. Then we have that

$$\sum_{v \nmid \infty} \log^+ \max_{i \leq m} \left| \left(\left(u^i \frac{\partial^i}{i!} q \right) X \right)_{<N+is} \right|_v$$

$$\leq \sum_{v \nmid \infty} \log^+ \left| q \right|_v + \sum_{v \nmid \infty} \log^+ \max_{i \leq m} \left| X_{<N+is} \right|_v$$

$$+ \sum_{v \nmid \infty} \log^+ \max_{i \leq m} \left| u^i \right|_v + \sum_{v \nmid \infty} \log^+ \max_{i \leq m} \gamma(\partial, \tilde{d}_x, 0, i, v).$$

Hence we have

$$\sum_{v \nmid \infty} \log^+ \max_{i \leq m} \left| (I,A)^{(i)}_\partial \right|_v \leq H_{f,N}(q) + H_{f,N+ms}(X)$$

$$+ m H_f(u) + m H_\infty(u) + (n-1) H_{f,N}(P)$$

$$+ n H_{\infty,N}(P) + 5n \log n!(N+1)$$

$$+ 2m \log(s+2) + 2 \sum_{v \nmid \infty} \log^+ \max_{i \leq m} \gamma(\partial, \max(d_x, \tilde{d}_x), d_y, i, v).$$

Since

$$H_{f,N}(q) + (n-1) H_{f,N}(P) + n H_{\infty,N}(P)$$

$$\leq H_{f,N}(q) + (n-1) H_{f,N}(q) + (n-1) H_{f,N}(X)$$

$$+ n H_{\infty,N}(q) + n H_{\infty,N}(X) + 3n \log(N+2)$$

$$\leq n H_N(q) + n H_N(X) + 3n \log(N+2),$$

we have

$$\sum_{v\nmid\infty} \max_{i\leq m} \log^+ \left|(I,A)_\partial^{(i)}\right|_v \leq nH_N(q) + nH_N(X)$$

$$+ H_{f,N+ms}(X) + mH(u) + 2m\log(s+2)$$
$$+ 5n\log n!(N+1) + 3n\log(N+2)$$
$$+ 2\sum_{v\nmid\infty} \log^+ \max_{i\leq m} \gamma(\partial, \max(d_x, \tilde{d}_x), d_y, i, v).$$

We assumed that $N + ms < L - m$, i.e., $N + ms < [(1 + \frac{1}{2}n^{-2})N] - d - m$. Now let m be the maximum integer such that $m \leq (N/(2n^2) - 1 - d)/(1 + s)$. Hence we find that $m/N \to (2n^2(1+s))^{-1}$, $\beta/N \to 0$ when $N \to \infty$, where β is as in Lemma 3. We put $\alpha' = \overline{\lim}_{N\to\infty} \alpha$. Then we have

$$\sigma_\partial(I,A) = \overline{\lim_{m\to\infty}} \frac{1}{m} \sum_{v\nmid\infty} \max_{i\leq m} \log^+ \left|(I,A)_\partial^{(i)}\right|_v$$

$$= \overline{\lim_{m\to\infty}} n\frac{L}{m}\frac{1}{L}\alpha(H_L(X) + \beta) + n\frac{N}{m}\frac{1}{N}H_N(X)$$

$$+ \frac{N+ms}{m}\frac{1}{N+ms}H_{f,N+ms}(X) + H(u) + 2\log(s+2)$$

$$+ \overline{\lim_{m\to\infty}} \frac{2}{m} \sum_{v\nmid\infty} \log^+ \max_{i\leq m} \gamma(\partial, \max(d_x, \tilde{d}_x), d_y, i, v)$$

$$\leq \left(2n^3(1+s)\left(1 + \frac{1}{2n^2}\right)\alpha' + 2n^3(1+s) + 2n^2(1+s) + s\right)\sigma(X)$$

$$+ H(u) + 2\log(s+2)$$

$$+ \overline{\lim_{m\to\infty}} \frac{2}{m} \sum_{v\nmid\infty} \log^+ \max_{i\leq m} \gamma(\partial, (2n^3(1+s) - 1)m, d_y, i, v).$$

Now we find that $\delta \to n^{-2}/2$ as $N \to \infty$ and $\alpha' = 1$. Hence we obtain (5).

Remark If we assume that $V \in GL_n(K[[y]])$ with $\det V = 1$ and $V_{|y=0} = I$ in (4), that is, V is not a matrix of polynomials, the inequality (5) holds for $d_y = (2n^3(1+s) - 1)m$.

2. Regular singularities of Fuchsian differential equations We consider the differential equation

$$\frac{d}{dx}X(x) = A(x)X(x) \tag{6}$$

with $xA(x) \in M_n(K[x]_{(x)})$. Throughout this section, we shall assume that

every eigenvalue of $\mathrm{Res}(A(x))$ is contained in $\mathbb{Q}$.

Proposition 4 *Let c be the common denominator of the eigenvalues o,
Res $(A(x))$ and let $\partial = (cx^{c-1})^{-1}\partial_x + (cx^c)^{-1}\partial_y$. For the differential equation (6) and its matrix solution $X(x)$, there exists $\tilde{T}(x) \in GL_n(K(x))$ whicl
satisfies the following properties: There exist $W(x) \in GL_n(K[[x]])$ and $V(y) \in
GL_n(K[y])$ with $W(0) = V(0) = I$ and with $\deg V(y) < n$, and there exist:
$B(x) \in M_n(K(x))$ which satisfies the partial differential equation:*

$$\partial(W(x)V(y)) = B(x)W(x)V(y),$$

where

$$B(x) = \tilde{T}(x)[A(x^c)]_\partial \quad and \quad W(x)V(\log x) = \tilde{T}(x)X(x^c).$$

Moreover the following inequality holds:

$$\sigma_{d/dx}(A(x)) \leq \sigma_\partial(B(x)) + nh \sum_{\substack{p:\text{prime} \\ (p,c)\neq 1}} \frac{\log p}{p-1} + nhc,$$

where $h = \max |c \cdot (eigenvalue\ of\ \text{Res}\,(A(x)))|$.

Proof First of all, we note that a differential equation $\partial \mathfrak{X} = \mathfrak{A}\mathfrak{X}$ is equivalent
to $\partial(\mathfrak{T}\mathfrak{X}) = \mathfrak{T}[\mathfrak{X}]_\partial(\mathfrak{T}\mathfrak{X})$ for any $\mathfrak{T} \in GL_n(K(x,y))$. After applying a Frobeniu:
transformation, $x \to x^c$, to the differential equation (6), we have

$$\frac{d}{dx}X(x^c) = cx^{c-1}A(x^c)X(x^c)$$

and $x \cdot cx^{c-1}A(x^c) \in M_n(K[x]_{(x)})$ and every eigenvalue of $\text{Res}\,(cx^{c-1}A(x^c))$
belongs to $\mathbb{Z}$. Applying several times the *sharing transformation* (see [4] anc
[9]), we find $T(x) \in GL_n(K(x))$ such that every eigenvalue of

$$\text{Res}\,(T(x)[cx^{c-1}A(x^c)]_{d/dx})$$

is zero. Since

$$T(x)[cx^{c-1}A(x^c)]_{d/dx} = cx^{c-1}T(x)[A(x^c)]_\partial,$$

we have

$$\frac{d}{dx}(T(x)X(x^c)) = cx^{c-1}T(x)[A(x^c)]_\partial(T(x)X(x^c)).$$

There exists $C \in GL_n(K)$ such that

$$C\text{Res}\,(cx^{c-1}T(x)[A(x^c)]_\partial)C^{-1} = C[\text{Res}\,(cx^{c-1}T(x)[A(x^c)]_\partial)]_\partial$$

is the Jordan form. Therefore we have by Proposition 1

$$\frac{d}{dx}(CT(x)X(x^c)) = cx^{c-1}(CT(x))[A(x^c)]_\partial(CT(x)X(x^c)).$$

We put $\tilde{T}(x) = CT(x)$ and $B(x) = \tilde{T}(x)[A(x^c)]_\partial$. In general, we know (see [1]) that there exists a unique solution $W(x) \in GL_n(K[[x]])$ with $W(0) = I$ of the differential equation

$$\frac{d}{dx}W(x) = cx^{c-1}B(x)W(x) - W(x)x^{-1}\mathrm{Res}\,(cx^{c-1}B(x)).$$

Moreover $W(x)x^{\mathrm{Res}\,(cx^{c-1}B(x))}$ is a matrix solution of $(d/dx)Z = cx^{c-1}B(x)Z$. We put $y = \log x$. Since

$$x^{\mathrm{Res}\,(cx^{c-1}B(x))} = \sum_{i=0}^{\infty} \frac{1}{i!}(\mathrm{Res}\,(cx^{c-1}B(x))y)^i$$

and since $(\mathrm{Res}\,(cx^{c-1}B(x)))^n = 0$, we find that $x^{\mathrm{Res}\,(cx^{c-1}B(x))} = I + J(y)$ where $J(y) \in M_n(K[y])$, $0 \le \deg J(y) < n$, $J(y)$ is a triangle matrix and all diagonal components of $J(y)$ are zeros. Let $V(y) := I + J(y)$. Hence we see that a matrix solution of $\partial Z(x,y) = B(x)Z(x,y)$ is given by $Z(x,y) = W(x)V(y)$. Moreover we have, as in [9, Section 3],

$$\sigma_\partial(A(x^c)) = \sigma_\partial(\tilde{T}(x), \tilde{T}(x)[A(x^c)]_\partial) \le \sigma_\partial(I, \tilde{T}(x)[A(x^c)]_\partial) + \sigma_\partial(\tilde{T}(x), 0)$$

by Propositions 2 and 3. Let F be the set

$$\{(a_{i,j})_{i,j=1,\dots,n} | a_{i,i} = 1, x \text{ or } x^{-1},\ a_{i,j} = 0\ (i \ne j).\}.$$

Following the proof of [9, Lemma 4.1], we see that there exist $C_1, \dots, C_{nh} \in GL_n(K)$ and $T_1, \dots, T_{nh} \in F$ such that $T(x) = T_{nh}C_{nh}\cdots T_1C_1$. Since

$$\sigma_\partial(\mathfrak{J}_1\mathfrak{J}_2, 0) \le \sigma_\partial(\mathfrak{J}_1, 0) + \sigma_\partial(\mathfrak{J}_2, 0),$$

we have

$$\sigma_\partial(\tilde{T}(x), 0) \le \sigma_\partial(C, 0) + \sum_{i=1}^{nh}(\sigma_\partial(T_i, 0) + \sigma_\partial(C_i, 0)).$$

An elementary computation gives

$$\sigma_\partial(T_i, 0) \le \varlimsup_{m \to \infty} \frac{1}{m} \sum_{\substack{(p,c) \ne 1, p \le m \\ p:\text{prime}}} m(\log|1/c|_p + (p-1)^{-1}\log p)$$

$$\le \sum_{\substack{(p,c) \ne 1 \\ p:\text{prime}}} \frac{\log p}{p-1} + c,$$

and $\sigma_\partial(C,0) = 0$, $\sigma_\partial(C_i,0) = 0$ for $i = 1,\dots,nh$. Then we obtain

$$\sigma_\partial(\tilde{T}(x),0) \le nh \sum_{\substack{(p,c)\ne 1 \\ p:\text{prime}}} \frac{\log p}{p-1} + nhc.$$

On the other hand, we have

$$(I, A(x^c))_\partial^{\langle 0\rangle} = (I, A(x))_{d/dx}^{\langle 0\rangle} = I;$$

and if $(I, A(x^c))_\partial^{\langle m\rangle} = ((I, A(x))_{d/dx}^{\langle m\rangle})_{|x\to x^c}$, then

$$\begin{aligned}
(m+1)(I, A(x^c))_\partial^{\langle m+1\rangle} &= \frac{1}{cx^{c-1}}\partial_x(I, A(x^c))_\partial^{\langle m\rangle} - A(x^c)(I, A(x^c))_\partial^{\langle m\rangle} \\
&= \frac{dx}{dx^c}\frac{d}{dx}(I, A(x^c))_\partial^{\langle m\rangle} - A(x^c)(I, A(x^c))_\partial^{\langle m\rangle} \\
&= \left(\frac{d}{dx}(I, A(x))_{d/dx}^{\langle m\rangle} - A(x)(I, A(x))_{d/dx}^{\langle m\rangle}\right)_{|x\to x^c} \\
&= (m+1)\left((I, A(x))_{d/dx}^{\langle m+1\rangle}\right)_{|x\to x^c}.
\end{aligned}$$

Then we find that $\sigma_\partial(A(x^c)) = \sigma_{d/dx}(A(x))$. Therefore we have

$$\sigma_{d/dx}(A) = \sigma_\partial(A(x^c)) \le \sigma_\partial(B) + nh \sum_{\substack{(p,c)\ne 1 \\ p:\text{prime}}} \frac{\log p}{p-1} + nhc,$$

which ends the proof.

Now we consider the estimation of the factor $\gamma(\partial,\dots,v)$ which we have introduced above; we may impose the condition $g \in K[x]$ for the sake of our application (see Section 1). For $\partial = a\partial_x + b\partial_y$ with any $a, b \in K(x,y)$, we have

$$\frac{\partial^m}{m!}(f/g) = \sum_{i+j=m}\left(\frac{\partial^i}{i!}f\right)\left(\frac{\partial^j}{j!}(1/g)\right).$$

Let γ_1, γ_2 be such that

$$\left|\frac{\partial^i}{i!}f\right|_v \le \gamma_1(\partial, \deg_x f, \deg_y f, i, v)\,|f|_v \quad (f \in K[x,y])$$

$$\left|\frac{\partial^j}{j!}(1/g)\right|_v \le \gamma_2(\partial, j, v)\,|1/g|_v \quad (g \in K[x]).$$

Then we may put

$$\gamma(\partial, \deg_x f, \deg_y f, m, v) = \max_{i+j=m} \gamma_1(\partial, \deg_x f, \deg_y f, i, v)\gamma_2(\partial, j, v).$$

Thus we need to estimate γ_1 and γ_2 for our particular differential operator $\partial = (cx^{c-1})^{-1}\partial_x + (cx^c)^{-1}\partial_y$:

First we shall estimate γ_1. Let integers $g_c(\cdot, \cdot)$ be defined by $g_c(s, 0) = 1$, $g_c(s, n) = (s - (n-1)c) \cdot g_c(s, n-1)$ recursively. Let $G(s, n, l)$ be the $\mathbb{Z}$-module generated by the numbers $g_c(s, n)/(a_1 \cdots a_l)$ where $a_i = 1$ or $s - jc$ for a certain $j = 0, \ldots, n-1$. It is easy to see that if $\tilde{g} \in G(s, n, l)$ then $(s - nc) \cdot \tilde{g} \in G(s, n+1, l)$ and thus $\tilde{g} = \tilde{g}(s - nc)/(s - nc) \in G(s, n+1, l+1)$. Hence $G(s, n, l) \subset G(s, n+1, l+1)$.

Lemma 6 *For $\partial = (cx^{c-1})^{-1}\partial_x + (cx^c)^{-1}\partial_y$, there exist $a_{s,k,i} \in G(s, k, t-i)$ such that*

$$c^k \partial^k (x^s y^t) = x^{s-kc} \sum_{i=0}^{t} a_{s,k,i} y^i$$

for $k, s, t \in \mathbb{Z}_{\geq 0}$.

Proof We use the induction with respect to the parameter k. For $k = 0$, we see that $0, 1 \in G(s, 0, i)$, $a_{s,0,t} = 1$ and $a_{s,0,i} = 0$ for $i = 0, \ldots, t-1$. Let us assume that there exist the coefficients $a_{s,l,i}$ for $l \leq k$. Increasing the value of k by 1, we put $b_{s,k+1,i} = (s - ck)a_{s,k,i} \in G(s, k+1, t-i)$, $c_{s,k+1,i-1} = ia_{s,k,i} \in G(s, k, t-i) \subset G(s, k+1, t-(i-1))$ and $c_{s,k+1,0} = c_{s,k+1,t} = 0$. Then we have

$$c^{k+1}\partial^{k+1} x^s y^t = x^{1-c}\partial_x \left(x^{s-kc} \sum_{i=0}^{t} a_{s,k,i} y^i \right) + x^{-c}\partial_y \left(x^{s-kc} \sum_{i=0}^{t} a_{s,k,i} y^i \right)$$

$$= x^{s-c(k+1)} \left(\sum_{i=0}^{t}(s - ck)a_{s,k,i} y^i + \sum_{i=1}^{t} ia_{s,k,i} y^{i-1} \right)$$

$$= x^{s-c(k+1)} \sum_{i=0}^{t}(b_{s,k+1,i} + c_{s,k+1,i}) y^i.$$

Since $b_{s,k+1,i} + c_{s,k+1,i} \in G(s, k+1, t-i)$, we end the proof.

For any $\tilde{g} \in \bigcup_{i=0}^{t} G(s, m, i) = G(s, m, t)$ and for v with $v \mid p$, we have, by a standard argument, that

$$\left| \frac{\tilde{g}}{m!} \right|_v \leq \begin{cases} (s + c + mc)^{td_v/d} & \text{if } (p, c) = 1, \ p \leq m, \\ |1/p|_v^{m/(p-1)} & \text{if } (p, c) \neq 1, \ p \leq m, \\ 1 & \text{otherwise,} \end{cases}$$

where $d = [K : \mathbb{Q}]$ and $d_v = [K_v : \mathbb{Q}_v]$. Thus by Lemma 6 we have

$$\gamma_1(\partial, n_1, n_2, m, v) \leq \begin{cases} (n_1 + c + n_1 m)^{n_2 d_v / d} & \text{if } (p, c) = 1, \ p \leq m, \\ |1/c|_v^m \, |1/p|_v^{m/(p-1)} & \text{if } (p, c) \neq 1, \ p \leq m, \\ 1 & \text{otherwise.} \end{cases}$$

Next, for $g \in K[x]$, $g \neq 0$, we have

$$\left| \frac{\partial^m}{m!} g \right|_v \leq \gamma_3(\partial, m, v) \, |g|_v \,,$$

where

$$\gamma_3(\partial, m, v) \leq \begin{cases} |1/c|_v^m \, |1/p|_v^{m/(p-1)} & \text{if } (p, c) \neq 1, \ p \leq m, \\ 1 & \text{otherwise.} \end{cases}$$

Since

$$\frac{\partial^m}{m!}(1/g) = -(1/g) \sum_{\substack{i+j=m \\ j<m}} \left(\frac{\partial^i}{i!} g \right) \left(\frac{\partial^j}{j!}(1/g) \right),$$

we have

$$\left| \frac{\partial^m}{m!}(1/g) \right|_v \leq |1/g|_v \max_{\substack{i+j=m \\ j<m}} \left| \frac{\partial^i}{i!} g \right|_v \left| \frac{\partial^j}{j!}(1/g) \right|_v.$$

Hence we may set

$$\gamma_2(\partial, 0, v) = 1, \quad \gamma_2(\partial, m, v) = \max_{\substack{i+j=m \\ j<m}} \gamma_3(\partial, i, v)\gamma_2(\partial, j, v).$$

In this way we obtain

Theorem 3 *Let $\partial = (cx^{c-1})^{-1}\partial_x + (cx^c)^{-1}\partial_y$. Let the valuation v divide a prime p. Let us put, for non-negative integers n_1, n_2, m,*

$$\gamma_1(\partial, n_1, n_2, m, v) = \begin{cases} (n_1 + c + n_1 m)^{n_2 d_v / d} & \text{if } (p, c) = 1, \ p \leq m, \\ |1/c|_v^m \, |1/p|_v^{m/(p-1)} & \text{if } (p, c) \neq 1, \ p \leq m, \\ 1 & \text{otherwise.} \end{cases}$$

Then for any $f \in K[x, y]$ with $\deg_x f \leq n_1$, $\deg_y f \leq n_2$ and for any $g \in K[x]$, $g \neq 0$, we have

$$\left| \frac{\partial^m}{m!}(f/g) \right|_v \leq \gamma_1(\partial, n_1, n_2, m, v) \, |f/g|_v \quad (m = 0, 1, 2 \dots).$$

Now let M and n_2 be finite and let n_1 be such that $n_1/m \leq M$ for any m. Then we find that

$$\varliminf_{m\to\infty} \frac{1}{m} \sum_{v\nmid\infty} \log^+ \max_{i\leq m} \gamma(\partial, n_1, n_2, i, v)$$

$$\leq \varliminf_{m\to\infty} \frac{1}{m} \sum_{\substack{p\leq m \\ p:\text{prime}}} n_2 \log(Mm + Mm^2 + c)$$

$$+ \sum_{\substack{(p,c)\neq 1 \\ p:\text{prime}}} m(\log|1/c|_p + (p-1)^{-1}\log p)$$

$$\leq 2n_2 + c + \sum_{\substack{(p,c)\neq 1 \\ p:\text{prime}}} \frac{\log p}{p-1},$$

for

$$\varliminf_{m\to\infty} \frac{1}{m} \sum_{\substack{p\leq m \\ p:\text{prime}}} \log m \leq \varliminf_{m\to\infty} \frac{1}{m}\pi(m)\log m = 1$$

with a common notation. Hence we obtain the following assertion via Theorem 2; this is the quantitative version of Theorem 1 that was mentioned in the introduction:

Corollary 1 *For the differential equation (6) and the corresponding solution $W(x)V(y)$ in Proposition 4, we have*

$$\sigma_{d/dx}(A) \leq ((1+s)n(4n^2 + 2n + 1) + s)\sigma(WV) + H(u) + 2\log(s+2)$$

$$+ 4n(n-1) + (2+nh)\sum_{\substack{(p,c)\neq 1 \\ p:\text{prime}}} \frac{\log p}{p-1} + (2+nh)c.$$

References

[1] Y. André. *G-functions and Geometry*. Max Planck Institut, Bonn, 1989.
[2] E. Bombieri. On *G*-functions. *Recent progress in Analytic Number Theory* (H. Halberstam and C. Hooley, Eds.), vol. 2., Academic Press, London, 1981, pp. 1–67.
[3] G. Christol and B. Dwork. Differential modules of bounded spectral norms. *Contemp. Math.*, **133** (1992), 39–58.
[4] —. Effective *p*-adic bounds at regular singular points. *Duke Math. J.*, **62** (1991), 689–720.

[5] D.V. Chudnovsky and G.V. Chudnovsky. Applications of Padé approximations to diophantine inequalities in values of G-functions. *Lect. Notes in Math.*, **1135**, Springer-Verlag, Berlin, 1985, pp. 9–51

[6] G.V. Chudnovsky. On applications of diophantine approximations. *Proc. Natl. Acad. Sci. USA*, **81** (1984), 1926–1930.

[7] P. Deligne. *Equations Différentielles á Points Singuliers Réguliers.* Lect. Notes in Math., **163**, Springer-Verlag, Berlin, 1970

[8] B. Dwork, G. Gerotto and F.J. Sullivan. *An Introduction to G-functions.* Annals of Math. Studies, **133**, Princeton Univ. Press, Princeton, 1994.

[9] M. Nagata. A generalization of the sizes of differential equations and its applications to G-function theory. Preprint of Department of Mathematics, Tokyo Institute of Technology.

[10] —. Sequences of differential systems. *Proc. Amer. Math. Soc.*, **124** (1996), 21–25.

[11] C.L. Siegel. Über einige Anwendungen diophantischer Approximationen. *Abh. Preuss. Akad. Wiss., Phys. Math. Kl. nr.1*, 1929.

Makoto Nagata
Research Institute for Mathematical Sciences, Kyoto University
Kyoto-606, Japan

20

Spectral Theory and L-functions
Talk Report

PETER SARNAK

1. Prime Geodesic Theorems There are some similarities between the Selberg Zeta Functions $Z_\Gamma(s)$, where Γ is a congruence subgroup of $SL_2(\mathbb{Z})$ and the Dedekind Zeta function $\zeta_K(s)$, where K is a number field. Recall that for such a group Γ whose set of primitive hyperbolic conjugacy classes $\{\gamma\}$ is denoted by P_Γ and whose norms we denote by $N(\{\gamma\})$ ($N(\gamma) = |\lambda|^2$ where γ is conjugate in $SL_2(\mathbb{R})$ to $\begin{bmatrix} \lambda & 0 \\ 0 & \lambda^{-1} \end{bmatrix}$, $|\lambda| > 1$), $Z_\Gamma(s)$ is defined to be the "Euler" like product

$$Z_\Gamma(s) = \prod_{\{\gamma\} \in P_\Gamma} \prod_{k=0}^{\infty} (1 - N(\gamma)^{-s-k}) \tag{1.1}$$

It is well known that these $\{\gamma\} \in P_\Gamma$ correspond to (primitive) closed geodesics on $X_\Gamma = \Gamma \setminus \mathbb{H}$ ($\mathbb{H}$ is the upper half plane) whose lengths are $2 \log N(\gamma)$. Like $\zeta_K(s)$, $Z_\Gamma(s)$ has a meromorphic continuation to $\mathbb{C}$ and a functional equation. Moreover it is expected to satisfy the analogue of the Riemann-Hypothesis: that $Z_\Gamma(s)$ has no zeros in $\text{Re}(s) > 1/2$, $s \neq 1$ (unlike $\zeta_K(s)$ it has a zero at $s = 1$ rather than a pole). Towards the latter we know much more about $Z_\Gamma(s)$ than for $\zeta_K(s)$. The reason is that the zeros $s_j = 1/2 + ir_j$ (in $\text{Re}(s) \geq 1/2$) of $Z_\Gamma(s)$ correspond via the trace formula to the eigenvalues of the Laplace–Beltrami operator Δ on $L^2(X_\Gamma)$, via the relation $\lambda_j = 1/4 + r_j^2$. Since λ_j is real and non-negative, the zeros of $Z_\Gamma(s)$ in $\text{Re}(s) > 1/2$ must lie in $(1/2, 1]$ and these will be there only if $\lambda_j < 1/4$. Thus the "Riemann Hypothesis" for $Z_\Gamma(s)$ is equivalent to the Selberg Conjecture: $\lambda_1(X_\Gamma) \geq 1/4$ ($\lambda_0(X_\Gamma) = 0$ corresponds to the zero of $Z_\Gamma(s)$ at $s = 1$). We will discuss this conjecture in detail in Section 2.

While there are the similarities described above, there are also major differences between $Z_\Gamma(s)$ and $\zeta_K(s)$. One difference is that, unlike the zeta and L-functions of number theory, $Z_\Gamma(s)$ is meromorphic of order 2. That is, the number of zeros of $Z_\Gamma(s)$ whose imaginary parts are of modulus at most T and whose real part lies between 0 and 1, is asymptotic to $C_\Gamma T^2$ as $T \to \infty$ ($C_\Gamma > 0$). This is in contrast to the $CT \log T$ behavior for L-functions. This difference leads to another. In the case of $\zeta_K(s)$, the Riemann Hypothesis implies (and is implied by) the following about counting of primes: For any

$\varepsilon > 0$

$$\Pi_K(x) := \sum_{N(P) \leq x} 1 = \text{Li}(x) + O_\varepsilon(x^{1/2+\varepsilon}) \qquad (1.2)$$

as $x \to \infty$, the sum being over the prime ideals of $\mathcal{O}_K$ and $\text{Li}(x) = \int_2^x \frac{dt}{\log t}$.

In the case of $Z_\Gamma(s)$ the preponderance of vertical zeros makes it much more difficult to estimate the remainder term in the counting of closed geodesics. If the possible exceptional zeros of $Z_\Gamma(s)$ in $(1/2, 1)$ are denoted by $s_1, \ldots, s_\nu$, then one can show using $Z_\Gamma(s)$ [2], or the trace formula directly [15], that the "Prime Geodesic Theorem" holds in the form

$$\Pi_\Gamma(x) := \sum_{\substack{\{\gamma\} \in P_\Gamma \\ N(\gamma) \leq x}} 1 = \text{Li}(x) + \sum_{j=1}^\nu \text{Li}(x^{s_j}) + O(x^{3/4}) \qquad (1.3)$$

In view of the estimates towards the Selberg Conjecture, established by Selberg – see (2.1) below, we have

$$\Pi_\Gamma(x) = \text{Li}(x) + O(x^{3/4}) \qquad (1.4)$$

We expect that (1.4) holds with the remainder term of size $O_\varepsilon(x^{1/2+\varepsilon})$ for any $\varepsilon > 0$ and to make any improvement of the exponent $3/4$ one needs to capture some cancellation in sums of the form

$$\sum_{|r_j| \leq T} X^{ir_j} \qquad (1.5)$$

for T and X in various ranges.

For $\Gamma = \text{SL}_2(\mathbb{Z})$, Iwaniec [5] introduced Rankin–Selberg L-functions into the picture in an ingenious way. This enabled him to obtain nontrivial upper bounds in (1.5) and hence that

$$\Pi_{\text{SL}_2(\mathbb{Z})}(x) = \text{Li}(x) + O_\varepsilon(x^{35/48+\varepsilon}). \qquad (1.6)$$

If ϕ_j is a basis of the Maass-Hecke cusp forms for $L^2(\text{SL}_2(\mathbb{Z}) \setminus \mathbb{H})$ we may write

$$\left. \begin{aligned} \Delta\phi_j + \lambda_j \phi_j &= 0 \\ T_n \phi_j &= \sqrt{n}\, \lambda_j(n) \phi_j \end{aligned} \right\} \qquad (1.7)$$

where T_n is the n-th Hecke operator on $L^2(\text{SL}_2(\mathbb{Z}) \setminus \mathbb{H})$. The Rankin–Selberg L-function associated with $\phi_j \otimes \phi_j$ is

$$L(s, \phi_j \otimes \phi_j) = \sum_{n=1}^\infty \frac{|\lambda_j(n)|^2}{n^s} \qquad (1.8)$$

In the same 1984 paper, Iwaniec notes that the following ("Mean Lindelöf") Conjecture would greatly simplify his proof and would also yield a better exponent.

$$\sum_{|r_j| \leq T} \left| L(\tfrac{1}{2} + it, \, \phi_j \otimes \phi_j) \right| \underset{\varepsilon}{\ll} (|t| + 1)^4 T^{2+\varepsilon} \qquad (1.9)$$

Recently Luo and the author [10], in the course of our investigations into the question of equidistribution the measures $|\phi_j(z)|^2 \, dx \, dy / y^2$ on X_Γ, established (1.9) for any congruence group Γ. Combining this with recent estimates on $\lambda_1(X_\Gamma)$, see Section 2, we proved: For any congruence subgroup of $SL_2(\mathbb{Z})$

$$\Pi_\Gamma(x) = \text{Li}(x) + O_\varepsilon(x^{\frac{7}{10}+\varepsilon}), \quad \varepsilon > 0 \qquad (1.10)$$

For more on the geometric and number theoretic ("class numbers") interpretation of the Prime geodesic Theorems, see [15]. The proof of (1.9) makes use of arithmetical ingredients such as Weil's bound [19] for Kloosterman sums and also analytic ingredients such as Kuznetsov's trace formula [9]. Another crucial more recent result which is used in the proof is the lower bound of Hoffstein and Lockhard [3]: For $\varepsilon > 0$, $\underset{s=1}{\text{Res}}\, L(s, \phi_j \otimes \phi_j) \underset{\varepsilon}{\gg} |r_j|^{-\varepsilon}$. We refer to [10] for details.

2. The Ramanujan Conjectures As we noted, the analogue of the Riemann Hypothesis for $Z_\Gamma(s)$ is equivalent to Selberg's eigenvalue conjecture. In the 1966 paper [16] in which Selberg formulates his conjecture, he proved that for Γ a congruence subgroup

$$\lambda_1(X_\Gamma) \geq \frac{3}{16} \qquad (2.1)$$

His proof of (2.1) is based on relating the spectrum of X_Γ to sums of Kloosterman sums. With this connection an application of Weil's bounds [19] yields (2.1). Even today we still don't know if it is possible to improve on (2.1) using Kloosterman sums. It is interesting to note the coincidence that in (1.3) the only zeros that are relevant are those with $s_j > 3/4$ which is exactly what $\lambda_1 \geq \frac{3}{16}$ eliminates, yielding (1.4). Another remark is that for Γ of small level – precisely for $\Gamma(N) = \left\{ \begin{pmatrix} a & b \\ c & d \end{pmatrix} \in SL_2(\mathbb{Z}) : \begin{pmatrix} a & b \\ c & d \end{pmatrix} \equiv \begin{pmatrix} 1 & 0 \\ 0 & 1 \end{pmatrix} \mod N \right\}$ and $N \leq 17$ – differential geometric methods can be used to prove $\lambda_1(X_\Gamma) \geq \frac{1}{4}$, see Huxley [4]. This is the reason that exceptional eigenvalues were not an issue in Iwaniec's result (1.6).

Recently Luo–Rudnick and the author [11] found a new way of using L-functions to give lower bounds on λ_1. This method finally goes beyond (2.1) and yields

$$\lambda_1(X_\Gamma) \geq \frac{21}{100} \qquad (2.2)$$

This corresponds to $Z_\Gamma(s)$ having no zeros in $\mathrm{Re}(s) > \frac{7}{10}$ $(s \neq 1)$, which by coincidence is again exactly the present limit of the remainder term in (1.10). This is really a coincidence since an application of our method to the symmetric square L-function (rather than the Rankin–Selberg L-function) leads to a slight improvement

$$\lambda_1(X_\Gamma) \geq \frac{171}{784} = 0.21811\ldots \tag{2.3}$$

The bound (2.2) is a part of some results towards the general Ramanujan Conjectures for GL_n over a number field, which we establish in [12]. Let F be a number field of degree d over Q. Let π be an automorphic cusp form on $\mathrm{GL}_n(\mathbb{A}_F)$ and denote by $\tilde\pi$ its contragredient. The standard L-function $L(s, \pi)$ associated to π [6] is given by

$$L(s, \pi) = \prod_v L(s, \pi_v) \tag{2.4}$$

the product being over all the places of F. At an unramified finite place v of F, $L(s, \pi_v)$ is of the form

$$\prod_{j=1}^n \left(1 - \alpha_{j,\pi}(v)N(v)^{-s}\right)^{-1} \tag{2.5}$$

$N(v)$ being the norm of v and $\alpha_{j,\pi} \in \mathbb{C}$. The local factor at an unramified archemedian place is of the form

$$\prod_{j=1}^n \Gamma_v\left(s - \alpha_{j,\pi}(v)\right) \tag{2.6}$$

where

$$\Gamma_v(s) = \begin{cases} \pi^{-s/2}\Gamma(s/2) & \text{if } v \text{ is real} \\ (2\pi)^{-s}\Gamma(s) & \text{if } v \text{ is complex.} \end{cases}$$

The general Ramanujan Conjectures assert that if π is unramified at v then

$$|\alpha_{j,\pi}(v)| = 1 \quad \text{if } v \text{ is finite}$$
$$\mathrm{Re}(\alpha_{j,\pi}(v)) = 0 \quad \text{if } v \text{ is archemedian.} \tag{2.7}$$

So if $F = Q$, $n = 2$ and π is the automorphic cusp form associated to a Maass eigenfunction of Δ on $\Gamma(N) \setminus \mathbb{H}$, then for $v = \infty$, $\lambda = \frac{1}{4} + (i\,\alpha_{1,\pi})^2$. Thus the general Conjecture 2.7 imply $\lambda \geq 1/4$.

Jacquet and Shalika [7], using local methods, proved that

$$\left|\log_{N(v)} |\alpha_{j,\pi}(v)|\right| < \frac{1}{2} \quad \text{for } v \text{ finite}$$

$$|\mathrm{Re}\left(\alpha_{j,\pi}(v)\right)| < \frac{1}{2} \quad \text{for } v\text{-archemedian} \tag{2.8}$$

Serre [17] observed that the Rankin–Selberg L-functions, $L(s, \pi \otimes \tilde{\pi})$ (assuming its analytic properties) can be used to show that for any <u>finite</u> v

$$|\log |\alpha_{j,\pi}(v)|| \leq \frac{1}{2} - \frac{1}{dn^2 + 1} \tag{2.9}$$

In [12] we establish the following:
Let F, n, π be as above and v a place of F at which π is unramified. Then

(a) $\qquad \left| \log_{N(v)} |\alpha_{j,\pi}(v)| \right| \leq \frac{1}{2} - \frac{1}{n^2 + 1} \qquad v$ finite

$$\left. \vphantom{\begin{array}{c}1\\1\\1\\1\end{array}} \right\} \tag{2.10}$$

(b) $\qquad |\mathrm{Re}\,(\alpha_{j,\pi}(v))| \leq \frac{1}{2} - \frac{1}{n^2 + 1} \qquad v$ archemedian

For $n \geq 3$ and v-archemedian this is the first result which goes beyond the local bounds. If we combine (2.10) in the case of $n = 3$ with the Gelbart–Jacquet lift [1] from GL_2 to GL_3 we get:
Let F, π be as above and let $n = 2$. If v is a place at which π is unramified then

(a) $\qquad \left| \log_{N(v)} |\alpha_{j,\pi}(v)| \right| \leq \frac{1}{5}, \quad v$ finite

$$\left. \vphantom{\begin{array}{c}1\\1\\1\\1\end{array}} \right\} \tag{2.11}$$

(b) $\qquad |\mathrm{Re}\,(\alpha_{j,\pi}(v))| \leq \frac{1}{5} \qquad$ if v is archemedian

Remark Part (a) of (2.11) is not new. Shahidi [18], using quite different methods, established this some time ago (in fact, in the somewhat stronger form which excludes the case of equality). As far as part (b) goes, previously such a bound was known with $1/5$ replaced by $1/4$ (Gelbart–Jacquet [1]), which, when $F = Q$, corresponds to Selberg's $3/16$ bound in (2.1). Applying (2.11) in this case of $F = Q$ yields $\lambda_1 \geq \frac{1}{4} - \left(\frac{1}{5}\right)^2 = \frac{21}{100}$.
The results (2.10) and (2.11) have a number of other applications of this type to bounding eigenvalues of Laplacians on arithmetic manifolds, see [12].
The proof of (2.10) makes essential use of the known analytic properties of the Rankin–Selberg L-functions $L(s, \pi_1 \otimes \pi_2)$, due to Jacquet, Piatetski-Shapiro and Shalika [8], Shahidi [18], and Moeglin–Waldspurger [13]. The proof is based on considering $L(s, \pi \otimes \tilde{\pi} \otimes \chi)$ for suitable ray class characters χ of F. In this connection the constructions of Rohrlich [14] are important.

References

[1] S. Gelbart and H. Jacquet. A relation between automorphic representations of GL(2) and GL(3). *Ann. Sci. Ecole Norm. Sup.*, 4e série, **11** (1978), 471–552.

[2] D. Hejhal. *The Selberg Trace Formula for* PSL(2,ℝ). II. Lect. Notes Math., **1001**, Springer-Verlag, Berlin, 1983.

[3] J. Hoffstein and P. Lockhard. Coefficients of Maass forms and the sieg zero. *Ann. Math.*, **140** (1994), 161–181.

[4] M. Huxley. Eigenvalues of congruence subgroups. *Contemp. Math.*, A.M.S 53 (1986), 341–349.

[5] H. Iwaniec. Prime geodesic theorem. *J. reine angew. Math.*, **349** (1984 136–159.

[6] H. Jacquet. Principal *L*-functions of the liner group. *Proc. Symp. Pu Math.*, A.M.S., Providence, 1979, pp. 63–86.

[7] H. Jacquet and J. Shalika. On Euler products and the classification automorphic representations. I. *Amer. J. Math.*, **103** (1981), 499–558.

[8] H. Jacquet, I. Piatetsky-Shapiro and J. Shalika. Rankin–Selberg convol tions. *Amer. J. Math.*, **105** (1983), 367–464.

[9] N.V. Kuznetsov. Petersson's conjecture for cusp forms of weight zero an Linnik's conjecture. Sums of Kloosterman sums. *Mat. Sbornik*, **111** (1980 334–383. (Russian)

[10] W. Luo and P. Sarnak. Quantum ergodicity of eigenfunctions on PSL₂(ℤ ℍ². *IHES Publ. Math.*, **81** (1995), 207–237.

[11] W. Luo, Z. Rudnick and P. Sarnak. On Selberg's eigenvalue conjectu Geom. Funct. Anal., **5** (1995), 387–401.

[12] —. On the generalized Ramanujan conjecture for GL(n). Preprint, 1996.

[13] C. Moeglin and L. Waldspurger. Le spectre residuel de GL(n). *Ann. S Ecole Norm Sup.*, 4ᵉ série, **22** (1989), 605–674.

[14] D. Rohrlich. Non vanishing of *L*-function for GL(2). *Invent. math.*, **9** (1989), 383–401.

[15] P. Sarnak. Prime geodesic theorems. Thesis. Stanford Univ., 1980.

[16] A. Selberg. On the estimation of Fourier coefficients of modular forms. *Prc Symp. Pure Math.*, A.M.S., Providence, 1965, pp. 1–15.

[17] J.P. Serre. Letter to J.-M. Deshouillers, 1981.

[18] F. Shahidi. On the Ramanujan conjecture and finiteness of poles for certa *L*-functions. *Ann. of Math.*, **127** (1988), 547–584.

[19] A. Weil. On some exponential sums. *Proc. Nat. Acad. Sci.*, **34** (1948 204–207.

Peter Sarnak
Deaprtment of Mathematics, Princeton University
Princeton, New Jersey 08544, U.S.A.

21

Irrationality Criteria for Numbers of Mahler's Type

Tarlok N. Shorey[1] and Robert Tijdeman[2]

Mahler, Bundschuh, Shan and Wang, Sander, and others have derived irrationality criteria for real numbers of the form

$$a(g) = 0.(g^{n_1})_h \; (g^{n_2})_h \; (g^{n_3})_h \cdots$$

where $g \geq 2$ and $h \geq 2$ are integers, $\{n_i\}_{i=1}^{\infty}$ is a sequence of non-negative
integers and $(m)_h$ denotes the finite sequence of digits of m written in h-adic
notation. In the present paper the results are extended to numbers of the form

$$a = 0.(g_1^{n_1})_h \; (g_2^{n_2})_h \; (g_3^{n_3})_h \cdots$$

where $h \geq 2$ is an integer, $\{g_i\}_{i=1}^{\infty}$ is any sequence of non-negative integers
and $\{n_i\}_{i=1}^{\infty}$ is a sequence of integers greater than 1. Furthermore, we give
an irrationality measure. We do so by exploiting estimates for linear forms in
logarithms of rational numbers.

1. Introduction In 1981 Mahler [5] showed that the number obtained by
the concatenation of the consecutive powers of two,

$$0.124816326412825651210 24..., \tag{1}$$

is irrational. For integers $m \geq 1$, $h \geq 2$, we write $(0)_h = 0$ and

$$(m)_h = m_1 m_2 \cdots m_r$$

when $m = m_1 h^{r-1} + m_2 h^{r-2} + \cdots + m_r$ for integers $r > 0$, $0 \leq m_i < h$ $(1 \leq i \leq r)$ with $m_1 \neq 0$. For a positive rational number m with denominator composed
only of prime factors of h, let ν be the least non-negative integer such that $h^\nu m$
is an integer and we extend the preceding definition by putting $(m)_h = (h^\nu m)_h$.
We call r the h-length of m. We express Mahler's above mentioned number (1)
by

$$0.(2^0)_{10} \; (2^1)_{10} \; (2^2)_{10} \; (2^3)_{10} \cdots .$$

This paper was written during a stay in Japan which was made possible by a grant from
Nihon University[1], Tokyo and a NISSAN-grant of NWO[2] (the Netherlands Organization
for the Advancement of Scientific Research).

We now know that the real number

$$a(g) = 0.(g^0)_h \ (g^1)_h \ (g^2)_h \ (g^3)_h \cdots$$

written in h-adic expansion is irrational for any integers $g \geq 2$ and $h \geq 2$. This was shown for $h = 10$ by Mahler and for general h by Bundschuh [4]. Bundschuh gave also an irrationality measure for $a(g)$ in case $\log g / \log h \in \mathbb{Q}$, but wrote not to have been able to give a similar result in case $\log g / \log h \notin \mathbb{Q}$. Niederreiter [6] gave a very simple proof of a more general qualitative irrationality result. Shan [9] gave another proof, using Kronecker's theorem.

Yu [12] studied classes of numbers of the form

$$a(g) = 0. \ (g^{n_0})_h \ (g^{n_1})_h \ (g^{n_2})_h \cdots,$$

where $\{n_i\}_{i=1}^{\infty}$ is a sequence of non-negative numbers. Shan and Wang [10] showed that $a(g)$ is irrational if $\{n_i\}_{i=1}^{\infty}$ is strictly increasing. Sander [7] observed that it suffices to require that the sequence $\{n_i\}_{i=1}^{\infty}$ is unbounded. We shall not restrict ourselves to powers of some fixed integer, but allow powers of different rational numbers whose denominators are composed of prime factors of h.

Theorem 1 *Let $h \geq 2$. Let $\{g_i\}_{i=1}^{\infty}$ with $g_i = A_i/B_i, B_i > 0$ and gcd $(A_i, B_i) = 1$ be a sequence of non-negative rational numbers such that all the B_i are composed solely of prime factors of h. Let $\{n_i\}_{i=1}^{\infty}$ be a sequence of integers each greater than one. If the real number*

$$a = 0. \ (g_1^{n_1})_h \ (g_2^{n_2})_h \ (g_3^{n_3})_h \cdots \qquad (2)$$

is rational, then the sequence $\{A_i^{n_i}\}_{i=1}^{\infty}$ is bounded such that each limit point is bounded by an effectively computable number depending only on h and the period r of a. Furthermore, except for finitely many i, $B_i^{n_i} = h^{\nu_i}/\lambda_i$ where ν_i is a non-negative integer and λ_i is a positive integer which is bounded by an effectively computable number depending only on h and r.

If $g_i = 0$, we understand that $A_i = 0$ and $B_i = 1$. Further we follow the convention that empty product is equal to one allowing g_i to be an integer in Theorem 1. If a is rational, then its h-adic expansion is periodic, that is, there exist positive integers r and M such that if the h-adic expansion of a is given by $a = 0.a_1a_2a_3...$, then $a_{m+r} = a_m$ for every $m \geq M$. We call r a period of a. If $M = 1$, we call the expansion purely periodic. The following example shows that the possibilities in Theorem 1 can occur. Let p and q be distinct primes and m any positive composite integer. Put $h = pq, A_i, B_i, n_i > 1$ such that $A_i^{n_i} = (h-1)^m$ and $B_i^{n_i} = h^{im}/p^m$ for all i. Then $(A_i/B_i)^{n_i} = p^m(h-1)^m/h^{im}$. Since $p^m(h-1)^m$ is not divisible by h, the h-adic expansion of $(A_i/B_i)^{n_i}$ is the

h-adic expansion of $p^m(h-1)^m$. Hence a is a rational number whose h-adic expansion consists of repetitions of the h-adic expansion of $p^m(h-1)^m$. Since this is an integer less than h^{2m}, the minimal period r of a is less than $2m$.

If $\{g_i\}_{i=1}^\infty$ is a sequence of non-negative integers, we note that it follows from Theorem 1 that this sequence itself is bounded by a number depending only on h and a. The method even enables us to give irrationality measures, although of a rather bad quality. As an example we prove the following result.

Theorem 2 *Let $g \geq 2$ and $h \geq 2$. Let $\{g_i\}_{i=1}^\infty$ be a sequence of integers greater than 1 and bounded by g. Let $\{n_i\}_{i=1}^\infty$ be a strictly increasing sequence of positive integers such that $n_i \leq C_0 n_{i-1}^2$ for $i = 2, 3, \ldots$. Let a be defined by (2). Then there exists an effectively computable positive number C depending only on C_0, g and h such that, for all $(P, Q) \in \mathbb{Z} \times \mathbb{N}$,*

$$\left| a - \frac{P}{Q} \right| \geq e^{-CQ^2}.$$

It is possible to derive an irrationality measure independent of g, but this measure might well be much worse. In order to compute such a measure one has to make all the estimates in the proof of Theorem 1 quantitative. It is interesting to compare Theorem 2 with the irrationality measure given in Theorem 3 of Becker [2].

Sander studied also the case that the sequence $\{n_i\}_{i=1}^\infty$ is non-periodic and bounded. He showed for this case that if $a(g)$ is rational, then $g^{N_1} \equiv g^{N_2}$ (mod h) for some pair of distinct limit points N_1, N_2 of $\{n_i\}_{i=1}^\infty$. In fact, there exist limit points $N_1 < N_2$ of $\{n_i\}_{i=1}^\infty$ such that $g^{N_1} \equiv g^{N_2} \pmod{h^l}$ where l is the h-length of g^{N_1}. This follows from part (a) of the following result.

Theorem 3 *Let $h \geq 2$. Let $\{g_i^{n_i}\}_{i=1}^\infty$ be a bounded, non-periodic sequence of perfect powers such that the resulting concatenation number a is rational.* (a) *Then there exists a pair of limit points $g_i^{n_i} < g_j^{n_j}$ such that*

$$g_i^{n_i} \equiv g_j^{n_j} \pmod{h^{L_0}} \tag{3}$$

where L_0 is the h-length of $g_i^{n_i}$. (b) *Then there exists a pair of limit points $g_i^{n_i} < g_j^{n_j}$ such that*

$$0 \leq h^{L_1} g_j^{n_j} - h^{L_2} g_i^{n_i} < h^{L_2} \tag{4}$$

where L_1 and L_2 are the h-lengths of $g_i^{n_i}$ and $g_j^{n_j}$, respectively.

Some further results of this type can be found in Becker and Sander [3].

A perfect power is an integer of the form b^n where b and n are integers with $b \geq 0$ and $n \geq 2$. Sander [7] and later Becker and Sander [3] gave necessary and sufficient conditions for the rationality of $a(g)$ in case $\{n_i\}_{i=1}^{\infty}$ has only two limit points. Again we shall extend the results to general perfect powers.

Theorem 4　Let $h \geq 2$. Let $\{g_i^{n_i}\}_{i=1}^{\infty}$ be a bounded, non-periodic sequence of perfect powers with exactly two limit points, $g_i^{n_i} < g_j^{n_j}$, both positive. Then the resulting number a is rational if and only if there exist positive integers b, k, l and r with $h^{r-1} \leq b < h^r$ such that

$$g_i^{n_i} = b\frac{h^{kr} - 1}{h^r - 1}, \quad g_j^{n_j} = b\frac{h^{lr} - 1}{h^r - 1}. \tag{5}$$

If $g_i = g_j = g$, then $g^{n_j - n_i} = (h^{lr} - 1)/(h^{kr} - 1) \in \mathbb{Z}$ which implies that $k|l$, whence h^{lr} is a power of h^{kr}. This is Sander's result. (Sander allowed exponents 1 and excluded 0.) He applied a result of Shorey and the author to show that in his result g, h, n_i and n_j can be bounded in terms of the smallest prime factor of g. His claim that they can also be bounded in terms of h only is unjustified, although probably true. Saradha and Shorey [8] have given a class of values of h for which they justified Sander's claim.

In Section 2 we apply results from the theory of linear forms in logarithms and its consequences. The elementary proofs of Theorems 3 and 4 are given in Sections 3 and 4. Section 3 contains two general lemmas. The authors thank the referee for pointing out a mistake in an earlier version of the paper.

2. Proofs of Theorems 1 and 2　We apply Theorem 12.1 and Corollary 1.1 of [11]. Let $P \geq 2$ and denote by S' the set of all integers which are composed of primes less than or equal to P. Let k be a non-zero integer. Then we have

Lemma 1　There exists an effectively computable number C depending only on k and P such that equation $ax^m - by^n = k$ in $a \in S', b \in S', x \in S', y \in \mathbb{Z}, m \in \mathbb{Z}$ and $n \in \mathbb{Z}$ with $m > 1, n > 1, x > 1, y > 1$ implies that

$$\max(|a|, |b|, m, n, x, y) \leq C.$$

Lemma 2　If $x > 0$ and $y > 0$ are elements of S' satisfying $x - y = k$, then $\max(x, y)$ is bounded by an effectively computable number depending only on k and P.

In order to prove Theorem 1 we assume that a in (2) is rational. Then its h-adic expansion is periodic with period r, say. Suppose $a_{m+r} = a_m$ for $m \geq M$. Choose N such that the representation of $(g_N^{n_N})_h$ in (2) starts with a_i where $i \geq M$. Consider some $g_i^{n_i}$ with $i \geq N$. Let ν be the least non-negative integer such that $h^\nu g_i^{n_i}$ is an integer. Then

$$h^\nu g_i^{n_i} = (b_{r-1}h^{r-1} + b_{r-2}h^{r-2} + \dots + b_0)(h^{r(k-1)+l} + h^{r(k-2)+l} + \dots + h^l) + c$$

where $(b_{r-1}b_{r-2}...b_0)$ with $b_{r-1} \neq 0$ is a period cycle of the h-adic representation of $g_i^{n_i}$ and $rk + l$ is the h-length of $g_i^{n_i}$ and $l < r$ and $c < h^l$. Hence

$$h^\nu g_i^{n_i} = bh^l \, \frac{h^{rk} - 1}{h^r - 1} + c \quad \text{with} \quad 0 < b < h^r. \tag{6}$$

Note that b, c and l are bounded in terms of h and r. We apply Lemma 1 and Lemma 2 to the resulting equation

$$(bh^\ell)(h^r)^k - (h^r - 1)\frac{h^\nu}{B_i^{n_i}}A_i^{n_i} = bh^\ell - c(h^r - 1)$$

where we observe that the right hand side is non-zero. It follows that $A_i^{n_i} \leq C_1$ and $1 \leq h^\nu/B_i^{n_i} \leq C_2$ where C_1 and C_2 are effectively computable numbers depending only on h and r.

The proof of Theorem 2 is based on the following estimate of Baker [1] for linear forms in logarithms of rational numbers .

Lemma 3 *Let* $b_1, b_2, ..., b_n$ *be integers with* $|b_j| \leq B$ *for* $j = 1, ..., n-1$ *and* $b_n \neq 0$ *where* $B \geq 4$. *Let* $a_1, a_2, ..., a_n$ *be positive rational numbers such that the numerator and denominator of* a_n *have modulus* $\leq A$ *where* $A \geq 2$. *Put* $\Lambda = b_1 \log a_1 + ... + b_n \log a_n$. *Then there exists an effectively computable number* C_3 *depending only on* $n, a_1, a_2, ..., a_{n-1}$ *such that, for any* δ *with* $0 < \delta < \frac{1}{2}$, *either*

$$\Lambda = 0 \quad \text{or} \quad |\Lambda| > \left(\frac{\delta}{|b_n|}\right)^{C_3 \log A} e^{-\delta B}.$$

In order to prove Theorem 2 we observe that Fermat's little theorem implies that the h-adic expansion of P/Q has a period r which divides $\varphi(Q)$. The preperiod of the expansion is at most $2\log Q$. Let i be so large that the number of h-adic digits of $(g_1^{n_1})_h(g_2^{n_2})_h...(g_{i-1}^{n_{i-1}})_h$ exceeds $2\log Q$ and that of $(g_i^{n_i})_h$ exceeds $4r$. Let m be the h-length of $(g_1^{n_1})_h(g_2^{n_2})_h...(g_i^{n_i})_h$ and $kr + l$ with $0 \leq l < r$ the h-length of $(g_i^{n_i})_h$. Hence we require

$$k \geq 4 \quad \text{and} \quad m > 2\log Q + kr + l. \tag{7}$$

We know by the periodicity of the h-adic expansion of P/Q that there exist integers d_1 and b with $0 \leq b < h^r$ such that

$$h^m \frac{P}{Q} = h^{kr+l}(d_1 + bh^{-r} + bh^{-2r} + bh^{-3r} + ...)$$

$$= d_1 h^{kr+l} + b\frac{h^{kr+l}}{h^r - 1}.$$

On the other hand, there exists an integer d_2 such that

$$0 < \mid h^m a - d_2 h^{kr+l} - g_i^{n_i} \mid < 1.$$

Notice that $h^{kr+l-1} \le g_i^{n_i} < h^{kr+l}$. Put $d = d_1 - d_2$. It follows that $|d| > 1$ implies

$$h^m \left| a - \frac{P}{Q} \right| \ge |d| h^{kr+l} - \left| b \frac{h^{kr+l}}{h^r - 1} - g_i^{n_i} \right| - 1 \ge h^{kr+l}. \qquad (8)$$

We now assume $|d| \le 1$. Then we have

$$h^m \left| a - \frac{P}{Q} \right| \ge \left| dh^{kr+l} - g_i^{n_i} + b \frac{h^{kr+l}}{h^r - 1} \right| - 1.$$

If $d + \frac{b}{h^r - 1} \le \frac{1}{h^2}$, then we obtain

$$h^m \left| a - \frac{P}{Q} \right| \ge g_i^{n_i} - h^{kr+l-2} - 1 \ge h^{kr+l-2} - 1. \qquad (9)$$

If $d + \frac{b}{h^r - 1} > \frac{1}{h^2}$, then we write

$$h^m \left| a - \frac{P}{Q} \right| \ge g_i^{n_i} \left| \frac{h^{kr+l}(dh^r - d + b)}{g_i^{n_i}(h^r - 1)} - 1 \right| - 1.$$

By $C_4, C_5, \ldots$, we shall denote effectively computable positive numbers depending only on g and h. By Lemma 3 with $\delta = 0.1$, we have

$$\left| (kr + l) \log h - n_i \log g_i + \log \frac{dh^r + b - d}{h^r - 1} \right| > (\frac{1}{10})^{C_4 \log A} e^{-B/10}$$

where $A < 2h^r$ and $B = \max(kr + l, n_i) \le (kr + l)\max(1, \log h / \log g_i)$. Hence, using that $|x - 1| \ge C_5 |\log x|$ for $x > \frac{1}{h^2}$,

$$h^m \left| a - \frac{P}{Q} \right| \ge C_5 g_i^{n_i} \left| \log \frac{h^{kr+l}}{g_i^{n_i}} \cdot \frac{dh^r - d + b}{h^r - 1} \right| - 1$$

$$\ge C_5 \, h^{kr+l-1} h^{-C_6 r} \, e^{-\frac{1}{10}(kr+l)\frac{\log h}{\log 2}} - 1 > C_5 \, h^{\frac{1}{2}(kr+l)-C_6 r} - 1.$$

Therefore

$$h^m \left| a - \frac{P}{Q} \right| > h^{\frac{1}{3}(kr+l)} \qquad (10)$$

provided that $k > C_7 \ge 4$. In view of (8) and (9) inequality (10) holds independently of the value of d whenever $k > C_7$.

Let i be the smallest integer > 1 such that $n_i > C_7 Q$. There exist positive numbers C_8 and C_9 such that the h-length of $g_j^{n_j}$ is in between $C_8 n_j$ and $C_9 n_j$ for $j = 1, 2, \ldots$. Hence the h-length of the concatenation $(g_1^{n_1})_h (g_2^{n_2})_h \cdots (g_{i-1}^{n_{i-1}})_h$ is at most $C_9 n_{i-1}^2$. Thus by the conditions on n_i,

$$m \leq C_9(n_{i-1}^2 + n_i) \leq C_9(1 + C_0)n_{i-1}^2 \leq C_{10}Q^2.$$

On substituting this estimate into (10) we obtain the inequality claimed in Theorem 2. Furthermore (7) is satisfied, since $k > C_7 \geq 4$ and $m - (kr + l) \geq n_{i-1} > (n_i/C_0)^{1/2} > C_{11}Q^{1/2} > 2 \log Q$ for $Q \geq Q_0$. For $Q < Q_0$ we can adjust the constant C, if necessary.

3. Some general results. Let a be a rational number in $(0, 1)$ and h an integer with $h \geq 2$. Let the h-adic expansion of a be given by

$$a = 0.a_1 a_2 a_3 \ldots .$$

Then the sequence $\{a_i\}$ is periodic. Let r be its minimal period. Let S be a set of positive integers. Let $\{s_i\}_{i=1}^{\infty}$ be a sequence with terms from S such that

$$a = 0.(s_1)_h (s_2)_h (s_3)_h \ldots . \tag{11}$$

Lemma 4 *If the sequence $\{s_i\}_{i=1}^{\infty}$ is not periodic, then there exist two strings $(t_1, t_2, \ldots, t_\kappa)$ and $(u_1, u_2, \ldots, u_\lambda)$ of elements of S with $t_1 \neq u_1$ and $t_\kappa \neq u_\lambda$ such that*

$$(t_1)_h (t_2)_h \cdots (t_\kappa)_h = (u_1)_h (u_2)_h \cdots (u_\lambda)_h. \tag{12}$$

Proof Choose M so large that $a_{m+r} = a_m$ for $m \geq M$. By the box principle there exists an l with $0 \leq l < r$ such that a_{kr+l} is the first digit of the h-adic expansion of some $(s_i)_h$ in (11) for infinitely many integers k. Let a_{Kr+l} with $Kr + l > M$ be the first digit of $(s_T)_h$. Let U be the smallest positive integer such that $(s_T)_h (s_{T+1})_h \cdots (s_{T+U-1})_h$ has h- length $L_1 r$ divisible by r. Since $\{s_i\}_{i=1}^{\infty}$ is not periodic, there exists an integer $K' > K$ such that $a_{K'r+l}$ is the first digit of $(s_V)_h$, but that not $s_T = s_V, s_{T+1} = s_{V+1}, \ldots, s_{T+U-1} = s_{V+U-1}$. Let W be the smallest positive integer such that $(s_V)_h (s_{V+1})_h \cdots (s_{V+W-1})_h$ has h-length $L_2 r$ divisible by r. Then the concatenation of L_2 strings $(s_T)_h (s_{T+1})_h \cdots (s_{T+U-1})_h$ equals the concatenation of L_1 strings $(s_V)_h (s_{V+1})_h \cdots (s_{V+W-1})_h$, since both are $L_1 L_2$ periodic parts of h-length r of a, but the corresponding strings are unequal,

$$(s_T, s_{T+1}, \ldots, s_{T+U-1}, s_T, s_{T+1}, \ldots) \neq (s_V, s_{V+1}, \ldots, s_{V+W-1}, s_V, s_{V+1}, \ldots).$$

By deleting on both sides the equal elements from the beginning as well as from the end in both strings, we obtain two strings as claimed in the lemma.

Lemma 5 *If in the above notation $S = \{\sigma, \tau\}$ with $\sigma < \tau$ and $\{s_i\}_{i=1}^{\infty}$ is not periodic, then the h-lengths of σ and τ are divisible by r.*

Proof By Lemma 4 there exist strings $(t_1, t_2, ..., t_\kappa)$ and $(u_1, u_2, ..., u_\lambda)$ of elements of S with $t_1 < u_1$ and $t_\kappa \neq u_\lambda$ such that (12) holds. Hence $t_1 = \sigma, u_1 = \tau$. If $(\sigma)_h = b_1 b_2 ... b_s$ and $(\tau)_h = c_1 c_2 ... c_t$, then apparently $b_1 = c_1$, $b_2 = c_2, ..., b_s = c_s$. Since t_2 is either σ or τ, we find that the digits of $(\tau)_h$ are purely periodic with period s. Let v be the minimal period of $(\sigma)_h$ with $v|s$. Then v is also a period of $(\tau)_h$. Let $t = Ks + L$ with $0 \leq L < s$. Suppose $L > 0$. Since u_2 is either σ or τ, we see by comparing both sides of (12) that $b_1 = b_{L+1}, b_2 = b_{L+2}, ..., b_s = b_{L+s}$ where the indices have to be read modulo s. This implies that L is a period of $(\sigma)_h$ and $v|L$. It follows that $v|t$. Since the h-adic expansion of a is a concatenation of σ's and τ's, v is also the minimal period of the h- adic expansion of a. Thus $v = r$ and the h-lengths of σ and τ are both divisible by r.

4. Proofs of Theorems 3 and 4 First Theorem 3. Let r be the minimal period of the h-adic expansion of a. Let M be so large that $a_{m+r} = a_m$ for $m \geq M$ and that every $g_i^{n_i}$ with $i \geq M$ is a limit point of $\{g_i^{n_i}\}_{i=1}^{\infty}$. According to Lemma 4 applied to $\{g_i^{n_i}\}_{i=M}^{\infty}$ there exist two strings

$$(g_T^{n_T}, g_{T+1}^{n_{T+1}}, ..., g_U^{n_U}) \quad \text{and} \quad (g_V^{n_V}, g_{V+1}^{n_{V+1}}, ..., g_W^{n_W})$$

with $M < T \leq U < V \leq W$ such that $g_T^{n_T} \neq g_V^{n_V}$ and $g_U^{n_U} \neq g_W^{n_W}$ whereas

$$(g_T^{n_T})_h \, (g_{T+1}^{n_{T+1}})_h \, \cdots \, (g_U^{n_U})_h = (g_V^{n_V})_h \, (g_{V+1}^{n_{V+1}}) \cdots (g_W^{n_W})_h.$$

Suppose $g_U^{n_U} < g_W^{n_W}$. (The case $g_U^{n_U} > g_W^{n_W}$ is similar.) Let L_0 be the h-length of $g_U^{n_U}$. Then

$$g_U^{n_U} \equiv g_W^{n_W} \quad (\mathrm{mod}\ h^{L_0}).$$

This proves (a). Now suppose $g_T^{n_T} < g_V^{n_V}$. Let L_1 be the h-length of $g_T^{n_T}$ and L_2 the h-length of $g_V^{n_V}$. Then

$$0 \leq g_V^{n_V} - h^{L_2 - L_1} \, g_T^{n_T} < h^{L_2 - L_1}.$$

Hence

$$0 \leq h^{L_1} \, g_V^{n_V} - h^{L_2} \, g_T^{n_T} < h^{L_2}$$

which is (4). This proves (b).

Finally, the proof of Theorem 4.

$\Longleftarrow$ From some M on $\{a_i\}_{i=M}^{\infty}$ is only composed of $(g_i^{n_i})_h$ and $(g_j^{n_j})_h$. By (5) both $g_i^{n_i}$ and $g_j^{n_j}$ have h-lengths divisible by r and consist of concatenations of $(b)_h$'s. It follows that $\{a_i\}_{i=M}^{\infty}$ is purely periodic with period r. Hence a is rational.

$\Longrightarrow$ Suppose a is rational. Choose M and r so that $\{a_i\}_{i=M}^{\infty}$ is purely periodic with minimal period r and is composed of $(g_i^{n_i})_h$ and $(g_j^{n_j})_h$ only. By Lemma 5 the h-lengths of both $(g_i^{n_i})_h$ and $(g_j^{n_j})_h$ are divisible by r. Suppose $(g_i^{n_i})_h$ consists of the concatenation of k periodic cycles $(b)_h$ of length r. Then $(g_j^{n_j})_h$ consists of an integral number of $(b)_h$'s too, l say. Now (5) follows. Obviously $b < h^r$. Furthermore $b \geq h^{r-1}$, since the h-length of $g_i^{n_i}$ is divisible by r.

References

[1] A. Baker. A sharpening of the bounds for linear forms in logarithms II. *Acta Arith.*, **24** (1973), 33–36.

[2] P.-G. Becker. Exponential diophantine equations and the irrationality of certain numbers. *J. Number Th.*, **39** (1991), 108–116.

[3] P.-G. Becker and J.W. Sander. Irrationality and codes. *Semigroup Forum*, **51** (1995), 117–124.

[4] P. Bundschuh. Generalization of a recent irrationality result of Mahler. *J. Number Th.*, **19** (1984), 248–253.

[5] K. Mahler. On some irrational decimal fractions. *J. Number Th.*, **13** (1981), 268–269.

[6] H. Niederreiter. On an irrationality theorem of Mahler and Bundschuh. *J. Number Th.*, **24** (1986), 197–199.

[7] J.W. Sander. Irrationality criteria for Mahler's numbers. *J. Number Th.*, **52** (1995), 145–156.

[8] N. Saradha and T.N. Shorey. The equation $\frac{x^n-1}{x-1} = y^q$ with x square, to appear.

[9] Z. Shan. A note on the irrationality of some numbers. *J. Number Th.*, **25** (1987), 211–212.

[10] Z. Shan and E.T.H. Wang. Generalization of a theorem of Mahler. *J. Number Th.*, **32** (1989), 111–113.

[11] T.N. Shorey and R. Tijdeman. *Exponential Diophantine Equations*. Cambridge Univ. Press, London, 1986.

[12] H. Yu. A note on a theorem of Mahler. *J. China Univ. Sci. Tech.*, **18**(3) (1988), 388–389.

Tarlok N. Shorey
School of Mathematics, Tata Institute of Fundamental Research
Homi Bhabha Road, Bombay 400 005, India

Robert Tijdeman
Mathematical Institute, Leiden University
P.O. Box 9512, 2300 RA Leiden, The Netherlands

22

Hypergeometric Functions and Irrationality Measures

CARLO VIOLA

1. Some arithmetical properties of the values of suitable hypergeometric functions at special rational points, and in particular the irrationality measures of such values, have been extensively studied during the last decades. The references at the end of this note represent a (largely incomplete) list of papers dealing with this or related subjects. Essentially, one considers Padé-type approximations to the hypergeometric functions involved yielding especially good rational approximations at suitable rational values of the variable. In several interesting cases, the irrationality results obtained through this general principle can be improved by eliminating common prime factors of the values of the approximating polynomials at the points considered. This elimination method originated in Siegel's work [18], and received new attention after Chudnovsky's paper [7].

The usual way of applying Siegel's elimination method is based on the analysis of the p-adic valuation of suitable binomial coefficients. Instead of this, the possibility of using the p-adic valuation of the gamma–factors occurring in the Euler–Pochhammer integral representation of the hypergeometric functions $_{n+1}F_n$, together with the symmetry properties of these functions, seems to have been generally overlooked. In Rhin and Viola's paper [16] this idea is combined with a group–theoretic approach, thus yielding symmetric statements about the p-adic valuation of rational approximations to $\zeta(2) = \sum_{n=1}^{\infty} n^{-2} = \pi^2/6 = {}_3F_2(1,1,1;2,2;1)$, and hence an improvement on the irrationality measure of this number previously obtained by Hata [11].

In this note we show how the analogue for one–dimensional Euler–Pochhammer integrals of the method developed in [16] can be applied to obtain very easily good irrationality measures for the values of the logarithm at rational points (see the inequality (16) below). Although the group structure underlying the one–dimensional case is almost trivial (see the remark at the end of this paper), our method yields the best known irrationality measures of a class of logarithms of rational numbers (roughly, when the rational numbers have either height or distance from 1 small enough). As a special instance, we get a simple proof of the best known irrationality measure of log 2, namely Rukhadze's result 3.89139978 (see [17]).

For special choices of our parameters, the estimate (16) below coincides with the inequality given in Theorem 2 of Heimonen, Matala–aho and Väänänen's recent paper [13]. These authors consider integrals involving Legendre–type polynomials, and apply Siegel's method using the p-adic valuation of products of binomial coefficients. The proof of our inequality (16) appears to be simpler than that of Theorem 2 of [13], and it is interesting to compare the two methods.

An alternative approach to the search for irrationality measures of logarithms of rational numbers is due to Amoroso [2]. Amoroso's paper is independent of Siegel's method, and combines the properties of weighted integer transfinite diameters of suitable real intervals with a method introduced by Dvornicich and Viola [8] and Rhin [15] independently.

2. Let h, j, l be integers satisfying $h > \max\{0, -l\}$, $j > \max\{0, l\}$, and let $z > -1$, $z \neq 0$. Define

$$I(h, j, l; z) = z^{h+j+1}(1 + z)^{\max\{0, -l\}} \int_0^1 \frac{x^h(1 - x)^j}{(1 + xz)^{j-l}} \frac{dx}{1 + xz}. \tag{1}$$

With the change of variable $1 + xz = t$ we obtain

$$I(h, j, l; z) = (1 + z)^{\max\{0, -l\}} \int_1^{1+z} \frac{(t - 1)^h(1 + z - t)^j}{t^{j-l}} \frac{dt}{t}.$$

Hence

$$I(h, j, l; z) = (1 + z)^{\max\{0, -l\}} \sum_{k=0}^{h} \sum_{m=0}^{j} (-1)^{h+j-k-m}$$

$$\times \binom{h}{k}\binom{j}{m}(1 + z)^m \int_1^{1+z} t^{k+l-m-1} dt$$

$$= a(1 + z) + b(1 + z) \log(1 + z),$$

where

$$a(1 + z) = \sum_{k=0}^{h} \sum_{\substack{m=0 \\ m \neq k+l}}^{j} (-1)^{h+j-k-m} \binom{h}{k}\binom{j}{m}$$

$$\times \frac{(1 + z)^{\max\{k, k+l\}} - (1 + z)^{\max\{m, m-l\}}}{k + l - m} \tag{2}$$

and

$$b(1 + z) = (-1)^{h+j-l}(1 + z)^{\max\{0, l\}}$$

$$\times \sum_{k=\max\{0, -l\}}^{\min\{h, j-l\}} \binom{h}{k}\binom{j}{k+l}(1 + z)^k. \tag{3}$$

Let $d_n = \text{l.c.m.}\{1, 2, \ldots, n\}$. If we define

$$M = \max\{j - l,\, h + l\},$$

we plainly have

$$d_M a(1 + z) \in \mathbb{Z}[1 + z], \quad b(1 + z) \in \mathbb{Z}[1 + z]. \tag{4}$$

Note that $b(1 + z)$ is easily expressed as a contour integral, since, for any $\varrho > 0$,

$$\frac{1}{2\pi i} \int_{|t| = \varrho} \frac{(t - 1)^h (1 + z - t)^j}{t^{j-l}} \frac{dt}{t}$$

$$= \sum_{k=0}^{h} \sum_{m=0}^{j} (-1)^{h+j-k-m} \binom{h}{k} \binom{j}{m} (1 + z)^m \frac{1}{2\pi i} \int_{|t| = \varrho} t^{k+l-m-1} dt$$

$$= (-1)^{h+j-l} (1 + z)^l \sum_{k=\max\{0, -l\}}^{\min\{h,\, j-l\}} \binom{h}{k} \binom{j}{k + l} (1 + z)^k.$$

Therefore

$$b(1 + z) = (1 + z)^{\max\{0, -l\}} \frac{1}{2\pi i} \int_{|t| = \varrho} \frac{(t - 1)^h (1 + z - t)^j}{t^{j-l}} \frac{dt}{t}. \tag{5}$$

We now apply the Euler–Pochhammer integral representation

$$_2F_1(\alpha, \beta; \gamma; y) = \frac{\Gamma(\gamma)}{\Gamma(\beta)\, \Gamma(\gamma - \beta)} \int_0^1 \frac{x^{\beta-1} (1 - x)^{\gamma-\beta-1}}{(1 - xy)^\alpha} \, dx,$$

valid for $\text{Re}\,\gamma > \text{Re}\,\beta > 0$, of the Gauss hypergeometric function

$$_2F_1(\alpha, \beta; \gamma; y) = \sum_{n=0}^{\infty} \frac{(\alpha)_n\, (\beta)_n}{(\gamma)_n} \cdot \frac{y^n}{n!},$$

where the Pochhammer symbols are defined by

$$(\xi)_0 = 1, \quad (\xi)_n = \xi(\xi + 1) \cdots (\xi + n - 1) \quad (n = 1, 2, \ldots).$$

Since

$$_2F_1(\alpha, \beta; \gamma; y) = {}_2F_1(\beta, \alpha; \gamma; y),$$

for $\text{Re}\,\gamma > \max\{\text{Re}\,\alpha,\, \text{Re}\,\beta\}$ and $\min\{\text{Re}\,\alpha,\, \text{Re}\,\beta\} > 0$ we obtain

$$\int_0^1 \frac{x^{\beta-1}(1 - x)^{\gamma-\beta-1}}{(1 - xy)^\alpha} \, dx = \frac{\Gamma(\beta)\, \Gamma(\gamma - \beta)}{\Gamma(\alpha)\, \Gamma(\gamma - \alpha)} \int_0^1 \frac{x^{\alpha-1}(1 - x)^{\gamma-\alpha-1}}{(1 - xy)^\beta} \, dx.$$

Choosing $\alpha = j - l + 1$, $\beta = h + 1$, $\gamma = h + j + 2$, $y = -z$, we get

$$I(h, j, l; z) = \frac{h!\,j!}{(j-l)!\,(h+l)!}\, I(j - l, h + l, l; z).$$ (6)

In order to use (6) we need

$$M^* = \max\{h,\,j\} \leq M = \max\{j - l,\, h + l\},$$ (7)

which we shall henceforth assume. Note that (7) implies

$$j - l < j \leq h + l, \quad \text{if } l > 0,$$
$$h + l < h \leq j - l, \quad \text{if } l < 0,$$

whence

$$M = \begin{cases} h + l, & \text{if } l > 0, \\ j - l, & \text{if } l < 0. \end{cases}$$ (8)

From (2), (3), (7), and (8) we obtain

$$\deg a(1 + z) \leq \max\{j - l,\, h + l\} = M$$ (9)

and

$$\deg b(1 + z) \leq \max\{h,\,j\} = M^* \leq M.$$ (10)

We now take a rational number $z = r/s$, with integers r and s satisfying $r \neq 0$, $s \geq 1$, $r > -s$, $(r, s) = 1$. By (4), (9), and (10) we have

$$s^M d_M a(1 + r/s) \in \mathbb{Z}, \quad s^M b(1 + r/s) \in \mathbb{Z}.$$ (11)

For $n = 1, 2, \ldots$ let

$$I_n = I(hn, jn, ln; r/s) = a_n + b_n \log(1 + r/s)$$

and

$$I_n^* = I\big((j - l)n,\, (h + l)n,\, ln;\, r/s\big) = a_n^* + b_n^* \log(1 + r/s).$$

The transformation formula (6) yields

$$\big((j - l)n\big)!\,\big((h + l)n\big)!\,a_n = (hn)!\,(jn)!\,a_n^*.$$

Multiplying by $s^{Mn} d_{Mn}$ we get

$$\big((j - l)n\big)!\,\big((h + l)n\big)!\,A_n = K(hn)!\,(jn)!\,A_n^*,$$

where, by (11), $A_n = s^{Mn} d_{Mn} a_n$, $A_n^* = s^{Mn} d_{M^* n} a_n^*$ and $K = d_{Mn}/d_{M^* n}$ are integers since $M^* \leq M$. By standard arguments (see [16], pp. 44–45), any prime $p > \sqrt{Mn}$ for which

$$[(j-l)\omega] + [(h+l)\omega] < [h\omega] + [j\omega], \tag{12}$$

where $\omega = \{n/p\}$, is such that $p|A_n$.

Let Ω be the set of $\omega \in [0, 1)$ satisfying (12), and let

$$\Delta_n = \prod_{\substack{p > \sqrt{Mn} \\ \{n/p\} \in \Omega}} p, \quad D_n = \frac{d_{Mn}}{\Delta_n}.$$

We have $s^{Mn} D_n a_n = A_n/\Delta_n \in \mathbb{Z}$. By (7), if $\omega < 1/M$ then $\omega < \min\{1/h, 1/j\}$, whence $[h\omega] = [j\omega] = 0$, $\omega \notin \Omega$. Thus $p|\Delta_n$ yields $\omega = \{n/p\} \in \Omega$, $n/p \geq \omega \geq 1/M$, $p \leq Mn$, $p|d_{Mn}$. Hence $D_n \in \mathbb{Z}$. Also, by (11), $s^{Mn} b_n \in \mathbb{Z}$. In order to get an irrationality measure of $\log(1 + r/s)$, we can apply Lemma 4.3 of [16] to

$$s^{Mn} D_n I_n = s^{Mn} D_n a_n + s^{Mn} D_n b_n \log(1 + r/s).$$

Again by standard arguments ([16], p. 51) we have

$$\lim_{n \to \infty} \frac{1}{n} \log\big(s^{Mn} D_n\big) = M \log s + M - \int_\Omega d\psi(x), \tag{13}$$

where $\psi(x) = \Gamma'(x)/\Gamma(x)$.

Let x_0 and x_1 be the stationary points $\neq 0, 1$ of the function

$$f(x) = \frac{x^h (1-x)^j}{\big(1 + (r/s)x\big)^{j-l}},$$

i.e., the solutions of

$$r(h+l)x^2 + \big(s(h+j) - r(h+l-j)\big)x - sh = 0,$$

with $0 < x_0 < 1$ and $1 + (r/s)x_1 < 0$. Then, by (1),

$$\lim_{n \to \infty} \frac{1}{n} \log |I_n| = (h+j)\log|r/s|$$
$$+ \max\{0, -l\}\log(1 + r/s) + \log f(x_0), \tag{14}$$

and, by (5),

$$\frac{1}{n} \log |b_n| \le \max\{0,\, -l\} \log(1 + r/s)$$
$$+ \min_{\varrho > 0} \log \frac{(1 + \varrho)^h (1 + r/s + \varrho)^j}{\varrho^{j-l}}.$$

With the change of variable $\varrho = -1 - (r/s)x$ we obtain

$$\min_{\varrho > 0} \log \frac{(1 + \varrho)^h (1 + r/s + \varrho)^j}{\varrho^{j-l}} = \log \left(|r/s|^{h+j} |f(x_1)| \right).$$

Hence

$$\limsup_{n \to \infty} \frac{1}{n} \log |b_n| \le \max\{0,\, -l\} \log(1 + r/s)$$
$$+ (h + j) \log |r/s| + \log |f(x_1)|. \tag{15}$$

Let $\mu(\alpha)$ denote the least irrationality measure of an irrational number α. Let

$$U = \log |f(x_1)| - \log f(x_0)$$

and

$$V = -\log f(x_0) + \int_\Omega d\psi(x) - M(1 + \log s)$$
$$- (h + j) \log |r/s| + \min\{0,\, l\} \log(1 + r/s).$$

From (13), (14), (15) and Lemma 4.3 of [16] we obtain

$$\mu\big(\log(1 + r/s)\big) \le \frac{U}{V}, \tag{16}$$

provided that $V > 0$.

In the case $r = s = 1$ we have

$$\mu(\log 2) \le \frac{\log |f(x_1)| - \log f(x_0)}{-\log f(x_0) + \int_\Omega d\psi(x) - h - l} \tag{17}$$

for any integers h, j, l such that $h > 0$, $j > l > 0$, $j \le h + l$ and

$$-\log f(x_0) > h + l - \int_\Omega d\psi(x).$$

With the choice $h = j = 7$, $l = 1$, the set Ω is the union of the intervals

$$\left[\tfrac{1}{7}, \tfrac{1}{6}\right), \quad \left[\tfrac{2}{7}, \tfrac{1}{3}\right), \quad \left[\tfrac{3}{7}, \tfrac{1}{2}\right), \quad \left[\tfrac{4}{7}, \tfrac{5}{8}\right), \quad \left[\tfrac{5}{7}, \tfrac{3}{4}\right), \quad \left[\tfrac{6}{7}, \tfrac{7}{8}\right),$$

and we have

$$\int_\Omega d\psi(x) = 2.31440700\ldots,$$

$$-\log f(x_0) = 11.98832512\ldots, \quad \log|f(x_1)| = 12.53812524\ldots.$$

Thus the right side of (17) is < 3.89139978, i.e., Rukhadze's irrationality measure of $\log 2$ ([17]).

In the special case $h = j > l \geq 0$, $|r| \leq s$, the above inequality (16) is a reformulation of Theorem 2 of [13] (the parameter α appearing in Theorem 2 of [13] is $1 - l/h$ in our notation). The table given on p. 186 of [13] for the irrationality measures of logarithms of several rational numbers should be compared with the numerical results obtained by Amoroso [2].

Remark With the change of variable $x = (1 - u)/(1 + uz)$ the integral (1) becomes

$$I(h, j, l; z) = z^{h+j+1}(1 + z)^{\max\{0,l\}} \int_0^1 \frac{u^j(1 - u)^h}{(1 + uz)^{h+l}} \frac{du}{1 + uz}$$

$$= I(j, h, -l; z).$$

From this and the transformation formula (6) we see that in the present case the analogues of the permutation groups $\mathbf{T}$ and $\mathbf{\Phi}$ considered in [16], Section 3, are isomorphic to the additive groups $\mathbb{Z}/2\mathbb{Z}$ and $\mathbb{Z}/2\mathbb{Z} \times \mathbb{Z}/2\mathbb{Z}$, respectively.

References

[1] K. Alladi and M.L. Robinson. Legendre polynomials and irrationality. *J. reine angew. Math.*, **318** (1980), 137–155.

[2] F. Amoroso. f-transfinite diameter and number theoretic applications. *Ann. Inst. Fourier*, **43** (1993), 1179–1198.

[3] A. Baker. Rational approximations to $\sqrt[3]{2}$ and other algebraic numbers. *Quart. J. Math. Oxford*, (2) **15** (1964), 375–383.

[4] M.A. Bennett. Simultaneous rational approximation to binomial functions. *Trans. Amer. Math. Soc.*, **348** (1996), 1717–1738.

[5] F. Beukers. A note on the irrationality of $\zeta(2)$ and $\zeta(3)$. *Bull. London Math. Soc.*, **11** (1979), 268–272.

[6] F. Beukers, T. Matala–aho and K. Väänänen. Remarks on the arithmetic properties of the values of hypergeometric functions. *Acta Arith.*, **42** (1983), 281–289.

[7] G.V. Chudnovsky. On the method of Thue–Siegel. *Ann. Math.*, (2) **117** (1983), 325–382.

[8] R. Dvornicich and C. Viola. Some remarks on Beukers' integrals. *Colloquia Math. Soc. János Bolyai*, **51** (1987), 637–657.

[9] M. Hata. Irrationality measures of the values of hypergeometric functions. *Acta Arith.*, **60** (1992), 335–347.

[10] —. Rational approximations to the dilogarithm. *Trans. Amer. Math. Soc.*, **336** (1993), 363–387.

[11] —. A note on Beukers' integral. *J. Austral. Math. Soc.*, (A) **58** (1995), 143–153.

[12] —. The irrationality of $\log(1 + 1/q)\log(1 - 1/q)$. *Trans. Amer. Math. Soc.*, in press.

[13] A. Heimonen, T. Matala-aho and K. Väänänen. On irrationality measures of the values of Gauss hypergeometric function. *Manuscripta Math.*, **81** (1993), 183–202.

[14] M. Huttner. Irrationalité de certaines intégrales hypergéométriques. *J. Number Th.*, **26** (1987), 166–178.

[15] G. Rhin. Approximants de Padé et mesures effectives d'irrationalité. *Progr. in Math.*, **71** (1987), 155–164.

[16] G. Rhin and C. Viola. On a permutation group related to $\zeta(2)$. *Acta Arith.*, **77** (1996), 23–56.

[17] E.A. Rukhadze. A lower bound for the approximation of $\ln 2$ by rational numbers. *Vestnik Moskov. Univ. Ser. I Mat. Mekh.*, 1987 (6), 25–29. (Russian)

[18] C.L. Siegel. Über einige Anwendungen diophantischer Approximationen. *Abh. Preuss. Akad. Wiss.*, **1** (1929).

Carlo Viola
Dipartimento di Matematica, Università di Pisa
Via Buonarroti 2, 56127 Pisa, Italy

23

Forms in Many Variables

TREVOR D. WOOLEY*

1. Introduction A system of homogeneous polynomials with rational coefficients has a non-trivial rational zero provided only that these polynomials are of odd degree, and the system has sufficiently many variables in terms of the number and degrees of these polynomials. While this striking theorem of Birch [1] addresses a fundamental diophantine problem in engagingly simple fashion, the problem of determining a satisfactory bound for the number of variables which suffice to guarantee the existence of a non-trivial zero remains unanswered in any but the simplest cases. Sophisticated versions of the Hardy–Littlewood method have been developed, first by Davenport [4] to show that 16 variables suffice for a single cubic form, and more recently by Schmidt [10] to show that $(10r)^5$ variables suffice for a system of r cubic forms. Unfortunately even Schmidt's highly developed version of the Hardy-Littlewood method is discouragingly ineffective in handling systems of higher degree (see [11, 12]). The object of this paper is to provide a method for obtaining explicit bounds for the number of variables required in Birch's Theorem. Our approach to this problem will involve the Hardy–Littlewood method only indirectly, being motivated by the elementary diagonalisation method of Birch. Although it has always been supposed that Birch's method would necessarily lead to bounds too large to be reasonably expressed, we are able to reconfigure the method so as to obtain estimates which in general are considerably sharper than those following from Schmidt's methods (see forthcoming work [15] for amplification of this remark). Indeed, for systems of quintic forms our new bounds might, at a stretch, be considered "reasonable".

In order to describe our conclusions we require some notation. When k is a field, d and r are natural numbers, and m is a non-negative integer, let $v_{d,r}^{(m)}(k)$ denote the least integer (if any such integer exists) with the property that whenever $s > v_{d,r}^{(m)}(k)$, and $f_i(\mathbf{x}) \in k[x_1, \ldots, x_s]$ $(1 \le i \le r)$ are forms of degree d, then the system of equations $f_i(\mathbf{x}) = 0$ $(1 \le i \le r)$ possesses a solution set which contains a k-rational linear space of projective dimension m. If no such integer exists, define $v_{d,r}^{(m)}(k)$ to be $+\infty$. We abbreviate $v_{d,r}^{(0)}(k)$ to $v_{d,r}(k)$, and define $\phi_{d,r}(k)$ in like manner, save that the arbitrary forms of degree d are restricted to be diagonal.

* Packard Fellow, and supported in part by NSF grant DMS-9622773

In view of the real solubility condition it is plain that $v_{d,r}^{(m)}(\mathbb{Q})$ can be finite only when d is odd. The simplest interesting examples to consider are therefore systems of cubic forms. In §3 we show how, for an arbitrary field k, one may bound $v_{3,r}^{(m)}(k)$ in terms of $\phi_{3,r}(k)$.

Theorem 1 *Let k be a field, let m and r be non-negative integers with $r \geq 1$, and suppose that $\phi_{3,r}(k)$ is finite. Then*

$$v_{3,r}^{(m)}(k) \leq r^3 (m+1)^5 \left(\phi_{3,r}(k) + 1\right)^5.$$

We remark that a modification of our method, which we outline in §3 below, yields a bound of the shape

$$v_{3,r}^{(m)}(k) \ll (m+1)^\alpha, \qquad (1.1)$$

where α is any number exceeding $\frac{1}{2}(5 + \sqrt{17}) = 4.56155\ldots$, and the implicit constant depends at most on k, r and α. Unfortunately the state of knowledge concerning upper bounds for $\phi_{d,r}(k)$ currently leaves much to be desired. Indeed, the only fields for which detailed investigations have thus far been executed are the local fields and $\mathbb{Q}$. Since we have considered local fields elsewhere (see [14]), we restrict attention to the case $k = \mathbb{Q}$, noting merely that recent developments in the theory of the Hardy-Littlewood method over algebraic number fields, when applied within Theorem 1, should yield useful bounds on $v_{3,r}^{(m)}(k)$ also when k is an algebraic field extension of $\mathbb{Q}$.

Corollary *Let r be a natural number, and let m be a non-negative integer. Then*

$$v_{3,r}^{(m)}(\mathbb{Q}) < (90r)^8 (\log(27r))^5 (m+1)^5.$$

For comparison, Lewis and Schulze-Pillot [7, equation (4)] have provided an estimate of the shape

$$v_{3,r}^{(m)}(\mathbb{Q}) \ll r^{11}(m+1) + r^3(m+1)^5, \qquad (1.2)$$

and have also indicated how to refine the latter bound for smaller m to obtain $v_{3,r}^{(m)}(\mathbb{Q}) \ll r^5(m+1)^{14}$. Thus the bound provided by the corollary to Theorem 1, which has strength

$$v_{3,r}^{(m)}(\mathbb{Q}) \ll_\varepsilon r^{8+\varepsilon}(m+1)^5,$$

is stronger than those of Lewis and Schulze-Pillot only for

$$r^{1/3+\varepsilon} \ll_\varepsilon m+1 \ll_\varepsilon r^{3/4-\varepsilon}.$$

Meanwhile, the improvement of our basic bound noted in (1.1) above yields a bound for $v_{3,r}^{(m)}(\mathbb{Q})$ superior to (1.2) whenever m is sufficiently large in terms of r (note, however, that a similar improvement may be put into effect in the work of Lewis and Schulze-Pillot). Also, when $r = 1$, work of Wooley [13, Theorem 2(b)] shows that $v_{3,1}^{(m)}(\mathbb{Q}) \ll (m+1)^2$, whence $v_{3,2}^{(m)}(\mathbb{Q}) \ll (m+1)^4$. Of course, when $m = 0$, so that one is seeking only the existence of rational points on the intersection of r cubic hypersurfaces, Schmidt's bound $v_{3,r}(\mathbb{Q}) < (10r)^5$ (see [10, Theorem 1]) is superior to the conclusion of the corollary.

In §4 we move on to the next most interesting class of examples, considering systems of quintic forms. Without any hypotheses concerning the behaviour of $\phi_{3,r}(k)$, unfortunately, the bounds on $v_{5,r}^{(m)}(k)$ stemming from our methods seem too complicated to merit mention. We therefore restrict attention to the rational field $\mathbb{Q}$. The sharpest estimates that we are able to derive for $v_{5,r}^{(m)}(\mathbb{Q})$ follow by exploiting an estimate of Lewis and Schulze-Pillot [7] for $v_{3,r}^{(m)}(\mathbb{Q})$ within the methods laid out in §2.

Theorem 2 *Let m and r be non-negative integers with $r \geq 1$. Then*

$$v_{5,r}^{(m)}(\mathbb{Q}) < \exp\left(10^{32}\left((m+1)r\log(3r)\right)^{\kappa}\log(3r(m+1))\right),$$

where

$$\kappa = \frac{\log 3430}{\log 4} = 5.87199\ldots.$$

In particular, $v_{5,r}(\mathbb{Q}) = o(e^{r^6})$.

For comparison, Schmidt [12, equation (2.5)] has shown that for a suitable positive constant A one has $v_{5,r}(\mathbb{Q}) \leq \exp(\exp(Ar))$. Thus our new result replaces a doubly exponential bound by one which is essentially single exponential.

We remark that when k is a field for which $\phi_i(k) < \infty$ for $2 \leq i \leq d$, as is the case, for example, for $\mathbb{Q}_p$ and its extensions, then Wooley [14, Theorem 2.4] has shown that

$$v_{d,r}^{(m)}(k) \leq 2(r^2\phi_d(k) + mr)^{2^{d-2}}\prod_{i=2}^{d-1}(\phi_i(k)+1)^{2^{i-2}}$$

(this sharpens an earlier result of Leep and Schmidt [6]; see also Schmidt [9] for a sharper conclusion for systems of cubic forms when $m = 0$). Thus the assumption of a suitable local to global principle would lead, for odd d, to the bound

$$v_{d,r}^{(m)}(\mathbb{Q}) = \sup_{p \text{ prime}} v_{d,r}^{(m)}(\mathbb{Q}_p) \ll_d (r^2 + mr)^{2^{d-2}},$$

a conclusion substantially stronger than those described in Theorems 1 and 2. In view of local obstructions, of course, one has the lower bound $v_{d,r}(\mathbb{Q}) \geq rd^2$, and some workers would even conjecture that the latter lower bound holds with equality for odd d.

For values of d larger than 5 our bound for $v_{d,r}(\mathbb{Q})$ is indescribably weaker, and our conclusions are considerably more complicated to explain. We will discuss such bounds, and the relevance of remarks of Schmidt [12, §2] in this context, on another occasion. Perhaps it is worth noting at this point, however, that subject to non-singularity conditions, stronger bounds are known for the number of variables required to solve systems of equations than have been derived herein (see Birch [2]). The point of the present paper, like that of Birch's original work [1], is to provide such conclusions without any hypotheses.

We describe our version of Birch's elementary diagonalisation argument in §2, this forming the core of our methods. In broad outline, our strategy is modelled on the original argument of Birch. Our superior conclusions stem from two sources. Firstly, by adapting an argument used by Lewis and Schulze-Pillot [7] to generate large dimensional linear spaces of rational solutions to systems of homogeneous cubic equations, we are able to efficiently generate large dimensional rational linear spaces on which a system of forms becomes diagonal. Roughly speaking, our argument doubles the dimension of the latter linear spaces with each iteration of the method, thereby leading to an exponential advantage over the methods available hitherto. Secondly, since we are able to apply this latter approach to a system of many forms simultaneously, we are able to exploit current knowledge concerning the solubility of systems of diagonal forms in order to avoid the inductive approach previously employed, in which large dimensional rational linear spaces of zeros of a single form are used to solve and remove one form at a time from the system. This second idea dramatically improves the quality of our conclusions.

Throughout, implicit constants in Vinogradov's notation $\ll$ and $\gg$ depend at most on the quantities occurring as subscripts to the notation.

The author gratefully acknowledges the extraordinary generosity and hospitality of the organizers and participants of The Taniguchi International Conference on Analytic Number Theory, during which time many of the ideas underlying this work were refined. The author is grateful to Professor Motohashi, in particular, for his benign tolerance and encouragement during the preparation of this paper.

2. Reduction to diagonal forms In this section we establish a reduction technique which, by rational change of variable, simplifies arbitrary systems of homogeneous polynomials into diagonal ones, albeit in far fewer variables. In order to fully implement our reduction argument we require some additional notation. Given an r-tuple of polynomials $\mathbf{F} = (F_1, \ldots, F_r)$ with coefficients in a field k, denote by $\nu(\mathbf{F})$ the number of variables appearing explicitly in $\mathbf{F}$. We are interested in the existence of solutions, over k, of systems of homogeneous polynomial equations with coefficients in k. When such a solution set contains a linear subspace of the ambient space, we define its dimension to be that when considered as a projective space. When d is a positive odd integer, denote by $\mathcal{G}_d^{(m)}(r_d, r_{d-2}, \ldots, r_1; k)$ the set of $(r_d + r_{d-2} + \cdots + r_1)$-tuples of homogeneous polynomials, of which r_i have degree i for $i = 1, 3, \ldots, d$, with coefficients in k, which possess no non-trivial linear space of solutions of dimension m over k. We define $\mathcal{D}_d^{(m)}(r_d, r_{d-2}, \ldots, r_1; k)$ to be the corresponding set of diagonal homogeneous polynomials. We then define $w_d^{(m)}(\mathbf{r}) = w_d^{(m)}(r_d, r_{d-2}, \ldots, r_1; k)$ by

$$w_d^{(m)}(r_d, r_{d-2}, \ldots, r_1; k) = \sup_{\mathbf{g} \in \mathcal{G}_d^{(m)}(r_d, r_{d-2}, \ldots, r_1; k)} \nu(\mathbf{g}),$$

and we define $\phi_d^{(m)}(\mathbf{r}) = \phi_d^{(m)}(r_d, r_{d-2}, \ldots, r_1; k)$ by

$$\phi_d^{(m)}(r_d, r_{d-2}, \ldots, r_1; k) = \sup_{\mathbf{f} \in \mathcal{D}_d^{(m)}(r_d, r_{d-2}, \ldots, r_1; k)} \nu(\mathbf{f}).$$

We observe for future reference that both $w_d^{(m)}(\mathbf{r})$ and $\phi_d^{(m)}(\mathbf{r})$ are increasing functions of the arguments m and r_i. For the sake of convenience we abbreviate $w_d^{(m)}(r, 0, \ldots, 0; k)$ to $v_{d,r}^{(m)}(k)$, and note that $w_d^{(0)}(r, 0, \ldots, 0; k) = v_{d,r}(k)$. We also abbreviate $\phi_d^{(m)}(r, 0, \ldots, 0; k)$ to $\phi_{d,r}^{(m)}(k)$, and write $\phi_{d,r}(k)$ for $\phi_{d,r}^{(0)}(k)$.

Next, when $m \geq 2$, we define $\mathcal{H}_d^{(m)}(r; k)$ to be the set of r-tuples, $(F_1, \ldots, F_r)$, of homogeneous polynomials of degree d, with coefficients in k, for which no linearly independent k-rational vectors $\mathbf{e}_1, \ldots, \mathbf{e}_m$ exist such that $F_i(t_1 \mathbf{e}_1 + \cdots + t_m \mathbf{e}_m)$ is a diagonal form in $t_1, \ldots, t_m$ for $1 \leq i \leq r$. We then define $\widetilde{w}_d^{(m)}(r) = \widetilde{w}_d^{(m)}(r; k)$ by

$$\widetilde{w}_d^{(m)}(r; k) = \sup_{\mathbf{h} \in \mathcal{H}_d^{(m)}(r)} \nu(\mathbf{h}).$$

Further, we adopt the convention that $\widetilde{w}_d^{(1)}(r; k) = 0$. Note that $\widetilde{w}_d^{(m)}(r; k)$ is an increasing function of the arguments m and r. Moreover, when $s > \widetilde{w}_d^{(m)}(r; k)$ and $F_1, \ldots, F_r$ are homogeneous polynomials of degree d with coefficients in k possessing s variables, then there exist linearly independent k-rational vectors $\mathbf{e}_1, \ldots, \mathbf{e}_m$ with the property that $F_i(t_1 \mathbf{e}_1 + \cdots + t_m \mathbf{e}_m)$ is a diagonal form in $t_1, \ldots, t_m$ for $1 \leq i \leq r$.

Lemma 2.1 *Let d be an odd integer with $d \geq 3$, and let r, n and m be natural numbers. Then*

$$\widetilde{w}_d^{(n+m)}(r; k) \leq s + w_{d-2}^{(M)}(\mathbf{R}; k),$$

where

$$M = \widetilde{w}_d^{(n)}(r;k), \quad s = 1 + w_{d-2}^{(N)}(\mathbf{S};k), \quad N = \widetilde{w}_d^{(m)}(r;k),$$

and for $0 \le u \le (d-1)/2$,

$$R_{2u+1} = r\binom{s+d-2u-2}{d-2u-1} \quad \text{and} \quad S_{2u+1} = r\binom{n+d-2u-2}{d-2u-1}.$$

Proof Write $\delta = (d-1)/2$, and take $\mathcal{N}$ to be an integer with $\mathcal{N} > s + w_{d-2}^{(M)}(\mathbf{R};k)$. For $1 \le j \le r$, consider forms $\mathcal{F}_j$ of degree d, all having $\mathcal{N}$ variables. For $1 \le j \le r$ and $0 \le u \le \delta$ define the polynomials $\mathcal{G}_{ju}$ and $\mathcal{H}_{ju}$ through the expansion

$$\mathcal{F}_j(\mathbf{y}+t\mathbf{x}) = \sum_{u=0}^{\delta}\left(\mathcal{G}_{ju}(\mathbf{y},\mathbf{x})t^{2u+1} + \mathcal{H}_{ju}(\mathbf{y},\mathbf{x})t^{d-2u-1}\right), \qquad (2.1)$$

valid for each $\mathbf{x},\mathbf{y} \in k^{\mathcal{N}}$. Notice that $\mathcal{G}_{ju}(\mathbf{y},\mathbf{x})$ is a form of degree $2u+1$ in $\mathbf{x}$, and of degree $d-2u-1$ in $\mathbf{y}$. Also, $\mathcal{H}_{ju}(\mathbf{y},\mathbf{x})$ is a form of degree $d-2u-1$ in $\mathbf{x}$, and of degree $2u+1$ in $\mathbf{y}$. Let T be an arbitrary, but fixed, k-linear subspace of $k^{\mathcal{N}}$ of affine dimension s, and let $\mathbf{a}_1,\dots,\mathbf{a}_s$ be a basis for T. Let U be any subspace of $k^{\mathcal{N}}$ such that $T \oplus U = k^{\mathcal{N}}$. Consider an arbitrary element of T, say $\mathbf{y} = u_1\mathbf{a}_1 + \cdots + u_s\mathbf{a}_s$, and substitute this expression into $\mathcal{G}_{ju}(\mathbf{y},\mathbf{x})$. We find that the latter polynomial becomes a form of degree $d-2u-1$ in $u_1,\dots,u_s$, whose coefficients are forms of degree $2u+1$ in $\mathbf{x}$. Moreover, following a simple counting argument, one finds that the number of such coefficients of degree $2u+1$ is

$$\binom{(d-2u-1)+(s-1)}{d-2u-1}.$$

Thus, as we consider all $\mathcal{G}_{ju}(\mathbf{y},\mathbf{x})$ with $1 \le j \le r$, we find that the total number of coefficients of degree $2u+1$ which arise is R_{2u+1} ($0 \le u \le \delta$). Then since $\mathcal{N} - s > w_{d-2}^{(M)}(\mathbf{R};k)$, we may conclude thus far that there exists a k-linear subspace, V, of U, with projective dimension M, on which all of the above coefficients of degrees $1,3,\dots,d-2$ vanish. Moreover, for each j one has $\mathcal{G}_{j\delta}(\mathbf{y},\mathbf{x}) = \mathcal{F}_j(\mathbf{x})$. Consequently, for each $\mathbf{x} \in V$ and each $\mathbf{y} \in T$ one has

$$\mathcal{F}_j(\mathbf{y}+t\mathbf{x}) = t^d\mathcal{F}_j(\mathbf{x}) + \sum_{u=0}^{\delta}\mathcal{H}_{ju}(\mathbf{y},\mathbf{x})t^{d-2u-1} \qquad (1 \le j \le r). \qquad (2.2)$$

Next, since $M+1 > \widetilde{w}_d^{(n)}(r;k)$, we deduce that there exist linearly independent vectors $\mathbf{b}_1,\dots,\mathbf{b}_n \in V$ with the property that for each $t_1,\dots,t_n \in k$ one has for $1 \le j \le r$ that $\mathcal{F}_j(t_1\mathbf{b}_1 + \cdots + t_n\mathbf{b}_n)$ is a diagonal form in $t_1,\dots,t_n$

Let W be the linear subspace of $k^{\mathcal{N}}$ spanned by $\mathbf{b}_1, \ldots, \mathbf{b}_n$, and consider an arbitrary element of W, say $\mathbf{x} = v_1 \mathbf{b}_1 + \cdots + v_n \mathbf{b}_n$. On substituting the latter expression into $\mathcal{H}_{ju}(\mathbf{y}, \mathbf{x})$, we find that the latter polynomial becomes a form of degree $d - 2u - 1$ in $v_1, \ldots, v_n$, whose coefficients are forms of degree $2u + 1$ in $\mathbf{y}$. Moreover, following a simple counting argument, one finds that the number of such coefficients of degree $2u + 1$ is

$$\binom{(d - 2u - 1) + (n - 1)}{d - 2u - 1}.$$

Thus, as we consider all $\mathcal{H}_{ju}(\mathbf{y}, \mathbf{x})$ with $1 \leq j \leq r$, we find that the number of coefficients of degree $2u + 1$ which arise is S_{2u+1} $(0 \leq u \leq \delta)$. Then since $s > w_{d-2}^{(N)}(\mathbf{S}; k)$, we may conclude that there exists a k-linear subspace, X, of T, with projective dimension N, on which all of the above coefficients of degrees $1, 3, \ldots, d - 2$ vanish. Moreover, for each j one has $\mathcal{H}_{j\delta}(\mathbf{y}, \mathbf{x}) = \mathcal{F}_j(\mathbf{y})$. Consequently, for each $\mathbf{y} \in X$ and each $\mathbf{x} \in W$ one has

$$\mathcal{F}_j(\mathbf{y} + t\mathbf{x}) = t^d \mathcal{F}_j(\mathbf{x}) + \mathcal{F}_j(\mathbf{y}) \quad (1 \leq j \leq r). \tag{2.3}$$

Thus, since the affine dimension of X is $N+1$ and $N+1 > \widetilde{w}_d^{(m)}(r; k)$, we deduce that there exist linearly independent vectors $\mathbf{c}_1, \ldots, \mathbf{c}_m \in X$ with the property that for each $s_1, \ldots, s_m \in k$ one has for $1 \leq j \leq r$ that $\mathcal{F}_j(s_1 \mathbf{c}_1 + \cdots + s_m \mathbf{c}_m)$ is a diagonal form in $s_1, \ldots, s_m$. Consequently, when $1 \leq j \leq r$,

$$\mathcal{F}_j(t_1 \mathbf{b}_1 + \cdots + t_n \mathbf{b}_n + s_1 \mathbf{c}_1 + \cdots + s_m \mathbf{c}_m)$$

is a diagonal form in $t_1, \ldots, t_n, s_1, \ldots, s_m$, whence $\widetilde{w}_d^{(n+m)}(r; k) < \mathcal{N}$, and the lemma follows immediately.

Recalling the trivial result $\widetilde{w}_d^{(1)}(r; k) = 0$, it is apparent that Lemma 2.1 may be exploited inductively to obtain bounds for $\widetilde{w}_d^{(m)}(r; k)$ for arbitrary m. We now indicate how to bound $v_{d,r}^{(m)}(k)$ in terms of $\widetilde{w}_d^{(n)}(r; k)$ for suitable n.

Lemma 2.2 *Let d be an odd positive number, let r be a natural number, and let m be a non-negative integer. Then*

$$v_{d,r}^{(m)}(k) \leq \widetilde{w}_d^{(M)}(r; k),$$

where $M = (m + 1)(\phi_{d,r}(k) + 1)$.

Proof Take N to be an integer with $N > \widetilde{w}_d^{(M)}(r; k)$, and for $1 \leq j \leq r$, consider forms $\mathcal{F}_j$ of degree d, all having N variables. By the definition of $\widetilde{w}_d^{(M)}(r; k)$, there exist linearly independent k-rational vectors $\mathbf{e}_1, \ldots, \mathbf{e}_M$ with the property that whenever $t_1, \ldots, t_M \in k$, one has for $1 \leq j \leq r$ that the form $\mathcal{F}_j(t_1 \mathbf{e}_1 + \cdots + t_M \mathbf{e}_M)$ is a diagonal form in $t_1, \ldots, t_M$. Let c_{ij} $(1 \leq i \leq r, 1 \leq j \leq M)$ be elements of k such that

$$\mathcal{F}_i(t_1 \mathbf{e}_1 + \cdots + t_M \mathbf{e}_M) = \sum_{j=1}^{M} c_{ij} t_j^d \quad (1 \leq i \leq r). \tag{2.4}$$

Write $\phi = 1 + \phi_{d,r}(k)$. We observe that, by the definition of $\phi_{d,r}(k)$, for $l = 0, 1, \ldots, m$, each of the systems of equations

$$\sum_{j=l\phi+1}^{(l+1)\phi} c_{ij} t_j^d = 0 \quad (1 \leq i \leq r)$$

possesses a non trivial k-rational solution. Consequently, there exist linearly independent k-rational vectors $\mathbf{a}_0, \ldots, \mathbf{a}_m$ such that for each $u_0, \ldots, u_m \in k$ one has

$$\mathcal{F}_j(u_0 \mathbf{a}_0 + \cdots + u_m \mathbf{a}_m) = 0 \quad (1 \leq j \leq r).$$

Thus the system of equations $\mathcal{F}_j(\mathbf{x}) = 0$ $(1 \leq j \leq r)$ possesses a linear space of solutions of projective dimension m, whence $v_{d,r}^{(m)}(k) \leq N$. This completes the proof of the lemma.

It is now clear how to bound $v_{d,r}^{(m)}(k)$ in terms of $w_{d-2}^{(n)}(\mathbf{s}; k)$, for suitable n and $\mathbf{s}$ depending on d, m and r, provided of course that we have sufficient knowledge concerning the solubility of systems of diagonal equations. We turn to the latter issue in §3. Our next lemma completes the preliminaries necessary to facilitate our induction by bounding $w_{d-2}^{(n)}(s_{d-2}, \ldots, s_1; k)$ in terms of $v_{d-2,r}^{(n)}(k)$ and $w_{d-4}^{(p)}(\mathbf{t}; k)$ for suitable r, p and $\mathbf{t}$.

Lemma 2.3 *Let d be an odd positive number with $d \geq 3$, and let $r_1, r_3, \ldots, r_d$ be non-negative integers with $r_d > 0$. Then for each non-negative integer m one has*

$$w_d^{(m)}(r_d, r_{d-2}, \ldots, r_1; k) \leq w_{d-2}^{(M)}(r_{d-2}, \ldots, r_1; k),$$

where $M = v_{d,r_d}^{(m)}(k)$.

Proof Take N to be an integer with $N > w_{d-2}^{(M)}(r_{d-2}, \ldots, r_1; k)$. Consider forms $\mathcal{F}_{ij}$ of degree i for $1 \leq j \leq r_i$ and $i = 1, 3, \ldots, d$, all having N variables and coefficients in k. By the definition of $w_{d-2}^{(M)}(\mathbf{r}; k)$, there exists a k-linear solution set of the system of equations

$$\mathcal{F}_{ij}(\mathbf{x}) = 0 \quad (1 \leq j \leq r_i, \ i = 1, 3, \ldots, d-2)$$

projective dimension M. Let $\mathbf{e}_0, \ldots, \mathbf{e}_M$ be a basis for the latter space of ions. Then for each $t_0, \ldots, t_M \in k$ one has

$$\mathcal{F}_{ij}(t_0\mathbf{e}_0 + \cdots + t_M\mathbf{e}_M) = 0 \quad (1 \le j \le r_i, \ i = 1, 3, \ldots, d - 2).$$

loreover, for $1 \le j \le r_d$, each of the forms $\mathcal{F}_{dj}(t_0\mathbf{e}_0 + \cdots + t_M\mathbf{e}_M)$ is a form of degree d in the $M + 1$ variables $t_0, \ldots, t_M$. Thus, since $M + 1 > v_{d,r_d}^{(m)}(k)$, there exists a k-linear solution set of the system of equations

$$\mathcal{F}_{dj}(t_0\mathbf{e}_0 + \cdots + t_M\mathbf{e}_M) = 0 \quad (1 \le j \le r_d)$$

with projective dimension m. Let $\mathbf{a}_0, \ldots, \mathbf{a}_m$ be a basis for the latter space of solutions. Then for each $u_0, \ldots, u_m \in k$ one has

$$\mathcal{F}_{ij}(u_0\mathbf{a}_0 + \cdots + u_m\mathbf{a}_m) = 0 \quad (1 \le j \le r_i, \ i = 1, 3, \ldots, d),$$

whence $w_d^{(m)}(r_d, r_{d-2}, \ldots, r_1; k) < N$. This completes the proof of the lemma.

3. Systems of cubic forms

Before embarking on our primary course, we detour in this section to discuss the existence of rational linear spaces in the solution set of systems of homogeneous cubic equations. This topic has been addressed in considerable generality by Lewis and Schulze-Pillot (see [7]), and more recently for a single equation in [13]. The conclusions of Lewis and Schulze-Pillot rest on the deep work of Schmidt [10]. Our methods, although elementary, yield superior conclusions to the aforementioned results whenever the dimension of the linear space lies in an interval intermediate in size in terms of the number of forms. We observe also that our methods apply in any field k for which suitable upper bounds are available for $\phi_{3,r}(k)$.

The proof of Theorem 1 We start by using Lemma 2.1 to bound $\widetilde{w}_3^{(n)}(r; k)$ as a function of n. Recall the notation of the statement of Lemma 2.1. Take $d = 3$, so that $R_1 = r\binom{s+1}{2}$ and $S_1 = r\binom{n+1}{2}$. We also take $m = 1$, so that $N = \widetilde{w}_3^{(1)}(r; k) = 0$ and

$$s = 1 + w_1^{(N)}(S_1; k) = 1 + S_1 = 1 + r\binom{n+1}{2}.$$

Notice that when $n \ge 2$, one has $s \le rn^2$. Further,

$$w_1^{(M)}(R_1; k) = R_1 + M = r\binom{s+1}{2} + \widetilde{w}_3^{(n)}(r; k).$$

On inserting these estimates into Lemma 2.1, we find that

$$\widetilde{w}_3^{(n+1)}(r; k) \le s + r\binom{s+1}{2} + \widetilde{w}_3^{(n)}(r; k), \tag{3.1}$$

and hence when $n \geq 2$ that

$$\widetilde{w}_3^{(n+1)}(r;k) \leq \widetilde{w}_3^{(n)}(r;k) + rn^2 + \tfrac{1}{2}r^2 n^2(rn^2 + 1)$$
$$\leq \widetilde{w}_3^{(n)}(r;k) + r^3 n^4. \tag{3 }$$

Moreover, $\widetilde{w}_3^{(1)}(r;k) = 0$, and hence by (3.1), on noting that when $n = 1$ one
has $s = r + 1$, we deduce that

$$\widetilde{w}_3^{(2)}(r;k) \leq 1 + r + r\binom{r+2}{2} + \widetilde{w}_3^{(1)}(r;k) < 5r^3 < 2^5 r^3. \tag{3.3}$$

On applying (3.2), we therefore deduce that when $n \geq 2$,

$$\widetilde{w}_3^{(n+1)}(r;k) < r^3 \left(5 + \sum_{m=2}^{n} m^4\right) < n^5 r^3. \tag{3.4}$$

On recalling (3.3), we conclude that $\widetilde{w}_3^{(n)}(r;k) < r^3 n^5$ for each positive integer
n.

Finally we apply Lemma 2.2, so that by (3.4) we arrive at the estimate

$$v_{3,r}^{(m)}(k) \leq r^3(m+1)^5 \left(\phi_{3,r}(k) + 1\right)^5.$$

This completes the proof of the theorem.

By altering the choice of m in the above argument we obtain the bound (1.1)
discussed in the introduction.

Theorem 3.1 *Let k be a field, let m and r be non-negative integers with
$r \geq 1$, and suppose that $\phi_{3,r}(k)$ is finite. Then whenever $\alpha > \tfrac{1}{2}(5 + \sqrt{17})$, one
has*

$$v_{3,r}^{(m)}(k) \ll_{k,r,\alpha} (m+1)^\alpha.$$

Proof We form the hypothesis that for some positive number β, with $\beta >
\tfrac{1}{2}(5 + \sqrt{17})$, one has

$$\widetilde{w}_3^{(n)}(r;k) \ll n^\beta, \tag{3.5}$$

where here, and throughout the rest of the proof of this theorem, the implicit
constant depends at most on k, r and β. We mimic the argument of the proof
of Theorem 1, but now take $m = [n^{2/\beta}] + 1$. Thus, in the notation of the
statement of Lemma 2.1, we have

$$s = 1 + w_1^{(N)}(S_1;k) = 1 + S_1 + \widetilde{w}_3^{(m)}(r;k),$$

e by (3.5) one obtains

$$s \ll n^2 + m^\beta \ll n^2.$$

we recall that

$$w_1^{(M)}(R_1; k) = r\binom{s+1}{2} + \widetilde{w}_3^{(n)}(r; k),$$

d thus deduce from Lemma 2.1 that

$$\widetilde{w}_3^{(n+m)}(r; k) - \widetilde{w}_3^{(n)}(r; k) \ll s^2 \ll n^4.$$

A trivial induction now reveals that for each positive integer n,

$$\widetilde{w}_3^{(n)}(r; k) \ll n^5/m \ll n^{5-2/\beta},$$

whence the hypothesis (3.5) holds with β replaced by $5 - 2/\beta$. In view of (3.4), we therefore conclude that the hypothesis (3.5) holds with β replaced by the exponent β_r, for any $r \in \mathbb{N}$, where β_r is defined by $\beta_1 = 5$, and $\beta_{r+1} = 5 - 2/\beta_r$ ($r \in \mathbb{N}$). After verifying that $\lim_{r\to\infty} \beta_r = \frac{1}{2}(5+\sqrt{17})$, the proof of the theorem is complete.

In order to establish the corollary to Theorem 1 we will require an estimate for $\phi_{3,r}(\mathbb{Q})$. We record for this and future use the following lemma.

Lemma 3.2 *Let d and r be natural numbers with d odd. Then*

$$\phi_{d,r}(\mathbb{Q}) + 1 \leq 48rd^3 \log\left(3rd^2\right).$$

Proof This is immediate from the corollary to Theorem 1 of Brüdern and Cook [3], the latter making fundamental use of the corresponding local results of Low, Pitman and Wolff [8].

We note that older results of Davenport and Lewis [5] would also yield reasonable, though somewhat weaker, conclusions when exploited within our methods.

We are now in a position to prove the corollary to Theorem 1, which provides an estimate for the number of variables required to guarantee the existence of a rational m-dimensional linear space of solutions on the intersection of a number of cubic hypersurfaces.

The proof of the corollary to Theorem 1 We apply Theorem 1, bounding $\phi_{3,r}(\mathbb{Q})$ by using Lemma 3.2. Thus

$$\phi_{3,r}(\mathbb{Q}) + 1 < 6^4 r \log(27r),$$

and hence

$$v_{3,r}^{(m)}(\mathbb{Q}) \le 6^{20} r^8 (\log(27r))^5 (m+1)^5,$$

and the corollary follows immediately.

Since it is useful for our discussion, in the following section, of system quintic forms, we record an explicit version of the bound of Lewis and Schu Pillot.

Lemma 3.3 *Let r be a natural number, and let m be a non-negative integ Then*

$$v_{3,r}^{(m)}(\mathbb{Q}) < (11r)^{11}(m+1) + 50r^3(m+1)^5.$$

Proof We employ the bounds on $v_{3,r}^{(m)}(\mathbb{Q})$ used by Lewis and Schulze-Pillo [7] in their proof of [7, inequality (4)], being careful to keep all intermediate estimates explicit. In combination with Schmidt's bound $v_{3,r}^{(0)}(\mathbb{Q}) < (10r)^5$ (see [10, Theorem 1]), the argument of Lewis and Schulze-Pillot yields

$$v_{3,r}^{(m)}(\mathbb{Q}) \le v_{3,r}^{(0)}(\mathbb{Q}) + 3r \sum_{j=1}^{m} \left(v_{3,r}^{(0)}(\mathbb{Q}) + 3rj^2 + 2 \right)^2$$
$$< (10r)^5 + 3r \sum_{j=1}^{m} \left((10r)^5 + 3rj^2 + 2 \right)^2,$$

and the desired conclusion follows with a modicum of computation.

4. Systems of quintic forms We now return to our major goal, that of bounding $v_{5,r}^{(m)}(\mathbb{Q})$. Once again the key to our argument is Lemma 2.1, and again we make use of the estimate for $\phi_{d,r}(\mathbb{Q})$ provided by Lemma 3.2. We begin with a lemma which bounds $\widetilde{w}_5^{(n)}(r;\mathbb{Q})$ as a function of n and r. The conclusion of the lemma represents a compromise between strength and simplicity. We remark on some possible improvements at the end of this section.

Lemma 4.1 *For each non-negative integer j one has*

$$\widetilde{w}_5^{(m_j)}(r;\mathbb{Q}) < (2rm_j^2)^{3\alpha_j}, \tag{4.1}$$

where

$$m_j = 4^j \quad and \quad \alpha_j = 3430^j.$$

Proof We use induction to establish that for each j one has

$$\widetilde{w}_5^{(m_j)}(r;\mathbb{Q}) < e^{\beta_j} r^{\gamma_j} m_j^{\delta_j}, \tag{4.2}$$

$m_j = 4^j$, and

$$3982 \cdot \frac{3430^j - 1}{3429}, \quad \gamma_j = 9261 \cdot \frac{3430^j - 1}{3429}, \quad \delta_j = 17836 \cdot \frac{3430^j - 1}{3429}. \quad (4.3)$$

first that (4.2) holds trivially when $j = 0$. We suppose next that (4.2) ds for a non-negative integer j, and aim to establish that (4.2) holds with j laced by $j + 1$. Let r, n and m be natural numbers. Recall the notation of statement of Lemma 2.1, and take $d = 5$. Then

$$S_1 = r \binom{n + 3}{4} \leq rn^4 \quad \text{and} \quad S_3 = r \binom{n + 1}{2} \leq rn^2,$$

whence by Lemma 3.3, on writing $N = \widetilde{w}_5^{(m)}(r; \mathbb{Q})$ and $N_1 = N + 1$, one has that

$$
\begin{aligned}
s &= 1 + w_3^{(N)}(S_3, S_1; \mathbb{Q}) = 1 + S_1 + v_{3,S_3}^{(N)}(\mathbb{Q}) \\
&< 1 + rn^4 + (11rn^2)^{11} N_1 + 50(rn^2)^3 N_1^5 \\
&< C_1(rn^2)^3 (N_1^5 + (rn^2)^{10}),
\end{aligned} \quad (4.4)
$$

where

$$C_1 = 11^{11} + 50 + 2 < e^{27}. \quad (4.5)$$

Next, on writing $M = \widetilde{w}_5^{(n)}(r; \mathbb{Q})$ and $M_1 = M + 1$, one has from Lemma 2.1,

$$
\begin{aligned}
\widetilde{w}_5^{(n+m)}(r; \mathbb{Q}) &\leq s + w_3^{(M)}(R_3, R_1; \mathbb{Q}) = s + R_1 + v_{3,R_3}^{(M)}(\mathbb{Q}) \\
&< s + rs^4 + (11rs^2)^{11} M_1 + 50(rs^2)^3 M_1^5 \\
&< C_1(rs^2)^3 (M_1^5 + (rs^2)^{10}).
\end{aligned} \quad (4.6)
$$

On substituting from (4.4) into (4.6), we find that

$$
\begin{aligned}
&\widetilde{w}_5^{(n+m)}(r; \mathbb{Q}) \\
&< C_1^7 r^{21} n^{36} \left(N_1^5 + (rn^2)^{10} \right)^6 \left(M_1^5 + C_1^{20} r^{70} n^{120} \left(N_1^5 + (rn^2)^{10} \right)^{20} \right) \\
&< C_2 \left(r^{351} n^{676} + r^{91} n^{156} N_1^{130} + r^{21} n^{36} M_1^5 N_1^{30} + r^{81} n^{156} M_1^5 \right), \quad (4.7)
\end{aligned}
$$

where

$$C_2 = 2^6 C_1^7 (1 + (2C_1)^{20}) < e^{748}. \quad (4.8)$$

First we take $n = m$ in (4.7) to obtain

$$\widetilde{w}_5^{(2m)}(r; \mathbb{Q}) < 4C_2 \left(r^{351} m^{676} + r^{91} m^{156} N_1^{130} \right), \quad (4.9)$$

where we recall that $N_1 = 1 + \widetilde{w}_5^{(m)}(r; \mathbb{Q})$. Next, on taking $n = 2m$ and [...] use of (4.9) in (4.7), we deduce that

$$\widetilde{w}_5^{(3m)}(r; \mathbb{Q}) < C_3 \left(r^{1836} m^{3536} + r^{536} m^{936} N_1^{680} \right),$$

where

$$C_3 = C_2(1 + 8C_2)^5 \left(2^{676} + 2^{157} + 2^{36} \right) < e^{4967}.$$

Finally, on taking $n = 3m$ and making use of (4.10) in (4.7), we conclude th[...]

$$\widetilde{w}_5^{(4m)}(r; \mathbb{Q}) < C_4 \left(r^{9261} m^{17836} + r^{2761} m^{4836} N_1^{3430} \right),$$

where

$$C_4 = C_2(1 + 2C_3)^5 \left(3^{676} + 3^{157} + 3^{36} \right) < e^{26330}.$$

Now recall the inductive hypothesis (4.2). We deduce from (4.12) and (4.13) that

$$\widetilde{w}_5^{(4m_j)}(r; \mathbb{Q}) < C_4 \left(r^{9261} m_j^{17836} + r^{2761} m_j^{4836} \left(1 + e^{\beta_j} r^{\gamma_j} m_j^{\delta_j} \right)^{3430} \right)$$
$$< C_4(1 + 2^{3430}) e^{3430\beta_j} r^{9261+3430\gamma_j} m_j^{17836+3430\delta_j}$$
$$< e^{3982+3430\beta_j} r^{9261+3430\gamma_j} m_{j+1}^{17836+3430\delta_j}.$$

Then in view of (4.3) we deduce that

$$\widetilde{w}_5^{(m_{j+1})}(r; \mathbb{Q}) < e^{\beta_{j+1}} r^{\gamma_{j+1}} m_{j+1}^{\delta_{j+1}},$$

whence the inductive hypothesis follows with j replaced by $j + 1$. We may therefore conclude that (4.2) holds for all non-negative integers j. Finally, (4.1) follows from (4.2) with a little calculation.

It is now possible to bound $v_{5,r}^{(m)}(\mathbb{Q})$ by combining the bound for $\phi_{5,r}(\mathbb{Q})$ provided by Lemma 3.2 together with Lemma 2.2.

The proof of Theorem 2   By Lemma 2.2 one has

$$v_{5,r}^{(m)}(\mathbb{Q}) \leq \widetilde{w}_5^{(M)}(r; \mathbb{Q}),$$

where $M = (m + 1)(\phi_{5,r}(\mathbb{Q}) + 1)$. We may therefore apply Lemma 4.1 with

$$j = \left[\frac{\log M}{\log 4} \right] + 1$$

in, in the notation of the statement of Lemma 4.1,

$$v_{5,r}^{(m)}(\mathbb{Q}) \le \widetilde{w}_5^{(m_j)}(r;\mathbb{Q}) < (2rm_j^2)^{3\alpha_j},$$

e

$$\alpha_j = 3430^j < \exp\left(\left(\frac{\log M}{\log 4} + 1\right)\log 3430\right) < 3430 M^\kappa,$$

_re $\kappa = (\log 3430)/(\log 4)$. Thus we deduce that

$$\log v_{5,r}^{(m)}(\mathbb{Q}) < 10290 M^\kappa \log(32rM^2).$$

ut by Lemma 3.2 one has

$$\phi_{5,r}(\mathbb{Q}) + 1 \le 6000r\log(75r),$$

whence, following a modicum of computation, one deduces that

$$\log v_{5,r}^{(m)}(\mathbb{Q}) < 10^{32}\left((m+1)r\log(3r)\right)^\kappa \log(3r(m+1)).$$

This completes the proof of the theorem.

If, in the proof of Lemma 4.1, the conclusion of the corollary to Theorem 1 is applied in place of Lemma 3.3 to obtain a substitute for (4.6), one arrives at an expression of the shape

$$\widetilde{w}_5^{(n+m)}(r;\mathbb{Q}) \ll_\varepsilon (rs^2)^{8+\varepsilon} M_1^5,$$

whence from (4.4) one deduces that

$$\widetilde{w}_5^{(2m)}(r;\mathbb{Q}) \ll_\varepsilon (rm)^{O(1)}\left(1 + \widetilde{w}_5^{(m)}(r;\mathbb{Q})\right)^{85+\varepsilon}.$$

On taking $n = 2m$ and making use of the latter bound in (4.7), one obtains

$$\widetilde{w}_5^{(3m)}(r;\mathbb{Q}) \ll_\varepsilon (rm)^{O(1)}\left(1 + \widetilde{w}_5^{(m)}(r;\mathbb{Q})\right)^{455+\varepsilon}.$$

Thus a theorem of similar shape to Theorem 2 may be established, save that κ is now replaced by any number exceeding

$$\frac{\log 455}{\log 3} = 5.57093\ldots.$$

A further modest reduction in the permissible value of κ would be made possible by a suitable version of the bound $v_{3,r}^{(m)}(\mathbb{Q}) \ll r^5(m+1)^{14}$ claimed by Lewis and Schulze-Pillot [7].

References

[1] B.J. Birch. Homogeneous forms of odd degree in a large number of var *Mathematika*, **4** (1957), 102–105.

[2] —. Forms in many variables. *Proc. Roy. Soc., Ser. A*, **265** (196ı 245–263.

[3] J. Brüdern and R. J. Cook. On simultaneous diagonal equations anɩ equalities. *Acta Arith.*, **62** (1992), 125–149.

[4] H. Davenport. Cubic forms in 16 variables. *Proc. Roy. Soc., Ser. A*, (1963), 285–303.

[5] H. Davenport and D.J. Lewis. Simultaneous equations of additive ty *Philos. Trans. Roy. Soc. London, Ser. A*, **264** (1969), 557–595.

[6] D.B. Leep and W.M. Schmidt. Systems of homogeneous equations. *Inver math.*, **71** (1983), 539–549.

[7] D.J. Lewis and R. Schulze-Pillot. Linear spaces on the intersection of cubi hypersurfaces. *Monatsh. Math.*, **97** (1984), 277–285.

[8] L. Low, J. Pitman and A. Wolff. Simultaneous diagonal congruences. *J. Number Th.*, **36** (1990), 1–11.

[9] W.M. Schmidt. On cubic polynomials III. Systems of p-adic equations. *Monatsh. Math.*, **93** (1982), 211–223.

[10] —. On cubic polynomials IV. Systems of rational equations. *Monatsh. Math.*, **93** (1982), 329–348.

[11] —. Analytic methods for congruences, Diophantine equations and approximations. *Proc. Intern. Cong. Math. Warszaw, 1983*, PWN, Warszaw, 1984, pp. 515–524.

[12] —. The density of integer points on homogeneous varieties. *Acta Math.*, **154** (1985), 243–296.

[13] T.D. Wooley. Linear spaces on cubic hypersurfaces, and pairs of homogeneous cubic equations. *Bull. London Math. Soc.*, in press.

[14] —. On the local solubility of diophantine systems. *Compositio Math.*, in press.

[15] —. An explicit version of Birch's theorem. In preparation.

Trevor D. Wooley
Department of Mathematics, University of Michigan
Ann Arbor, Michigan 48109-1109, U.S.A.

24

Remark on the Kuznetsov Trace Formula

Eiji Yoshida

e aim of this note is to relax the condition for test functions appearing in
e Kuznetsov trace formula over the full modular group $\Gamma = PSL(2, \mathbb{Z})$. Non-
vial bounds for Kloosterman sums play a crucial role in our method, and in
is respect our theorem is a consequence of the Weil estimate.

1. Statement of the result Let $\mathcal{H} = \{z = x + iy \in \mathbb{C} : y > 0\}$ be
the complex upper half plane equipped with the hyperbolic measure $d\mu(z) =
dxdy/y^2$. Let $L^2(\Gamma\backslash\mathcal{H})$ be the set of all Γ-automorphic functions which are
square integrable with respect to $d\mu$ over the quotient $\Gamma\backslash\mathcal{H}$. This is a Hilbert
space with the inner-product

$$\langle f, g \rangle = \int_{\Gamma\backslash\mathcal{H}} f(z)\overline{g(z)}d\mu(z).$$

Let $\{f_j(z)\}_{j\geq 1}$ be an orthonormal basis of the subspace composed of all cusp
forms in $L^2(\Gamma\backslash\mathcal{H})$. We have the Fourier expansion

$$f_j(z) = y^{\frac{1}{2}} \sum_{n\neq 0} d_j(n) K_{ir_j}(2\pi|n|y)e(nx),$$

where $e(x) = \exp(2\pi ix)$, and $\frac{1}{4} + r_j^2 (r_j > 0)$ is the corresponding eigenvalue
of the Laplacian. Let $E(z, s)$ be the Eisenstein series. We have the Fourier
expansion

$$E(z, s) = y^s + \pi^{\frac{1}{2}}\frac{\Gamma(s - \frac{1}{2})\zeta(2s - 1)}{\Gamma(s)\zeta(2s)}y^{1-s} + y^{\frac{1}{2}}\sum_{n\neq 0}\varphi_n(s)K_{s-\frac{1}{2}}(2\pi|n|y)e(nx).$$

Here $K_\nu(y)$ is the K-Bessel function of order ν, and

$$\varphi_n(s) = 2\pi^s|n|^{s-\frac{1}{2}}\frac{\sigma_{1-2s}(|n|)}{\Gamma(s)\zeta(2s)},$$

where $\sigma_\nu(|n|)$ is the sum of the νth powers of divisors of $|n|$, and $\zeta(s)$ is the
Riemann zeta function.

This work was partially supported by a Grant-Aid for General Scientific Research from
the Ministry of Education, Science and Culture.

The Kloosterman sum is defined by

$$S(m,n,c) := \sum_{0 \le a,d < c} e((ma+nd)/c), \quad ad \equiv 1 \bmod c,$$

with non-zero integers m, n. We have the bound due to Weil:

$$|S(m,n,c)| \le (|m|,|n|,c)^{1/2}\sigma_0(c)c^{1/2}.$$

We then have

The Kuznetsov trace formula *Let m, n be non-zero integers. Let $h(r)$ a function of a complex variable r satisfying certain conditions. Then*

$$\sum_{j \ge 1} \frac{\overline{d_j(m)}d_j(n)}{\cosh(\pi r_j)}h(r_j) + \frac{1}{\pi}\int_{-\infty}^{\infty} \varphi_m(\tfrac{1}{2} - ir)\varphi_n(\tfrac{1}{2} + ir)\frac{h(r)}{\cosh(\pi r)}dr$$

$$= \frac{\delta_{m,n}}{\pi^2}\int_{-\infty}^{\infty} r\tanh(\pi r)h(r)dr$$

$$+ \sum_{c=1}^{\infty} \frac{S(m,n,c)}{c}\frac{2i}{\pi}\int_{-\infty}^{\infty} rM_{2ir}(4\pi|mn|^{\frac{1}{2}}/c)\frac{h(r)}{\cosh(\pi r)}dr, \qquad (2)$$

where M_ν stands for either J_ν or I_ν according as $mn > 0$ or $mn < 0$.

The formula (2) was first established by Kuznetsov [5] (see [1], [4], [6] for alternative proofs). He proved it for $h(r)$ satisfying the condition:

$(\mathcal{C}_1)$ $\begin{cases} h(r) \text{ is even and holomorphic in the strip } |\operatorname{Im} r| < \frac{1}{2} + \varepsilon; \text{ and there} \\ |h(r)| \ll (1+|r|)^{-2-\delta} \text{ as } |r| \to \infty, \end{cases}$

where ε and δ are arbitrary small positive constants.

We shall show that

Theorem *The trace formula (2) holds if $h(r)$ satisfies the condition:*

$(\mathcal{C}_2)$ $\begin{cases} h(r) \text{ is even and holomorphic in the strip } |\operatorname{Im} r| < \frac{1}{4} + \varepsilon; \text{ and there} \\ |h(r)| \ll (1+|r|)^{-2-\delta} \text{ as } |r| \to \infty, \end{cases}$

where ε and δ are as above.

Remark Our argument gives, more precisely, that if we have $S(m,n,c) \ll c^{1-\delta_0}$ for a certain $\delta_0 \ge 0$ then the width of the relevant strip can be taken to be $\frac{1}{2}(1 - \delta_0) + \varepsilon$. Thus the Weil estimate (1) gives our theorem. Also we see that the condition $(\mathcal{C}_1)$ corresponds to the trivial bound $S(m,n,c) \ll c$. The foresees that this sort of correspondences should hold for any Fuchsian s of the first kind with cusps.

3asic identity   In this section we state an intermediate trace formula which our theorem follows. We first introduce the Poincaré series P_m and series F_m defined by Niebur [6](see also [2], [3]). Let Γ_∞ be the stabilizer group in Γ of the cusp at infinity. Then we have

$$P_m(z,s) = \sum_{\gamma \in \Gamma_\infty \backslash \Gamma} e^{-2\pi|m|y(\gamma z)} y(\gamma z)^s e(mx(\gamma z)),$$

$$F_m(z,s) = \sum_{\gamma \in \Gamma_\infty \backslash \Gamma} y(\gamma z)^{\frac{1}{2}} I_{s-\frac{1}{2}}(2\pi|m|y(\gamma z)) e(mx(\gamma z))$$

with common abuse of notation. Both of them converge absolutely for $\mathrm{Re}\,(s) >$ 1. We note that P_m belongs to $L^2(\Gamma \backslash \mathcal{H})$ but F_m does not.

The inner product $\langle F_m(\cdot,s), P_n(\cdot,\overline{w})\rangle$ has no sense because the relevant integral is divergent. We can, however, mimic the procedure of taking the inner-product of these series. To indicate it we invoke the relation:

$$F_m(z,s) = \frac{2^{1-2s}}{\Gamma(s+\frac{1}{2})} \sum_{k=0}^{\infty} \frac{(s)_k}{(2s)_k} \frac{(4\pi|m|)^{s-\frac{1}{2}+k}}{k!} P_m(z,s+k),$$

(see [9, Proposition 2]). This suggests that we should investigate, instead, the expression

$$\pi^{-\frac{1}{2}}(4\pi|n|)^{w-\frac{1}{2}}\Gamma(w)\frac{2^{1-2s}}{\Gamma(s+\frac{1}{2})}$$

$$\times \sum_{k=0}^{\infty} \frac{(s)_k}{(2s)_k}\frac{(4\pi|m|)^{s-\frac{1}{2}+k}}{k!}\langle P_m(\cdot,s+k), P_n(\cdot,\overline{w})\rangle. \tag{3}$$

Here, for an obvious reason, we are not able to exchange the order of the summation over k and the integration involved in the inner-products. Thus it is remarkable that the expression is indeed convergent, and the sum can be expressed in a compact way.

To show this fact, we use on one side the spectral decomposition for each of the inner-products; it converges uniformly for $\mathrm{Re}\,s, \mathrm{Re}\,w > \frac{1}{2}$. On the other side we use an expression for the inner-product which was recently obtained by Motohashi [6, Lemma 9, 10, and 11]; it holds for $\mathrm{Re}\,s, \mathrm{Re}\,w > \frac{3}{4}$.

Then the computation of (3) readily reduces to verifying the relation

$$\frac{2^{1-2s}}{\Gamma(s+\frac{1}{2})\Gamma(s)}\Gamma(s-\tfrac{1}{2}+ir)\Gamma(s-\tfrac{1}{2}-ir)F\big(s-\tfrac{1}{2}+ir, s-\tfrac{1}{2}-ir; 2s; 1\big)$$

$$= \frac{\pi^{-\frac{1}{2}}}{(s-\frac{1}{2})^2 + r^2},$$

where F is the hypergeometric function. In this way we are led to ou
identity:

Lemma *For any non-zero integers m, n, and arbitrary complex number.
satisfying*

$$\tfrac{3}{4} < \operatorname{Re} s, \quad \tfrac{3}{4} < \operatorname{Re} w < 1,$$

we have

$$\frac{\delta_{m,n}}{\pi} \frac{\Gamma(w)\Gamma(s + w - 1)\Gamma(1 - w)}{\Gamma(1 + s - w)}$$

$$+ \sum_{c=1}^{\infty} \frac{S(m,n,c)}{c} \frac{2i}{\pi} \int_{-\infty}^{\infty} r M_{2ir}(4\pi|mn|^{\frac{1}{2}}/c)\Psi(w,r)\frac{1}{(s - \frac{1}{2})^2 + r^2}\,dr$$

$$= \sum_{j \geq 1} \overline{d_j(m)}d_j(n)\Psi(w, r_j)\frac{1}{(s - \frac{1}{2})^2 + r_j^2}$$

$$+ \frac{1}{\pi} \int_{-\infty}^{\infty} \varphi_m(\tfrac{1}{2} - ir)\varphi_n(\tfrac{1}{2} + ir)\Psi(w, r)\frac{1}{(s - \frac{1}{2})^2 + r^2}\,dr, \qquad (4)$$

where $M_{2ir}(y)$ is as in (2), and

$$\Psi(s, r) = \Gamma(s - \tfrac{1}{2} + ir)\Gamma(s - \tfrac{1}{2} - ir).$$

The condition $\operatorname{Re} w < 1$ can not be removed. For otherwise the spectral sum
would not converge absolutely; here we need Kuznetsov's spectral mean square
estimate of the Fourier coefficients $d_j(n)$:

$$\sum_{0 < r_j \leq X} |d_j(n)|^2 e^{-\pi r_j} \ll_n X^2.$$

Obviously the formula (4) is a special case of the trace formula (2); we have
only to put

$$h(r) = \cosh(\pi r)\Psi(w, r)\frac{1}{(s - \frac{1}{2})^2 + r^2}.$$

But the point is that this choice of $h(r)$ is not included in the class induced by
$(\mathcal{C}_1)$.

3. Proof of Theorem We put, in (4),

$$s = \alpha - it, \quad w = \alpha + it \quad (\tfrac{3}{4} + 2\delta < \alpha < \tfrac{3}{4} + 4\delta, t \in \mathbb{R}),$$

with $\delta = \varepsilon/4$ (see $(\mathcal{C}_2)$). Then we shift the path in the integral on the left side
of (4) to $\operatorname{Im} r = -\tfrac{1}{4} - \delta$, which is to ensure absolute convergence. Let $h(r)$ be a

nction satisfying the condition $(\mathcal{C}_2)$. We multiply the resulting identity
factor

$$\Theta(h;\alpha,t) = -i\pi^{-\frac{3}{2}}(\alpha-\tfrac{1}{2})2^{2-2\alpha-2it}\frac{\Gamma(1-\alpha-it)}{\Gamma(\alpha-\tfrac{1}{2}+it)}h(t-i(\alpha-\tfrac{1}{2}))t$$

d integrate with respect to t over the real axis. The verification of abso-
e convergence is immediate. After changing the order of the sums and the
egrals, we see that our problem is reduced to the evaluation of

$$h^*(\alpha,r) = \int_{-\infty}^{\infty} ((\alpha-\tfrac{1}{2}-it)^2+r^2)^{-1}\Psi(\alpha+it,r)\Theta(h;\alpha,t)dt$$

with Im $r = 0$ or $-\tfrac{1}{4}-\delta$. Shifting the path to Im $t = |\text{Im } r| + \alpha + \delta - \tfrac{1}{2}$, we
get

$$h^*(\alpha,r) = h(r)/\cosh(\pi r) + h_1^*(\alpha,r),$$

where h_1^* is the integral over the new path. We have that $h_1^*(\alpha,r)$ is regular
for Re $\alpha > \tfrac{1}{2}(1-\delta)$, and moreover

$$h_1^*(\alpha,r) \ll (1+|r|)^{2\alpha-4}e^{-\pi|r|}.$$

Thus by analytic continuation with respect to α we see that we may put $\alpha = \tfrac{1}{2}$
in the integrated trace formula. This obviously ends the proof of the theorem,
since we have $h_1^*(\tfrac{1}{2};r) = 0$.

There is another approach to the theorem. It depends on Fay's functional
equation for the Kloosterman zeta function, and is a minor modification of the
argument developed in, e.g., Iwaniec [4, §9.3]. Suffice to say, we use the Weil
estimate in his procedure.

The functional equation in question was proved by using the properties of
the resolvent kernel (see [2], [3], and [4]). Concerning this, it is worth remarking
that it is also possible to derive the equation from the formula (4).

In addition to these proofs, recently Motohashi [7, Chap.2] obtained an
alternative proof of the theorem. His proof is based on his formula for the
inner-product $\langle P_m(\cdot,s), P_n(\cdot,\overline{w})\rangle$ which we have mentioned prior to the lemma
in §2. His argument had been developed, independently from ours, in his
extension of Kuznetsov's trace formulas to the three dimensional hyperbolic
space.

Acknowledgement The present article is an improved version of author's
draft. The improvement was kindly shown to the author by the referee.

382   *E. Yoshida*

References

[1] R.W. Bruggemàn. Fourier coefficients of cusp forms. *Invent. mai* (1978), 1–18.

[2] J.D. Fay. Fourier coefficients of the resolvent for a Fuchsian group. *J. angew. Math.*, **294** (1977), 143–203.

[3] D.A. Hejhal. *The Selberg Trace Formula for* PSL$(2,\mathbb{R})$. Lect. Note: Math., **1001**, Springer, Berlin–New York, 1983.

[4] H. Iwaniec. *Introduction to the Spectral Theory of Automorphic Forms.* E Rev. Mat. Iberoamericana, Madrid, 1995.

[5] N.V. Kuznetsov. Petersson's conjecture for cusp forms of weight zero a Linnik's conjecture: Sums of Kloosterman sums. Preprint: Habarovsk Co plex Res. Inst., East Siberian Branch Acad. Sci. USSR, 1977 (Russian); s also *Math. Sb.*, **39** (1981), 299–342.

[6] Y. Motohashi. On real-analytic Poincaré series. Manuscript.

[7] —. *Spectral Theory of the Riemann Zeta-Function.* Cambridge Univ. Press, in press.

[8] D. Niebur. A class of nonanalytic automorphic functions. *Nagoya Math. J.*, **52** (1973), 133–145.

[9] E. Yoshida. On Fourier coefficients of non-holomorphic Poincaré series. *Mem. Fac. Sci. Kyushu Univ.*, (A)**45** (1991), 1–17.

Eiji Yoshida
Graduate School of Mathematics, Kyushu University
Fukuoka-812, Japan

Printed in the United States
By Bookmasters